CONCRETE AND CONCRETE STRUCTURES: NUMERICAL MODELLING AND APPLICATIONS

M. Y. H. BANGASH

*Middlesex Polytechnic
Faculty of Engineering, Science and Mathematics
London, UK*

ELSEVIER APPLIED SCIENCE
LONDON and NEW YORK

ELSEVIER SCIENCE PUBLISHERS LTD
Crown House, Linton Road, Barking, Essex IG11 8JU, England

Sole Distributor in the USA and Canada
ELSEVIER SCIENCE PUBLISHING CO., INC.
655 Avenue of the Americas, New York, NY 10010, USA

WITH 40 TABLES AND 319 ILLUSTRATIONS

© 1989 ELSEVIER SCIENCE PUBLISHERS LTD

British Library Cataloguing in Publication Data
Bangash, M. Y. H.
  Concrete and concrete structures.
  1. Reinforced concrete structures. Analysis. Finite
  element
  I. Title
  624.1'8341

ISBN 1-85166-294-4

Library of Congress Cataloging in Publication Data
Bangash, M. Y. H.
    Concrete and concrete structures: numerical modelling and applications / M. Y. H. Bangash.
    p.   cm.
    Bibliography: p.
    Includes index.
    1. Concrete construction—Mathematical models. I. Title.
  TA439.B27 1989
  624.1'834—dc20   89-33644
                                                                  CIP

No responsibility is assumed by the Publisher for any injury and/or damage to persons or property as a matter of products liability, negligence or otherwise, or from any use or operation of any methods, products, instructions or ideas contained in the material herein.

Special regulations for readers in the USA

This publication has been registered with the Copyright Clearance Center Inc. (CCC), Salem, Massachusetts. Information can be obtained from the CCC about conditions under which photocopies of parts of this publication may be made in the USA. All other copyright questions, including photocopying outside the USA, should be referred to the publisher.

All rights reserved. No part of this publication may be reproduced, stored in a retrieval system, or transmitted in any form or by any means, electronic, mechanical, photocopying, recording, or otherwise, without the prior written permission of the publisher.

Printed in Great Britain by Galliard (Printers) Ltd, Great Yarmouth

Preface

The problem of modelling the constitutive equations representing the isotropic and anisotropic behaviour of concrete in the analysis of concrete structures remains one of the most difficult tasks in the field of structural engineering. At present analyses are being carried out on concrete structures for elastic, plastic and cracking conditions under static and dynamic loads. The behaviour of concrete under such complex loads depends on the state of stress, relative Poisson's ratios and Young's moduli, and the nature of cracking. All these vary in three dimensions. The state of stress or cracking may also depend on the instant load combinations such as compression–compression–compression, tension–compression–compression, tension–tension–tension, etc. All these cases at present cannot easily be handled by codes of practice or simplified empirical formulae.

The basic aim of this book is to present a unified approach for the available mathematical models of concrete. These are then linked to finite element analysis and to a computer program in which special provision is made for concrete plasticity, cracking and crushing with and without concrete aggregate interlocking. Creep, temperature and shrinkage formulations are included and also geared to various concrete constitutive models. Their influence is taken into consideration in the operational and overloading behaviour of concrete structures.

The book is divided into three parts. In the first part Chapter I gives a literature review of the basic properties of concrete as seen by various researchers. Useful data are included on the stress–strain criteria of concrete under monotonic, cyclic and thermal loads. The effects of creep, shrinkage and cryogenic temperatures on the material properties of concrete are clearly established. Equations and graphs are provided to support the data for concrete under uniaxial, biaxial and triaxial conditions with and without the changing influence of Young's modulus and Poisson's ratio. They are arranged in such a way that they can be used in the proposed finite element analysis.

Chapter II gives various mathematical models for concrete. They include the one-parameter to five-parameter models; the hypoelastic model; the shear and bulk moduli models; the endochronic theory; cracking criteria for single and multiple cracks; blunt crack band propagation; and concrete penetration, perforation, scabbing and spalling. Tables and graphs are provided along with diagrams to support these numerical models. Chapter III gives several constitutive creep models including time and strain hardening concepts related to creep models. They are followed by a comprehensive treatment of the three-dimensional thermal creep model. Shrinkage and thermal models are discussed under separate headings within this chapter. The chapter is supported by relevant data. All concrete models in these chapters are presented in such a way that they can also be easily incorporated into the finite element analysis.

The bond and bond–slip phenomenon is one of the most difficult concepts in the field of concrete. Many areas have not been properly understood because of several variables and problems encountered in the measurement of bond. A comprehensive treatment of the bond and bond–slip phenomenon is given in Chapter IV with an emphasis on models which are related to the finite element analysis. Both conventional steel and prestressed steel are examined for bond and bond–slip conditions. For the first time a three-dimensional bond linkage element has been developed by Dr M. Ahmed under the supervision of the author. The element is known as the Ahmlink element. Various constitutive equations and the final stiffness matrix are given for this element to be included in the global finite element analysis. Analytical bond–slip curves obtained have been validated by experimental results and by those available in published data. Chapter V gives a summary of the empirical formulae for the fatigue of concrete. A number of computer subroutines have been developed on the basis of these formulae. They are linked to the global finite element analysis and the computer program ISOPAR.

Part II contains two major chapters on finite element analysis and the simulation techniques of various concrete numerical models. Chapter VI gives a step-by-step formulation of the finite element analysis of concrete structures in which provision is made for elastic, elasto-plastic cracking and crushing of concrete structures. The influence of creep, shrinkage, bond and fatigue is included. The analysis caters for both static and dynamic loads with special emphasis on the effects of impact, blast and earthquakes. Solid isoparametric elements, panel/shell elements and line elements are included to represent concrete, steel liners/membranes and reinforcement/prestressing tendons respectively. Other popular elements are introduced as an alternative choice for finite element analysis. This chapter includes details of solution procedures for the finite element equations applied to a number of concrete structures. Chapter VI is supported by a number of appendices which include additional analyses, element types, material compliance matrices and some computer subroutines.

Chapter VII gives material modelling of major constitutive equations given in earlier chapters. A procedure is given for the simulation of these models in the finite element analysis. The numerical models described in Chapter II along with additional numerical models are expanded and a link is established between them and the relevant equations of Chapter VI. This linking procedure forms the basis on which various subroutines are written for the computer program ISOPAR. The linking procedure is also appropriate for other similar finite element packages and paves the way for the effective use of this technique in other research areas for concrete.

Part III is devoted to the application of this work. Case studies have been selected from various areas of civil engineering. They include structures for building works, hydro-electric and nuclear power stations, offshore, defence, shelters and storage facilities. Care is taken to include those existing structures which are initially supported by experimental test results and/or site monitoring. In this way the work reported in the text can assume a degree of validation.

Chapter VIII covers various case studies which include beams, slabs, concrete-filled tubes, tunnels, nuclear reactor pressure and containment vessels, underground nuclear shelters, gravity platforms and liquefied natural gas tanks. The analytical results compared well with those obtained from the experiments, site monitoring and some published results.

In order to acquaint the reader with such specialised topics, each case study is supported by a brief introduction to the background of the relevant areas. In this way the reader becomes familiar with the problem prior to its solution. In addition, the text gives a comprehensive bibliography for those who wish to study these topics in greater depth.

When writing the book the author relied on a wide range of published references and data provided by organisations, but he also sought to incorporate the results of his own research papers which have appeared in various journals, periodicals and conference proceedings but have not been published in book form. The book has the flexibility to include additional topics and case studies. The book can be used by engineers, technologists, mathematicians and specialists in computer-aided techniques who are involved in numerical modelling of materials such as concrete and the finite element modelling of concrete structures. This book can also be used by practising engineers and designers in the field of concrete for nuclear power, concrete in the ocean, concrete under extreme temperatures and concrete in conventional structures such as buildings, bridges, dams and heavy foundations. The book is also relevant to postgraduate courses on concrete structures subject to static, impact, dynamic, blast and earthquake loads.

M. Y. H. Bangash

Acknowledgements

The author wishes to express his appreciation and gratitude to many civil/structural, nuclear, offshore, petroleum and aerospace engineers, engineering consortia and research establishments, whose names are mentioned in the text, for helpful discussions, criticisms and guidance. The author is also indebted to all the authors of the research papers and reports mentioned in the Bibliography for generous help in the acquisition of data, graphs and computer outputs.

Special thanks to Professor Z. P. Bazant, Professor F. H. Wittmann, Dr G. L. England, Dr J. Stevenson, Dr J. Swanson, Dr D. W. Murray, Professor O. C. Zienkewitz, Professor J. H. Argyrus and Dr N. S. Ottoson.

Thanks to Professor E. Wilson, Dr K. J. William, Mr Janson, Mr A. Nielsen and Mr D. Olsen for generously providing their research materials.

It will not be out of place to mention the help of many students notably, Dr M. Ahmad, Mr A. Aziz, Mr C. Lai, Mr N. Y. Lau, Mr V. Salashoori, Mr J. Tang, Mr A. Ghorbani and Mr Thirugonosothy, in the development of some computer subroutines.

The author acknowledges that owing to a unified approach, many symbols given in original sources had to be replaced by new ones.

Contents

Preface v

Acknowledgements. ix

Notation xvii

PART I

Chapter I   ***Material Properties and Strength of Concrete — A Literature Survey of Numerical Modelling*** . . 3
- I.1   Introduction 3
- I.2   Stress–Strain Curves 3
  - I.2.1   Uniaxial Compression . . 3
  - I.2.2   Uniaxial Tension . . . 11
- I.3   Behaviour of Concrete Under Biaxial Stress   11
  - I.3.1   Biaxial State Adopting the Octahedral Stress Approach 16
- I.4   The Incremental Constitutive Laws for Concrete Under a Multiaxial State of Stress   19
  - I.4.1   General Review of Several Failure Criteria and Strength Theories (Existing Failure Theories) 20
- I.5   Data on Poisson's Ratio and Young's Modulus 36
  - I.5.1   Poisson's Ratio 36
  - I.5.2   Young's Modulus 38
  - I.5.3   Temperature Effects on Young's Modulus 39

Chapter II	**Numerical Modelling of Concrete Strength and Failure**		41
	II.1 Failure Models for Concrete . . .		41
		II.1.1 One-Parameter Model . . .	41
		II.1.2 Two-Parameter Model . . .	42
		II.1.3 Three-Parameter Model . .	45
		II.1.4 Four-Parameter Model . . .	47
		II.1.5 Five-Parameter Model . . .	51
		II.1.6 Hypoelastic Model	55
		II.1.7 Shear and Bulk Moduli Models .	66
		II.1.8 Concrete Model Based on Endochronic Theory	69
	II.2 Concrete Cracking, Penetration, Perforation, Scabbing and Spalling . . .		74
		II.2.1 Concrete Tension and Cracking .	74
		II.2.2 The Strain Softening Rule . .	75
		II.2.3 A Step-by-Step Formulation for Cracking	77
		II.2.4 Blunt Crack Band Propagation .	80
		II.2.5 Penetration, Perforation, Scabbing and Spalling	84
Chapter III	**Numerical Models for Creep, Shrinkage and Temperature**		91
	III.1 Introduction		91
	III.2 Creep		91
		III.2.1 A Brief Survey of Creep . .	91
		III.2.2 Creep Constitutive Models . .	96
	III.3 Shrinkage		106
		III.3.1 Shrinkage Models	107
	III.4 Temperature		109
Chapter IV	**Numerical Modelling of Bond and Bond–Slip** .		112
	IV.1 Introduction		112
	IV.2 Simple Models for Bond between Steel and Concrete		113
		IV.2.1 BS 8110 Model for Bond . .	113
		IV.2.2 Brice, Saliger and Wastlund Models	114
	IV.3 Somayaji and Shah Model . . .		116
	IV.4 Marshall and Krishnamurthy Model .		118
	IV.5 Bond and Bond–Slip Models . . .		118
		IV.5.1 Bond Stress–Slip Constitutive Models	127

Contents xiii

| Chapter V | **Numerical Models for Fatigue of Concrete** | 135 |

V.1 Introduction 135
V.2 Fatigue Models 136
    V.2.1 Kakuta *et al.* Model . . . 136
    V.2.2 Aas-Jacobson Model . . . 136
    V.2.3 Miner Model 136
    V.2.4 Waagaard Model 137
    V.2.5 JSCE Model 137
    V.2.6 CEB–FIP Model 138
    V.2.7 AASHTO Model 139
    V.2.8 NCHR Model 139
    V.2.9 Wilson Model 139

PART II

| Chapter VI | **Finite Element Modelling of Concrete Structures** | 145 |

VI.1 Introduction 145
VI.2 Solid Isoparametric Element Representing Concrete 145
    VI.2.1 The Shape Function . . . 146
    VI.2.2 Derivatives and the Jacobian Matrix 148
    VI.2.3 Determination of Strains . . 150
    VI.2.4 Determination of Stresses . . 151
    VI.2.5 Load Vectors and Material Stiffness Matrix 151
    VI.2.6 The Superelement and Substructuring 155
VI.3 The Membrane Isoparametric Elements . 157
    VI.3.1 Derivatives 159
VI.4 Isoparametric Line Elements . . 161
    VI.4.1 Two-, Three- and Four-Noded Elements 162
    VI.4.2 The Strain–Displacement Relation . 163
    VI.4.3 Line Element in the Body of a Solid Element 164
VI.5 Plastic Flow Rule and Stresses during Elastoplastic Straining . . . 166
    VI.5.1 Sample Cases 172
VI.6 Dynamic Analysis 173
    VI.6.1 Nonlinear Transient Dynamic Analysis 173
    VI.6.2 Reduced Linear Transient Dynamic Analysis 174
    VI.6.3 Mode Frequency Analysis . . 175
    VI.6.4 Spectrum Analysis . . . 178
    VI.6.5 Impact/Explosion . . . 180
VI.7 Buckling Analysis 181

Chapter VII Material Modelling Simulation for Finite Element Formulation 182

- VII.1 Introduction 182
- VII.2 Uniaxial Stress–Strain Curve . . . 182
  - VII.2.1 Solution Technique for Descending Stress–Strain Curve . . 184
- VII.3 Biaxial Stress Envelope 186
- VII.4 Simulation Procedures for One- to Five-Parameter Models 191
  - VII.4.1 One-Parameter Model . . . 191
  - VII.4.2 Two-Parameter Model . . . 193
  - VII.4.3 Three-Parameter Model . . 194
  - VII.4.4 Four-Parameter Model . . . 196
  - VII.4.5 Five-Parameter Model . . . 201
- VII.5 Creep Model 202
- VII.6 Thermal Stress–Strain Model . . . 204
- VII.7 Nonlinear Bond Linkage Element . . 205
- VII.8 Bulk and Shear Moduli Model . . . 206
- VII.9 Endochronic Cracking Model . . . 206
- VII.10 Heat-Conduction Model 207
  - VII.10.1 Steady State 208
  - VII.10.2 Transient 208
- VII.11 Miscellaneous Orthotropic Constitutive Models 212
  - VII.11.1 Liu, Nilson and Slate . . . 212
  - VII.11.2 Darwin and Pecknold . . . 214
  - VII.11.3 Elwi and Murray 215
  - VII.11.4 Strain Hardening Models . . 216

PART III

Chapter VIII Application to Engineering Problems . . . 221

- VIII.1 Introduction 221
- VIII.2 Case Study No. 1—Reinforced and Prestressed Concrete Beams 221
  - VIII.2.1 Reinforced Concrete Beams . . 221
  - VIII.2.2 Prestressed Concrete Beams . . 227
- VIII.3 Case Study No. 2—Reinforced Concrete Slabs 231
  - VIII.3.1 Reinforced Concrete Rectangular Slabs Subject to Concentrated Loads 231

	VIII.3.2	Reinforced Micro-Concrete Square Slabs Subject to Patch Loads	237
	VIII.3.3	Reinforced Concrete Slabs Under Impact Loads	238
VIII.4	Case Study No. 3—Bonded Prestressed Concrete Specimens and Slabs		244
	VIII.4.1	Bond–Slip Test Specimen	244
	VIII.4.2	Details of the Bonded Slab	250
VIII.5	Case Study No. 4—Concrete Filled Tubular Columns		270
VIII.6	Case Study No. 5—Concrete Power and Diversion Tunnels for Hydro-Electric Schemes		280
	VIII.6.1	Introduction	280
	VIII.6.2	Diversion, Power and Pressure Tunnels	280
VIII.7	Case Study No. 6—Prestressed Concrete Nuclear Reactor Vessels		293
	VIII.7.1	General	293
	VIII.7.2	Historical Development	294
	VIII.7.3	Problems Associated with Vessels	296
	VIII.7.4	Vessel Layouts and Finite Element Mesh Schemes	299
	VIII.7.5	Design Analysis	299
VIII.8	Case Study No. 7—Concrete Containment Vessels		356
	VIII.8.1	Introduction	356
	VIII.8.2	Loads and Stresses	364
	VIII.8.3	Containment Vessel Analysis	368
VIII.9	Case Study No. 8—Concrete Nuclear Shelters		399
	VIII.9.1	Introduction	399
	VIII.9.2	Characteristics of the Blast Wave in Air	400
	VIII.9.3	Air Blast Loading and Target Response	407
	VIII.9.4	Loading Versus Structural Characteristics	412
	VIII.9.5	Damage Classification	413
	VIII.9.6	Blast Loads and Stresses	413
	VIII.9.7	Finite Element Analysis of a Domestic Nuclear Shelter	413
VIII.10	Case Study No. 9—Silos and Bunkers		418
	VIII.10.1	Introduction	418

		VIII.10.2 Loads on Silos	419
		VIII.10.3 Material Properties	419
		VIII.10.4 Application to Silos	430
	VIII.11	Case Study No. 10—Offshore Structures	430
		VIII.11.1 Concrete Gravity Platforms	430
		VIII.11.2 Design of the Condeep Platform	431
		VIII.11.3 Data for Condeep	436
		VIII.11.4 Ship Impact on Concrete Gravity Platforms	438
		VIII.11.5 Dynamic Finite Element Analysis	442
	VIII.12	Case Study No.11—Liquefied Natural Gas (LNG) Tanks	451
		VIII.12.1 Introduction	451
		VIII.12.2 Methods of Analysis and Design	453
		VIII.12.3 Finite Element Analysis of LNG Tanks	463

Bibliography 471

Appendix I	**Relevant Element Library**	527
Appendix II	**Transformation Matrices**	563
Appendix III	**Material and Cracking Matrices**	567
Appendix IV	**Gaussian Quadrature**	575
Appendix V	**Acceleration and Convergence**	579
Appendix VI	**Concrete Cubes and Steel Bars—Experimental and Finite Element Results**	587
Appendix VII	**Stress–Strain Curves for Reinforcements and Prestressing Tendons (Data for the Program ISOPAR)**	591
Appendix VIII	**Concrete Cracking—Step-by-Step Formulation and Computer Sub Routines**	603
Appendix IX	**Data and Results for Offshore Structures**	615
Appendix X	**Geometry and Forces for Domes for Offshore Cells and LNG Tanks**	649

Index 653

Notation

Major Notation

$[B]$	Strain–displacement matrix
$[B_i]$	Strain–displacement matrix at node i
$[C]$	Inverse of material matrix
d, ϕ	Diameter of the line element
d, D	Missile diameter or diameter for structures
$\det J$	Determinant of Jacobian
ds	Differential surface area
$d\,\text{vol}$	Differential volume
$[D]$	Material compliance matrix
$[D_e]$	Elastic material matrix for steel
$[D_{ep}]$	Elasto-plastic material matrix for steel
$[D_p]$	Plastic material matrix for steel
$\{D_E\}$	Elastic material matrix for membrane element
$[D_T]$	Tangent constitutive matrix in global co-ordinate system
$[D^*]$	Constitutive matrix in crack co-ordinate system
E_b	Bond linkage constitutive matrix
E_c	Modulus of elasticity for concrete
E_1, E_2, E_3	Moduli of elasticity in three principal directions
F	Vector of external element loads
$\{F\}$	Global load vector
G	Shear modulus
G'	Shear modulus for cracked concrete
$[J]$	Jacobian matrix
k_H, k_V, k_E (or k_h, k_v, k_e)	Bond–slip moduli in horizontal, vertical and lateral directions
$[K]$	Global structural stiffness matrix

$[K_b]$	Bond linkage stiffness matrix
$[K_T]$	Total stiffness matrix
$[K_0]$	Effective stiffness matrix
$[K^e]$	Element stiffness matrix
l, m, n	Direction cosines relating local to global axes
L	Length of the line elements
n_1, n_2	Unit vectors
N	Element shape function matrix
N_i	Shape function of node i
$\{P_G\}$	Vector of body forces per unit volume
P_s	Vector of surface pressure
q	Starting iteration vector
$\{R_i\}$	Residual force vector
$\{R_t\}$	External load vector
$\{R_t\}$	Projected load vector
s	Spalling thickness/scabbing thickness
t	Time during analysis
$[T]''$	Transformation matrix (relating local to global displacements at nodes)
$\{T_\varepsilon''\}$	Strain transformation matrix
$\{T_\sigma''\}$	Stress transformation matrix
$\{U\}$	Global nodal displacement vector
U_i, V_i, W_i	Displacements at node i
$\{U^e\}$	Element nodal displacement vector
V	Impact velocity
w, W	Missile weight or any other weights
W_e	External work
w_i	Nodal displacement in the z-direction at node i
X	Constant
X, Y, Z	Global co-ordinate system
X', Y', Z'	Local Cartesian co-ordinate system
X^*, Y^*, Z^*	Crack co-ordinate system
α_{ij}	Total translation of the yield surface
β'	Shear interlocking factor
δ, u	Displacement
$\dot{\delta}, \dot{u}$	Velocity
$\ddot{\delta}, \ddot{u}$	Acceleration
δ_{ij}	Kronecker delta
$\{\delta\}_0$	First displacement vector

$\delta_1, \delta_2, \delta_3$	Displacement increments
δ^*	Virtual displacement
$\{\Delta S\}$	Incremental slips
$\Delta\varepsilon$	Incremental uniaxial strain
$\Delta\varepsilon_i, \Delta\sigma_i$	Incremental strain/stress vectors
$\{\Delta\varepsilon_0\}$	Plastic strain increment
$\{\Delta\varepsilon_t\}$	Incremental thermal strains
$\{\Delta\varepsilon^*\}\{\Delta\sigma^*\}$	Incremental strain/stress vectors in crack directions
$\{\Delta\sigma_b\}$	Incremental bond stress vector
ε	Equivalent strain
$\{\varepsilon_{cu}\}$	Concrete crushing strain
ε_i	Vector of Cartesian strain components at point i
ε_0	Initial strain vector
ε^c	Creep strain vector
$\{\varepsilon^*\}$	Strain vector in crack directions
ν_c	Poisson's ratio for concrete
ξ, η, ζ	Local curvilinear co-ordinates
σ	Equivalent stress
$\{\sigma_b\}$	Bond stress vector
σ_c	Concrete cylinder compressive strength
σ_i	Principal stresses, $i = 1, 2, 3$
σ_i	Vector of Cartesian stress components at point i
σ_{ij}	Stress tensor
σ_{ij}''	Deviatoric component of the stress tensor
σ_m	Mean stress
σ_0	Initial stress
σ_0	Initial stress vector
$\{\sigma^*\}$	Stress vector in crack directions
σ_t	Concrete limiting tensile strength
τ	Increase in time t

PART I

This part contains a literature review of the material properties and strengths of concrete. Detailed work is presented on the numerical modelling of concrete strength and the failure of concrete. Numerical modelling of creep, shrinkage temperature, bond and fatigue is also covered in Parts II and III for the finite element analysis of concrete structures.

Chapter I

Material Properties and Strength of Concrete—A Literature Survey of Numerical Modelling

I.1 INTRODUCTION

This chapter includes a literature survey of the material properties and strength of concrete. Various concrete strength models are discussed. Material data collected for concrete subjected to monotonic loads, cyclic loads and temperature has been used elsewhere in the analysis of concrete structures. The effects of creep and shrinkage are discussed within the text. A comparative study of the behaviour of concrete is given for such numerical models.

I.2 STRESS–STRAIN CURVES

I.2.1 Uniaxial Compression

Experimental tests show that concrete behaves in a highly nonlinear manner in uniaxial compression. Figure I.1 shows a typical stress–strain relationship subjected to uniaxial compression. This stress–strain curve is linearly elastic up to 30% of the maximum compressive strength. Above this point the curve increases gradually up to about 70–90% of the maximum compressive strength. Eventually it reaches the peak value which is the maximum compressive strength σ_{cu}. Immediately after the peak value, this stress–strain curve descends. This part of the curve is termed *softening*. After the curve descends, crushing failure occurs at an ultimate strain ε_{cu}. A numerical expression has been developed by Hognestad [1] which treats (Fig. I.1) the ascending part as a parabola and the descending part as a straight line. This numerical expression is given as:

for $0 < \varepsilon < \varepsilon_0'$
$$\frac{\sigma}{\sigma_{cu}} = 2\frac{\varepsilon}{\varepsilon_0'}\left(1 - \frac{\varepsilon}{2\varepsilon_0'}\right) \tag{I.1}$$

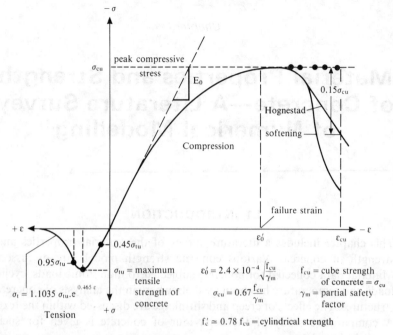

Fig. I.1. Typical uniaxial compression or tension stress–strain curve for concrete.

for $\varepsilon_0' < \varepsilon < \varepsilon_{cu}$
$$\frac{\sigma}{\sigma_{cu}} = 1 - 0.15\left(\frac{\varepsilon - \varepsilon_0'}{\varepsilon_{cu} - \varepsilon_0'}\right) \tag{I.2}$$

where ε_0' is the strain at the peak stress, ε_{cu} is the ultimate strain and σ is the stress.

The initial tangent modulus E_0 is then computed as

$$E_0 = \left.\frac{d\sigma}{d\varepsilon}\right|_{\varepsilon'=0} = 2\frac{\sigma_{cu}}{\varepsilon_0'} \tag{I.3}$$

and equals twice the secant modulus taken at the peak stress σ_{cu}.

Hognestad suggested that the modulus of elasticity E_0 in psi is

$$E_0 = 1\cdot 8 \times 10^6 + 460(0\cdot 85 \times \text{the compressive strength of standard cylinders}) \tag{I.4}$$

Desayi and Krishnan [2] give a general expression for both the ascending and descending parts of the curve as

$$\sigma = \frac{E_0 \varepsilon}{1 + \left(\dfrac{\varepsilon}{\varepsilon_0'}\right)^2} \tag{I.5}$$

The European Concrete Committee (CEB) for short-term loading gives a parabola and a straight line up to ultimate strain ε_{cu} as shown in Fig. I.1. In numerical terms this is given as

$$\frac{f_c'}{\sigma_{cu}} = \frac{k\eta - \eta^2}{1 + (k-2)\eta} \tag{I.6}$$

where

$$\eta = \frac{\varepsilon_{cu}}{0 \cdot 002} \qquad k = \frac{[0 \cdot 0022(1 \cdot 1 E)]}{\sigma_{cu}}$$

The value of ε_{cu} is given as between 0·003 and 0·0035.

The concept of equivalent uniaxial strain for any stress has been developed [3] so as to assess, at any stage, the degradation of stiffness and strength of plain concrete and also to allow the actual biaxial stress–strain

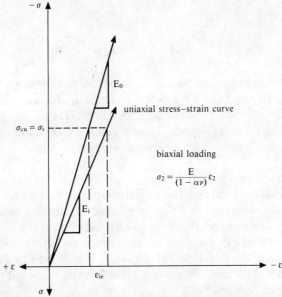

Fig. I.2.   Stress–strain curve and equivalent strain.

curves to be duplicated from such uniaxial curves. This concept provides a method of separating the Poisson effect from the cumulative strain. Figure I.2 shows the equivalent uniaxial strain ε_{ie} for concrete as a linear material. The value of ε_{ie} for nonlinear material is written as

$$\varepsilon_{ie} = \sum d\varepsilon_{ie} = \int \frac{d\sigma_i}{E_{ti}} = \sum \frac{\Delta \sigma_i}{E_{ti}} \tag{I.7}$$

where $d\varepsilon_{ie}$ and $d\sigma_i$ are, respectively, differential changes in stress and equivalent uniaxial strain in the ith direction. $\sum$ is for all load increments. E_{ti} is tangent modulus in the ith direction.

Saenz [4] selected curves for compressive loading and has defined them numerically (Fig. I.3) as

$$\sigma_{cu} \simeq \sigma_i = \frac{\varepsilon_{ie} E_0}{1 + \left(\dfrac{E_0}{E_{sec}} - 2\right)\dfrac{\varepsilon_{ie}}{\varepsilon_0'} + \left(\dfrac{\varepsilon_{ie}}{\varepsilon_0'}\right)^2} \tag{I.8}$$

where E is the second modulus. Elwi and Murray [5] have expressed eqn (I.8) as

$$\varepsilon_{ie} E_0 \bigg/ 1 + \left(\frac{E_0}{E_{sec}} + R - 2\right)\frac{\varepsilon_{ie}}{\varepsilon_0'} - (2R-1)\left(\frac{\varepsilon_{ie}}{\varepsilon_0'}\right)^2 + R\left(\frac{\varepsilon_{ie}}{\varepsilon_0'}\right)^3 \tag{I.8a}$$

where

$$R = \frac{E_0}{E_{sec}}\left(\frac{R_\sigma - 1}{R_\varepsilon - 1}\right) - \frac{1}{R_\varepsilon} \qquad R_\sigma = \frac{1}{0.85} \quad R_\varepsilon = \frac{\varepsilon_0'}{\varepsilon_{cu}}$$

Typical stress–strain curves for concrete in compression have been produced by Bangash [6] for different concrete grades, and are shown in Fig. I.4 along with the BS 8110 actual stress curves. The idealised curves are also plotted in the same figure. Figure I.5 shows a comparative study of concrete stress–strain diagrams in compression under various age conditions.

Figure I.6 shows a typical stress–strain curve for concrete undergoing several cycles of load reversals in uniaxial compression. In such a case many values of strain occur for a single value of stress. An idealised model curve is proposed by Darwin and Pecknold [7a,b] and is plotted on such cyclic curves as given in Fig. I.6. This curve is based on several experimental results from Karson and Jirsa [8]. Equation (I.8) is used to describe the ascending part of the idealised curve, and the descending part of the curve is a straight line ending at $0.2 f_c'$, where f_c' is the cylindrical compressive strength of concrete. Figure I.7 shows some of Karson and Jirsa's

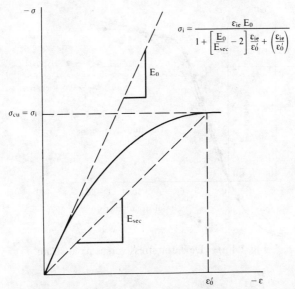

Fig. I.3. Equivalent uniaxial stress–strain curve.

experimental results. An empirical relation is established between ε_{en} (the strain on the envelope curve prior to unloading) and the residual strain remaining after the entire load is released, ε_p (the plastic strain). This is given as

$$\frac{\varepsilon_p}{\varepsilon_{cu}} = 0 \cdot 145 \left(\frac{\varepsilon_{en}}{\varepsilon_{cu}}\right)^2 + 0 \cdot 13 \left(\frac{\varepsilon_{en}}{\varepsilon_{cu}}\right) \qquad (I.9)$$

Karson and Jirsa found a band of points on the stress–strain plane which controls the degradation of the concrete under continued load cycles. Below this band is the 'stability limit' and above it is the 'common point' limit. The band is reduced to a single curve which is then called the *locus of common points*. As shown in Fig. I.6, if the load is cycled above the stability limit and peak stress maintained between cycles, additional permanent strain would accumulate. The common point limit shows the maximum stress at which a reloading curve may intersect the original unloading curve. Each new envelope point defines a common point and a new plastic strain. The location of the common point with respect to the envelope curve can be adjusted in order to control the number of cycles leading to failure. Continued cycles above this common point can eventually intersect the envelope curve. If the locus of common points is lowered, fewer cycles are

Concrete and Concrete Structures

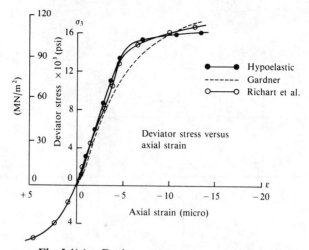

Fig. I.4(a). Deviator stress versus axial strain.

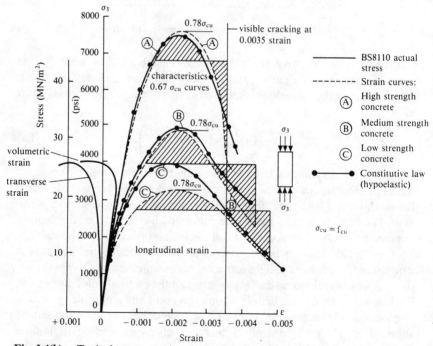

Fig. I.4(b). Typical stress–strain curves for concrete compression: comparison with BS 8110.

Material Properties and Strength of Concrete

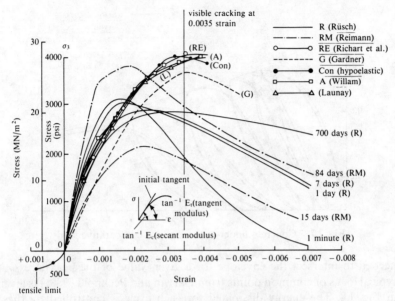

Fig. I.5. Stress–strain curve for concrete.

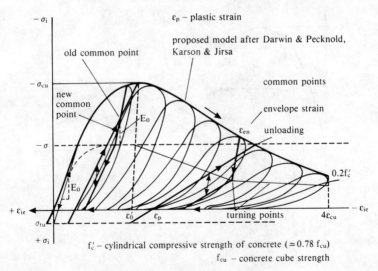

Fig. I.6. Cyclic loading of concrete.

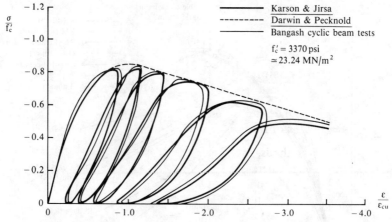

Fig. I.7. Concrete under cyclic load—a comparative study.

needed to intersect the envelope for a given value of the peak stress. A typical locus of common points from Darwin and Pecknold [7a,b] is given in Fig. I.8. The energy dissipated for each cycle is controlled by these turning points. The lower the turning point, the greater the energy dissipated per cycle. Figure I.8 gives a simple satisfactory energy for all but

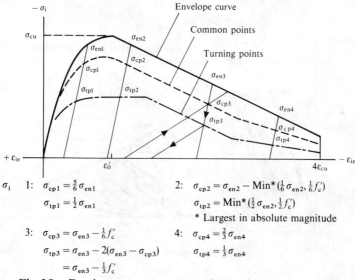

σ_i 1: $\sigma_{cp1} = \frac{5}{6}\sigma_{en1}$ 2: $\sigma_{cp2} = \sigma_{en2} - \text{Min}^*(\frac{1}{6}\sigma_{en2}, \frac{1}{6}f'_c)$

       $\sigma_{tp1} = \frac{1}{2}\sigma_{en1}$ $\sigma_{tp2} = \text{Min}^*(\frac{1}{2}\sigma_{en2}, \frac{1}{2}f'_c)$

                                             * Largest in absolute magnitude

    3:   $\sigma_{cp3} = \sigma_{en3} - \frac{1}{6}f'_c$ 4: $\sigma_{cp4} = \frac{2}{3}\sigma_{en4}$

       $\sigma_{tp3} = \sigma_{en3} - 2(\sigma_{en3} - \sigma_{cp3})$ $\sigma_{tp4} = \frac{1}{3}\sigma_{en4}$

         $= \sigma_{en3} - \frac{1}{3}f'_c$

Fig. I.8. Envelope curve, common points and turning points [7].

low values of ε_{en}. For low values of ε_{en} the unloading and reloading take place on a single line when the slope is E_0 and when

$$|\varepsilon_{en}| \leq \tfrac{1}{4}|\varepsilon_{cu}| \tag{I.10}$$

I.2.2 Uniaxial Tension

Tensile behaviour is evaluated by either

(a) the split cylinder test or
(b) the modulus of rupture or bending test

In both these methods, when the tensile strength σ_{cu} decreases, the compressive strength increases. For any stress less than $0.6\sigma_{cu}$, microcracks are ignored. Welch [9] and Evans [10] give a value of $0.75\sigma_{cu}$ for the onset of unstable cracks. The tensile strength can be expressed in terms of the cylindrical compressive strength of concrete. Hughes and Chapman [11] have carried out several tests. They found that the ratio between uniaxial tensile strength and compressive strength ranges from 0·05 to 0·1. The modulus of elasticity under uniaxial tension is higher, and Poisson's ratio lower, than those obtained from the uniaxial compression curves. For the tensile splitting case, σ_{tu} is given by

$$\sigma_{tu} = 0.55\sqrt{f'_c} \quad \text{(MPa)} \tag{I.11}$$

For the modulus of rupture case, the tensile strength σ_t is given by

$$\sigma_{tu} = 0.95\sqrt{f'_c} \quad \text{(MPa)} \tag{I.12}$$

where f'_c is the cylindrical strength of concrete in compression. Figures I.1 and I.5 show the tensile 'cut-off'. Data on the complete stress–strain curves for concrete subjected to direct tension have been reported by Evans [10].

Johnson [13] carried out tests and found that the ratio of uniaxial tensile strength to compressive strength ranges from 0·05 for high-strength concrete to about 0·1 for a concrete of medium strength, thus endorsing the figures given by Hughes and Chapman. He also found that the type of aggregate influences the tensile strength of concrete. Siliceous aggregates decrease the tensile strength, and calcareous aggregates increase the tensile strength of concrete.

A general form of the tensile stress–strain curve is given in Fig. I.1.

1.3 BEHAVIOUR OF CONCRETE UNDER BIAXIAL STRESS

A number of investigators [1–31] have attempted to evaluate concrete strength under biaxial compression, biaxial tension–compression, or under any combination of loads. Kupfer *et al.* [29] and Liu *et al.* [30] have conducted biaxial loading tests. Three detailed experiments have been carried out and the results are summarised in Figs I.9 and I.10. For biaxial compression, Kupfer and Gerstle [31] have closely approximated the strength curve by eqn (I.13). Where $\sigma_1 \geq \sigma_2$,

$$\left(\frac{\sigma_1}{f'_c} + \frac{\sigma_2}{f'_c}\right)^2 - \left(\frac{\sigma_2}{f'_c}\right) - 3\cdot 65\left(\frac{\sigma_1}{f'_c}\right) = 0 \tag{I.13}$$

Equation (I.13) can be rewritten for $\alpha = \sigma_{1c}/\sigma_{2c}$ to give σ_{2c}. The maximum compressive strength of concrete as a function of the uniaxial compressive strength f'_c and α (Fig. I.11) is

$$\sigma_{2c} = \frac{1 + 3\cdot 65\alpha}{(1 + \alpha)^2} f'_c \tag{I.14}$$

The peak stress σ_{cu} is given by

$$\sigma_{1c} = \sigma_{cu} = \alpha\sigma_{2c} \tag{I.15}$$

For the tension–compression case, Kupfer and Gerstle suggest a straight

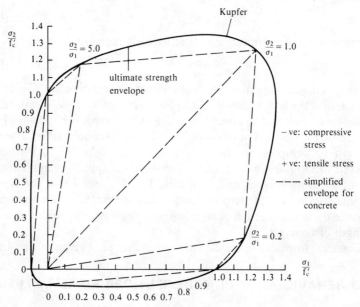

Fig. I.9. Concrete under biaxial stress state in terms of ultimate uniaxial cylinder crushing strength.

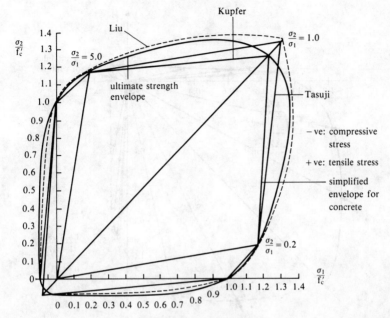

Fig. I.10. Concrete under biaxial stress state in terms of ultimate uniaxial cylinder crushing strength.

line reduction in tensile strength, taking into consideration the increased compressive stress

$$\sigma_{pt} = \left(1 - 0.8 \frac{\sigma_2}{f'_c}\right)\sigma_{tu} \tag{I.16}$$

For tension–compression, Darwin and Pecknold [7a] modified eqn (I.14) and presented it as

$$\sigma_2 = \left[\frac{1 + 3.28\alpha}{(1 + \alpha)^2}\right] f'_c \tag{I.17}$$

For tension–tension, Kupfer and Gerstle suggest a constant tensile strength equal to the uniaxial tensile strength of concrete.

The biaxial compression stress–strain curve ($\alpha = 1$) can determine the effect on increased ductility. In this case the real strain at maximum compressive stress, i.e. the peak stress σ_{cu}, is converted to $\varepsilon'_0/(1-v)$, thus eliminating the Poisson effect from the strain value, as stated in Section I.2.

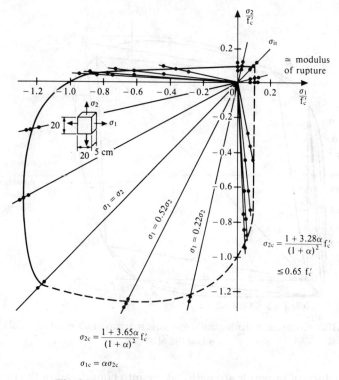

Fig. I.11. Biaxial stress state strength [7a,b].

If the value of ε'_0 varies with σ_{cu}, then ε'_0 is given by

$$\varepsilon'_0 = \varepsilon_{cu}\left[1\cdot 6\left(\frac{\sigma_{cu}}{f'_c}\right)^3 + 2\cdot 25\left(\frac{\sigma_{cu}}{f'_c}\right) + 0\cdot 35\left(\frac{\sigma_{cu}}{f'_c}\right)\right] \qquad (I.18)$$

The tensile ductility is greater under biaxial compression than under uniaxial compression. It is estimated that a 25% increase in maximum strength occurs if $\alpha = 2$, and a decrease of approximately 15·5% occurs when $\alpha = 1$. It is interesting to note that when a failure point is approached, with increasing compressive stress, an increase in volume occurs; this inelastic volume is known as *dilatancy*. This is the onset of progressive growth of microcracks in concrete.

Many attempts have been made to produce graphically the biaxial ultimate strength envelope. Figure I.9 shows the biaxial stress state in terms of ultimate cylindrical strength. Using data provided by Kupfer and Gerstle [31], Fig. I.10 shows a comparative study of the biaxial ultimate strength envelopes of Kupfer and Gerstle [31], Tasuji and co-workers [73, 74], and Liu and co-workers [69, 70].

Failure occurs when non-dimensionalised stress lies outside the failure envelope. The maximum compressive stress σ_{cu} in directions 1 and 2 is then determined by

$$\sigma_{cu1} = f'_c \left[1 + \frac{\left(\frac{\sigma_2}{\sigma_1}\right)}{1 \cdot 2 - \left(\frac{\sigma_2}{\sigma_1}\right)} \right] \quad \text{when } \frac{\sigma_2}{\sigma_1} < 0 \cdot 2 \tag{I.19}$$

$$\sigma_{cu1} = 1 \cdot 2 f'_c \quad \text{when } 1 \cdot 0 \geq \frac{\sigma_2}{\sigma_1} \geq 0 \cdot 2$$

$$\sigma_{cu2} = f'_c \left[1 + \frac{\left(\frac{\sigma_1}{\sigma_2}\right)}{1 \cdot 2 - \left(\frac{\sigma_1}{\sigma_2}\right)} \right] \quad \text{when } \frac{\sigma_2}{\sigma_1} > 5 \cdot 0 \tag{I.20}$$

$$\sigma_{cu2} = 1 \cdot 2 f'_c \quad \text{when } 1 \cdot 0 \leq \frac{\sigma_2}{\sigma_1} \leq 5 \cdot 0$$

Throughout the analysis, tensile stresses are considered to be positive, and compressive stresses to be negative. For tensile failure it is assumed that the load may be carried in the direction of the crack and may not be transferred across the crack. For a case of compressive failure it is assumed that no load may be carried in any direction. In order to comply with these limitations, the following boundaries for corresponding regions are established:

(a) Tension–tension region: σ_{tu}/f'_c as non-dimensionalised values for directions 1 and 2.

(b) Compression–tension region: the boundary is defined as

$$\frac{\sigma_1}{f'_c} = \frac{\sigma_{tu}}{f'_c}\left(1 - \frac{\sigma_1}{f'_c}\right) \quad \frac{\sigma_2}{f'_c} = \frac{\sigma_{tu}}{f'_c}\left(1 - \frac{\sigma_2}{f'_c}\right) \tag{I.21}$$

(c) Compression–compression region:

Direction 1

$$\frac{\sigma_1}{f'_c} = 1 + \left[\frac{\left(\frac{\sigma_2}{\sigma_1}\right)}{1\cdot 2 - \left(\frac{\sigma_2}{\sigma_1}\right)}\right] \qquad \text{when } \frac{\sigma_2}{\sigma_1} < 0\cdot 2$$

$$\frac{\sigma_2}{f'_c} = 1 + \left[\frac{1}{1\cdot 2\left(\frac{\sigma_1}{\sigma_2}\right) - 1}\right]\left(\frac{\sigma_2}{\sigma_1}\right) \qquad \text{when } \frac{\sigma_2}{\sigma_1} < 0\cdot 2 \qquad (I.22)$$

$$\frac{\sigma_1}{\sigma_2} = 1\cdot 2 \qquad \text{when } 1\cdot 0 \geq \frac{\sigma_2}{\sigma_1} \geq 0\cdot 2$$

$$\frac{\sigma_2}{f'_c} = 1\cdot 2\left(\frac{\sigma_2}{\sigma_1}\right) \qquad \text{when } 1\cdot 0 \geq \frac{\sigma_2}{\sigma_1} \geq 0\cdot 2$$

Direction 2

$$\frac{\sigma_2}{f'_c} = 1 + \left(\frac{\sigma_1}{\sigma_2}\right) \bigg/ 1\cdot 2\left(\frac{\sigma_1}{\sigma_2}\right) \qquad \text{when } \frac{\sigma_2}{\sigma_1} > 5\cdot 0$$

$$\frac{\sigma_1}{\sigma_2} = \left[1 + \frac{1}{1\cdot 2\left(\frac{\sigma_2}{\sigma_1}\right) - 1}\right]\left(\frac{\sigma_1}{\sigma_2}\right) \qquad \text{when } \frac{\sigma_2}{\sigma_1} > 5\cdot 0 \qquad (I.23)$$

$$\frac{\sigma_2}{\sigma} = 1\cdot 2 \qquad \text{when } 5\cdot 0 > \frac{\sigma_2}{\sigma_1} > 1\cdot 0$$

$$\frac{\sigma_1}{\sigma} = 1\cdot 2\left(\frac{\sigma_1}{\sigma_2}\right) \qquad \text{when } 5\cdot 0 > \frac{\sigma_2}{\sigma_1} > 1\cdot 0$$

I.3.1 Biaxial State Adopting the Octahedral Stress Approach

In the biaxial state of stress, the value of the shear modulus is assumed to be a function of the confined pressure and the second invariants of the stress and strain. The yield criterion in general is represented by $\tau_0 = f(\sigma_0)$, in which the hydrostatic pressure σ_0, as explained later, is given by $\sigma_0 = \frac{1}{3}J_1$. Again using Kupfer et al.'s experimental results [29], a good approximation is obtained from a yield function with a linear expression of the form

$$\tau_0 + A\sigma_0 - B = 0 \qquad (I.24)$$

in which A and B are constants, and τ_0 and σ_0 are the corresponding octahedral stresses. As a result, eqn (I.24) will have two forms—one representing the biaxial compression region and the other representing the biaxial tension and compression–tension regions. It is interesting to note that eqn (I.24) can now be extended to the triaxial state as well, in which case a surface of revolution with an axis $\sigma_1 = \sigma_2 = \sigma_3$ can easily be formed. The linear form of eqn (I.24) is a cone. Equation (I.24) has two forms, and hence two different cones can be plotted.

I.3.1.1 Determination of Constants A and B

(a) Biaxial compression stress state. Under this condition, if $\sigma_1 = \sigma_2$, the concrete, as non-homogeneous material, will yield when it reaches its biaxial yield strength in compression $R_c f_c'$. The numerical value of R_c will simply be the ratio of biaxial/uniaxial strength of the concrete.

Using
$$\tau_0 = \frac{\sqrt{2}}{3} \sigma_{cu}^2$$
together with $R_c f_c'$, eqn (I.24) becomes
$$+ \tfrac{\sqrt{2}}{3} R_c f_c' (A - 1) - B = 0 \tag{I.25}$$
For a uniaxial value at yield with $\sigma_1 = \sigma$ and $\sigma_2 = \sigma_3 = 0$ eqn (I.24) leads to
$$+ \tfrac{1}{3} f_c' (A - \sqrt{2}) - B = 0 \tag{I.26}$$
Solving eqns (I.24) and (I.26), the values for the constants A and B are
$$A = \sqrt{2} L_1 \qquad B = \frac{\sqrt{2}}{3} L_1 f_c' \tag{I.27}$$
where
$$L_1 = \left(\frac{R_c - 1}{2R_c - 1} \right)$$

Equation (I.24) now represents a simplified function of the type given in eqn (I.28):
$$\tau_0 + \sqrt{2} L_1 \sigma_0 + \frac{\sqrt{2}}{3} L_1 f_c' = 0 \tag{I.28}$$

The intersection of the yield surface, based on eqn (I.28) for the σ_1–σ_2 plane, is now represented by
$$\sigma_1^2 L_1^2 - \sigma_1^2 + 2\sigma_1 \sigma_2 L_1^2 + \sigma_1 \sigma_2 + \sigma_2^2 L_1^2 - \sigma_2^2 \\ - 2L_1^2 \sigma_1 f_c' - 2L_1^2 \sigma_2 f_c' + L_1^2 f_c'^2 = 0 \tag{I.29}$$
Equation (I.28) is a general equation for the intersection of yield surfaces.

The value of R_c can be obtained by experiment or from the test results of the several investigators mentioned earlier. This general equation can become the basis of comparison and can test the sensitivity of the investigator's results. For example, Kupfer's tests give the value of $R_c = 1\cdot16$, therefore eqn (I.28) leads to an expression which indicates that eqn (I.28) represents an ellipse:

$$\left(\frac{2L_1^2 + 1}{2}\right)^2 - (L_1^2 - 1)^2 < 0 \qquad (I.30)$$

The construction of such an ellipse is a sure test of the authenticity of these equations.

At the centre of the ellipse it is obvious that $\sigma_1 = 0 = \sigma_2$. The direction of axes of such an ellipse can be determined in the manner given below.

If the inclination of σ_1, say, is α', then

$$\tan 2\alpha' = \frac{(2L_1^2 + 1)}{(L_1^2 - 1) - (L_1^2 - 1)} = \infty \qquad (I.31)$$

from which $\alpha' = \pi/4$ or $3/4\pi$.

(b) Biaxial tension and combined compression–tension stress states. In this case two of the three stresses are less than zero and the third stress is zero. For example, $\sigma_1 < 0$, $\sigma_2 < 0$ and $\sigma_3 = 0$ will give a biaxial tension state of stress. Under the uniaxial tension condition the concrete will reach its cracking condition with a limiting value of $R_t f_c''$, where $R_t = \sigma_{tu}/f_c''$. When f_c'' is less than zero and σ_{tu} is greater than zero, the value of R_t is normally around 10%, which is in full agreement with the findings of several investigators [29–39]. As with eqn (I.25) and eqn (I.26), the yield function becomes

$$-\tfrac{1}{3}R_t f_c''(A + \sqrt{2}) - B = 0 \qquad (I.32)$$

Equation (I.29) is available for uniaxial compression. The second set of values for the constants A and B is determined simultaneously:

$$A = \sqrt{2}\,L_3 \qquad B = \frac{2\sqrt{2}}{3} L_4 \qquad (I.33)$$

The yield function for a single constitutive equation defined by eqn (I.24) can now be expressed as

$$\tau_0 + \sqrt{2L_3}\sigma_0 + \frac{2\sqrt{2}}{3}L_4 = 0 \qquad (I.34)$$

where

$$L_3 = \frac{1-R_t}{1+R_t} \qquad L_4 = \left(\frac{R_t}{1+R_t}\right) f'_c$$

The yield surface intersection of eqn (I.35) for the σ_1–σ_2 plane becomes

$$\sigma_1 L_3^2 - \sigma_1 + 2L_3^2 \sigma_1 \sigma_2 + \sigma_1 \sigma_2 + \sigma_2 L_3^2 - \sigma_2 + 2\sigma_1 L_3 L_4 \qquad (I.35)$$

The polynomial of eqn (I.35) reduces to a simplified equation, eqn (I.36), by using the accepted value of $R_t = 0.1$:

$$\left[\left(\frac{2L_3^2+1}{2}\right)^2 - (L_3-1)^2\right] > 0 \qquad (I.36)$$

Equation (I.36) is very similar to eqn (I.30), but represents a hyperbola for which σ_1 and σ_2 at the centre are

$$\sigma_1 = \frac{4L_3 L_4}{(1-4L_3^2)} = \sigma_2 \qquad (I.37)$$

The direction of this hyperbola with respect to the σ_1 axis, say at an angle of α', is still $\pi/4$ or $\tfrac{3}{4}\pi$ since

$$\tan 2\alpha' = \frac{2L_1^2+1}{(L_3^2-1)-(L_3^2-1)} = \infty \qquad (I.38)$$

I.4 THE INCREMENTAL CONSTITUTIVE LAWS FOR CONCRETE UNDER A MULTIAXIAL STATE OF STRESS

Numerous attempts have been made [32–125] to define the mechanics of failure of concrete under a complex state of stress. Some investigators [57–125] have also made attempts to derive a single failure criterion for concrete. Due to crude assumptions and limited boundary conditions, most of these laws or criteria lack consistency in their results and, in some cases, they are diametrically opposed to one another. Although most of them [32–125] do recognise concrete as a redundant structure, able to carry loads even when cracks have extended through the mortar mix, advances in the use of prestressing techniques for prestressed concrete reactor vessels (PCRV) have necessitated a thorough investigation of the existing failure criteria. In PCRV, due to variations in loading conditions, the principal stresses at any time and any point may vary from biaxial and triaxial compression to compression–tension–tension is any combination. These

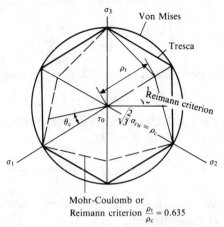

Fig. I.12a. View along space diagonal. Principal stress space for different failure theories. (Note: $\sigma_{cu} = f_{cu}$.)

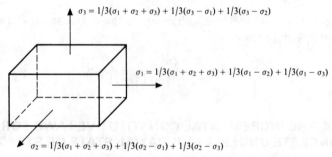

Fig. I.12b. Values of stress resultants.

instant changes in the principal stress profile are not directly covered by these laws or criteria. However, with the advent of computers, these instant changes can be detected and simulated by iteration procedures. In the following pages a systematic review is given of the work of some well-known investigators.

I.4.1 General Review of Several Failure Criteria and Strength Theories (Existing Failure Theories)

Failure criteria proposed for various materials have been reviewed by Nadai [18], Timoshenko [101] and Marin [102].

Many of the proposed criteria are given in empirical form, e.g. the maximum principal stress of Lamé [103], Clapeyron [104] and Rankine

Material Properties and Strength of Concrete

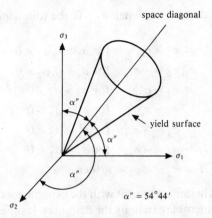

Fig. I.12c. Von Mises yield surface—cohesionless material.

[105]; the maximum shear stresses of Tresca [108], Coulomb [109], Navier [110] and Mohr [111]; and the octahedral shear and normal stresses of Balmer [16], Nadai [18], Wastlung [22], Bresler and Pister [23, 42], Kupfer and Gerstle [31], Vile [40] and Bellamy [43], etc. Another independent theory expressed in terms of shear strain energy has been suggested by Von Mises [35]. Most of the elasto-plastic and crack analyses do make use of these criteria. Nevertheless, they are better suited to metallic structures because they do not predict characteristics such as concrete in tension and compression, and material instability leading to a cracking phenomenon. In order to understand such characteristics, some investigators [48–125] modify their failure criteria in order genuinely to predict the behaviour of concrete under a complex state of stress.

The Tresca criterion is a one-parameter theory based on the assumption that failure occurs when the maximum shear stress reaches a limiting value. The yield surface for a cohesive material is represented by a regular hexagon with its axis on the space diagonal. For concrete with cohesionless ingredients the Tresca criterion gives a yield surface which is represented by a regular hexagonal pyramid with its axis on the space diagonals and its apex at the origin. Both these cases are shown in Fig. I.12a. The maximum shear stress τ_{max} is generally $(\sigma_3 - \sigma_1)/2$. For simple tension when $\sigma_2 = \sigma_1 = 0$ the maximum shear stress at yield will be $\frac{1}{2}\sigma_{cu}$, where σ_{cu} is the ultimate or yield stress. The Tresca criterion then asserts that yielding will occur when any one of the following conditions is met:

$$\sigma_3 - \sigma_2 = \pm\sigma_{cu} \qquad \sigma_2 - \sigma_1 = \pm\sigma_{cu} \qquad \sigma_1 - \sigma_3 = \pm\sigma_{cu} \qquad (I.39)$$

For the biaxial state of stress when $\sigma_3 = 0$, the following conditions hold good:

$$\begin{aligned}
\sigma_1, \sigma_2 &= \sigma_{cu} & \text{for } \sigma_1 > 0, \sigma_2 < 0 \\
\sigma_1, \sigma_2 &= -\sigma_{cu} & \text{for } \sigma_1 < 0, \sigma_2 > 0 \\
\sigma_2 &= \sigma_{cu} & \text{for } \sigma_2 > \sigma_1 > 0 \\
\sigma_1 &= \sigma_{cu} & \text{for } \sigma_1 > \sigma_2 > 0 \\
\sigma_1 &= -\sigma_{cu} & \text{for } \sigma_1 < \sigma_2 < 0 \\
\sigma_2 &= -\sigma_{cu} & \text{for } \sigma_2 < \sigma_1 < 0 \\
\end{aligned}$$
$$\text{For } \sigma_1 = -\sigma_2 = \bar{y}, \text{ then } \bar{y} = \tfrac{1}{2}\sigma_{cu} \qquad (\text{I.39a})$$

As this criterion is in fair agreement with the experimental results [48–125] it is used to a considerable extent by the designers. It suffers, however, from one major drawback, i.e. it will always be necessary to know in advance the maximum and minimum principal stresses. It is extremely difficult to forecast these stresses in advance in structures.

Von Mises theory, on the other hand, is a distortion energy theory, and failure is assumed to occur when the shear strain energy density exceeds a critical value or when the octahedral shear stress exceeds a limiting value. For three-dimensional systems the three principal stresses are expressed in Fig. I.12b, and the first term in each expression is the hydrostatic mean stress:

$$\sigma_0 = \tfrac{1}{3}(\sigma_1 + \sigma_2 + \sigma_3) = \tfrac{1}{3}J_1 \qquad (\text{I.40})$$

which acts on all faces of the cube and which produces a volumetric strain only. The second and third terms are proportional to the maximum shear stress values and it is these stresses that cause distortion or energy of distortion. The volumetric strain ε_v and volumetric strain energy U_v can easily be computed from Figs I.12b and I.12c, and are given in eqn (I.41):

$$\varepsilon_v = \left(\frac{\sigma_1 + \sigma_2 + \sigma_3}{E}\right)(1 - 2v) \qquad U_v = \frac{\varepsilon_v}{6} \qquad (\text{I.41})$$

For a three-dimensional system the strains in three directions, $\varepsilon_1, \varepsilon_2, \varepsilon_3$, which are dependent on the Poisson value of v and Young's modulus E, are difficult to determine. Using constant values of v and E as assumed by Von Mises, the shear strain energy can be computed by subtracting the volumetric strain energy from the total strain energy:

$$\tfrac{1}{2}/E[\sigma_1^2 + \sigma_2^2 + \sigma_3^2 - 2v(\sigma_1\sigma_2 + \sigma_2\sigma_3 + \sigma_1\sigma_3)]$$
$$- \tfrac{1}{6}/E[(\sigma_1 + \sigma_2 + \sigma_3)^2(1 - 2v)] \qquad (\text{I.42})$$

Substituting $(1 + v) = E/2G$, eqn (I.42) becomes

$$\tfrac{1}{12}/G[(\sigma_1 - \sigma_2)^2 + (\sigma_2 - \sigma_3)^2 + (\sigma_3 - \sigma_1)^2] \tag{I.43}$$

or

$$(\sigma_1 - \sigma_2)^2 + (\sigma_2 - \sigma_3)^2 + (\sigma_3 - \sigma_1)^2 = \text{a constant}$$

For a case of pure shear when $\sigma_3 = -\sigma_1$ and $\sigma_2 = 0$, Von Mises general yield condition becomes (Figs I.12a and I.12c)

$$\tfrac{1}{2}[(\sigma_1 - \sigma_2)^2 + (\sigma_2 - \sigma_3)^2 + (\sigma_3 - \sigma_1)^2] = \sigma_{cu}^2 = \bar{y}^2 \tag{I.44}$$

Using tensor notation, the second invariant I_2 of the stress deviator at the yield point becomes

$$I_2 = \tfrac{1}{3}\sigma_{cu}^2 = \tfrac{1}{6}(\sigma_1 - \sigma_2)^2 + (\sigma_2 - \sigma_3)^2 + (\sigma_3 - \sigma_1)^2 = \bar{y}^2 \tag{I.45}$$

A yield stress in pure shear $\bar{y}$ becomes

$$\bar{y} = \frac{1}{\sqrt{3}} \sigma_{cu} \tag{I.46}$$

Von Mises criterion predicts a pure shear yield stress which is about 15% higher than predicted by the Tresca criterion. The criterion is widely used since no knowledge is needed of the relative magnitude of the principal stresses. Equations (I.40) and (I.46) are unable to cover areas such as instant tension–compression and unstable cracking phenomena. However, for elasto-plastic problems involving nonlinearity arising from the nonlinear material and geometric behaviour, the stress–strain relation can be expressed in incremental form by combining the yield criteria of the Von Mises and Prandtl–Reuss [35, 36] equations. This idea is used in the following chapters for the solution of various problems.

Hencky and Nadai [37] later showed that it was convenient to assume that yielding can take place when the octahedral shear stress τ_0 reaches the octahedral shear stress at yield in simple tension. Referring to eqn (I.44) this then becomes

$$\tau_0 = \frac{\sqrt{2}}{3}\sigma_{cu}^2 = \tfrac{1}{3}\sqrt{[(\sigma_1 - \sigma_2)^2 + (\sigma_2 - \sigma_3)^2 + (\sigma_3 - \sigma_1)^2]} \tag{I.47}$$

For a simple tension at yield, Fig. I.12a shows the value of τ_0, which is given by

$$\tau_0 = \frac{\sqrt{2}}{3}\sigma_{cu}^2 = \frac{\sqrt{2}}{3}J_2 \tag{I.48}$$

Alternatively, this can be looked upon as stating that yielding will occur when the second invariant J_2 of the stress deviator tensor reaches a critical value. This is the stage where the importance of the octahedral shear stress becomes imminent. If a suitable relationship between a normal stress and this octahedral shear stress can be found, then it could be possible to define the mechanics of failure criteria. In general terms this relationship can be written as

$$F(\sigma_0, \tau_0) = 0 \qquad (I.49)$$

$$\tau_0 = f(\sigma_0) \qquad (I.50)$$

Whatever the type of loading on structures, using eqns (I.49) and (I.50) should give rise to normal or shear stresses, or both, in the concrete of structures. As concrete is a non-homogeneous and anisotropic material, the simulation of some boundary conditions under a complex state of stress in eqns (I.49) and (I.50) is, by itself, a complex phenomenon as these boundary conditions depend on several dependent variables. The important variables amongst them are the loading history and time, temperature and creep dependent values of E and v at any axis of the failure surface.

Mohr [111] simplifies eqns (I.49) and (I.50) by stating a failure criterion in which the limiting shear stress τ_0 in any plane is dependent upon the normal stress σ_0 of eqn (I.50) in the same plane. This means that the intermediate principal stress has no influence on the failure. This may present problems if it is applied to the behaviour of concrete in complex structures. A special version of the Mohr criterion is Coulomb's criterion, which is a two-parameter theory in which the coefficient of friction μ and cohesion value c are used. All this involves is the achievement of a limiting Mohr envelope for all possible Mohr circles that can be drawn for different states of stress. The value of τ_0 will then become

$$|\tau_0| = c + \mu\sigma \qquad (I.51)$$

It is interesting to note that for $\mu = 0$ Coulomb's criterion corresponds to Tresca's criterion when $\sigma_3 \geq \sigma_2 \geq \sigma_1$. This theory, as is now evident, can be the basis of a satisfactory criterion for triaxial compression only when the two lesser principal stresses are equal. However, this fails when one of them is zero or becomes negative, which may well be the case in complex structures under extreme loads.

This point has been supported by Bresler and Pister [23], Bellamy [43], Balmer [16], and Karni and McHenry [83]. Vile [40] later took this further to indicate that the limiting tensile strain at discontinuity depends on the state of stress, and increases with the degree of hydrostatic compression shown by the value of mean normal stress. This means that the tensile failures due to compressive forces (mainly shear bond failures) occur at higher strains than those due to direct tensile forces (mainly tensile bond failures). This observation is supported by Sturman *et al.* [38]. It is interesting to note that, although failure due to slip lines in the Mohr–Coulomb case is widely accepted because of, amongst other things, the creation of cones in the uniaxial compressive stress, it was found by Wastlung [22] and Kupfer [113] that the cone creation can easily be explained by the friction on the loaded surface, and hence that the compressive failure in the biaxial state of stress is generally followed by splitting perpendicular to the direction of the greater principal strain.

Rüsch [55] indicated the significance of creep and creep strain, and stated that the behaviour and ultimate strength of concrete are affected by the rate of loading. At slow rates the creep strain becomes significant at a small proportion of ultimate load. He therefore suggested that the static modulus of elasticity should be defined in terms of secant value at an appropriate stress level. Using such a relation he showed that the ultimate compressive strength of concrete under long-term loading is around 80% of the short-term ultimate strength. This is relevant to the short- and long-term concrete behaviour in some structures. It was not clear whether or not his results took into consideration short- and long-term thermal effects—a vital case in certain complex structures. In any event, the behaviour of concrete under creep and thermal effects, which is very much the environment in reactor vessels, is something that was not considered in earlier criteria.

Newman and Newman [21] have dealt more fully with the relationship expressed in eqns (I.43), (I.48) and (I.50) by averaging data obtained by Richart *et al.* [14], Balmer [16], Bresler and Pister [23], Karni and McHenry [83], Bellamy [43], and Weigher and Becker [57], and have plotted the values at failure of mean normal stress σ_0 against the value of shear strain. They have concluded that a small tensile stress combined with compression can, to a greater extent, reduce the strength of concrete to a value well below the biaxial compression line. The value of octahedral normal stress was kept as σ_0, as given in eqn (I.47). They have discovered that a short cylindrical specimen will fail under a triaxial state of stress if the height/width ratio is less than about $1\frac{1}{2}$ and will fail under a uniaxial compressive state of stress ($\sigma_1 = \sigma_2 = 0$) if the height/width ratio is greater than $2\frac{1}{2}$. They consider a true uniaxial tensile stress to be equal to one of the values of σ. Throughout their work they have considered concrete to be a homogeneous and isotropic material.

Hannant and Frederick [58] have suggested a failure criterion for concrete which is valid only for multiaxial compressive stresses. Simple equations are presented which take into consideration the form of the failure surface as a three-lobed curve which is not rotationally axisymmetric.

Two cases have been reviewed by Hannant and Frederick. For isotropic material conditions these cases are

$$\sigma_3 > \sigma_1 = \sigma_2$$
$$\sigma_3 = \sigma_{cu} + 4\sigma_2 \quad \text{(Case 1)} \tag{I.52}$$
$$\sigma_3 = 1{\cdot}75\sigma_{cu} + 3\sigma_2 \quad \text{(Case 2)}$$

Based on short-term loading conditions, they arrived at the conclusion that for a safe design when the maximum principal stress is increased by a factor of two while other stresses remain constant, failure will not occur when

$$\sigma_1 = \sigma_2 \quad \text{and} \quad \sigma_3 = \tfrac{1}{2}\sigma_{cu} + 2\sigma_2$$
$$\sigma_3 = \sigma_1 \quad \text{and} \quad \sigma_2 = \tfrac{1}{2}\sigma_{cu} + 2\sigma_2 \tag{I.53}$$
$$\sigma_3 = \sigma_2 \quad \text{and} \quad \sigma_1 = \tfrac{1}{2}\sigma_{cu} + 2\sigma_2$$

For a maximum stress of $4\sigma_{cu}$, the factor of safety is found to be 1·75. The introduction of a safety factor in the stress space is plausible. This work is partially recommended in the new British code of practice [89]. Figure I.13 shows a failure surface covering both the cases given above.

Although this criterion is quite an improvement upon the previous criteria nevertheless, concrete has been considered as a material behaving isotropically. It might be interesting to look at the numerical work done by Link [59], who has presented a numerical analysis of the failure criterion of concrete incorporating a material behaviour law for concrete, in the elastic range. The analysis is based only on the biaxial test results of Kupfer and co-workers [29, 31] and Rüsch [55].

An isotropic stress–strain relation has been derived with two Poisson's ratios v_1 and v_2 and two Young's moduli E_1 and E_2 in the entire biaxial stress field covering the domains of tension–tension, tension–compression and compression–compression. Although such an analysis is best suited to ultimate load analyses using finite elements, the relationship of initial and new values of v and E are based on lightweight concrete. The use of a true relationship between v and E would certainly produce an accurate failure criterion under a biaxial state of stress.

A criterion for concrete failure suggested by Baker [60] is based on the fact that concrete has a heterogeneous system, and the principal causes of

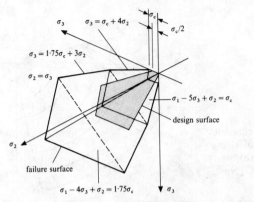 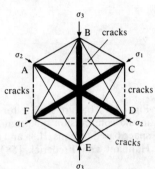

Fig. I.13. Concrete failure surface under triaxial compression (after Hannant and Frederick [58]). (Note: $\sigma_c = f'_c \sigma_c = \sigma_{cu}$.)

Fig. I.14. Isometric of tetrahedral model (after Baker).

cracking and failure are the differential stiffness between the aggregate matrix and interface, and the bond interaction and the weakness in tension of the mortar mix. He has modelled the failure surface by using a tetrahedron (Fig. I.14) in which the Poisson thrust ring, plastic flow and crack effects are represented by the relative stiffnesses of rods, which, in turn, depend on the changing load behaviour. He has developed expressions assuming that cracks form first in mortar pockets and then eventually extend around the stone interfaces. It is an impressive contribution to improving representation of the situation in which even the Poisson's ratio and the Young's modulus E change rapidly as the failure approaches. Using the convention for co-ordinate axes adopted by Baker, the two equations obtained are

$$\frac{\sigma_3 + \sigma_1}{f'_c} = 1 + 4 \cdot 5 \left(\frac{\sigma_2}{f'_c}\right) \quad \text{for } \sigma_3 > \sigma_1 = \sigma_2 \qquad (I.54)$$

$$\frac{\sigma_3 + \sigma_1}{f'_c} = 2 \cdot 2 + 10 \left(\frac{\sigma_2}{f'_c}\right) \quad \text{for } \sigma_3 = \sigma_1 > \sigma_2 \qquad (I.55)$$

For $\sigma_2 = 0$, the factor of safety

$$\frac{\sigma_3}{f'_c} = \frac{\sigma_1}{f'_c} = 1 \cdot 1 \qquad (I.56)$$

The test case in eqn (I.56) proves what Vile [40] has established. It is interesting to note that, using eqns (I.54) and (I.55) throughout the triaxial

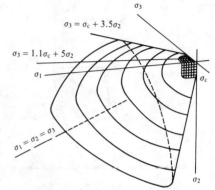

Fig. I.15. Failure surface for concrete in the compression quadrant of stress space based on eqns (II.8a and 8b) (after Hannant and Frederick [58]). (Note: $\sigma_c = f_c''$.)

compression quadrant of stress space, the following relationships are achieved, respectively:

$$\sigma_3 = f_c'' + 3\cdot 50\sigma_2 \qquad \sigma_3 = 1\cdot 1 f_c'' + 5\sigma_2 \qquad (I.57)$$

When they are plotted in the stress space, as shown in Fig. I.15, the failure envelope agrees both in shape and magnitude, as suggested by Hannant and Frederick [58] in eqns (I.56) and (I.57). These relationships are extremely sensitive to the assumed values of v and E. The safety factors thus depend both on values of v and values of E, and could vary from 1·1 to 2·5. Further research is needed in Baker's equations for $(\sigma_1 + \sigma_2 + \sigma_3)$ in excess of 16. Chawalla [61], very similarly to other investigators, shows the three-dimensional envelope which defines the onset of failure as a surface of revolution having an axis of symmetry, the meridian of which can be represented by σ_0–τ_0. The surface is similar to a nose-cone of a rocket with a peak partially flattened to a lesser or greater extent, depending upon material properties. The only difference is that $\sigma_1 = \sigma_2 = \sigma_3 = \sigma_{cu}$ defines the stress in the material at rupture caused by equal tensile forces, and the failure is anticipated if the point is outside the σ_0–τ_0 envelope.

Chinn and Zimmerman have reported their own work on [62] this subject, and it appears they are the only researchers who have investigated the failure of concrete when the cases of both axial pressure and confining pressure predominate. The outcome of their investigation is that the maximum principal stress, say σ_3, is equal to $\sigma_{cu} + 4\sigma_1$ (or σ_2). Akroyd [48] has found that this is only true up to about $\sigma_{cu} + 3\sigma_1$ (or σ_2). These findings again are very much in line with those found by Hannant and Frederick [58], Baker [60] and Garas [68].

Reimann [63] gives a failure criterion, and using Fig. I.12a the compressive meridian ρ_c is expressed by

$$\frac{\frac{1}{\sqrt{3}}J_1}{f'_c} = A\frac{\rho_c^2}{f'_c} + B\frac{\rho_c}{f'_c} + C \tag{I.58}$$

Equation (I.58) is a parabolic equation where A, B and C are constants, and $J_1 = \sigma_1 + \sigma_2 + \sigma_3$ is the first invariant of the stress tensor. The other meridians are assumed to be affined with the compressive meridians $\rho = \phi(\theta_c)\rho_c$ for $-60° < \theta_c < 60°$. This corresponds to a failure curve in the deviatoric plane consisting of straight lines which are tangents to a circle of which the radius corresponds to the tensile meridian. As shown in Fig. I.12a, besides having a constant value, the surface is not smooth and it is only valid for compressive stresses.

Argyris [64] suggested the following three-parameter criterion involving all the stress invariants:

$$A\frac{J_1}{f'_c} + (B - C\cos 3\theta_c)\left(\frac{\sqrt{I_2}}{f'_c}\right) - 1 = 0 \quad \text{(straight meridians)} \tag{I.59}$$

where A, B and C are constants, and f'_c is the uniaxial compressive cylindrical strength, i.e. straight meridians. The failure curves are similar in all deviatoric planes and concavity exists when $\rho_t/\rho_c < 0.777$.

Willam and Wanke [130] suggested a five-parameter criterion in which the tensile and compressive meridians are expressed by

$$\sqrt{I_{2,t}} = A_{01} + A_{11}J_1 + A_{22}\left(\frac{J_1^2}{f'_c}\right) \quad \text{tensile meridian } (\theta_c = 0°)$$

$$\sqrt{I_{2,c}} = B_{01} + B_{11}J_1 + B_{22}\left(\frac{J_1^2}{f'_c}\right) \quad \text{compressive meridian } (\theta_c = 60°) \tag{I.60}$$

where A_{01}, A_{11}, A_{22}, B_{01}, B_{11} and B_{22} are constants. The hydrostatic axis intersects these two meridians at the same point. They succeeded in finding an expression for the failure curve in the deviatoric plane, which is convex for all ratios $\frac{1}{2} < \rho_t/\rho_c < 1$. This was obtained by prescribing this curve as part of an elliptic curve.

Referring to Fig. I.12a, Launay and Gachon [66] suggested the following relation to express the failure curve in the deviatoric plane:

$$\rho^2\left[\frac{\cos^2 \frac{3}{2}\theta_c}{\rho_c\bigg/\left(\frac{J_1^2}{3}\right)} + \frac{\sin^2 \frac{3}{2}\theta_c}{\rho_t\bigg/\left(\frac{J_1^2}{3}\right)}\right] = 1 \tag{I.61}$$

Table I.1
Pressure vessel parameters

O		
P = gas pressure = 2.45 MN/m² Oldbury	P = 2.64 MN/m² Wylfa	P = 4.03 MN/m² Hinkley point
H	**D**	**FT**
P_G = 4.03 MN/m² Hartlepool	P_G = 3.30 MN/m² Dungeness B	P_G = 4.86 MN/m² Fort St Vrain
P = 3.04 MN/m² Chinon	P = 2.76 MN/m² St Laurent	P = 4.48 MN/m² Bugey

Parameter	O	D	H	FT
$2R_1$ = Internal diameter	18.3m	19.95m	13.1048m	10.67m
$2H_1$ = Internal height	23.5m	17.70m	29.261m	21.95m
t = Wall thickness	4.6m	3.82m	6.40m	2.5m
D_1 or D_2 = Cap thickness	6.7m	6.2T 5.94B	5.5m	5.0m

Table I.1—contd.

Concrete
- σ_{cu} = minimum cube strength of concrete at 28 days — $41.34\,\text{kN/mm}^2$
- σ_y = yield strength — $0.66\sigma_{cu}$
- σ_t = tensile strength — $+0.1\sigma_{cu}$
- E_e = elastic modulus — $+41.4\,\text{kN/mm}^2$
- E_p = plastic modulus — $+0.476E$
- ε_{cu} = ultimate strains — 0.0035
- v = Poisson's ratio — 0.18
- α_T = coefficient of linear thermal expansion — $8.0\,\mu\text{M/m}\,°\text{C}$
- K = thermal conductivity — $1.75\,\text{W/m}\,°\text{C}$
- a' = coefficient for aggregates — $0.65, 0.68, 0.87, 0.87$

Conventional steel
- $\bar{y} = \sigma_y$ = yield strength — $4516\,\text{N/mm}^2\ (50\,\varnothing)$
  $4400\quad (25\,\varnothing)$
- E_e = elastic modulus — $200\,\text{kN/mm}^2$
- E_p = plastic modulus — $0.1E = 20\,\text{kN/mm}^2$

Liner
- t_s = thickness — $12\,\text{mm} + 10\%,\ 12\,\text{mm} + 10\%,\ 19\,\text{mm} + 5\%,\ 19\,\text{mm} + 5\%$
  Up to 40 mm max, up to 40 mm, up to 25 mm, up to 25 mm
- $\bar{Y} = \sigma_y$ = yield strength — $3.4 \times 10^5\,\text{kN/m}^2$
- α_T = coefficient of linear thermal expansion — $10\,\mu\text{M/m}\,°\text{C}$
- k = thermal conductivity — $41.6\,\text{W/m}\,°\text{C}$

Loadings
- Case 1 Prestress and ambient temperature
- Case 2 Prestress + 1.15 × design pressure + ambient temperature
- Case 3[a] Prestress + design pressure + temperature
- Case 4[a] Overload (prestress + increasing pressure + temperature)

[a] Short- and long-term conditions apply [126]

where

$$\rho = \frac{\sqrt{2\rho_c \rho_t}}{\sqrt{(\rho_t^2 + \rho_c^2) + (\rho_t^2 - \rho_c^2)\cos 3\theta_c}} \quad \text{(I.61a)}$$

where

$$\cos 3\theta_c = \frac{3\sqrt{3}}{2}[(I_3)/(I_2)^{3/2}]$$

and

$$I_3 = \tfrac{1}{3}(\sigma_1^3 + \sigma_2^3 + \sigma_3^3)$$

However, this gives convexity only for $\rho_t/\rho_c > 0.745$, and for nearly all practical applications $\rho_t/\rho_c > 0.745$. The tensile and compressive meridians have not been prescribed in this equation.

The details of these several parameter models have been discussed later on in relation to numerical modelling. Bangash has reported [126] triaxial stress results using statistical results based on the mutual relationship of stresses at various incremental stages of deformations of several existing vessels listed in Table I.1. A computer program was developed to compare these stresses individually or in combination and evaluate them in terms of a uniaxial compressive stress. Figures I.15 and I.16 show a relationship between these stresses for a case of triaxial compression–compression–tension. They also indicate a relationship for various increments of the stress components. Results can be similarly obtained for triaxial

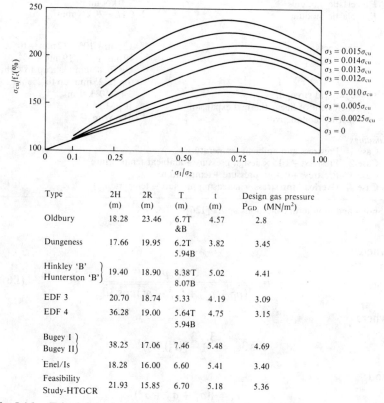

Type	2H (m)	2R (m)	T (m)	t (m)	Design gas pressure P_{GD} (MN/m^2)
Oldbury	18.28	23.46	6.7T &B	4.57	2.8
Dungeness	17.66	19.95	6.2T 5.94B	3.82	3.45
Hinkley 'B' Hunterston 'B'	19.40	18.90	8.38T 8.07B	5.02	4.41
EDF 3	20.70	18.74	5.33	4.19	3.09
EDF 4	36.28	19.00	5.64T 5.94B	4.75	3.15
Bugey I Bugey II	38.25	17.06	7.46	5.48	4.69
Enel/Is	18.28	16.00	6.60	5.41	3.40
Feasibility Study-HTGCR	21.93	15.85	6.70	5.18	5.36

Fig. I.16. Triaxial compression/compression/compression from the vessel operational analysis.

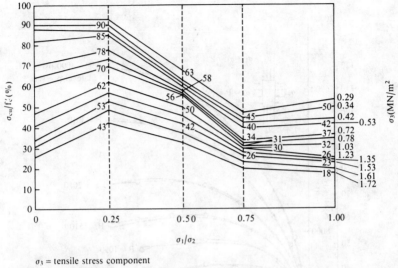

σ_3 = tensile stress component
σ_1, σ_2 = compressive stresses $\sigma_1 < \sigma_2$
σ_c = uniaxial strength of concrete = f'_c
σ_{cu} = computed ultimate load in the direction of high compressive stress = f_{cu}

Fig. I.17. Triaxial compression/compression/tension with specified tensile stress.

compression. If any two stresses are equal and the third is decreased by 1/10 to 3/10 of the value of the others, the values of (maximum stress)/(uniaxial compressive stress 100%) = 200 to 260%.

The triaxial compression results have been reported by Bangash [126] using the hypoelastic concept discussed later in this book. These results have been compared with those obtained by Richart *et al.* [14] in Fig. I.18. Similarly, hypoelastic analysis is compared with the experimental results of Kupfer [113] and summarised in Fig. I.19.

Hobbs [131] has carried out a series of tests for the failure criteria of concrete under triaxial conditions for various ages of concrete. He has obtained an ultimate strength equation of

$$\frac{\sigma_1}{f'_c} = 1 + \frac{4 \cdot 8 \sigma_3}{f'_c} \tag{I.62}$$

where σ_1 and σ_3 are the principal stresses at failure along the 1, 3 axes.

Hobbs has carried out a special triaxial test, contrary to the conventional triaxial tests in which intermediate principal stress is equal to the minor

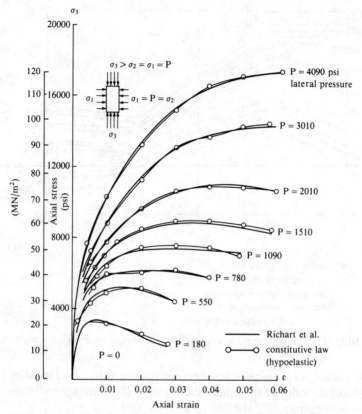

Fig. I.18. Stress–strain curves for concrete in triaxial compression.

principal stress, in which he has managed to equate the intermediate principal stress to the major principal stress. This is known as the 'extension test'. This test has the flexibility so that the biaxial compression plus tension can be included. The results of Hobbs [131] and Kotsovos [132] are summarised in Fig. I.20. Hobbs comes to the following major conclusions:

$$\sigma_u = 0.45 f_{cu} + 3\sigma_3$$

for the ultimate collapse condition where $\sigma_1 > \sigma_2 > \sigma_3 \geq 0$ \hfill (I.63)

$$\sigma_s = 0.3 f_{cu} + 2\sigma_3$$

for the serviceability condition where $\sigma_1 > \sigma_2 > \sigma_3 \geq 0$ \hfill (I.64)

The other important theories, namely hypoelastic [91, 96, 5], bulk

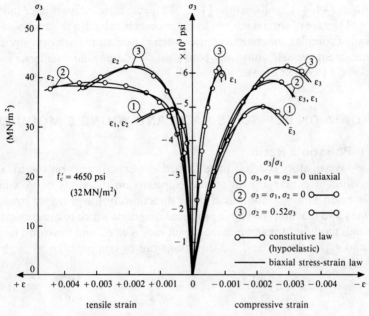

Fig. I.19. Biaxial stress–strain curves.

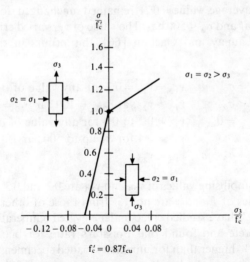

Fig. I.20. Concrete under triaxial stress—extension test results (after Hobbs [131] and Kotsovos [132]).

modulus [44, 47], endochronic [172, 388], etc., are discussed later in this book. They are of special interest to researchers involved in the analysis and design of complex concrete structures, such as nuclear reactor pressure and containment vessels, offshore platforms, underground shelters, silos, bunkers, LPG/LNG tanks, etc.

I.5 DATA ON POISSON'S RATIO AND YOUNG'S MODULUS

I.5.1 Poisson's Ratio

Some investigators [68–87] have carried out experiments on the triaxial behaviour of concrete, with special emphasis on the study of Poisson's value v and Young's modulus E in the directions of the principal stresses. Launay and Gachon [66] have produced diagrams which relate stresses to Young's moduli; for example, the secant modulus E_3 and Poisson's ratios v_{13} and v_{23} in the 1, 3 and 2, 3 directions can be computed as

$$E_{\text{sec}} = E_3 = \frac{\sigma_3}{\varepsilon_3} \qquad v_{13} = -\frac{\varepsilon_1 \sigma_3}{\varepsilon_3 \sigma_3} \qquad v_{23} = -\frac{\varepsilon_2 \sigma_3}{\varepsilon_3 \sigma_3} \qquad (\text{I.65})$$

The values in eqn (I.65) were derived for $\sigma_3 > \sigma_2 > \sigma_1$ on 140-mm cubes subject to temperatures of 20, 40 and 60°C at pressures up to 500 bar. The value of v_{23} (average value = 0·2) remained unchanged for any value $\sigma_1 \leq \sigma_2 \leq 400$ bar and $\sigma_3 \leq 500$ bar. The value of v_{13} varied erratically and, according to Launay and Gachon [66], this occurs in the following manner:

$$\begin{aligned}
&\text{for } \sigma_1 = \sigma_2, && v_{13} = v_{23} = 0\cdot 2 \text{ for any value of } \sigma_3 \\
&\text{for } \sigma_1 < \sigma_2, && v_{13} > v_{23}; \text{ at } \sigma_{1_{\max}} = 100 \text{ bar } v_{13} = v_{23} \\
&\text{for } \sigma_{1_{\max}} = 0, && v_{13} = 0\cdot 2 \text{ to } 0\cdot 4 \text{ for any value of } \sigma_2; \qquad (\text{I.65a})\\
&&& v_{13} = 0\cdot 4 \text{ for } \sigma_3 \text{ up to } 500 \text{ bar} \\
&\text{for } \sigma_2 = \sigma_3 > \sigma_1, && v_{13} > v_{23}
\end{aligned}$$

The suggested stabilising values of v_{23} and v_{13} are 0·2 and 0·3, respectively. The above values of E and v are not far from those of other researchers.

York et al. [67] have studied creep under elevated temperatures suitable for vessel concrete and found that Poisson's ratio for biaxially loaded specimens is 55% higher than for uniaxially loaded specimens. Garas [68] found on wire-wound cylindrical specimens that concrete fails under equal biaxial compression at a stress equivalent to 1·2 to 1·5f'_c. With no axial restraints, it fails at a stress equal to between 0·5 and 0·62f'_c, which in turn is equivalent to a limiting tensile strain of about 150 microstrains. In these experiments Poisson's ratio increases from an initial value of 0·15 to a maximum value of 0·42. Hussain and Saugy [76], whilst assuming the concrete surface at failure to be a cone in tension and a truncated cone in compression, arrived at a figure of 1·25 as the ratio of the biaxial compression strength to uniaxial compression, and $v = 0·25$.

Linse et al. [77] have investigated the strength of concrete under multiaxial all-state stresses of σ_1, σ_2 and σ_3, and under elevated temperatures. They have found that Poisson's ratio of creep at a temperature of 80°C is about 35–50% higher than at normal temperatures. The values suggested range between 0·21 and 0·235.

Gardner tested specimens using the theory developed by Row [78], which relates the stress in the concrete to the instantaneous value of v. The theory relates the ratio of the minimum work done by the axial stress to the instantaneous geometry of a regular packing of uniform spheres. Comparison of this theory with triaxial test results on sand showed good agreement.

Under monotonic loading in tension–tension and compression, Poisson's ratio of 0·2 is quite effective. This value of 0·2 is adequate for uniaxial compression and compression–tension at very low values of stress. Around 0·8f'_c, the value of 0·2 (as pointed out) is too low, and according to Darwin and Pecknold [7a,b] should be computed as

$$v = 0·2 + 0·6\left(\frac{\sigma_2}{f'_c}\right)^4 + 0·4\left(\frac{\sigma_1}{f'_c}\right)^4 \quad (I.66)$$

for uniaxial compression and compression–tension cases.

Elwi and Murray [5, 135], using a least square fit of a cubic polynomial, evaluated an expression for Poisson's ratio. They used the uniaxial compression data of Kupfer [113]. This expression is

$$v = v_0[1 + 1·3763(\varepsilon_R) - 5·36(\varepsilon_R)^2 + 8·586(\varepsilon_R)^3] \quad (I.67)$$

where

v_0 = initial value of Poisson's ratio

$\varepsilon_R = \varepsilon_i/\varepsilon'_0$

ε_i = uniaxial strain in the ith direction

Poisson's ratio under cryogenic temperatures for various conditions [136] are given in Table I.2.

Table I.2
Modulus of elasticity and Poisson's ratio at cryogenic temperatures

Mix no.	Aggregates	Cement content (kg/m³)	Condition	\multicolumn{6}{c}{E_0 (kN/mm²)}	v
				24°C	2°C	−18°C	−59°C	−101°C	−157°C	
1	sand/gravel	415	moist	45·5	45·5	48·3	55·2	62·1	62·1	0·22
2	sand/gravel	326	moist	37·9	40·0	45·5	52·5	58·0	58·7	0·22
2	sand/gravel	326	50% relative humidity	34·5	35·2	35·2	35·2	36·6	37·3	0·23
2	sand/gravel	326	dry	29·7	29·7	29·7	29·7	29·7	29·7	0·25
3	sand/gravel	237	moist	35·9	35·9	42·8	46·2	56·6	57·3	0·18
4	expanded shale	355	moist	20·7	20·7	22·8	26·2	29·0	29·7	0·21

The German Specification DIN 4227 gives the following Poisson ratios:

$$E_0 \text{ (kN/mm}^2\text{):} \quad 24 \quad\quad 30 \quad\quad 35 \quad\quad 40$$
$$v: \quad\quad 0{\cdot}15\text{–}0{\cdot}18 \quad 0{\cdot}17\text{–}0{\cdot}20 \quad 0{\cdot}20\text{–}0{\cdot}25 \quad 0{\cdot}25\text{–}0{\cdot}30 \quad (\text{I.68})$$

1.5.2 Young's Modulus

Young's modulus does not change substantially with time. After 28 days, in general, about 86% of the final value is reached and, after three months, it attains 99·7% of its final value. The values of E_0 defined in some codes are given below:

British Standard BS 8110:
$$E_0 = 5{\cdot}5\sqrt{f_{cu}/\gamma_m} \quad \text{kN/mm}^2 \tag{I.69}$$

Indian Standard IS 456–1979:
$$E_0 = 5688\sqrt{f_{cu}} \quad \text{N/mm}^2 \tag{I.70}$$

The European Concrete Committee (CEB):
$$E_0 = 6000\sqrt{f_{cu}} \quad \text{N/mm}^2 \tag{I.71}$$

The American Concrete Institute (ACI):
$$E_0 = 4800\sqrt{f_{cu}} \quad \text{N/mm}^2 \tag{I.72}$$

For lightweight concrete these values are multiplied by $(D_c/2300)^2$, where D_c is the density of lightweight concrete in kg/m³.

The German Code DIN 4227 gives E_0 values already given in eqn (I.68). The E_0 values for concrete under cryogenic conditions are given in Table I.2.

The US Bureau of Reclamation [102] quoted by Marin expresses E_0 values in the form of

$$E_0 = E(\bar{\tau}) = [4{\cdot}4 \times 10^6 (1 - e^{-0{\cdot}07\bar{\tau}})\,\text{lbf/in}^2] \times 6{\cdot}895\,\text{kN/m}^2 \tag{I.73}$$

Hobbs reported a theoretical expression for E_0:

$$E(\bar{\tau}) = E_{cm}(\bar{\tau})\left\{1 + \frac{2V_a[E_a - E_{cm}(\tau)]}{E_a + E_{cm} - V_a[E_a - E_{cm}(\tau)]}\right\} \quad (I.74)$$

where $\bar{\tau}$ = number of days, E_{cm} = Young's modulus of cement paste, E_a = Young's modulus of aggregate, and V_a = aggregate volume concentration. Green and Swanson [145] give an empirical equation:

$$E(\bar{\tau}) = E\left\{0.4 + 0.6\left[\frac{f(\bar{\tau})}{f_{cu}}\right]\right\} \quad (I.75)$$

where E = Young's modulus at 28 days, $f(\bar{\tau})$ = cube strength of concrete at time $\bar{\tau}$, and f_{cu} = cube strength of concrete at 28 days.

I.5.3 Temperature Effects on Young's Modulus

The influence of temperature distribution on the Young's modulus is much more substantial.

Directly predicted thermal stresses are dependent on the values assumed for Young's modulus and, therefore, the stresses are altered whether or not the change in the elastic properties is the same throughout. Therefore Young's moduli used in constitutive equations need modification in order to suit a particular temperature distribution for given vessel parameters. Three variations of modulus with temperature are plotted in Fig. I.21 for

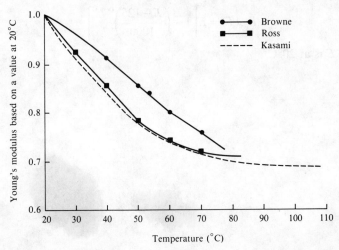

Fig. I.21. Variation of Young's modulus with temperature for concrete.

data from Ross [92, 141], Browne [84] and Kasami [80]. In the analyses that follow these variations are considered for specific temperature distributions.

Equations (I.71) and (I.72) also give expressions for Young's modulus. From the work of Elwi and Murray [5, 135] eqn (I.8a) is differentiated. The tangent moduli with respect to equivalent uniaxial strains are given by

$$E = \frac{d\sigma_i}{d\varepsilon_{0_i}} = \frac{E_0[1 + (2R-1)\varepsilon_R^2 - 2R\varepsilon_R^3]}{\left\{1 + \left[\left(\frac{E_0}{E_{\text{sec}}}\right) + R - 2\right]\varepsilon_R - (2R-1)\varepsilon_R^2 + R\varepsilon_R^2\right\}^2}$$

$$(i = 1, 2, 3) \quad (I.76)$$

where $\varepsilon_R = \varepsilon_i/\varepsilon_0'$.

Bangash and England [94] suggested two simplified expressions for temperature-dependent E:

$$E_T = E\left(1 - \frac{T - 20°C}{137}\right) \quad \text{for } 20°C \le T \le 50°C$$

$$E_T = 0.78E\left(1 - \frac{T - 50°C}{341}\right) \quad \text{for } 50°C \le T \le 85°C$$

(I.77)

where E is taken to be at 20°C and T is any temperature not exceeding 85°C.

A similar equation has been produced by Dibsi [147] describing E_{sec} due to temperature effects as

$$E_{\text{sec}}(T) = E\left[1 - \left(\frac{T - 125}{1000}\right)^{0.48}\right] \quad (I.78)$$

where E is at 20°C.

Chapter II

Numerical Modelling of Concrete Strength and Failure

A literature review is given for the constitutive and numerical modelling of concrete. Concrete is characterised by a materially nonlinear deformation behaviour. The material nonlinearity is assumed to occur due to cracking of concrete in tension and plasticity of concrete in compression. All numerical models discussed in this chapter include the possibility of such effects. They are written specifically in a format suitable for finite element analysis. They have the flexibility to include creep, shrinkage and temperature changes, and also fatigue—discussed later in this book.

II.1 FAILURE MODELS FOR CONCRETE

II.1.1 One-Parameter Model

The Tresca and Von Mises shear stress criteria are one-parameter models and have already been discussed in Chapter I, Section I.4. For numerical analysis the Tresca model is written as

$$f(J_2, \theta) = \sqrt{J_2} \sin\left(\theta + \frac{\pi}{3}\right) - \bar{y} = 0 \qquad (\text{II.1})$$

In terms of invariants J_2 and J_3, eqn (II.1) can also be written as

$$f(J_2, J_3) = J_2^3 - 6{\cdot}75 J_3^2 - 9\bar{y}^2 J_2^2 + 24\bar{y}^4 J_2 - 16\bar{y}^6 = 0 \qquad (\text{II.2})$$

If the material is isotropic and equal yield stress exists in tension and in compression, eqn (II.2) converges to

$$f(J_2, J_3) = J_2^3 - 2{\cdot}25 J_3^2 - \bar{y}^6 = 0 \qquad (\text{II.3})$$

Here $\bar{y}$ can also be approximated as $(1/\sqrt{3})f_{cu}$.

An alternative to the yield stress in pure shear, the octahedral shear stress concept, can be substituted, which is given in the form

$$f(J_2) = J_2 - \bar{y}^2 = 0 \qquad \text{(II.4)}$$

where J_2 is related to τ_0 as given in eqn (I.48).

Using finite element analysis, the Von Mises surface can be combined with the tension cut-off surface in order to take into consideration the limited tensile capacity of concrete. Similarly the same Von Mises surface can be combined with the ultimate compressive strain criterion ε_{cu} for the compressive ductility and crushing of concrete.

II.1.2 Two-Parameter Model

Equation (II.5) assumes that yielding will occur when the shear stress τ_0 reaches the value that depends on the cohesion stress C and the normal stress σ. Figure II.1 shows a yield surface in a three-dimensional stress space. This is graphically represented by a hexagonal pyramid, the axis of which lies along the space diagonal, known as the hydrostatic stress axis. The deviatoric plane (π-plane) passes through the origin and perpendicular to this hydrostatic axis. Graphical representation shows that it is an

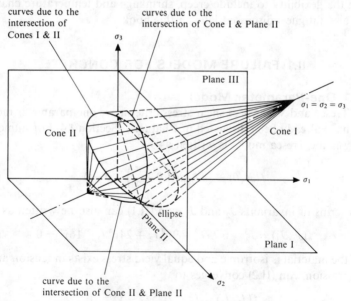

Fig. II.1. Graphical representation of octahedral shearing stress yield criteria of concrete.

irregular right hexagon and the yield stress in tension differs from that in compression. In numerical form the yield criterion of eqn (II.5) can be expressed in a number of ways. Chen [148] expresses it as

$$f(I_1, J_2, \theta) = \tfrac{1}{3} I_1 \sin\phi + \sqrt{J_2} \sin\left(\theta + \frac{\pi}{3}\right)$$

$$+ \frac{\sqrt{J_2}}{\sqrt{3}} \cos\left(\theta + \frac{\pi}{3}\right) \sin\phi - C\cos\phi = 0 \quad \text{(II.5)}$$

The final expressions developed by Kojic and Cheathem [149] are written as

$$f(I_1, J_2, \theta) = \tfrac{1}{3} I_1 \sin\phi + \sqrt{J_2} \cos\theta - \frac{\sqrt{J_2}}{\sqrt{3}} \sin\theta + \sin\phi - C\cos\phi = 0$$

$$\text{(II.6)}$$

in which I_1 is the first invariant of stresses and J_2 is the second invariant of stress deviations. The angle θ is the lode angle which is an alternative form of the third stress invariant.

Bangash [126, 750] has analysed a containment vessel using a simplified version of the two-parameter model as a function of three co-ordinates

$$f(z, \gamma, \delta) = 0 \quad \text{(II.7)}$$

where

$$z = \frac{1}{\sqrt{3}}(\sigma_1 + \sigma_2 + \sigma_3)$$

$$\gamma = \frac{1}{\sqrt{3}}\sqrt{(\sigma_1 - \sigma_2)^2 + (\sigma_2 - \sigma_3)^2 + (\sigma_1 - \sigma_3)^2}$$

$$\delta = \cos^{-1}[\sigma_1 + \sigma_2 - 2\sigma_3/\sqrt{2}\sqrt{3}\gamma]$$

Suh [152] modified the Mohr–Coulomb yield criterion to take into account both dilatancy and compaction using a closed surface of the failure cone. Mathematically, this is represented as

$$f(I_1, J_2, \theta) = \left[\left(-\frac{I_1}{3} + K\right)^2 + \gamma J_2\right]^{1/2} - \beta\left[\cos\frac{\pi\theta}{2\phi}\right]^n = 0 \quad \text{(II.8)}$$

where

$$\tan\theta = \sqrt{\gamma J_2}\Big/\left(\frac{I_1}{3} + K\right) \quad \text{(II.8a)}$$

γ is generally taken as 2/3, and K, n, β are positive constants, depending on concrete properties.

The yield surface is the revolution of a lemniscate. Equation (II.8) does include the effects of strain hardening and porosity. Since the Mohr–Coulomb surface is angular in the plane and has a singular vertex point in the principal stress space, a greater computational advantage can be achieved by eliminating angularity and vertex singularity. Zienkiewicz and Pande [153] have introduced hyperbolic and parabolic approximations of Mohr–Coulomb with an added strain dependent elliptical surface. A general surface without corners has been developed and is given in the form

$$f(I_1, J_2, \theta) = \alpha' \left(\frac{I_1}{3}\right)^2 + \beta \frac{I_1}{3} + \gamma' + \left\{\frac{J_2}{g(\theta)}\right\}^2 = 0 \qquad (II.9)$$

where α', β', γ' are material constants. The function $g(\theta)$ is such that $g(\pi/6) = 1$ so that it determines the shape of the π-section. Various forms of $g(\theta)$ have been suggested in order to avoid singularities.

In the Drucker–Prager criterion [154] it was discovered that in the case of a Mohr–Coulomb surface where information is not available in advance about the use of any one of the six sides or the non-fixation of the principal directions at each point, numerical computation presents problems. It gives a smooth approximation of the Mohr–Coulomb surface and presents it in a mathematical form:

$$f(I_1, J_2) = \alpha I_1 + \sqrt{J_2} - K = 0 \qquad (II.10)$$

where α and K are positive and constant at each point of the material. Drucker and Prager have shown that eqn (II.9) gives a right circular cone which possesses a vertex and in which the corners have been smoothed. The axis of the cone, as shown in Fig. (II.1), lies along the hydrostatic stress. If $\alpha = 0$, eqn (II.10) reduces to the Von Mises yield criterion stress. Where $\alpha \neq 0$ the plastic deformation is accompanied by an increase in volume. The values of α and K are given as

$$\alpha = 2 \sin \phi / \sqrt{3(3 \pm \sin \phi)}$$
$$K = 6C \cos \phi / \sqrt{3(3 \pm \sin \phi)} \qquad (II.10a)$$

ϕ is used when the inner cone passes through the tensile meridian ρ_t when $\theta = 0$; ϕ is used when two surfaces of the model agree along the compressive meridian ρ_c when $\theta = 60°$.

II.1.3 Three-Parameter Model

Three-parameter models developed by Reimann [63] and Argyris [64] are briefly described in Chapter I. Willam [65] has given a three-parameter failure surface for concrete in a low compression regime and in tension cross-sections. The failure model (Fig. II.2) which is convex and smooth has been achieved by the curve fitting of an ellipse, which is a portion of the failure curve. Owing to symmetry, smoothness and convexity, the elliptic form can easily degenerate to a circle which, in turn, shows that both the Von Mises and the Drucker–Prager criteria are special cases of this model.

The standard form of an ellipse is

$$\frac{x^2}{a^2} + \frac{y^2}{b^2} = 1 \tag{II.11}$$

where a and b are half axes.

Sampling this equation at a point $P_2(m, n)$ yields

$$\frac{m^2}{a^2} + \frac{n^2}{b^2} = 1$$

where these values have been defined as

$$m = \frac{\sqrt{3}}{2}\rho_c \qquad n = b - (\rho_t - \tfrac{1}{2}\rho_c)$$

$$a = \sqrt{\frac{\rho_c(\rho_t - 2\rho_c)^2}{(5\rho_c - 4\rho_t)}} \qquad b = (2\rho_t^2 - 5\rho_t\rho_c + 2\rho_c^2)/(4\rho_t - 5\rho_c) \tag{II.12}$$

Using $x = \rho \sin\theta$, $y = \rho \cos\theta - (\rho_t - b)$, the final surface form in polar co-ordinates ρ, θ for $0 \le \theta \le 60°$ is given as

$$\rho(\theta) = \frac{2\rho_c(\rho_c^2 - \rho_t^2)\cos\theta + \rho_c(2\rho_t - \rho_c)[\alpha' + 5\rho_t^2 - 4\rho_t\rho_c]^{1/2}}{\alpha' + (\rho_c - 2\rho_t)^2} \tag{II.12a}$$

where

$$\alpha' = 4(\rho_c^2 - \rho_t^2)\cos^2\theta$$

The angle of similarity θ is computed as

$$\theta = \cos^{-1}\left[\frac{2\sigma_1 - \sigma_2 - \sigma_3}{\sqrt{2}[(\sigma_1 - \sigma_2)^2 + (\sigma_2 - \sigma_3)^2 + (\sigma_3 - \sigma_1)^2]^{1/2}}\right] \tag{II.12b}$$

If $\tfrac{1}{2} \le \rho_t/\rho_c \le 1$ satisfies the position vector ρ, it shows that the failure curve has both convexity and smoothness. This three-parameter model

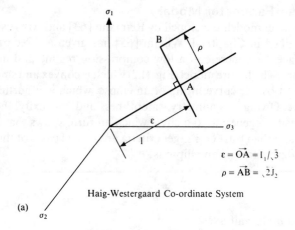

(a) Haig-Westergaard Co-ordinate System

$\varepsilon = \overrightarrow{OA} = I_1/\sqrt{3}$
$\rho = \overrightarrow{AB} = \sqrt{2J_2}$

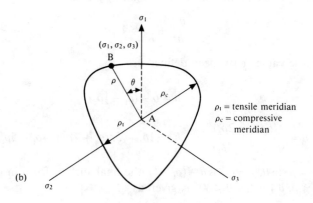

(b)

ρ_t = tensile meridian
ρ_c = compressive meridian

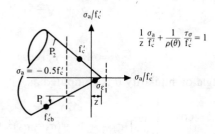

(c) Hydrostatic section ($\theta = 0°$)

$$\frac{1}{z}\frac{\sigma_a}{f'_c} + \frac{1}{\rho(\theta)}\frac{\tau_\sigma}{f'_c} = 1$$

Fig. II.2. Three-parameter model strength ratios, $\alpha_u = 1\cdot 3$, $\alpha_z = 0\cdot 1$ [63, 64].

predicts failure if the average stresses σ_a, τ_a, and the angle of similarity θ satisfy the boundary conditions of the following equation:

$$f_c(\sigma_a, \tau_a, \theta) = \frac{1}{z}\frac{\sigma_a}{f'_c} + \frac{1}{\rho(\theta)}\frac{\tau_a}{f'_c} - 1 = 0 \tag{II.13}$$

where

$$\sigma_a = \tfrac{1}{3}I_1$$
$$\tau_a = \sqrt{\tfrac{3}{5}} \quad \tau_0 = \tfrac{2}{5}J_2 = \tfrac{1}{5}\rho^2 \tag{II.13a}$$
$$z = \frac{\alpha_u \alpha_z}{\alpha_u - \alpha_z}$$

$$\alpha_z = \sigma_t/f'_c \quad \alpha_u = \frac{\sigma_{cb}}{f'_c} \tag{II.13b}$$

σ_{cb} = stress at biaxial conditions

The values for the constituents are given as:

$\dfrac{\sigma_a}{f'_c}$	$\dfrac{\tau_a}{f'_c}$	θ	$\rho(\theta)$
$\tfrac{1}{3}\alpha_z$	$\sqrt{\tfrac{2}{15}}\alpha_z$	$0°$	$\rho_1 = \dfrac{\alpha_u \alpha_z}{2\alpha_u - \alpha_z}\sqrt{\tfrac{6}{5}}$
$-\tfrac{1}{3}$	$\sqrt{\tfrac{2}{15}}$	$60°$	$\rho_2 = \dfrac{\alpha_u \alpha_z}{3\alpha_u \alpha_z + \alpha_u - \alpha_z}\sqrt{\tfrac{6}{5}}$
$-\tfrac{2}{3}\alpha_u$	$\sqrt{\tfrac{2}{15}}\alpha_u$	$0°$	ρ_1

$$\tag{II.13c}$$

II.1.4 Four-Parameter Model

A failure criterion proposed by Ottoson [24–26], known as the four-parameter model, is given here. This failure surface contains all three stress invariants and has the following characteristics:

(a) The surface is smooth and convex with curved meridians, determined by the constants a and b.
(b) It is open in the negative direction of the hydrostatic axis.
(c) The trade in the deviatoric plane changes from an almost triangular to a circular shape with increasing hydrostatic pressure, defined by λ on the deviatoric plane.
(d) The surface is in good agreement with experimental results over a wide range of stress states, including those where tensile stresses occur.

Figures II.3 and II.4 show the surface in the principal stress co-ordinate system in which the compressive meridian $\rho_c(\theta = 60°, \sigma_1 = \sigma_2 > \sigma_3)$ and the tensile meridian $\rho_t(\theta = 0°, \sigma_1 = \sigma_2 < \sigma_3)$ are defined. The meridians are curved, smooth and convex, and ρ increases with increasing hydrostatic pressure.

An analytical failure surface containing all the above characteristics is defined in the following form by Ottoson

$$f(I_1, J_2, J) = a\frac{J_2}{(f_c')^2} + \lambda\frac{\sqrt{J_2}}{f_c'} + b\frac{I_1}{f_c'} - 1 = 0 \quad \text{(II.14)}$$

where

$I_1 = \sigma_x + \sigma_y + \sigma_z =$ the first invariant of the stress tensor (II.14a)

$J_2 =$ the second invariant of the stress deviator tensor
$= \frac{1}{2}(S_x^2 + S_y^2 + S_z^2) + \tau_{xy}^2 + \tau_{yz}^2 + \tau_{zx}^2$ (II.14b)

$J = \cos 3\theta = 1\cdot 5\sqrt{3}\dfrac{J_3}{\sqrt{J_2}}$ (II.14c)

$J_3 =$ the third invariant of the stress deviator tensor
$= S_x S_y S_z + 2\tau_{xy}\tau_{yz}\tau_{zx} - S_x \tau_{yz}^2 - S_y \tau_{xz}^2 - S_z \tau_{xy}^2$ (II.14d)

$S_x = \sigma_x - I_1/3$
$S_y = \sigma_y - I_1/3$ (II.14e)
$S_z = \sigma_z - I_1/3$

$\lambda = \lambda(\cos 3\theta) > 0 \quad a$ and b are constant

$\lambda = K_1 \cos(\frac{1}{3}\cos^{-1}(K_2 \cos 3\theta))$ for $\cos 3\theta > 0$

$\lambda = K_1 \cos(\pi/3 - \frac{1}{3}\cos^{-1}(-K_2 \cos 3\theta))$ for $\cos 3\theta \leq 0$

K_1, K_2, a and b are material parameters $(0 \leq K_2 \leq 1)$

$f_c' =$ uniaxial compressive cylinder strength for concrete

$\sigma_t =$ uniaxial tensile strength for concrete

For $a > 0, b > 0$, the meridians become curved (non-affine), smooth and convex and the surface opens in a negative direction of the hydrostatic axis. When $a = 0, \lambda =$ constant, this criterion comes closer to the Drucker–Prager criterion. When $a = 0, b = 0$ and $\lambda =$ constant, it represents the Von Mises criterion. The relationship of $\sqrt{J_2}/f_c'$ is given by

$$\frac{\sqrt{J_2}}{f_c'} = \frac{1}{2a}\left[-\lambda + \sqrt{\lambda^2 - 4a\left(b\frac{I_1}{f_c'} - 1\right)}\right] \quad \text{(II.15)}$$

Numerical Modelling of Concrete Strength and Failure

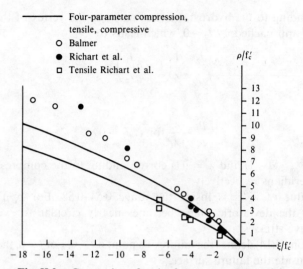

Fig. II.3. Comparison for the four parameter model.

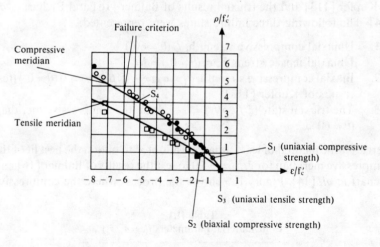

Fig. II.4. Determination of material parameters (S_1, S_2, S_3, S_4 = failure stresses).

Corresponding to the hydrostatic tension, when the vertex of the failure surface is approached, $\sqrt{J_2} \to 0$, which leads to

$$\frac{\sqrt{J_2}}{f'_c} \to \frac{1}{\lambda}\left(1 - b\frac{I_1}{f'_c}\right) \tag{11.16}$$

i.e.

$$\frac{\rho_t}{\rho_c} \to \frac{\lambda_c}{\lambda_t} \text{ for } \sqrt{J_2} \to 0$$

in which $\lambda_c = \lambda(-1)$ and $\lambda_t = \lambda(1)$ corresponding to the compressive and tensile meridians respectively.

The value of λ_c/λ_t is inside the range 0·54–0·58. For $\rho_t/\rho_c \to 1$ for $I_1 \to -\infty$, the deviatoric plane becomes nearly circular for very high compressive stresses.

Equation (II.14) defines the failure at a point if $f \geq 0$. For $f < 0$, the point is situated inside the failure surface.

II.1.4.1 Parameters a, b, K_1 and K_2 (Tables II.1 and II.2)

The four material parameters are determined using the biaxial test results of Kupfer [113] and the triaxial results of Balmer [16] and Richart et al. [14]. The following three failure states were represented:

1. Uniaxial compressive strength, f'_c ($\theta = 60°$).
   Uniaxial tensile strength, σ_t ($\theta = 0°$) = Kf'_c.
2. Biaxial compressive strength, $\sigma_1 = \sigma_2 = -1·16\sigma_c$; $\sigma_3 = 0$ ($\theta = 0°$) (test results of Kupfer [113]).
3. The triaxial state $(\xi/f'_c, \rho/f'_c) = (-5, 4)$ on the compressive meridian ($\theta = 60°$).

Hereafter the method of least squares is adopted to obtain the best fit of the compressive meridian for $\xi/f'_c = -5·0$ to test the results of Balmer [16] and Richart et al. [14]. Figure II.3 shows the process where the compressive

Table II.1
Four material parameters ($k = \sigma_t/\sigma_c$)

k	a	b	k_1	k_2
0·08	1·807 6	4·096 2	14·486 3	0·991 4
0·10	1·275 9	3·196 2	11·736 5	0·980 1
0·12	0·921 8	2·596 9	9·911 0	0·964 7

Table II.2
Values of the function ($k = \sigma_t/\sigma_c$)

k	λ_t	λ_c	λ_c/λ_t
0·08	14·492 5	7·783 4	0·737 8
0·1	11·710 9	6·531 5	0·557 7
0·12	9·872 0	5·697 9	0·577 2

meridian passes through a point (where ξ/σ_c, $\rho/\sigma_c = (-5, 4)$). With this procedure, the values of material parameters are determined as given in Tables II.1 and II.2. From these tables it is clear that the material parameters show considerable dependence on $k = \sigma_t/f_c'$, but the failure stresses in the compressive regime are only slightly affected.

II.1.4.2 σ_{cu} value

It remains now to assess σ_{cu} ($i = 1, 2, 3$), the peak stress, for the calculation of tangent moduli (eqn (I.75)) for various principal stress ratios. Under uniaxial conditions σ_{cu} is equal to the compressive cylinder strength. However, under multiaxial stress conditions the compressive strength of concrete increases. To obtain σ_{cu} in three directions for the principal stress ratio, a surface in the stress space is used. First of all, current principal stresses are established: let these be σ_{p1}, σ_{p2}, σ_{p3}, where $\sigma_{p1} > \sigma_{p2} > \sigma_{p3}$. It is then assumed that σ_{p1} and σ_{p2} are held constant while the third principal stress is changed such that it reaches to the failure surface. This establishes that the ultimate stress is σ_{p3}^c. Similarly, σ_{p1}^c and σ_{p2}^c are calculated by increasing their values while the other two stresses remain constant. This means that the principal stresses are substituted in the failure surface (eqn II.14) and then one of the stresses (the more compressive) is increased while the other two remain constant until eqn (II.14) is satisfied.

II.1.5 Five-Parameter Model

The three-parameter model developed by Willam [65] has been extended by adding two additional degrees of freedom for describing curved meridians so that the failure surface model can be applied to low as well as high compression regions. The more general failure condition is suggested in contrast to eqn (II.13) which determines the average stress components using a linear relationship. This failure condition is given by

$$f(\sigma) = f(\sigma_a, \tau_a, \theta) = \frac{1}{\rho(\sigma_a, \theta) f_c''} \tau_a - 1 \qquad (II.17)$$

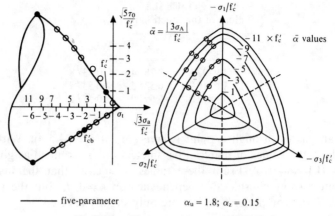

five-parameter $\alpha_u = 1.8$; $\alpha_z = 0.15$

Fig. II.5. Five-parameter model [65].

The average shear stress is restricted for $f(\sigma) = 0$ to

$$\frac{\tau_a}{f'_c} = \rho(\sigma_\theta, \theta)$$

such that τ_a is now a single function of σ_a and θ. The proposed model now removes the affinity of deviatoric planes and sections built into previous theories.

The failure model is constructed by first obtaining the two meridians at $\theta = 0°$ and $\theta = 60°$ by two second-order parabolas which are connected by an ellipsoidal surface (Fig. II.5). These parabolas, whose tensile and compressive meridians exist, are given by

$$\rho'_1(\sigma_a) = \frac{\rho_t}{\sqrt{5}f'_c} = a_0 + a_1 \frac{\sigma_a}{f'_c} + a_2 \left(\frac{\sigma_a}{f'_c}\right)^2 \quad \text{at } \theta = 0° \qquad (II.18)$$

$$\rho'_2(\sigma_a) = \frac{\rho_c}{\sqrt{5}f'_c} = b_0 + b_1 \frac{\sigma_a}{f'_c} + b_2 \left(\frac{\sigma_a}{f'_c}\right)^2 \quad \text{at } \theta = 60° \qquad (II.18a)$$

The surface is defined on the basis of the average normal stress σ_a ($0° \leq \theta \leq 60°$) as

$$\rho(\sigma_a, \theta) = \frac{2\rho_c(\rho_c^2 - \rho_t^2)\cos\theta + \rho_c(2\rho_t - \rho_c)[\sqrt{\alpha' + 5\rho_t^2 - 4\rho_t\rho_c}]}{\alpha' + (\rho_c - 2\rho_t)^2} \qquad (II.19)$$

which is the right-hand-side of eqn (II.12a).

The six degrees of freedom a_0, a_1, a_2, b_0, b_1 and b_2 require evaluation.

When the following adjustments are made, the Von Mises, Drucker–Prager, Willam three-parameter and Ottoson criteria are established

$a_0 = b_0$ and $a_1 = b_1 = b_2 = a_2 = 0$ leads to Von Mises

$a_0 = b_0$ $\quad a_1 = b_1 \quad a_2 = b_2 = 0$ leads to Drucker–Prager

$\dfrac{a_0}{b_0} = \dfrac{a_1}{b_1} \quad a_2 = 0 = b_2$ leads to Willam three-parameter

$\dfrac{a_0}{b_0} = \dfrac{a_1}{b_1} = \dfrac{a_2}{b_2}$ leads to Ottoson

(II.20)

When the two parabolas pass through a common apex $\bar{\sigma}_a$ at the equisectrix or hydrostatic axis, this imposes a constraint

$$\rho_t(\xi_0) + \rho_c(\xi'_0) = 0 \tag{II.21}$$

where

$$\xi'_0 = \frac{\bar{\sigma}_a}{f'_c}$$

The stress state of five sets together with the constraint conditions of the common apex for $\theta = 0°$ and $\theta = 60°$ are given in Table II.3.

Substituting the three strength values in Table II.3 into eqn (II.18a) at the tensile meridian $\theta = 0°$, the three parameters a_0, a_1 and a_2 are given as

$$a_0 = \tfrac{2}{3}\alpha_u a_1 - \tfrac{4}{9}\alpha_u^2 \alpha_z + \sqrt{\tfrac{2}{15}}\alpha_u$$

$$a_1 = \tfrac{1}{3}(2\alpha_u - \alpha_z)a_2 + \sqrt{\tfrac{6}{5}}\frac{\alpha_z - \alpha_u}{2\alpha_u + \alpha_z} \tag{II.22}$$

$$a_2 = \frac{\sqrt{\tfrac{6}{5}}\zeta'(\alpha_z - \alpha_u) - \sqrt{\tfrac{6}{5}}\alpha_z\alpha_u + \rho_1(2\alpha_u + \alpha_z)}{(2\alpha_u + \alpha_z)(\zeta'^2 - \tfrac{2}{3}\alpha_u\zeta' + \tfrac{1}{3}\alpha_z\zeta' - \tfrac{2}{9}\alpha_z\alpha_u)}$$

The apex of the failure surface follows from the condition $\rho_t(\bar{\sigma}_a) = 0$. Hence the following equation is obtained

$$a_2\xi'^2_0 + a_1\xi'_0 + a_2 = 0 \quad \text{and} \quad \xi'^2_0 = -\frac{a_1 - \sqrt{a_1 - 4a_0 a_2}}{2a_2} \tag{II.23}$$

Substituting the second set of three strength values of Table II.3 into eqn

Table II.3
Parameters and their values

Test	σ_a/f_c'	τ_a/f_c'	$\theta°$	$\rho(\sigma_a, \theta)$
$\sigma_1 = \sigma_t$	$\tfrac{1}{3}\alpha_z$	$\sqrt{\tfrac{2}{15}}\alpha_z$	0	$\rho_t(\sigma_a)$
$\sigma_2 = \sigma_3 = \sigma_{cb}$	$-\tfrac{2}{3}\alpha_u$	$\sqrt{\tfrac{2}{15}}\alpha_u$	0	$\rho_t(\sigma_a)$
	$-\zeta'^a$	$\rho_1^{\ b}$	0	$\rho_t(\sigma_a)$
$\sigma_3 = f_c'$	$-\tfrac{1}{3}$	$\sqrt{\tfrac{2}{15}}$	60	$\rho_c(\sigma_a)$
	$-\zeta'$	$\rho_2^{\ c}$	60	$\rho_c(\sigma_a)$
	ξ_0'	0	60	$\rho_c(\sigma_a)$

$^a\ \zeta' = -\sigma_a/f_c'$.
$^b\ $ At $\theta = 0°$; $\rho_1 = \tau_a/f_c'$.
$^c\ \rho_2 = \tau_a/f_c'$; at $\theta = 60°$.

(II.18a) at the compressive meridian $\theta = 60°$, the remaining three parameters b_0, b_1 and b_2 are established:

$$b_0 = -\xi_0' b_1 - \xi_0'^2 b_2$$

$$b_1 = (\zeta' + \tfrac{1}{3})b_2 + \frac{\sqrt{\tfrac{6}{5}} - 3\rho_2}{3\zeta' - 1} \tag{II.23a}$$

$$b_2 = \frac{\rho_2(\xi_0' + \tfrac{1}{3}) - \sqrt{\tfrac{2}{15}}(\xi_0' - \zeta')}{(\zeta' + \xi_0')(\zeta' - \tfrac{1}{3})(\xi_0' + \tfrac{1}{3})}$$

The five-parameter model is illustrated in Fig. II.5 and is corroborated by the experimental data from Launay and Gachon [66]. The five-parameter model features a smooth surface and produces the principal features of the triaxial failure surface of concrete. It is periodical in the deviatoric plane (with a period of 120°) and has symmetry at 60°. The cross-section is non-circular and the meridians are second-order parabolas with a failure surface in the deviatoric plane prescribed by a part of an elliptic curve. In order to produce convexity in both the deviatoric plane and along the meridians, the following boundary conditions or constraints are applied:

$$a_0 > 0 \quad \text{and} \quad a_1 \leq 0 \quad a_2 \leq 0$$
$$b_0 > 0 \quad \text{and} \quad b_1 \leq 0 \quad b_2 \leq 0 \tag{II.23b}$$
$$\rho_t(\sigma_a)/\rho_c(\sigma_a) > \tfrac{1}{2}$$

The failure surface intersects the hydrostatic axis if, as in eqn (II.23b), $a_2 \leq 0$, $b_2 \leq 0$.

II.1.6 Hypoelastic Model

The hypoelasticity element may be a specific material which can undergo softening or hardening in strain, but, at the same time, has neither a preferred state nor a preferred stress and its response to deformation is smooth and gradual. The hypoelastic material differs from most materials employed in plasticity theory in that it is dynamically sound. If it is possible to judge the behaviour of the hypoelastic material under large strains, large rotations and abrupt stress changes, then the Truesdell [90], Rivlin and Erickson [91] and Noll [166] constitutive equations will be subject to certain additional restrictions.

Such a suggestion obviously frees concrete from the need to choose any of the specific failure criteria discussed earlier. It is admitted in the development of such a constitutive law that the hypoelastic element is appropriate only to the isotropic case unless a preferred initial state is specified. Here, for the sake of argument, the anisotropy concept is appropriate only to materials which have a preferred state. The equations to be developed in this case must be supported by experimental results in order to given an initial preferred state. Since, in hypoelasticity, equations of motion are involved and initial stresses are satisfied by these equations, the situation herein is not the same as that of ordinary elasticity in which special classes of initial stresses obtained from elastic strains are admissible. Using the equations of motion the concept of hypoelasticity offers better treatment of initially stressed configurations of models or prototypes under large deformations. Truesdell [90] has shown that his equations do give large deformations for various specimens. Since these deformations are not covered in the classical theory of finite strains, the opportunity can be taken to extend the existing work to include terms for elasto-plasticity.

In addition, the model contains stability criterion and provides the most general representation for time independent incremental behaviour and accommodates stress-induced anisotropy. More discussions are given by Coons and Evans [159] and Bangash [126].

II.1.6.1 Concrete Stress–Strain Criteria Using Hypoelasticity

A time dependent law by which an increment of stress is related to the stress is given by Truesdell [90] as

$$\text{rate of stress} = f(\text{stress, rate deformation}) \qquad (\text{II}.24)$$

where f is the tensor function. The purpose is to obtain a concept of the elastic behaviour of the material expressed in terms of rates. In other words, this means that the initial state of the material is arbitrarily stressed, which is different from what was believed earlier, and that the unstrained state of material can also be considered as unstressed. If eqn (II.24) is given a three-dimensional status, the difference between the results based on the constitutive law concept of expressing material behaviour in rates, and the concept of material being unstressed whilst unstrained, will be greater. This is due to the fact that the three-dimensional equations bring about the inter-relationship of several phenomena and parameters, noted amongst these are: extension, shear, hydrostatic pressure, variable elastic properties, etc.

In eqn (II.24) Green and Swanson [145] retained the features of plasticity when they applied them to the deformation of metals. This idea is extremely attractive for predicting the short-term response of the frictional material. Equations developed on this basis, when stability criteria are included, can reasonably predict failure criteria for plain concrete. This is the theme which will be followed throughout in the development of the theoretical model.

Equation (II.24) can easily be written in a combined tensor form in three dimensions as

$$f(\sigma, \varepsilon, \dot{\sigma}, \dot{\varepsilon}) = 0 \qquad (II.25)$$

in which $f(\sigma, \varepsilon)$ and $f(\dot{\sigma}, \dot{\varepsilon})$ are related to the elastic behaviour, and $f(\sigma, \dot{\varepsilon})$ defines the onset of plasticity zones. It is now clearly established that eqn (II.25) is a general representation of the constitutive relationship in which classical stress (σ) and strain (ε) tensors and their scalar parameters represented by dots are included. It is initially assumed that the increment of stress in eqn (II.25) is not linearly dependent on the increment of strain. Within eqn (II.25) the incremental stress $d\sigma_{ij}$ is given as in a 'trilinear' relationship as

$$d\sigma_{ij} = S_{ijkl}(\sigma_{ij}) \, d\sigma_{kl} \qquad (II.26)$$

where S_{ijkl} is the material property tensor, σ_{ij} is the stress, and $d\sigma_{kl}$ is the rate of deformation as stated in eqn (II.24).

Truesdell [90] suggests for isotropic material that the values of S_{ijkl} and $d\sigma_{ij}$ are

$$\begin{aligned}
S_{ijkl} &= (A_{01} + A_{11}\sigma_{rr})\delta_{ij}\delta_{kl} + \tfrac{1}{2}(A_{02} + 2A_{11}\sigma_{rr}) \\
&\quad \times (\delta_{ik}\delta_{jl} + \delta_{jk}\delta_{il}) + A_{13}(\sigma_{ij}\delta_{kl} + \sigma_{kl}\delta_{ij}) \\
&\quad + \tfrac{1}{2}A_{14}(\sigma_{jk}\sigma_{li} + \sigma_{jl}\delta_{ki} + \sigma_{ik}\delta_{lj} + \sigma_{il}\delta_{kj})
\end{aligned} \qquad (II.26a)$$

$$\begin{aligned}
d\sigma_{ij} &= (A_{01}\dot{\varepsilon}_{kk}\delta_{ij} + A_{02}\dot{\varepsilon}_{ij} + A_{11}\sigma_{pp}\dot{\varepsilon}_{kk}\delta_{ij}) \\
&\quad - 2A_{11}\sigma_{kk}\dot{\varepsilon}_{ij} + A_{13}(\sigma_{ij}\dot{\varepsilon}_{kk} + \sigma_{kl}\dot{\varepsilon}_{kl}\delta_{ij}) \\
&\quad + A_{14}(\sigma_{jk}\dot{\varepsilon}_{ik} + \sigma_{ik}\dot{\varepsilon}_{jk})
\end{aligned} \qquad (II.26b)$$

where δ represents Kronecker deltas and A represents various hydrostatic constants to be determined by experiments on the lines suggested by Gardner [78]. Hence eqn (II.26a) exhibits a stress-induced anisotropicity if its last three terms are retained. This phenomenon occurs in most nonlinear isotropic behaviour. As stated earlier, Richart *et al.*'s data [14] suggest a small amount of asymmetry about the hydrostatic stress axis, and Bresler and Pister [42] have indicated that the failure of concrete is affected by the third invariant of stress J_3; it therefore becomes necessary to retain the constant A_{14} in eqn (II.26a) and eqn (II.26b) in order to represent the asymmetry. Coons and Evans [159], in their general failure equations, have seven material properties: $A_{01}, A_{11} \ldots A_{13} \ldots$ etc. These have been reduced to five in eqns (II.26a) and (II.26b) by making $A_{13} = A_{15}$ and $A_{12} = A_{11}$ in order to guarantee the entire behaviour of concrete as hypoelastic. It is interesting to note the two constants A_{01} and A_{02}. The hypoelastic constant A_{01} is incorporated in the equations (II.26a and II.26b) so that, for its zero value, it should give a zero value for Poisson's ratio at very small loads. Similarly, the constant A_{02} is essential in order to give the slope at the origin of the stress–strain curve for zero lateral pressure. The values of the remaining constants can easily be obtained by using failure data from a number of investigations [77–176] mentioned earlier.

For any combination of material properties, a stress state is sought for which an incremental deformation may be obtained without changing the stress state itself. Coons and Evans [159] suggest that, if constants other than A_{01} and A_{02} are zero, the hypoelastic case will be brought back to an elastic case represented by Hooke's law with an additional freedom that the initial stress can be prescribed for the zero strain rate. For the hypoelastic case suggested by eqns (II.26a) and (II.26b) the five sets of conditions suggested for the kind of stress state for which the incremental deformation is required, are written in modified forms as

$$\begin{aligned} A_{14} &\neq 0 \\ -2A_{11}(3A_{11} - A_{11}) \text{ or } -2A_{11}^2 &= 0 \\ -2A_{11}A_{13} &= 0 \\ A_{13}(3A_{11} + A_{13}) &= 0 \\ -6A_{01}A_{11} + A_{02}(-2A_{11} - A_{13}) &= 0 \end{aligned} \qquad \text{(II.27)}$$

Equation (II.27) must therefore be satisfied if the stress–strain criterion assumed is initially independent of the loading path. The stress condition for failure after fully satisfying eqn (II.27) will be

$$S_{ijkl}\, d\varepsilon_{kl} = 0 \qquad \text{(II.28)}$$

Equation (II.28) is satisfied for the incremental deformation and condition of failure when the determinant of the material constants is zero, i.e.

$$|S_{ijkl}| = 0 \qquad (II.29)$$

In order to evaluate the determinant in eqn (II.29), eighty-one components of the material tensor for a homogeneous anisotropic concrete element must have 36 coefficients. From considerations based on the law of conservation of energy and on the study of potential energy stored in the concrete element by the strains, Love [160] and Lekhnitskii [161] have shown that the relation $K_{ij} = K_{ji}$ of $|S_{ijkl}|$, for $i,j = 1, 2, \ldots, 6$, holds between these constants. Since the original matrix $[S_{ijkl}]$ is symmetric around the main diagonal, all 36 coefficients of this matrix therefore involve a maximum of 21 constants. Nye [162] thus concludes that the relationship given in eqn (II.29) also exists amongst the material stiffnesses, and hence it is easy to write

$$S_{ijkl} = S_{ijlk} = S_{jikl} = S_{klij} \qquad (II.29a)$$

Consider a concrete element that is symmetric in properties with respect to one plane, say x_1–x_2. This symmetry can be expressed by transforming the invariants such that $x_1 = x'_1$, $x_2 = x'_2$ and $x_3 = x'_3$. The direction cosines given in Table II.4 are used for this transformation. The stresses and strains of the transformed co-ordinate systems are related to those of the original co-ordinate system by:

$$\sigma'_{kl} = c_{ki}c_{lj}\sigma_{ij} \qquad \varepsilon'_{kl} = c_{ki}c_{lj}\varepsilon_{ij} \qquad (II.29b)$$

Therefore, for $i = 1, 2, 3, 6$, $\sigma'_i = \sigma_i$ and $\varepsilon'_i = \varepsilon_i$. However, $\varepsilon'_{23} = c_{22}c_{33}$, $\varepsilon_{23} = -\varepsilon_{23}$; therefore $\varepsilon'_4 = -\varepsilon_4$. Similarly, $\sigma'_4 = -\sigma_4$. Likewise $\varepsilon'_{31} = c_{33}c_{11}$, $\varepsilon_{31} = -\varepsilon_{31}$. Therefore $\varepsilon'_5 = -\varepsilon_5$ and also $\sigma'_5 = -\sigma_5$.

Now eqn (II.26) can be written both in terms of the original co-ordinate system and the transformed co-ordinate systems as:

$$\sigma_1 = D'_{11}\varepsilon_1 + D'_{12}\varepsilon_2 + \cdots + D'_{16}\varepsilon_6$$
$$\sigma'_1 = D'_{11}\varepsilon'_1 + D'_{12}\varepsilon'_2 + \cdots + D'_{16}\varepsilon'_6 \qquad (II.29c)$$

Since $\varepsilon'_4 = -\varepsilon_4$ and $\varepsilon'_5 = -\varepsilon_5$, then $\sigma'_1 = \sigma_1$ requires that $D'_{14} = D'_{15} = 0$. By similarly considering the other equations of eqn (II.26) such as $\sigma'_2 = \sigma_2$; $\sigma'_3 = \sigma_3$; $\sigma'_4 = \sigma_4$; $\sigma'_5 = \sigma_5$; and $\sigma'_6 = \sigma_6$, it is found that $D'_{24} = D'_{25} = D'_{34} = D'_{35} = D'_{64} = D'_{65} = 0$.

Knowing that Lekhnitskii [161] has established the relation $D_{ij} = D_{ji}$, and with the material having one plane of symmetry, the elasticity tensor

Table II.4
Direction cosines C

Materials with one plane of symmetry				Orthotropic materials			
	x_1	x_2	x_3		x_1	x_2	x_3
x'_1	0	1	0	x'_1	0	1	0
x'_2	1	0	0	x'_2	−1	0	0
x'_3	0	0	−1	x'_3	0	0	1

$[S_{ijkl}] = [D]$ has the following array which involves 13 independent components:

$$[D] = \begin{bmatrix} D'_{11} & D'_{12} & D'_{13} & 0 & 0 & D'_{16} \\ D'_{21} & D'_{22} & D'_{23} & 0 & 0 & D'_{26} \\ D'_{31} & D'_{32} & D'_{33} & 0 & 0 & D'_{36} \\ 0 & 0 & 0 & D'_{44} & D'_{45} & 0 \\ 0 & 0 & 0 & D'_{54} & D'_{55} & 0 \\ D'_{61} & D'_{62} & D'_{63} & 0 & 0 & D'_{66} \end{bmatrix} \quad \text{(II.29d)}$$

Concrete as a material is generally orthotropic, i.e. it contains three mutually orthogonal planes of symmetry. This also means that this material must be symmetric with respect to the x_1–x_3 and the x_2–x_3 planes. Proceeding as before, the diagram shown in Table II.4 for this case can be made for symmetry with respect to the x_1–x_3 plane. Using the transformation, it becomes clear that $D'_{16} = D'_{26} = D'_{36} = D'_{45} = 0$ from the x_1–x_3 plane of symmetry. For concrete as an orthotropic (orthogonally anisotropic) material the S_{ij} tensor can be modified by substituting these values into eqn (II.29d), and thus the property of this material is described by nine independent components.

Sometimes concrete as a material is assumed to be *transversely isotropic*, i.e. like the orthotropic case, it also has three mutually orthogonal planes but in addition is *isotropic* with respect to one of the three planes. For example, if the material isotropicity is in the x_1-x_2 plane, then this means that $D'_{11} = D'_{22}$, $D'_{13} = D'_{23}$, $D'_{44} = D'_{55}$. Similarly, the tensor coefficients for the x_2-x_3 planes are: $D'_{13} = D'_{12}$, $D'_{31} = D'_{21}$ and $D'_{23} = D'_{32}$.

II.1.6.2 Concrete as an Orthotropic Material

After modification of the tensor array given in eqn (II.29d)

$$[D] = \begin{bmatrix} D'_{11} & D'_{12} & D'_{13} & 0 & 0 & 0 \\ D'_{21} & D'_{22} & D'_{23} & 0 & 0 & 0 \\ D'_{31} & D'_{32} & D'_{33} & 0 & 0 & 0 \\ 0 & 0 & 0 & D'_{44} & 0 & 0 \\ 0 & 0 & 0 & 0 & D'_{55} & 0 \\ 0 & 0 & 0 & 0 & 0 & D'_{66} \end{bmatrix} \quad \text{(II.29e)}$$

If the co-ordinates are aligned with the direction of the principal stresses and using specific equations for the invariants of the stress tensor and stress deviator tensor, the *determinant* in eqn (II.29) will simply be the determinant of eqn (II.29e)

$$I_1 = \sigma_1 + \sigma_2 + \sigma_3$$
$$J_2 = \tfrac{1}{2}(\sigma_1^2 + \sigma_2^3 + \sigma_3^2) \quad \text{(II.30)}$$
$$J_3 = \tfrac{1}{3}(\sigma_1^3 + \sigma_2^3 + \sigma_3^3)$$

Using eqns (II.26), (II.26a), (II.26b), (II.27) and (II.30), the values of the tensor coefficients of eqn (II.29) are written as:

$$D'_{11} = A_{01} + A_{02} - A_{11}I_1 + 2(A_{13} + A_{14})\sigma_1$$

D'_{22}, D'_{33} have all the terms of D_{11} except σ_1 is changed to σ_2 for D'_{22} and σ_3 for D'_{33}

$$D'_{44} = (A_{02} - 2A_{11}I_1) + A_{14}(J_1 - \sigma_1)$$
$$D'_{55} = (A_{02} - 2A_{11}I_1) + A_{14}(J_1 - \sigma_2)$$
$$D'_{66} = (A_{02} - 2A_{11}I_1) + A_{14}(J_1 - \sigma_3)$$
$$D'_{12} = (A_{01} + A_{11}I_1) + A_{13}(\sigma_1 + \sigma_2) = \bar{D} + A_{13}(\sigma_1 + \sigma_2)$$
$$D'_{13} = \bar{D} + A_{13}(\sigma_1 + \sigma_3) \quad \text{(II.30a)}$$
$$D'_{23} = \bar{D} + A_{13}(\sigma_2 + \sigma_3)$$
$$D'_{21} = \bar{D} + A_{13}(\sigma_2 + \sigma_1) \quad \text{where } \bar{D} = A_{01} + A_{11}I_1$$
$$D'_{31} = \bar{D} + A_{13}(\sigma_3 + \sigma_1)$$
$$D'_{32} = \bar{D} + A_{13}(\sigma_3 + \sigma_2)$$

The determinant in eqn (II.30) after expansion leads to:

$$[[(A_{02} - 2A_{11}I_1)]\{-3A_{13}^2(2I_2 - J_1^3/3) + [3A_{01} + A_{02} + (A_{11} + 2A_{13})I_1]\}$$
$$+ 2A_{14}\{(A_{02} - 2A_{11}I_1)^2 I_1 + 2(A_{02} - 2A_{11}I_1)I_1(A_{01} + A_{11}I_1)$$
$$+ (2J_1^2/3 - (J_2 - I_1^3/3))[A_{14}(A_{01} + A_{11}I_1) + (A_{02} - 2A_{11}I_1)(2A_{13} + A_{14})]$$
$$+ ((J_{3/3} - I_1(J_2 - I_1^3/3))/2)[2A_{14}(6A_{13} + 2A_{14}) + 6A_{13}^2]$$
$$- A_{13}^2[J_1^3/3 + I_1(J_2 - I_1^3/3) - J_3]\}]]$$
$$\times [(A_{02} - 2A_{11}I_1) + A_{14}(I_1 - \sigma)] \times [(A_{02} - 2A_{11}I_1) + A_{14}(I_1 - \sigma_2)]$$
$$\times [(A_{02} - 2A_{11}I_1) + A_{14}(I_1 - \sigma_3)] = 0 \quad \text{(II.31)}$$

Equation (II.31) in the expanded form geometrically represents a surface in the principal stress space comprising two shapes—the first is pyramidal resulting from the last three products of eqn (II.31); whilst the second, the curved shape, results from the determinant whose elements are D_{12}, D_{13}, D_{23}, D_{21}, D_{31} and D_{32} in eqn (II.30a). In terms of invariants of the stress deviator tensor, I_1, J_2 and J_3 values are substituted into eqns (II.30), (II.30a) and (II.31).

Looking at eqn (II.30) and (II.31) it is clear that there is a class of deformation associated with stresses, and when plastic yield occurs two important properties emerge.

(a) The incremental plastic strain vector $d\varepsilon_p$ is normal everywhere to the yield surface except the pyramidal part.
(b) On the pyramidal part, the determinant in eqn (II.29), the directions of the principal stresses in this case do not coincide with those of the principal strains.
(c) The principal directions of the stress–strain rate do coincide for the part of the failure surface where $D_{ij} = 1, 2, 3$, $i \neq j$.

If a strain history is regarded as a strain path drawn in this space (Fig. II.6) the values of ε_1, ε_2, and ε_3 may all be varied.

At any point A the equivalent strain ε_e is given by the radius as shown in Fig. (II.6), and $d\varepsilon_p$ is its incremental value along the strain path. For example, for two intervals it can be computed as follows:

$$\frac{d\varepsilon_{p2}}{d\varepsilon_{p3}} = \frac{-D'_{11}D'_{23} + D'_{21}D'_{13}}{D'_{12}D'_{23} - D'_{22}D'_{13}}$$

and (II.32)

$$\frac{d\varepsilon_{p1}}{d\varepsilon_{p2}} = \frac{-D'_{12}D'_{21} + D'_{11}D'_{22}}{D'_{12}D'_{23} - D'_{22}D'_{13}}$$

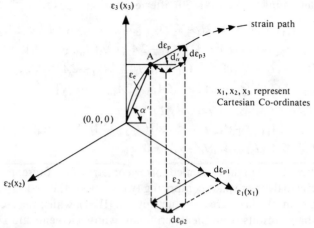

Fig. II.6. Strain space.

The inclination $d\alpha'$ of $d\varepsilon_p$ at any instant is computed from

$$d\alpha' = \tan^{-1}\frac{1}{\sqrt{3}}\frac{(d\varepsilon_{p_1} - d\varepsilon_{p_2}) + (d\varepsilon_{p_1} - d\varepsilon_{p_3})}{d\varepsilon_{p_2} - d\varepsilon_{p_3}} \quad \text{(II.33)}$$

A_{01} (at infinitesimal loads for zero Poisson's ratio) $= 0$
A_{02} (determined from the slope at the origin of the stress–strain curve with zero lateral pressure) $= 8\cdot99 \times 10^5$
A_{14} (for symmetry of hydrostatic stresses) $= 0$

The coefficients A_{11} and A_{13} are found from the stress–strain increment vector and are plotted on the basis of data from Richart et al. [14] and Gardner [78]. Again for this specific case the following values are evaluated:

$$A_{11} = 43\cdot5 \qquad A_{13} = -1200$$

These values can be substituted into equations for orthotropic, transversely isotropic, isotropic and one-dimensional tension or compression cases. Similar data can easily be obtained from the stress–strain curves of other experimental tests.

Similarly, by using eqns (II.30), (II.31) and (II.32) and following the above principles, $d\sigma_3/d\varepsilon_3$ for an *orthotropic* case can be computed. The procedure for evaluating $d\sigma_1/d\varepsilon_1$ and $d\sigma_2/d\varepsilon_2$ for orthotropic and transversely isotropic cases is exactly the same. Only the appropriate tensor coefficients from the respective matrices $[D]$ have to be chosen to replace those of $d\sigma_3/d\varepsilon_3$. For any iteration procedure the tangent moduli can be determined by holding lateral stresses σ_1 and σ_2 constant for prescribed loading conditions. In the case of $d\sigma_3/d\varepsilon_3$ it will give the tangent modulus E_3, which can be prescribed as a function of the current values of σ_1, σ_2 and σ_3. This function must reflect not only the magnitudes of σ_1, σ_2 and σ_3, but also their signs.

The procedure for evaluating $d\sigma_1/d\varepsilon_1$ and $d\sigma_2/d\varepsilon_2$, and the corresponding tangent moduli E_1 and E_2, is exactly the same for both orthotropic and transversely isotropic cases, except that the appropriate tensor coefficients from the respective matrices $[D]$ will replace those of $d\sigma_3/d\varepsilon_3$.

II.1.6.3 Isotropic Case

Having established values of constants consistent with both initial and ultimate loading, it is then possible to compare observed and predicted behaviour of the concrete over the complete load range of the concrete structure vessel. Results from the standard triaxial tests of various authors [35–133] can be used for this purpose. For any prescribed loading path of the vessel, the above hypoelasticity equations can be solved by integration. Where integration is avoided, initial slopes of the stress–strain curves are used to provide the relationship between various material constants in eqns (II.29)–(II.32). For example, for an isotropic case, the strain increments are given as

$$\frac{d\varepsilon_2}{d\varepsilon_3} = \frac{[A_{01} + (A_{11} + A_{13})\sigma_3 + (2A_{11} + A_{13})]}{2[A_{01} + \tfrac{1}{2}A_{02} + (2A_{13} + A_{14})]} \quad (II.34)$$

For $\sigma_3 > \sigma_1$ or σ_2, the stress–strain criterion will simply be $d\sigma_3/d\varepsilon_3$, and

$$\frac{d\sigma_3}{d\varepsilon_3} = \Big\{ [(A_{01} + A_{02}) + (-A_{11} + 2A_{13} + 2A_{14})\sigma_3 - 2A_{11}\sigma_1]$$
$$\times [A_{01} + \tfrac{1}{2}A_{02} + (2A_{13} + A_{14})\sigma_1]$$
$$- \frac{[A_{01} + (A_{11} + A_{13})\sigma_3 + (2A_{11} + A_{13})\sigma_1]^2}{A_{01} + \tfrac{1}{2}A_{02} + (2A_{13} + A_{14})} \Big\} \quad (II.35)$$

$$\frac{d(\sigma_3 - \sigma)}{d\varepsilon_3} = \Big\{ A_{01} + A_{02} + [-A_{11} + 2(A_{13} + A_{14})\sigma_3 - 2A_{11}\sigma]$$
$$- \frac{[A_{01} + (A_{11} + A_{13})\sigma_3 + (2A_{11} + A_{13})\sigma]^2}{A_{01} + \tfrac{1}{2}A_{02} + (2A_{13} + A_{14})\sigma} \Big\} \quad (II.36)$$

$$\frac{d\varepsilon_v}{d\varepsilon_3} = \frac{\tfrac{1}{2}A_{02} + (-A_{11} + A_{13})\sigma_3 + (-2A_{11} + A_{14} + A_{13})\sigma}{A_{01} + \tfrac{1}{2}A_{02} + A_{13}\sigma} \quad (II.37)$$

where ε_v is *the volumetric strain under the triaxial state of stress.*

Table II.5
Isotropic compression and plane strain cases

(a) Isotropic condition

$$\sigma_1 = \sigma_2 = \sigma_3 = \sigma \qquad \varepsilon_1 = \varepsilon_2 = \varepsilon_3 = \varepsilon$$

Final differential equations:

$$d\sigma = (C_1 + C_2\sigma)d\varepsilon$$
$$C_1 = 3A_{01} + A_{02}$$
$$C_2 = 3A_{11} + 6A_{13} + 2A_{14}$$

Upon integration, the stress–strain relation becomes

$$\sigma = (C_1/C_2 + \sigma_c)e^{C_2\varepsilon} - C_1/C_2$$

(b) Plain strain condition

Constant pressure (confined) σ.
Differential equations pertaining to this condition:

$$\frac{d\varepsilon_v}{d\varepsilon_3} = \frac{A_{02} + (A_{12} + A_{13})\sigma_3 - 2A_{11}\sigma_2 + (-2A_{11} + A_{13} + 2A_{14})\sigma}{A_{01} + A_{02} - A_{11}(\sigma_3 + \sigma_2) + (-A_{11} + 2A_{13} + 2A_{14})}$$

$$\frac{d\sigma_3}{d\varepsilon_3} = A_{01} + A_{02} + (-A_{11} + 2A_{13} + 2A_{14})\sigma_3 + (-A_{11})(\sigma_2 + \sigma)$$

$$- [A_{01} + (A_{11} + A_{13})\sigma_3 + A_{11}\sigma_2 + (A_{11} + A_{13})\sigma]$$

$$\times \frac{[A_{01} + (A_{13} + A_{11})\sigma_3 + A_{11}\sigma_2 + (A_{11} + A_{13})\sigma]}{A_{01} + A_{02} - A_{11}(\sigma_3 + \sigma_2) + (-A_{11} + 2A_{13} + 2A_{14})}$$

$$\frac{d\sigma_2}{d\varepsilon_3} = A_{01} + (A_{11} + A_{13})\sigma_3 + A_{11}\sigma_2 + (A_{11} + A_{13})\sigma$$

$$- [A_{01} + A_{11}\sigma_3 + (A_{11} + A_{13})\sigma_2 + (A_{11} + A_{13})\sigma]$$

$$\times \frac{[A_{01} + (A_{11} + A_{13})\sigma_3 + A_{11}\sigma_2 + (A_{11} + A_{13})\sigma]}{A_{01} + A_{02} - A_{11}(\sigma_3 + \sigma_2) + (-A_{11} + 2A_{13} + 2A_{14})}$$

Initial slopes are obtained by setting $\sigma_3 = \sigma$ in eqns (II.36) and (II.37). These initial slopes are related to the constants C_1 and C_2 evaluated in Table II.5 for the isotropic compression case. Hence

$$\left[\frac{d(\sigma_3 - \sigma)}{d\varepsilon_3} \bigg/ \frac{d\varepsilon_v}{d\varepsilon_3}\right]_{\sigma_3 = \sigma} = (C_1 + C_2\sigma) \qquad (II.38)$$

A simple case for the material constants is to use only the initial slope of eqn (II.36). For isotropic conditions eqn (II.36) may be written as

$$A_{01} + (3A_{11} + 2A_{13})\sigma = \frac{[d(\sigma_3 - \sigma)/d\varepsilon_3]_{\sigma_3 = \sigma} - (A + B\sigma)}{\left[\dfrac{d(\sigma_3 - \sigma)}{d\varepsilon_3}\right]_{\sigma_3 = \sigma} \sigma - 3(A + B\sigma)}(C_1 + C_2\sigma)$$

(II.39)

Equation (II.39) gives a relationship between the lateral pressure σ, material constants and the initial slope of the stress–strain curves. At the peak stress condition under the triaxial state of stress, the following relations between the material parameters are obtained by equating eqn (II.36) equal to zero:

$$[3C_1 + C_2(\sigma_3 + 2\sigma)][3A_{02} + (-6A_{11} + 2A_{14})(\sigma_3 + 2\sigma) + 2A_{14}(\sigma_3 - \sigma)$$
$$- 2(\sigma_3 - \sigma)^2(3A_{13} + 2A_{14})^2] = 0 \quad (II.40)$$

Equation (II.40) is a simplified version under isotropic conditions of the general eqn (II.31).

II.1.6.4 Deduction of a One-Dimensional Tension or Compression

The application of the boundary conditions and the loading path corresponding to one-dimensional tension or the compression test gives the two related differential equations

$$\frac{d\sigma_3}{d\varepsilon_3} = A_{01} + A_{02} + (-A_{11} + 2A_{13} + 2A_{14})\sigma_3 + (-A_{11})\sigma \quad (II.41)$$

$$\frac{d\sigma}{d\varepsilon_3} = A_{01} + (A_{11} + A_{13})\sigma_3 + (2A_{11} + A_{13})\sigma \quad (II.42)$$

assuming $K_0\sigma$ is related to σ_3 by

$$\sigma = K_0\sigma_3$$
$$= (1 - \sin\phi)\sigma_3 \quad (II.43)$$

where ϕ is the angle of frictional resistance for concrete.

Using eqns (II.41)–(II.43), two relations are obtained between the material parameters

$$A_{01} = K_0(A_{01} + A_{02}) \quad (II.44)$$

$$A_{11}(1 - K_0)(1 + 2K_0) - K_0(1 + 2K_0)(-2A_{11})$$
$$- 2K_0A_{14} + (1 + K_0)A_{13} = 0 \quad (II.45)$$

Substituting eqn (II.43) into eqn (II.41) gives the general equation

$$\frac{d\sigma_3}{d\varepsilon_3} = (\bar{\alpha} + \bar{\beta}\sigma_3) \tag{II.46}$$

where

$$\bar{\alpha} = A_{01} + A_{02}$$
$$\bar{\beta} = -A_{11}(1 + 2K_0) + 2(A_{13} + A_{14}) \tag{II.47}$$

Integrating eqn (II.46) leads to a stress–strain relation similar to the one obtained in Table II.5 for the isotropic case:

$$\sigma_3 = \left(\frac{\bar{\alpha}}{\bar{\beta}} + \sigma_{3_c}\right)e^{\bar{\beta}\varepsilon_3} - \bar{\alpha}/\bar{\beta} \tag{II.48}$$

The constants $\bar{\alpha}$ and $\bar{\beta}$, as usual, are computed from the initial slope and one point of the experimental stress–strain curve from a one-dimensional test.

Table II.5 gives differential equations pertaining to plain strain conditions. A similar integration procedure is required in order to arrive at suitable relations for the stress–strain conditions. For example, the test conducted by Gardner [78] gives good agreement with the experimental data of Richart et al. [14]. Table I of Gardner [78] with an average of four tests, and Tables 8 and 9 of Richart et al. [14] with an average of four tests each, can now be correlated for the mix 1:1:2. Data from Richart et al. [14] for this mix gives a good agreement with those of Gardner [78].

II.1.7 Shear and Bulk Moduli Models

Nonlinear incremental elastic models are proposed by Phillips et al. [167] in which the bulk modulus K is assumed constant, and the tangential shear modulus G is assumed to be a function of the octahedral shear stress only. A similar approach has been adopted by Cedolin et al. [169] who consider the bulk and shear moduli to depend on all the stress invariants. The proposed model is applicable to triaxial compressive states only.

Another triaxial model of nonlinear type was proposed by Saugy et al. [170] in which the bulk modulus was considered as constant while the shear modulus was assumed to vary as a logarithmic function of the second stress invariant. This model has been used for three-dimensional work.

In this work the elastic bulk and shear moduli are treated as scalar functions of the invariants of the stress or strain tensor. The deviatoric and hydrostatic (volumetric) components are written as

$$S_{ij} = 2G\varepsilon_{ij} = 2G\,\Delta\varepsilon_{ij} \tag{II.49a}$$
$$\sigma_m = \sigma_0 = K\varepsilon_{kk} \tag{II.49b}$$

in which

$$\sigma_0 = \frac{\sigma_{kk}}{3} \quad \text{or} \quad \Delta\sigma_m = 3K\varepsilon_{kk}$$

The stress–strain relation is then

$$\Delta\sigma_{ij} = \Delta\delta_{ij} + \delta_{ij}\Delta\sigma_m$$
$$\Delta\sigma_m = (\Delta\sigma_{11} + \Delta\sigma_{22} + \Delta\sigma_{33})/3$$
$$\Delta\varepsilon_{ij} = \Delta\varepsilon_{ij} - \delta_{ij}\Delta\varepsilon_m \tag{II.50}$$
$$\Delta\varepsilon_m = \frac{(\Delta\varepsilon_{11} + \Delta\varepsilon_{22} + \Delta\varepsilon_{33})}{3}$$

δ_{ij} (the Kronecker delta) $= 1$ for $i = j$ and $= 0$ for $i \neq j$.

Equation (II.50) can also be written as

$$\begin{aligned}\Delta\sigma_{ij} &= 2G\,\Delta\varepsilon_{dij} + K\Delta\varepsilon_{kk}\delta_{ij} \\ &= 2\varepsilon\,\Delta\varepsilon_{dij} + (K - \tfrac{2}{3}G)\,\Delta\varepsilon_{kk}\delta_{ij}\end{aligned} \tag{II.51}$$

In the case of the biaxial model and using the approach adopted by Kupfer and Gerstle [31], K and G can be expressed as a function of octahedral shear strain $\gamma_0 = \tau_0/G$. Using E and v as

$$E = \frac{9KG}{(3K + G)} \qquad v = \frac{(3K - 2G)}{2(3K + G)} \tag{II.52}$$

Equation (II.51) is written in matrix form as

$$\{\sigma\} = [D]\{\varepsilon\} \tag{II.53}$$

where $[D]$ for the *biaxial case* is given below

$$[D] = 4GD_2 \begin{bmatrix} D_1 & D_3/2 & 0 \\ D_3/2 & D_1 & 0 \\ 0 & 0 & D_2/4 \end{bmatrix} \tag{II.54}$$

where

$$D_1 = G + 3K \qquad D_2 = 4G + 3K \qquad D_3 = 3K - 2G$$

In three dimensions Bangash [126] adopts the stress–strain relation for the triaxial case as:

$$\begin{pmatrix} \Delta\sigma_x \\ \Delta\sigma_y \\ \Delta\sigma_z \\ \Delta\tau_{xy} \\ \Delta\tau_{yz} \\ \Delta\tau_{zx} \end{pmatrix} = \begin{bmatrix} (K+\tfrac{4}{3}G), & (K-\tfrac{2}{3}G), & (K-\tfrac{2}{3}G) & 0 & 0 & 0 \\ & (K+\tfrac{4}{3}G), & (K-\tfrac{2}{3}G) & 0 & 0 & 0 \\ & & (K+\tfrac{4}{3}G) & 0 & 0 & 0 \\ & & & G_{12} & 0 & 0 \\ & \text{sym} & & & G_{23} & 0 \\ & & & & & G_{13} \end{bmatrix} \begin{pmatrix} \Delta\varepsilon_x \\ \Delta\varepsilon_y \\ \Delta\varepsilon_z \\ \Delta\gamma_{xy} \\ \Delta\gamma_{yz} \\ \Delta\gamma_{zx} \end{pmatrix}$$

(II.55)

or in short

$$\{\Delta\sigma\} = [D]\{\Delta\varepsilon\}$$

where $[D]$ is the required material matrix.

During loading of the structure, the shear modulus of eqn (II.55) is assumed to vary as a logarithmic function of the second stress invariant and is expressed as follows:

$$G_{12} = G_{23} = G_{13} = G = G_e - \alpha \log_e \frac{J_2}{J_2^e} \quad \text{for } J_2 > J_2^e \quad \text{(II.56)}$$

$$G = G_e \quad \text{for } J_2 \le J_2^e \quad \text{(II.57)}$$

and $K = K_e$ is constant throughout, and where

$$G_e = \frac{E}{2(1+v)} \quad \text{is the linear elastic shear modulus} \quad \text{(II.58)}$$

$$K_e = \frac{E}{3(1-2v)} \quad \text{is the linear elastic bulk modulus} \quad \text{(II.59)}$$

$$J_2 = \tfrac{1}{2}(S_x^2 + S_y^2 + S_z^2) + \tau_{xy}^2 + \tau_{yz}^2 + \tau_{zx}^2 \text{ is the second stress invariant}$$

(II.60)

$$S_x = \sigma_x - \sigma_m \quad S_y = \sigma_y - \sigma_m \quad S_z = \sigma_z - \sigma_m \quad \text{(II.61)}$$

J_2^e is the limit of linearity $= \tfrac{1}{3}f_c'$.

When E, v and G change, the material matrix (II.55) is modified on the lines suggested by Bangash [126] and in Appendix I. The values of G_{12}, G_{23}

and G_{13} are given as

$$G_{12} = \frac{1}{2}\left[\frac{E_1}{2(1+v_{12})} + \frac{E_2}{2(1+v_{21})}\right]$$

$$G_{23} = \frac{1}{2}\left[\frac{E_2}{2(1+v_{23})} + \frac{E_3}{2(1+v_{32})}\right] \quad \text{(II.62)}$$

$$G_{13} = \frac{1}{2}\left[\frac{E_3}{2(1+v_{31})} + \frac{E_1}{2(1+v_{13})}\right]$$

The values of v are given in Chapter I.

Due to symmetry

$$E_1 v_{21} = E_2 v_{12} \qquad E_2 v_{32} = E_3 v_{23} \qquad E_3 v_{13} = E_1 v_{31} \quad \text{(II.62a)}$$

In terms of the uniaxial stress–strain relation in compression

$$\varepsilon_{cu} = \frac{\sigma_{cu}}{3}\left[\frac{1}{3K_e} - \frac{1}{G}\right] \le 0.0035 \qquad \sigma_{cu} = f_{cu}$$

Using eqn (II.56) Figs II.7a and b show variation of G with E and with J_2.

II.1.8 Concrete Model Based on Endochronic Theory

This type of model was initially developed for steel by Valanis [171] and has subsequently been modified by Bazant and Bhat [172] and Bazant and Shieh [173]. In this theory the inelastic strains are shown by means of increments of intrinsic time, a non-decreasing scalar variable whose increments depend on strain increments. The difference between the formulations for metal and steel is only due to the hydrostatic pressure sensitivity of inelastic strain from inelastic dilatancy caused by large deviatoric strains. The theory applicable to concrete takes into consideration improved description of strain softening with and without cyclic loading, unloading and reloading. Bazant and Bhat [172] give this theory in an incremental form. Bazant and Shieh [173], using further test data, give a significant refinement of this theory.

The basic incremental constitutive equations of the endochronic theory are:

$$d\varepsilon_{dij} = d\varepsilon_{dij}^E + d\varepsilon_{dij}^P \quad \text{(II.63)}$$

$$= \frac{dS_{ij}}{2G} + \frac{S_{ij}}{2G}dz \quad \text{(II.63a)}$$

$$d\varepsilon_m = d\varepsilon_m^E + d\varepsilon_m^P \qquad (II.63b)$$

$$= \frac{d\sigma_m}{3K} + d\lambda \qquad (II.63c)$$

$$d\varepsilon_{ij} = d\varepsilon_{dij} + \delta_{ij} d\varepsilon_m$$

where

$d\varepsilon_{ij}$ = strain increment
$d\varepsilon_{dij}$ = deviatoric strain increment
$d\varepsilon_m$ = volumetric strain increment
$d\sigma_m$ = volumetric stress increment
dS_{ij} = deviatoric stress increment
$d\varepsilon_{dij}^E, d\varepsilon_{dij}^P$ = elastic and inelastic deviatoric strain increments
$d\varepsilon_m^E, d\varepsilon_m^P$ = elastic and inelastic volumetric strain increments
δ_{ij} = Kronecker delta ($\delta_{ij} = 1$ for $i = j$, otherwise $\delta_{ij} = 0$)
K = bulk modulus
G = shear modulus
dz = intrinsic time increment
$d\lambda$ = inelastic dilatancy increment

$$dS_{ij} = d\sigma_{ij} - \delta_{ij} d\sigma_m \qquad (II.63d)$$

The increment of intrinsic time is defined by Bazant and Bhat [172] as

$$dz = \frac{d\eta}{z_1 F_2 \left\{ 1 + \dfrac{\beta_1 \eta + \beta_2 \eta^2}{1 + F_1 a_7} \right\}} \qquad (II.64)$$

Bazant and Shieh give the value of dz as

$$dz = \left[\left(\frac{d\zeta}{z_1}\right)^2 + \left(\frac{dt}{\tau_1}\right)^2 \right]^{1/2}$$

$$d\zeta = d\eta / f(\eta, \varepsilon\sigma) = \frac{d\eta}{1 + \dfrac{\beta_1 \eta + \beta_2 \eta^2}{1 + F_2/a_7}} F_3 \qquad (II.65)$$

For eqn (II.64) the parameters defined are:

$$d\eta = \left\{ \frac{a_0}{1 - (a_6 I_3(\sigma))^{1/3}} + F_1 \right\} d\xi \qquad (II.65a)$$

$$d\xi = \sqrt{\tfrac{1}{2} d\varepsilon_{dij} d\varepsilon_{dij}} \tag{II.65b}$$

$$F_1 = \frac{a_2[1 + a_5 I_2(\sigma)]\sqrt{J_2(\varepsilon)}}{\{1 - a_1 I_1(\sigma) - (a_3 I_3(\sigma))^{1/3}\}(1 + a_4 I_2(\sigma)\sqrt{J_2(\varepsilon)})} \tag{II.65c}$$

$$F_2 = 1 + \frac{a_8}{\left(1 + \dfrac{a_9}{\eta^2}\right) J_2(\varepsilon)} \tag{II.65d}$$

$I_1(\sigma)$ = first stress invariant
$I_2(\sigma)$ = second stress invariant
$I_3(\sigma)$ = third stress invariant
$J_2(\varepsilon)$ = second invariant of the deviatoric strain tensor

In the case of Bazant and Shieh additional factors are included

$$F_3 = 1 + \frac{a_{10}}{(1 + a_9/\eta^2) J_2(\varepsilon)} \tag{II.66}$$

$$F_1 = a_0(1 - g_1)/[1 - a_5(I_3(\sigma)^{1/3}(1 + g_2)] \tag{II.67}$$

$$F_2 = [a_2\sqrt{J_2(\varepsilon)}(1 + [a_6 I_2(\sigma)]^{1/4} + F_5]/[1 - a_1 I_1(\sigma) + \bar{L}] \tag{II.68}$$

$$\bar{L} = [[a_8 I_2(\sigma)^{1/4} F_4 - a_3 I_3(\sigma)[J_2(\sigma)]^{1/8}(1 + g_2)]$$

$$g_1 = g_{11} g_{12}$$

$$g_2 = g_{21} g_{22} g_{23}$$

$$g_{12} = 1 - \left[1 + \left(\frac{\sigma_{min}}{a_{17}(\sigma_{max} - a_{23})}\right)^4\right]^{-1} \tag{II.69}$$

$$g_{11} = a_{14}[J_2(\varepsilon)]^{1/4} \frac{\sigma_{med} - \sigma_{min}}{\sigma_{max} - a_{23}} \times \left[a_{15}\left(\frac{\sigma_{med} - \sigma_{min}}{\sigma_{max} - a_{23}}\right)^{4/3} - a_{16}\right]$$

$$g_{21} = \left[\frac{a_{18}\left(\dfrac{\sigma_{med} - \sigma_{min}}{\sigma_{med} - a_{23}}\right) - 1}{a_{19}\left(1 - a_{20}\dfrac{|\sigma_{min}|}{\sigma_{max} - a_{23}}\right)(\sigma_{min} - a_{23})}\right]^{5/4}$$

$$g_{22} = \left[1 + a_{21}\left(\frac{\sigma_{min}}{\sigma_{max} - a_{23}}\right)^4\right]^{-1}$$

$$g_{23} = \left[\frac{[J_2(\varepsilon)]^{1/4}}{a_{22} + [J_2(\varepsilon)]^{1/2}}\right]^3$$

$$F_4 = \left[\frac{[J_2(\varepsilon)]^{1/4}}{a_4 + [J_2(\varepsilon)]^{1/2}}\right]^3$$

$$F_5 = a_{11}\sigma_{\min}(1 + a_{12}\sigma_{\min})\left[\frac{[J_2(\varepsilon)]^{1/4}}{|a_{13}\sigma_{\min}|^{1/4} + [J_2(\varepsilon)]^{1/2}}\right]^3 \quad \text{(II.70)}$$

The values of G and K are

$$G = \frac{E_0}{2(1+v)}\left(\frac{1}{1+c_5\lambda}\right) \qquad K = \frac{E_0}{3(1-2v)}\left(\frac{1}{1+c_5\lambda}\right)$$

The values of a, β and c are given in Table II.6.

The shear and bulk moduli are assumed to be dependent on λ (inelastic dilatancy) and are given in the case of Bazant and Bhat as

$$G = \frac{E_0}{2(1+v)}\left(1 - 0.25\frac{\lambda}{\lambda_0}\right) \quad \text{(II.71)}$$

$$K = \frac{E_0}{3(1-2v)}\left(1 - 0.25\frac{\lambda}{\lambda_0}\right) \quad \text{(II.72)}$$

E_0 = initial modulus of elasticity

$$l(\lambda) = 1 - \frac{\lambda}{\lambda_0} \quad \text{(II.72a)}$$

The increment of the inelastic dilatancy variable is defined as:

$$\frac{1 - \lambda/\lambda_0}{1 - c_1 I_1(\sigma)}\left\{\left(\frac{J_2(\varepsilon)}{c_2^2 + J_2(\varepsilon)}\right)^3 + \left(\frac{\lambda}{\lambda_0}\right)^2\right\}d\xi \quad \text{(II.73)}$$

and $d\lambda$ is accumulated along the load path as

$$\lambda = \sum_{\text{all load increments}} d\lambda \quad \text{(II.73a)}$$

Bazant and Shieh give λ' (shear compaction), and the corresponding intrinsic time for compaction is z'.

Similar to eqns (II.65a), (II.73) and (II.73a), the new expressions are:

$$dz' = \left[\left(\frac{d\zeta'}{z_2}\right)^2 + \left(\frac{dt}{\tau_1}\right)^2\right]^{1/2} \quad \text{(II.74)}$$

$$d\xi' = \frac{d\eta'}{h(\eta')} \qquad d\eta' = H(\sigma)\,d\xi'$$

Table II.6
Coefficients for endochronic theory

	1^a	2^b		1^a	2^b
a_0	0·7	0·7	a_{20}	—	14
a_1	$0·6/f'_c$	$0·6/f'_c$	a_{21}	—	1 000
a_2	1 400	$1400\left(\dfrac{f'_c}{4650\text{ psi}}\right)^{1/2}$	a_{22}	—	0·04
			a_{23}	—	$0·2f'_c$
a_3	$500/f'^3_c$	$\dfrac{90}{f'_c}\left(\dfrac{3600\text{ psi}}{f'_c}\right)$	b_1	—	$9·1\left(\dfrac{f'_c}{7200\text{ psi}}\right)$
a_4	$475/f'^2_c$	0·045	b_2	—	$1·0f'_c$
			c_1	$100/f'_c$	$2·0/f'_c$
a_5	$0·8/f'^2_c$	$\dfrac{0·6}{f'_c}\left(\dfrac{3600\text{ psi}}{f'_c}\right)$	c_2	0·000 5	0·003
			c_3	—	0·5
a_6	$0·055/f'^3_c$	$0·15/f'^2_c$	c_4	—	2·0
a_7	20·0	0·05	c_5	—	150
			β_1	30	30
a_8	0·000 125	$\dfrac{15}{f'_c}\left(\dfrac{f'_c}{3600\text{ psi}}\right)^{1·5}$	β_2	3 500	3 500
			β_3	—	0·08
			β_4	—	0·23
a_9	0·001 5	0·001 5	z_1	0·001 5	0·001 5
a_{10}	—	0·000 125	z_2	—	0·012 5
a_{11}	—	$0·2/f'_c$	λ_0	0·001	0·003
a_{12}	—	$0·8/f'_c$	λ'_0	—	0·003
a_{13}	—	$2·2 \times 10^{-5}/f'_c$			
a_{14}	—	25			0·002
a_{15}	—	1·095	c_6	—	psi
a_{16}	—	1·216			
a_{17}	—	0·055			$1·05 \times 10^6$
a_{18}	—	0·94	c_7	—	psi
a_{19}	—	$\dfrac{1·0}{f'_c}\left(\dfrac{3600\text{ psi}}{f'_c}\right)$	c_8	—	0·001
			E_0	$(0·565 + 0·000\,1f'_c) \times 5\,700\sqrt{f'_c}$	

[a] 1: Bazant and Bhat—f'_c shall be in kN/m2, except E_0 where f'_c shall be in N/mm².
[b] 2: Bazant and Shieh—f'_c shall be in kN/m², except where the 6·89 factor is applied and f'_c shall be in lbs/in².

$$d\zeta' = \sqrt{[I_1(d\varepsilon)]^2} = [d\varepsilon_{11} + d\varepsilon_{22} + d\varepsilon_{33}] \qquad (II.75)$$

$$h(\eta') = 1 + \frac{\eta'}{\beta_3} + \left(\frac{\eta'}{\beta_4}\right)^2 \qquad H(\sigma) = b_1 \left[\frac{I(\sigma)}{b_2 - I_1(\sigma)}\right]^2 \qquad (II.76)$$

$l(\lambda)$ is as in eqn (II.72a):

$$l(\lambda, \varepsilon, \sigma) = \frac{c_3}{1 - c_1 I_1 \sigma} \left[\left(\frac{\lambda}{\lambda_0}\right)^2 + \left(\frac{c_4 J_2(\varepsilon)}{c_2^2 + J_2(\varepsilon)}\right)^2\right] \qquad (II.76a)$$

$$d\lambda' = l'(\lambda) L'(\lambda', \varepsilon, \sigma)$$

$$l'(\lambda') = c_6 \left(1 - \frac{|\lambda'|}{\lambda_0'}\right) \qquad L'(\lambda', \varepsilon, \sigma) = \frac{\sigma_{\min} g_3^{1/3}}{1 + |g_3/c_8|^3} \qquad (II.77)$$

$$g_3 = |c_7 \sigma_{\min}| \times 0.93 - [J_2(\varepsilon)]^{1/2} \qquad (II.78)$$

J_2 is the second invariant of the deviator of the tensor; J_3 is the third invariant of the tensor; $\sigma_{\max}$, $\sigma_{\min}$ and σ_{med} are the maximum minimum and medium principal stresses; and $H(\sigma)$ is the softening function.

The additional parameters defined above are also given in Table II.6.

II.2 CONCRETE CRACKING, PENETRATION, PERFORATION, SCABBING AND SPALLING

II.2.1 Concrete Tension and Cracking

Concrete in tension is modelled as a linear elastic strain softening material. The principal stresses and their directions are computed initially in uncracked concrete. If the maximum principal stress for some reason exceeds a limiting value, a crack is assumed to form in a plane orthogonal to this stress. After this, the behaviour of that zone of concrete becomes orthotropic. The local material axes in the given zone coincide with the principal stress direction. If this direction is fixed, the procedure is known as 'fixed crack approach'. If the principal material axes can be given a rotation to coincide with the principal stress and strain directions, the model is known as the 'rotating crack model'. It is also possible in computational models to keep the directions of the first set of cracks fixed, a search is then performed to determine the maximum stress value in the plane parallel to the existing crack. If this maximum stress exceeds the limiting value, then new sets of cracks can be formed which may be perpendicular to the previous one. The limiting values to define the onset of cracking are given as follows:

(a) $\quad \sigma_{pi} = \sigma_{tui} \quad i = 1, 2, 3$ for triaxial tension $\quad$ (II.79)

(b) tension–tension–compression and tension–compression–compression, linear decreasing tensile strength expressions

$$\sigma_{pi} = \sigma_{tui}\left(1 + \frac{\sigma_{i+1}}{f'_c}\right) \quad \text{for } \sigma_{i+1} \leq 0$$
$$\sigma_{pi} = \sigma_{tui}\left(1 + \frac{\sigma_{i+1}}{f'_c}\right)\left(1 + \frac{\sigma_{i+2}}{f'_c}\right) \quad \text{for } \sigma_{i+1}, \sigma_{i+2} \leq 0$$
(II.80)

where σ_{tui} is the tensile strength of concrete.

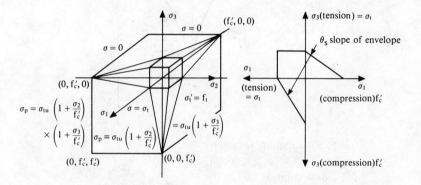

(a) Three-dimensional cracking criteria (b) Criterion in $\sigma_3 - \sigma_1$ plane

Fig. II.7. Concrete tensile strength and cracking.

Figure II.7 shows a typical three-dimensional tensile failure envelope of a concrete model [751].

II.2.2 The Strain Softening Rule

Concrete can be considered as an elastic strain softening material in tension. The problem with this concept is the extent to which the tension stiffening can be incorporated. Alternatively concrete can be assumed to be elastic–brittle in tension. When a crack occurs, the stress normal to it can be immediately released and drop to zero.

The general strain softening model for concrete is given by Petersson [752] as a bilinear curve. Figure II.8 shows the strain softening model. The stress σ is defined as

$$\sigma = \sigma_{tu}[\exp[-(\varepsilon - \varepsilon_{cu})/\alpha]] \quad (II.81)$$

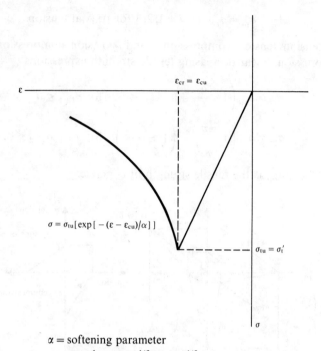

Fig. II.8. Tensile strain softening model for concrete.

α = softening parameter
$= G_f - \tfrac{1}{2}\sigma_{tu}\varepsilon_0(dv)^{1/3} / \sigma_{tu}(dv)^{1/3} > 0$
dv = volume;
$\varepsilon_0 = \varepsilon_{cr}$ = strain at cracking;
σ_{tu} = tensile strength of concrete;
$(dv)^{1/3}\varepsilon_{eq}$ = crack opening;
ε_{eq} = equivalent crack strain, based on smeared crack width.

where ε is the nominal tensile strain in the cracked zone, ε_{cu} is the cracking, and α, the softening parameter, is given by

$$\alpha = (G_f - \tfrac{1}{2}\sigma_{tu}\varepsilon_{cu}l_c)/\sigma_{tu}l_c > 0 \qquad (\text{II.82})$$

where

$$l_c = \frac{\text{volume containing a crack}}{\text{area of crack}} = \frac{V}{A_{cs}}$$
$$\approx (dV)^{1/3}$$

and dV is the volume of concrete represented by the sampling point

$$G_f = \int_0^\infty \sigma(\omega)\,d\omega = \text{fracture energy needed to separate two cracks} \quad (11.83)$$

$\approx 50\text{–}200 \text{ N/m}$

$\omega = \text{crack width} = l_c \varepsilon_{cu}$

It is important to note that a considerable amount of shear can be transferred across the surfaces of cracked concrete. The shear transfer mechanism is aggregate interlock which depends on the aggregate size and grading. On the basis of this concept the next section gives a step-by-step procedure for crack initiation, closure and opening.

II.2.3 A Step-by-Step Formulation for Cracking

Bangash [126] has presented cracking criteria, assuming that under incremental loads progressive and well-disposed cracks occur. The following assumptions are made concerning the initiation, closure and re-opening of cracks.

(a) Where cracks occur the direct tensile stress cannot be supported in the direction normal to the cracks.
(b) A concrete element parallel to the crack is assumed to carry stress given by the constitutive relationship consistent with biaxial or triaxial conditions prevailing in the plane parallel to the crack. On further loading new cracks occur at an angle to the previous ones.
(c) The opposite faces of a crack opening in a normal direction will interlock when they are subjected to a parallel differential movement. As a result, restraint of movement and volume changes will cause forces which will only be transmitted across the crack. With large crack widths, these surfaces would completely separate and the aggregate interlock phenomenon would cease. If the strain across the crack is positive, the crack is considered open and rigidity for that direction is kept zero.
(d) Where interlocking exists, the shear stress τ_0 along the crack will not be zero and instead it is linearly related to the strain caused by the differential movement.
(e) During the unloading procedure of the structure the interlocking faces of a crack and the surface deterioration will not interfere with the closing of a crack. A crack closes if the normal (negative quantity) across the crack is compressive.

The best way to handle these cracks is to adjust material laws in zones where they occur with their natural directions. The material matrice $[D]$ can easily be modified to initiate cracks in specific directions using direction cosines.

II.2.3.1 Step-by-Step Analysis of Single and Multiple Cracks

The cracks are initiated, closed and re-opened using the following steps:

(i) Assess initial strains $\{\varepsilon_0\}$, initial stress $\{\sigma_{in}\}$ and adjusted material property matrix $[D]$.

(ii) Compute incremental stresses

$$\delta\{\sigma_i\} = [D]\delta\{\varepsilon_i\} \quad \text{for no crack}$$
$$\delta\{\sigma_i\}^* = [D]\delta\{\varepsilon_i\} \quad \text{for an already initiated crack}$$

(II.84)

(iii) Calculate total stresses and strains

$$\{\sigma_i\}_{tot} = \{\sigma_{in}\}^* + \delta\{\sigma\}_i^* \qquad \{\varepsilon\}_{tot} = \{\varepsilon_{in}\}^* + \delta\{\varepsilon\}_i^* \qquad (II.85)$$

(* is excluded if cracks do exist).

(iv) For the initially uncracked condition, compute principal stresses $\{\sigma\}$ corresponding to total stresses in point step (iii) and check for the initiation of cracking

$$\{\sigma\}_{pi} \geq \sigma_{tu} \quad i = 1, 2, 3, \text{ triaxial tension zone} \qquad (II.86)$$

If the above inequality is satisfied, calculate the angle of cracks and their directions. Principal stresses are calculated and converted into a global system using a standard transformation matrix $[T]$

$$\{\sigma_g\} = [T]\{\sigma\}_{pi} \qquad (II.87)$$

(v) The stresses are converted into nodal loads

$$\{F\} = \int [B]^{T^{11}}\{\sigma_g\} \, d \, vol \qquad (II.88)$$

(vi) Residual forces are examined and accumulated for elements and are checked at the end of each iteration. The final iteration terminates when the values of these forces are less than a predetermined constant of 0·001 or any other acceptable value.

(vii) At this stage a thorough check is required of the crushing of concrete and the yielding of steel. When the principal compressive strain $\varepsilon_i \geq \varepsilon_{cu}$, ε_i is such that where ε_{cu} is the limiting compressive strain (assumed 0·0035), concrete is assumed to crush in the compressive zones at the integration points. The material matrix $[D]^*$ is set to zero. The stresses are then zero. For subsequent load increments these points are ignored. Stresses at these points are converted into equivalent nodal forces as shown in eqn (II.88). The steel elements, whether lying within or on the solid element, yield on the basis of a uniaxial stress–strain relationship. The crushing failure condition of concrete can also be on the basis of

$$3J'_2 = \varepsilon_{cu}^2$$

where J'_2 is the second deviatoric strain invariant.

(viii) This criterion for the crack closure is based on the inequality of strains

$$\varepsilon_{cu} \leq \varepsilon_c \qquad (II.89)$$

where

ε_{cu} = strain normal to the crack

$\varepsilon_{cu} = \dfrac{v^*}{E_c^*}\sigma_{cu}$ or strain from the stress–strain curve

v^* and E_c^* = relevant concrete Poisson's ratio and Young's modulus

If $\varepsilon_{cu} < \varepsilon_c$, compressive strains are induced which may not be compatible with the concrete strains. When cracks are closed at an intermediate load within the increment then at the end of the load increment stresses may occur at the contact faces of the crack.

(ix) The opening of the crack is similar to new crack initiation. The only difference is the tensile resistance offered by concrete. The closed crack will re-open if

$$\sigma_{cu} \geq 0 \qquad (II.90)$$

The shear stresses τ_{cu} on the crack interface and the normal stresses σ_{cu} must be redistributed such that

$$\hat{t}_{fv} = \sigma_{cu} \qquad \hat{t}_{f\tau} = \tau_{cv} \qquad (II.91)$$

When these stresses are equated, the rest of the procedure should then be the same as given earlier.

(x) The crack nodes are not identical from one iteration to another at any specific Gaussian point. The most obvious reasons are the improvement in $[D]^*$ and the effect of interlocking with the shear stress along the crack.

These changes from one crack node to another can be included. The following linear relationship is adopted for shear τ^*

$$\tau^* = \beta' G \gamma^* \tag{II.92}$$

where G is the shear modulus of the uncracked material; β' is the interlocking factor $= 1 - (\varepsilon_t/0\cdot005)^{K_1}$; ε_t is the tensile strain normal to the crack plane; and K_1 ranges from $0\cdot3-1$.

II.2.4 Blunt Crack Band Propagation

The smeared crack concept is far simpler than the alternative approach of isolated sharp inter-element cracks in which one has to cope with topological changes in element connectivity. The smeared crack requires an adjustment of element stiffness. Cedolin et al. [169] have developed the blunt crack approach and a refined version is given by Pfiffer et al. [175]. The smeared crack band of a blunt front is that in which one can easily choose cracks of any direction without paying a penalty if the crack

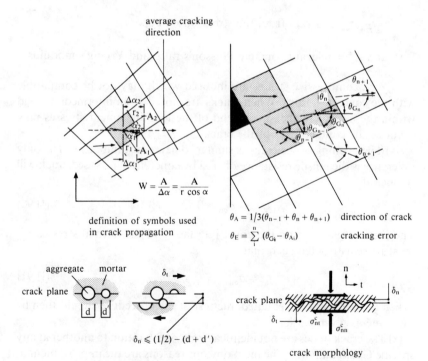

Fig. II.9. Blunt crack propagation.

direction is unknown. On the contrary, a change in the direction of a sharp inter-element crack requires changes in the co-ordinates of the finite element mesh. The smeared crack concept suffers from the drawback that the magnitude of stress depends on the sharpness of the front of the crack band, particularly the width of the front element of the blunt crack band. The smaller the width, the larger is the stress in the element just ahead of the front under smaller applied load. If the element width is zero, the stresses ahead of the crack front become infinite for any finite load value. This indicates that cracks would propagate at nearly zero load, which cannot be true. To overcome this problem of fracture mechanics, a criterion in conjunction with the highly efficient smeared crack concept is adopted.

The equivalent strength and energy variation are utilised for crack propagation once it is initiated within the element. The equivalent strength criterion is used for crack propagation by specifying an equivalent stress within the surrounding elements of an existing crack at which cracking should be propagated. The expression for the equivalent strength σ_{eq} is given as

$$\sigma_{eq} = C[EG_f/W(1 - 2v\sigma_2^0/\sigma_1^0)]^{1/2} \qquad (II.93)$$

where

C = a constant dependent on the choice of elements
E = elastic modulus
v = Poisson's ratio
$W = A/\Delta a = A/r \cos \alpha$
A = area of the element at the front (Fig. II.9)
r = centroidal distance from the last cracked elements in the front element
α = angle measured from the established crack to the line between the centroids
G_f = the energy release rate = $\partial U(P_i, a)/\partial a = \Delta U(P_i, a)/\Delta a$
U = total strain energy in a cracked body
a = cracked area
P_i = loads
ΔU = energy released by the structure into the element which cracks

The band length is specified as $a + \Delta a/2$.

In the initial state prior to cracking, the strain energy U_0 is based on the principal stresses σ_2^0, where σ_1^0 is the largest tensile stress. After cracking σ_1^1 becomes zero and $\sigma_2^1 \neq 0$, which is used for the current value of U_1. The change of strain energy $\Delta U = U_0 - U_1$ is equated to crack length $\Delta a \times G_f$

defined above. If $\sigma_1^1 = 0$, cracks relieve the σ_1 stress. The energy flow into the crack front depends on γ and α as indicated in Fig. II.9.

II.2.4.1 Formulation of Cracking Directions

In this method one specifies the cracking direction within an arbitrary grid, given by

$$\theta_A = \tfrac{1}{3}(\theta_{n-1} + \theta_n + \theta_{n+1}) \qquad (II.94)$$

where θ_A is the average crack direction, θ_{n-1} and θ_n are the cracking angles of the next to last cracked element, and θ_{n-1} is the impending cracking angle of the element adjacent to the crack front. Since in the arbitrary grid the cracking direction is specified by the accumulated error of the cracked direction, the accumulated cracking error θ_E is given by

$$\theta_E = \sum_i^n (\theta_{Gi} - \theta_{Ai}) \qquad (II.95)$$

where n is the number of cracked elements and θ_G is the actual average crack propagation angle within each element.

Gambarova and Karakoc [177] introduce the relation between the interface shear and the crack displacements in Bazant's formulations. A better description is given of the effects of aggregate size, and a better formulation is given of the tangent shear modulus G^{CR} whenever cracking is initiated.

$$G^{CR} = \frac{\sigma_{nt}^c}{\varepsilon_{nn}^{CR}} k \frac{1}{r(a_3 + a_4|r|^3)} \qquad (II.96)$$

or

$$G^{CR} = \frac{\tau_0}{\varepsilon_{nn}^{CR}} k \{1 - [2(P/D_a)\varepsilon_{nn}^{CR}]^{1/2}\}$$

where

$$K = \frac{a_3 + 4a_4|\gamma|^3 - 3a_3 a_4 \gamma^4}{(1 + a_4 \gamma^4)^2} \qquad (II.97)$$

and

a_3, a_4 = coefficients as a function of the standard cylindrical strength f_c''
τ_0 = crack shear strength (ranging from 0·25 to 0·7f_c'')
D_a = maximum aggregate size (up to 4 mm)

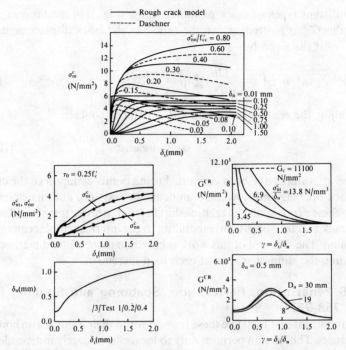

Fig. II.10. Crack displacement versus tangent shear modulus [169, 175, 177].

P = large percentage of crack asperities
$\gamma = \delta_t/\delta_n$
δ_t, δ_n = crack displacements along the normal and tangential directions (Figs II.9 and II.10)

Curves have been plotted showing a decrease in the value of G^{CR} when a crack opening linearly increases at increasing shear. The paper [177] gives a constitutive law in which a confinement stress within the rough crack model is given by

$$\sigma_{nn}^c = -a_1 a_2 \frac{\delta_t \sigma_{nt}^c}{(\delta_n^2 + \delta_t^2)^q} \qquad \text{(II.98)}$$

where

σ_{nn}^c = interface normal stress
σ_{nt}^c = interface shear stress
a_1, a_2 = constants ($a_1 a_2 = 0.62$)
q = a function of crack opening; taken to be 0.25.

For different types of crack dilatancy δ_n/σ_{nt}^c (Fig. II.9) the tangent shear modulus G^{CR} is plotted against the ratio r of the crack displacement. The value of σ_{nt}^c is given by

$$\sigma_{nt}^c = \tau_0 \left(1 - \sqrt{\frac{2p}{D_a}\varepsilon_{nn}^{CR}}\right) r \frac{a_3 + a_4 |r|^3}{1 + a_4 r^4} \qquad r = \gamma_{nt}^{CR}/\varepsilon_{mn}^{CR} \qquad (II.99)$$

where p is the crack spacing and CR is cracked concrete

$$\varepsilon_{nn}^{CR} = \frac{\delta n}{p} = \text{strain against } \sigma_{nn}^c \qquad (II.100)$$

The σ_{nn}^c values have been computed using points common to the curves of crack opening and constant confinement stress. The authors claim that using these expressions the relationship between stresses transferred across the crack and the strains corresponding to the interface displacements can be found. The solution of this work is based on finite element methods, by updating the stiffness matrix at each load increment.

II.2.5 Penetration, Perforation, Scabbing and Spalling [178, 179, 753–757]

It is important first to define these terms prior to their influence on concrete structures. This section pertains only to local effects largely independent of the dynamic characteristics of the concrete structures. Local effects consist of missile penetration into and perforation through concrete structures, and spalling and scabbing of such concrete structures. They are defined below:

(a) Penetration — The measure of the depth of the crater formed at the level of impact.
(b) Perforation — Full penetration through the structure.
(c) Scabbing — The peeling off of the material from the opposite face to that at which the impact occurred.
(d) Spalling — The ejection of target material from the face at which the impact occurs.

Analytical and rational mathematical prediction of these effects is extremely difficult. This is due to the complex nature of the transient loads. Various papers have been published in recent years which have described new ideas and methods for analysing and designing against impact on concrete structures connected with nuclear-plant facilities. Owing to complexities in evaluating structural damage due to impact loading the design criteria so far developed have been mainly dependent upon experimental tests and empirical formulae.

There are various empirical formulae that used to be employed in connection with missile impact problems on nuclear-plant structures, but most of them have been discarded because over the years new more relevant test data have been made available and these empirical formulae did not give good or realistic results compared with these.

However, there are some older and newer formulae which are still very much applicable. Of these, some empirical formulae are given in the next few paragraphs.

II.2.5.1 The Modified National Defence Research Formula (NDRC)

This formula is really the only one of the older formulae still applicable to nuclear-plant facilities for design against impacts. It was put forward in 1946 by the National Defence Research Committee (USA), and proposed a theory of penetration for a non-deforming missile penetrating a massive concrete target. This theory enabled one to calculate the penetration, scabbing and perforation thickness for given missile data, and also gave the thickness that would be required to prevent scabbing and perforation.

The modified NDRC formula for penetration is given below

$$x = \left[4KNWd \left(\frac{V_0}{1000d} \right)^{1.8} \right]^{1/2} \quad \text{for } \frac{x}{d} \leq 2 \cdot 0 \quad \text{(II.101)}$$

or

$$x = \left[KNW \left(\frac{V_0}{1000d} \right)^{1.8} + d \right] \quad \text{for } \frac{x}{d} \geq 2 \cdot 0 \quad \text{(II.102)}$$

where all the symbols have been defined earlier.

The modified NDRC formula for perforation thickness is given by

$$\frac{e}{d} = 3 \cdot 19 \left(\frac{x}{d} \right) - 0 \cdot 718 \left(\frac{x}{d} \right)^2 \quad \text{for } \frac{x}{d} \leq 1 \cdot 35 \quad \text{(II.103)}$$

This formula is applicable for $x:d$ ratios less than 1·35 only, and for other $x:d$ ratios the results are very conservative. For other $x:d$ ratios the US Army Corps of Engineers put forward the formula

$$\frac{e}{d} = 1 \cdot 32 + 1 \cdot 24 \left(\frac{x}{d} \right) \quad \text{for } 1 \cdot 35 \leq \frac{x}{d} \leq 13 \cdot 5 \quad \text{(II.104)}$$

It should be stressed that the above formulae for perforation within the stated $x:d$ ratios are not strictly NDRC formulae but are based on the penetration thickness x, as obtained from the original NDRC formula.

The values of e and d in the formulae are the calculated values of the target thickness to prevent perforation and scabbing, respectively. It has been suggested that a factor of safety be used with these values so that the design thickness of the concrete target is 20% or 30% greater than the value calculated.

For example, if a 20% safety factor is used the design thickness to prevent perforation, t_4 would be given by

$$t_4 = 1\cdot 2e \tag{II.105}$$

Similarly, the thickness to prevent scabbing is given by

$$t_4 = 1\cdot 2s \tag{II.106}$$

where s = scabbing thickness.

It is up to the designer what he considers most important for design, the perforation thickness or the scabbing thickness. Usually the larger of the two values obtained is used.

In most cases if the structure is designed against perforation then it is automatically made adequately safe against punching, and further analysis for punching shear is not necessary. This is strictly true for non-deformable missiles such as steel rods but also holds good for most deformable missiles.

II.2.5.2 The Bechtel Formula

This formula is sometimes used to determine the thickness to prevent scabbing. It was developed by the Bechtel Corporation and is based on recent test data applicable to missile impacts on nuclear-plant structures. However, the formula is restricted to essentially non-deformable missiles such as solid steel slugs and rods. It is only moderately applicable to hollow pipe missiles. There are two formulae for scabbing thickness which are given below:

The Bechtel formula for scabbing thickness for solid steel missiles

$$s = \frac{15\cdot 5}{\sqrt{f'_c}} \left(\frac{W^{0\cdot 4} V_0^{0\cdot 5}}{d^{0\cdot 2}} \right) \tag{II.107}$$

The Bechtel formula for steel pipe missiles (scabbing)

$$s = \frac{5\cdot 42}{\sqrt{f'_c}} \left(\frac{W^{0\cdot 4} V_0^{0\cdot 65}}{d^{0\cdot 2}} \right) \tag{II.108}$$

The variables given above are in Imperial units. The symbols again are those defined earlier. d is the nominal missile diameter. The formula was strictly developed with 200 mm (8 in) diameter pipes but also holds true for other diameters.

Again s obtained from the formula is the target thickness to prevent scabbing. A factor of safety may also be used to convert to design-target thickness.

II.2.5.3 The IRS Formulae for Penetration and Complete Protection

The IRS formula for penetration is expressed as

$$x = 1183 f_c'^{-0.5} + 1038 f_c'^{-0.18} \exp(-0.82 f_c'^{0.18}) \qquad \text{(II.109)}$$

The IRS formula for total protection of a target against penetration, perforation and scabbing is:

$$\text{SVOLL} = 1250 f_c'^{-0.5} + 1673 f_c'^{-0.18} \exp(-0.82 f_c'^{0.18}) \qquad \text{(II.110)}$$

where SVOLL is the minimum wall thickness to provide complete protection. (Note: The units of the penetration and concrete strength are not Imperial. The penetration is in units of centimetres and f_c' is in the units kg force/cm^2—1 kg force/cm$^2 \approx 100$ N/m^2.)

In the IRS formula the value of SVOLL is the equivalent of design thickness as obtained from the other formulae. The penetration depth is the same as those in the other above-mentioned formulae.

II.2.5.4 The ACE Formulae to Prevent Penetration or Perforation

The formulae for penetration and perforation are as follows:

(i) The formula for penetration

$$x = \frac{282}{\sqrt{f_c'}} \left(\frac{W}{d^2}\right) d^{0.215} \left(\frac{V_0}{1000}\right)^{1.5} + 0.5d \qquad \text{(II.111)}$$

(ii) The formula for perforation

$$e = 1.23d + 1.07x \qquad \text{(II.112)}$$

The symbols used are those previously defined. The units of the variables are Imperial. The ACE formula for perforation gives the thickness of target to prevent perforation. Together with a factor of safety of say 1·2, the value of e can be used to determine the target design thickness.

$$t_a = 1.2e \qquad \text{(II.112a)}$$

This is because e is the minimum wall thickness of the target to prevent perforation.

II.2.5.5 The Stone and Webster Formula

Scabbing thickness $(s) = (Wv_0/C)^{1/3}$ C is a coefficient (Table II.7)

(II.113)

II.2.5.6 The CKW–BRL Formula for Penetration and Perforation

$$x = \frac{6Wd^{0.2}}{d^2}\left(\frac{V_0}{1000}\right)^{4/3} \quad \text{(Imperial units)} \quad \text{(II.114)}$$

The perforation thickness is given by

$$e = 1\cdot3x \quad \text{(II.114a)}$$

e is the minimum thickness to prevent perforation.
Table II.7 shows comparisons of scabbing thicknesses.

Table II.7
Comparison of predicted scabbing thickness and known low velocity test results

Missile	Missile velocity (m/s)	Test results scabbing thickness (mm)	Calculated scabbing thickness	
			NDRC (mm)	Bechtel (mm)
200 mm slug 0·95 kN	60	305	360	315
75 mm pipe schedule-40 0·35 kN	61	300	285	343
300 mm pipe 3·3 kN	60	450–600	607	627
75 mm slug 0·06 kN	30	127	129	129
75 mm diameter pipe 0·05 kN	59	133	141	151

Stone and Webster: C coefficients lb/in/s

s/d	1	1·5	2·0	2·5	3·0
Solid C	←	900	—	950	→
Hollow C					
$2t/d = 0\cdot125$	2 250	2 450	2 500		2 550
$= 0\cdot08$	3 000	3 250	3 350		3 400
$= 0\cdot06$	3 600	3 750	3 900		4 050

II.2.5.7 The Barr, Carter, Howe and Neilson Formulae

Experiments were undertaken on circular reinforced concrete slabs at three different scales (1×, 0·38× and 0·128× prototype). Transient loads, deflections and impactor velocities were measured during impact. The bending reinforcement in the targets varied between 0 and 0·5% each way on each face and the missile velocity varied from 50 m/s to 300 m/s.

Based on these tests, a modification of the CEA–EDF formula was suggested. Thus, the velocity to cause perforation V, can be calculated by

$$V = 1\cdot 3\sigma^{1/2}D_c^{1/6}(de^2/m_s)^{2/3}r_b^{0\cdot 27} \qquad \text{(II.115)}$$

Because of the limited test data, it was proposed that the formulae be restricted to the following range:

$$160 \text{ kg/m}^3 < \text{reinforcing steel}$$
$$20 \text{ m/s} < \text{impact velocity} < 230 \text{ m/s}$$
$$0\cdot 5 \leq \text{slab thickness/projectile diameter}$$

II.2.5.8 The Kar Formula

The equation for predicting the penetration depth, X, in concrete structures is given as:

$$G(X/d) = [\bar{\alpha}/(f_c')^{0\cdot 5}]N_2(E^m/E_s)^{1\cdot 25}[W/D(d^{1\cdot 8})(V/1000)^{1\cdot 8}] \qquad \text{(II.116)}$$

where

$$G(X/d) = (X/2d)^2 \quad \text{for } X/d \leq 2\cdot 0$$
$$G(X/d) = X/d - 1 \quad \text{for } X/d > 2\cdot 0$$

$N =$ 0·72 for flat-nosed solid bodies, and
$N =$ 0·72 + 0·25 $(n - 0\cdot 25)^{0\cdot 5} \leq 1\cdot 17$ for missiles with special nose shapes; n is the ratio of the radius of the nose to the diameter of the missile, and
$N =$ 0·72 + $[(D/d) - 1](0\cdot 0306) \leq 1\cdot 17$, for hollow pipe or irregular sections
E^m is the modulus of elasticity of the material of the missile (psi)
E_s is the modulus of elasticity of mild steel (psi)
f_c' the ultimate compressive strength of the concrete test cylinder (psi)
d is the projectile diameter (in.)
D is the outside diameter of a circular section (in.)
W is the weight of the missile (lbs)
V is the impact velocity of the missile (ft/s)
$\bar{\alpha}$ is a constant depending on the type of units used ($\bar{\alpha} = 180\cdot 0$ when using imperial units)

The thickness t_a to prevent perforation, and the thickness t_s to prevent backface scabbing are determined by

$$(t_a - a_1)/d = 3\cdot19(X/d) - 0\cdot718(X/d)^2 \quad \text{for } X/d < 1\cdot35 \quad \text{(II.117)}$$

$$\rho(t_s - a_1)/d = 7\cdot91(X/d) - 5\cdot06(X/d)^2 \quad \text{for } X/d \leq 0\cdot65 \quad \text{(II.118)}$$

in which $\rho = (E_s/E^m)$ ($\rho = 1$ for steel missiles), and a_1 is half the aggregate size in concrete (in.) $t_a = e$ in this case.

For X/d ratios larger than those shown in the previous equations, the perforation depth and the scabbing thickness can be calculated using the following equations:

$$(t_a - a_1)/d = 1\cdot32 + 1\cdot24(X/d) \quad \text{for } 3 \leq t_a/d \leq 18 \quad \text{(II.119)}$$

$$\beta'(t_s - a_1)/d = 2\cdot12 + 1\cdot36(X/d) \quad \text{for } 3 \leq t_s/d \leq 18 \quad \text{(II.120)}$$

Chapter III

Numerical Models for Creep, Shrinkage and Temperature

III.1 INTRODUCTION

This chapter deals with various constitutive models for creep, shrinkage and temperature. As far as possible up-to-date information is included for all three areas so that it can be used in the finite element analyses presented later in the text.

Concrete exhibits time-dependent strains due to creep shrinkage and temperature, which profoundly affect the properties of concrete as a structural material. The literature covering concrete creep is comprehensive and it is not therefore intended to review the entire literature in great depth here. However, a brief discussion of the results obtained by various researchers is included.

III.2 CREEP

III.2.1 A Brief Survey of Creep

Creep in concrete represents the dimensional change in the material under the influence of sustained mechanical loading. Quite small loads will cause the concrete to deform. The phenomenon of creep occurs at elevated and at ambient temperatures. The rate of creep is increased at elevated temperatures. Various experimental tests [204–228] have been conducted to identify the effect of temperature on concrete creep. England and Phok [86] and Ross [207] have presented results for sealed and unsealed cylinders up to a temperature of 140°C and for a test duration of 60 days. The results for sealed cylinders show that the creep at 80 and 140°C was about 3·5 and 4·2 times the value at 20°C. Nasser and Neville [204] reported their observations from experimental tests at temperatures ranging from 21·1 to 96·1°C with stress/strength ratios from 0·35 to 0·7. The concrete was cured at 41·34 N/mm² (6000 psi) and tested after 24 h of casting. They found that the pronounced maximum for the creep rate is at a temperature of about 71°C. This creep rate was based on creep measurements made during a period from 21 to 91 days after loading. Hannant [205] conducted creep tests on sealed 104·775 mm (4⅛ in) by 305 mm (12 in) cylinders of (approximately 62 N/mm² (9000 psi)) limestone aggregate after curing them for five months in water and for an additional month in a sealed and saturated condition. Results showed a nearly linear increase of specific creep with a range of 27 to 77°C for loading periods of two years. The creep at 77°C was approximately 4–4·8 times that at 27°C. Up to stresses of 12·78 N/mm² (2000 psi) creep remained proportional to stress. The Poisson

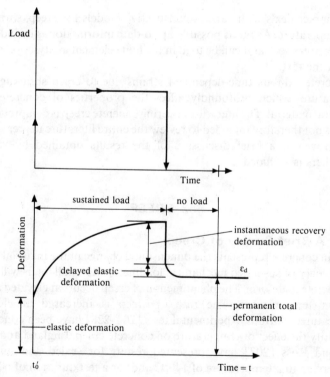

Fig. III.1. Typical deformation versus time curves for concrete under constant load.

ratios of creep determined for sealed specimens were similar in magnitude to their elastic values.

In order to understand the influence of creep, a typical deformation–time curve is shown in Fig. III.1. A concrete specimen loaded under uniaxial compression shows an immediate elastic deformation. If this load is sustained, additional deformation due to creep occurs. The rate of the deformation decreases with time (Fig. III.1). If this load is removed, there is an immediate recovery of deformation and, following this, a recovery of creep deformation (delayed recovery) rate occurs, which rapidly decreases with time, as shown in Fig. III.2. At the end of this, a residual deformation is left which is greater than the initial elastic deformation. Figure III.2 shows a time-dependent strain curve for ambient and elevated temperatures between T_1 and T_2. The creep recovery strain occurs immediately after the instantaneous elastic strain, which is extensive at first but reduces after a short period of time. The creep recovery is essentially independent of temperature [204].

At a low level of stress, the concept of specific creep is introduced. Specific creep or creep strain per unit of stress is a useful indicator of creep effects. It is also sometimes useful to normalise creep strain data with

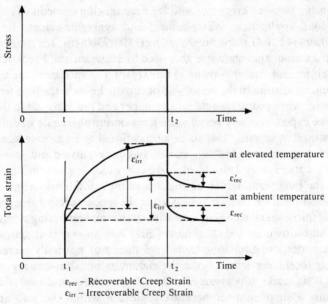

ε_{rec} – Recoverable Creep Strain
ε_{irr} – Irrecoverable Creep Strain

Fig. III.2. Time-dependent strains in concrete.

respect to stress and temperature. This quantity is known as specific thermal creep.

The creep rate, creep–time curves and the Young's modulus of concrete are all temperature dependent. The creep rate is also dependent on the type of aggregate and the mix proportions, and the age of the concrete. The average values and increases in creep rate are subject to $+30\%$ due to the different qualities of aggregate. The creep of concrete at normal temperatures has been investigated by several authors [80–96, 207–234, 418]. Some tests indicate that even very old concrete is still liable to creep of considerable magnitude. The creep and time curves of old concrete approach very nearly those of young concrete [80–96, 207–234]. This is a very important phenomenon and is very relevant to the analysis and design of the reactor vessels since these vessels are loaded when concrete has attained several years of age. For prototype vessels the maximum temperature level permitted on concrete [82, 94] is about 80°C, and hence creep is generally high, as suggested by Ross [207]. Bangash and England have carried out an optimum analysis, based on a statistical approach, to plot long-term specific creep values. The long-term specific creep values have been determined by extrapolating short-term (Fig. III.3) tests by Hannant and Frederick [58] and Browne [84] based on a linear relationship between creep rate and log time existing immediately after the initial load application. Water/cement and aggregate/cement ratios are $(0.47:0)$ and $(4.5:4.4)$, respectively, giving $(7800:7300)$ psi. The difference lies in the maximum aggregate size, that used by Hannant and Frederick being 9·5 mm ($\frac{3}{8}$ in) and that of Browne 39 mm ($1\frac{1}{2}$ in). The specimens are sealed to represent conditions in the vessel wall or cap thickness in the long term. The two curves when plotted would give an upper and lower bound to the creep loss to be expected in any vessel when a maximum aggregate size of 18 mm ($\frac{3}{4}$ in) is used. Assuming loss to be proportional to aggregate size, linear interpolation between the two curves has been carried out in order to obtain a curve for 19·3 mm ($\frac{3}{4}$ in) aggregate which is given in Fig. III.3.

For the long term, under normal operating conditions, creep loss has been determined for nuclear vessels using the specific creep value from this curve at thirty years in conjunction with the normal operating stresses. For a hot shut-down period, Hannant [205] has shown that temperature cycling under constant long-term load does not normally increase the resulting total creep strain. Since the maximum hot shut-down period for vessels is six weeks, it is assumed to be a short-term shut-down, which will not cause a significant temperature reduction within the wall and slab thicknesses of the vessel from that of normal operation.

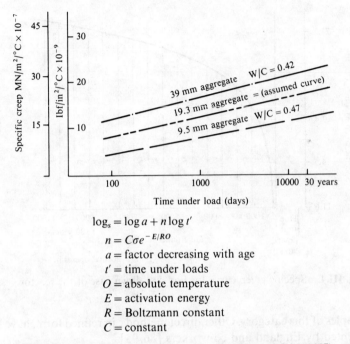

$\log_s = \log a + n \log t'$
$n = C\sigma e^{-E/RO}$
a = factor decreasing with age
t' = time under loads
O = absolute temperature
E = activation energy
R = Boltzmann constant
C = constant

Fig. III.3. Specific creep curves (after Refs [84, 88, 205, 909]).

The shut-down periods last for six weeks and occur every second year after the first four years throughout the vessel life. During these six-week periods there is an increase in stress in the vessel and a corresponding increase in creep strain. These increases in creep strain have been superimposed on the 30-year normal operating creep strain assuming that there is no creep recovery on return to normal operating conditions. This, therefore, gives a conservative estimate of creep loss. Figure III.4 gives the relationship between specific creep and time for a hot shut-down case for up to the entire predicted life of the vessel. Similar curves are used for structures in different operating temperature conditions.

Various methods of creep analysis have been used over the last 50 years; they may be classified into two main categories:

(a) Direct methods
(b) Iterative or step-by-step methods

The direct methods allow the calculation of creep effects in a single time step. The effective modulus [222] and steady-state [222] methods are

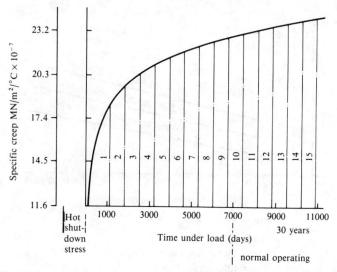

Fig. III.4. Specific creep curve for hot shut-down case of the reactor vessel.

examples of this category. Other direct methods, in refined form, have been presented by England and co-workers [86, 93].

In the iterative methods, the period of time (over which creep is sought) is divided into a number of steps, and separate calculations are carried out for each step. An assumption is made that stress is constant during each time step while strain is being calculated. The accuracy and stability of the solution depends on the length of the time step chosen and successive calculations depend on those in previous time steps. The types of iterative creep solution methods are:

1. Method of superposition
2. Rate of creep
3. Strain hardening
4. Rate of flow

III.2.2 Creep Constitutive Models
III.2.2.1 Indian Standard Code IS 1343 [212]
The ultimate creep strain ε_r^u is computed as

$$\varepsilon_r^u = 17 \cdot 5 \left(\frac{1 \cdot 25 - H}{0 \cdot 25} \right) \left(\frac{400}{f_{cu}} \right) \left(\frac{f_{cu}}{f_{ci}} \right) \times 10^{-6} \qquad \text{(III.1)}$$

where

H = the average ambient humidity of the environment (0·5–0·2 dry; 0·75–0·9 humid; 1·0 average)

f_{cu} = cube strength of concrete at 28 days

f_{ci} = cube strength of concrete at transfer

III.2.2.2 The Hilsdorf and Muller Method [213]

This method gives the general creep function $\varepsilon_c(t, t')$:

$$\varepsilon_c(t, t') = \frac{1}{E_c(t')} + \frac{\phi(t, t')}{E_c(t')} \quad \text{(III.2)}$$

where

$E_c(t')$ = modulus of elasticity at time t'

$\phi(t, t')$ = creep coefficient = $\varepsilon_c/\varepsilon_e$ for concrete at time t

III.2.2.3 The ACI Method

The American Concrete Institute ACI 209 method is based mainly upon the method proposed by Branson [214]. The hyperbolic function $\phi(t, t')$ is given by

$$\phi(t, t') = \frac{(t - t')^{0·6}(\phi_\infty)}{10 + (t - t')^{0·6}} \quad \text{(III.3)}$$

where ϕ_∞, the creep coefficient, is given by

$$\phi_\infty = 2·35 \beta_{t'} \beta_H \beta_d \beta_S \beta_F \beta_{AC}$$

$\beta_{t'}$ = a coefficient for the age of the concrete t' when load is applied

$$\beta_{t'} = \{1·25 t'\}^{-0·118} \quad \text{for } t' \geq 7 \text{ days} \quad \text{(III.3a)}$$

β_H = a coefficient to estimate the relative humidity H

$$\beta_H = 1·27 - 0·0067 H \quad \text{for } H \geq 0·4 \quad \text{(III.3b)}$$

β_d = a coefficient for the average thickness of the concrete member

(i) for $50 \leq d \leq 150$ mm

d (mm):	50	75	100	125	150
β_d:	1·3	1·17	1·11	1·04	1·0

(III.3c)

(ii) for $150 \leq d < 380$ (all in mm)

$$\beta_d = 1·14 - 2·093 \times 10^{-5} \to t - t' \leq 365 \text{ days}$$

or (III.3d)

$$\beta_d = 1·10 - 1·139 \times 10^{-5} d^2 \to t - t' > 365 \text{ days}$$

(iii) $d \geq 375$ mm

$$\beta_d = 0.67(1 + 1.13e^{-0.212V/A})(1.13e^{-0.54V/A}) \quad \text{(III.3e)}$$

(V = volume of the member/structure and A = area of the member/structure)

β_S = a coefficient for the consistency of fresh concrete

$$\beta_S = 0.82 + 1.7688 \times 10^{-4} S \quad \text{(III.3f)}$$

(S = slump in concrete (mm))

β_F = a coefficient for fine aggregate particles

$$\beta_F = 0.88 + 24 \times 10^{-4} F \quad \text{(III.3g)}$$

(F = particles/total aggregate content $\times$ 100)

β_{AC} = a coefficient for the air content of the concrete

$$\beta_{AC} = 0.46 + 0.09 AC \geq 1.0 \quad \text{(III.3h)}$$

(AC is the percentage air content (in volume of concrete))

III.2.2.4 The CEB–FIP Method [215]

This method suggests the following creep function for the initial load application at 28 days:

$$\phi(t, t') = \frac{1}{E_c(t')} + \frac{\alpha_{28}(t, t')}{E_c(t_{28})} \quad \text{(III.4)}$$

where

$$\alpha_{28}(t, t') = \beta'_H \beta'_{t'} \beta'_F \beta'_d \beta_{td} \quad \text{(III.4a)}$$

(i) β'_H = a creep coefficient as a function of relative humidity

H:	0.3	0.4	0.5	0.6	0.7	0.8	0.9	1.0
β_H:	3.25	3.15	2.90	2.70	2.35	1.90	1.5	1.0

(III.4b)

(ii) $\beta'_{t'}$ = an age coefficient at loading of normal cement (III.4c)

t' (days):	1	3	7	14	28	90	180	360
$\beta'_{t'}$:	1.8	1.6	1.4	1.2	1.0	0.75	0.65	0.3

(iii) β'_F = a coefficient for concrete composition (Table III.1) (III.4d)

(iv) β'_d = a coefficient based on cross-sectional dimension (III.4e)

Effective thickness, d (cm):	100	200	300	400	500
β'_d:		1.0	0.85	0.75	0.71	0.7

(v) β_{td} = a coefficient based on theoretical thickness versus age of concrete at loading (Table III.2) (III.4f)

Table III.1
β'_F values for different water/cement ratios and unit weights

Water/cement ratio	Unit weight (kg/m³)			
	500	400	300	200
0·4	1·0	0·8	0·45	—
0·5	1·4	1·2	0·9	—
0·6	—	1·6	1·19	0·75
0·7	—	—	1·5	1·0
0·8	—	—	—	1·3

Table III.2
β_{td} values for different ages and values of d

d (mm)	Age (days)				
	1	10	100	1000	10 000
50	0·15	0·38	0·8	0·99	—
100	—	0·21	0·65	0·95	—
200	—	0·1	0·43	0·90	—
400	—	—	0·24	0·78	—
800	—	—	0·1	0·5	0·85

III.2.2.5 Hansen's Formula as Expressed by Nasser and Neville [81, 216]

$$\frac{C}{\sigma} = \frac{B(0.31h_K + W/C)V_1}{(NK_1 + 0.31)h_K}(1 - e^{-mt}) + \alpha \frac{W}{C} V_{cp} \log_e \frac{t + t'}{t'} \quad \text{(III.5)}$$

where

C = creep value

σ = applied stress

W/C = water/cement ratio by weight (correction made for bleeding)

V_{cp} = volume concentration of cement paste

$= \frac{1}{27}\left[\frac{\text{weight of cement}}{\rho_{cm} \times \rho_w} + \frac{\text{weight of water}}{\rho_w}\right]$

(ρ_w = specific gravity of water, ρ_{cm} = specific gravity of cement)

h_K = degree of hydration for loading

$K_1 = \dfrac{\text{weight of non-evaporable water}}{\text{weight of cement at full hydration}}$

t' = aggregate loading

$N = 0.75(1 + 4t')$

t = time under load

Hansen reported the values of α, B and m to be

$$\alpha = 5.7 \times 10^{-6} \qquad B = 1.76 \times 10^{-6} \qquad m = 0.33$$

The first right-hand term in eqn (III.5) is the term for delayed elastic deformation and the second right-hand term is the term for viscous (long-term) deformation. The value of $C \times 10^{-6}$ gives creep in inches per inch, and this is then converted to SI units.

III.2.2.6 The Dibsi Strain Hardening Model [147]

The creep rate at any instant of time is defined as

$$d\varepsilon_c/dt = f(\bar{\sigma}, T, \varepsilon_c) \qquad (\text{III.6})$$

where

$d\varepsilon_c/dt$ is the creep rate

$\bar{\sigma}$ = effective stress $\qquad \varepsilon_c$ = thermal creep = $(1/10)\sigma_t^{0.31} F(T)$ (III.6a)

$$F(T) = \left\{ 1 \pm \left[\dfrac{125 - T)}{1000}(t) \right]^{0.6} \right\} \qquad (\text{III.6b})$$

The creep rate at any instant of time is written as (time hardening)

$$d\varepsilon_c/dt = 0.31 \sigma_t^{-0.61} F(T) \qquad (\text{III.7})$$

In terms of strain hardening, eqn (III.7) is written as

$$d\varepsilon_c/dt = 0.31[\sigma F(T)]^{3.225} \varepsilon_c^{-2.225} \qquad (\text{III.8})$$

Equation (III.8) can be used successfully in the finite element creep analysis of structures.

III.2.2.7 The Zienkiewicz and Watson Creep Formulation [87]

England and Phok [86], for variable stress conditions, suggested a strain rate equation for total strain including elastic and creep. Zienkiewicz and Watson [87] give a modified version of the equation of England and Phok

[86] in which the value of E is considered as a function of time and temperature, and is written as

$$\varepsilon_s = \phi(T,t')\log[1+(t-t')] + (1/E)(T,t') \tag{III.9}$$

where

ε_s = specific strains
$\phi(T,t')$ = constant depending upon time and temperature
t' = age of loading
t = time at which strains are to be considered
T = temperature

The last term in eqn (III.9) is the elastic component. The specific value of the creep strain increment can be written in terms of elastic and creep strains:

$$\{\varepsilon_{0,c}\} = [D_0]^{-1}\left[\sum_{t=0}^{t=t_n} \frac{1}{E(T,t')}\frac{\Delta\{\sigma\}}{\Delta t'}\Delta t'\right] \quad \text{(elastic strains)}$$

$$+ \sum_{t=0}^{t=t_n} \phi(T,t')\log[1+(t-t')]\frac{\Delta\{\sigma\}}{\Delta t'}\Delta t' \quad \text{(creep strains)}$$

$$[D_0]\{\varepsilon_c\}_1 = \Delta\{\sigma_0\}\log(1+t_1-t_0') \tag{III.10a}$$

$$[D_0]\{\varepsilon_c\}_2 = \Delta\{\sigma_0\}\phi(T_1,t_0')\log(1+t_2-t_0') \\ + \Delta\{\sigma_1\}\phi(T_2,t_1')\log(1+t_2-t_1') \tag{III.10b}$$

$$[D_0]\{\varepsilon_c\}_{n-1} = \Delta\{\sigma_0\}\phi(T_{n-1},t_0')\log(1+t_{n-1}-t_0') \\ + \Delta\{\sigma_1\}\phi(T_{n-1},t_1')\log(1+t_{n-1}-t_1') \\ + \Delta\{\sigma_{n-2}\}\phi(T_{n-2},t_{n-2}')\log(1+t_{n-1}-t_{n-2}') \tag{III.10c}$$

$$[D_0]\{\varepsilon_c\}_n = \Delta\{\sigma_0\}\phi(T_n,t_0')\log(1+t_n-t_0') \\ + \Delta\{\sigma_1\}\phi(T_n,t_1')\log(1+t_n-t_1') \\ + \Delta\{\sigma_{n-1}\}\phi(T_{n-1},t_{n-1}')\log(1+t_n-t_{n-1}') \tag{III.10d}$$

Thus any specific value of creep strain increment can be computed from eqns (III.10c) and (III.10d) as

$$\Delta\{\varepsilon_c\}_n = \{\varepsilon_c\}_n - \{\varepsilon_c\}_{n-1} \tag{III.11}$$

The time intervals between $t_1,\ldots,t_n$ may be chosen quite arbitrarily.

III.2.2.8 The Bangash Incremental Creep Formulation

The creep strain increment $\Delta\varepsilon_c$ is written in the form

$$\phi_1(\Delta\varepsilon_c)\phi'(t,t') \tag{III.12}$$

where ϕ' is a function which determines whether time hardening, strain hardening or strain dependent relations, respectively, are being used; and ϕ_1 is a standard function defining stress.

Two forms of eqn (III.12) are given below:

Time hardening:

$$\frac{\Delta\varepsilon_{cx}}{\Delta t} = \bar{L}(\sigma_x - \sigma_y) + (\sigma_x - \sigma_z) \qquad \frac{\Delta\varepsilon_{xy}}{\Delta t} = \bar{L} \times 2\tau_{xy}$$

$$\frac{\Delta\varepsilon_{cy}}{\Delta t} = \bar{L}(\sigma_y - \sigma_x) + (\sigma_y - \sigma_x) \qquad \frac{\Delta\varepsilon_{xz}}{\Delta t} = \bar{L} \times 2\tau_{xz} \tag{III.13}$$

$$\frac{\Delta\varepsilon_{cz}}{\Delta t} = \bar{L}(\sigma_z - \sigma_y) + (\sigma_z - \sigma_x) \qquad \frac{\Delta\varepsilon_{zy}}{\Delta t} = \bar{L} \times 2\tau_{zy}$$

where $\bar{L} = \tfrac{1}{2}\phi_1[(\sigma_{eq})/\sigma_{eq}]\phi'(t)\phi(t')$, and where equivalent stress σ_{eq} is defined by

$$\sigma_{eq} = \sigma_{cu} + \sqrt{3\tau^2}$$
$$\sigma_{cu} = f_{cu} = \sqrt{(\sigma_x - \sigma_z)^2 + (\sigma_x - \sigma_y)^2 + (\sigma_y - \sigma_z)^2} \tag{III.13a}$$

If σ_{eq} and t' are independent of time then

$$\sigma_x = \sigma_z = \sigma_{xy} = \sigma_{zx}$$

Equation (III.13) reduces to eqn (III.12) since $\varepsilon_c t = 0$, and hence the field of equivalent creep strain can be written as

$$\varepsilon_c = \sum_0^{t_n} \left\{ \frac{9}{2} \left[\left(\frac{\Delta\varepsilon_{cx}}{\Delta t} - \frac{\Delta\varepsilon_{cy}}{\Delta t}\right)^2 + \left(\frac{\Delta\varepsilon_{cx}}{\Delta t} - \frac{\Delta\varepsilon_{cz}}{\Delta t}\right)^2 \right. \right.$$

$$\left. \left. + \left(\frac{\Delta\varepsilon_{cz}}{\Delta t} - \frac{\Delta\varepsilon_{cy}}{\Delta t}\right)^2 + \frac{1}{2}\left(\frac{\Delta\varepsilon_{cx}}{\Delta t} + \frac{\Delta\varepsilon}{\Delta t} + \frac{\Delta\varepsilon_{cz}}{\Delta t}\right) \right] \right\} \tag{III.14}$$

Strain hardening:

The term $\phi'(t)$ in eqn (III.13) is replaced by $\phi'(t'')$ such that

$$\frac{\Delta\varepsilon_{cx}}{\Delta t} = \frac{1}{2} \frac{\phi'(\sigma_{eq})}{\sigma_{eq}} \phi'(t'')\phi(t')[(\sigma_x - \sigma_y) + (\sigma_x - \sigma_z)], \text{ etc.} \tag{III.15}$$

The equivalent time t'' is calculated from the creep strain

$$\phi(t'') = \varepsilon_c/\phi_1(\sigma_{eq})\phi(\sigma) \tag{III.16}$$

For a large range of stresses a suitable creep strain function $\varepsilon_{c_{eq}}$ can be chosen from any creep model described above.

III.2.2.9 The Three-Dimensional Thermal Creep Model

Bangash and England [1004] have presented a three-dimensional formulation for creep which allows independent variation of irreversible and delayed recoverable strains with respect to temperature and time. Simultaneous adoption of a pseudo-time concept simplifies the mathematical form of the modelling and allows the four-element Burger representation to be employed in pseudo-time. When the loads are sustained for a long time, the solution of the equation can be governed by the Maxwell elements of the model at elevated temperatures with small creep recovery.

The constitutive relation for concrete under uniaxial stress is established by using the rate of flow method. The total time dependent strain in concrete may be written (Fig. III.5) as

$$\varepsilon(t) = \varepsilon_e(t) + \varepsilon_f^e(t) + \varepsilon_d^e(t)$$
$$= \frac{\sigma(t')}{E(t')} + \int_0^{t'} \frac{\partial \sigma(\tau)}{\partial \tau} J_f(t' - \tau) d\tau + \int_0^{t'} \frac{\partial \sigma(\tau)}{\partial \tau} J_d(t' - \tau) d\tau \tag{III.17}$$

where

$\varepsilon'(t)$ = total strain

$\varepsilon_f(t')$ = irreversible or flow component of creep strain
 = $\sum_0^{t'} J_f(t' - \tau) \Delta \sigma^*$

$\varepsilon_e(t')$ = elastic strain

$\varepsilon_d(t)$ = reversible or delayed elastic component of creep strain
 = $\sum_0^{t'} J_d(t' - \tau) \Delta \sigma^*$

$J_f(t')$ = specific flow (flow strain per unit of stress)

$J_d(t')$ = delayed elastic strain per unit of stress

t' = age of concrete

τ = time under load

$\Delta \sigma^*$ represents incremental stress

Creep compliance for concrete is written as

$$J(t', \tau) = (1/E) + J_f(t' - \tau) + J_d(t' - \tau) \tag{III.18}$$

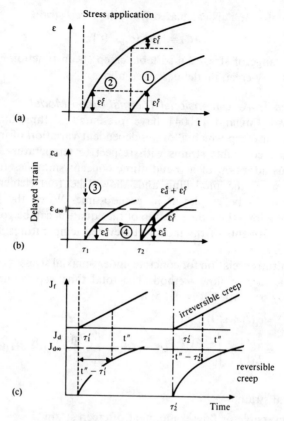

Fig. III.5. Creep components. (a) Influence of age at loading on flow component of strain; (b) creep strain at different age of loading; (c) component of creep strain in pseudo-time.

A parameter is introduced at this stage known as pseudo-time, t'', which itself is a specific flow component; and this may be used in place of actual time, t', together with the representation of a non-ageing visco-elastic material. The pseudo-time concept transforms the age dependent creep relationship in real time to the simpler non-ageing Maxwell law in pseudo-time. The time transformation eases the analytical or numerical computation which leads to a solution without changing the basic creep equation. Now, with this transformation, J_f may be written as

$$J_f(t'' - \tau') = t'' - \tau' \tag{III.19}$$

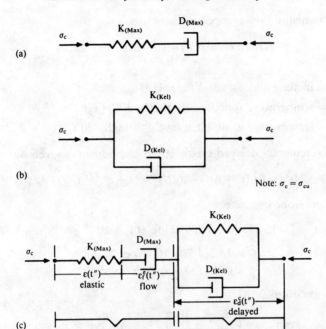

Fig. III.6. Visco-elastic models for concrete creep. (a) Model A—the Maxwell model of creep strain (flow component); (b) Model B—the Kelvin model of creep strain (delayed elastic component); (c) Model 3—the Burger model (models A + B).

and also the delayed elastic strain:

$$J_d(t'' - \tau') = J_{d\alpha_i}[1 - e^{-\beta(t'' - \tau')}] \tag{III.20}$$

When a Maxwell unit is connected in series with a Kelvin unit (Fig. III.6), a Burger model is obtained. In this model the dashpot and the Kelvin unit correspond to flow and delayed elastic component. Equation (III.18) is written as

$$J(t'' - \tau') = (1/E) + (t'' - \tau') + J_{d\alpha_i}[1 - e^{-\beta(t'' - \tau')}] \tag{III.21}$$

The components of creep strains in pseudo-time having special provisions for reversible and irreversible creep effects can now be written as

$$J_d(t'') = \sum_{i=1}^{n} J_{d\alpha_i}(1 - e^{-J_\tau/Q_i}) = Q[1 - e^{-\beta(t'' - \tau')}] \quad \text{mm}^2/\text{N} \tag{III.22}$$

The incremental flow component is written as

$$\Delta J_f(t'') = \Delta t'' \rho(T) = [(t'' + \Delta t'') - t'']\rho(T) \qquad (III.23)$$

where

$\Delta t''$ = incremental pseudo-time = $1/50$
$\rho(T)$ = temperature function = $1/50(T + 30°)\rho(T_\phi)$
T_ϕ = temperature at ambient level (normally 20°C)

The incremental delayed elastic strain component is given by

$$\Delta J_d(t'') = [J_d(t'' + \Delta t'') - J_d(t'')] = (1 - e^{-\beta \Delta t''})[Q - J_d(t'')] \qquad (III.24)$$

For limestone concrete

$$J_f = t'' = [-15\cdot373\,46 + 7\cdot889\,77 \log_e(t')] \times 10^{-6} \text{ per N/mm}^2 \qquad (III.25)$$

$$\begin{aligned} Q &= 14\cdot214\,57 \times 10^{-6} \text{ per N/mm}^2 \\ \beta &= 0\cdot281\,357\,8 \times 10^6 \text{ N/mm}^2 \end{aligned} \qquad (III.26)$$

Young's modulus is

$$\begin{aligned} E(T) &= E_{20}\left[1 - \frac{T - 20°}{137}\right] \quad \text{for } 20° \leq T \leq 50° \\ E(T) &= E_{20}(0\cdot78)\left[1 - \frac{T - 50°}{341}\right] \quad \text{for } 50° \leq T \leq 85° \end{aligned} \qquad (III.27)$$

This concept is used for many problems solved in this text. Further explanations are given later on in the finite element formulations (Chapters VI and VII).

III.3 SHRINKAGE

Shrinkage is considered to be a time dependent deformation in concrete which occurs due to the movement of the gel water from within the gel pores. The evaporation process of the drying out of the concrete is sufficient to cause such shrinkage strains. Most of the time, shrinkage due to drying out and creep due to stress occur in the same concrete. As moisture seepage takes place in the case of both shrinkage and creep, it is difficult to observe the two cases separately within the same concrete specimen. The type of cement definitely has an influence on the shrinkage of concrete. Although the function of shrinkage is to reduce the volume of the unloaded concrete at constant temperature, it induces stresses acting over extended time periods. These stresses are generally reduced by creep. Hence this is the reason why the shrinkage phenomenon is taken into account with the problem of creep.

III.3.1 Shrinkage Models
III.3.1.1 The Hilsdorf Model [218]
Hilsdorf found that as the water/cement ratio increases and the relative humidity drops, shrinkage increases. His relationship can be expressed as

$$\varepsilon_{sh\infty} = \beta_1 p \sqrt{\frac{100-H}{100}} \qquad (III.28)$$

where

$\varepsilon_{sh\infty}$ = final shrinkage of concrete
β_1 = proportionality coefficient
p = total porosity
H = relative humidity

III.3.1.2 The ACI 209 Method [219]
The final shrinkage strain is estimated to be

$$\varepsilon_{sh\infty} = 780 \times 10^{-6} (\beta^s_{cp} \beta^s_H \beta^s_d \beta^s_S \beta^s_F \beta^s_{ce} \beta^s_{AC}) \qquad (III.29)$$

(i) β^s_{cp} = a coefficient for curing periods different from seven days (III.29a)

Days:	1	3	7	14	28	90
β^s_{cp}:	1·2	1·1	1·0	0·93	0·86	0·75

(ii) β^s_H = a coefficient to estimate the contribution from the relative humidity (III.29b)

$\beta^s_H = 1\cdot40 - 0\cdot01H \qquad 40 \le H \le 80\%$
$\beta^s_H = 3\cdot0 - 0\cdot03H \qquad 80 \le H \le 100\%$

(iii) β^s_d = a coefficient for the average thickness of the concrete member (III.29c)

when $50 \le d \le 150$ (all mm):

d (mm):	50	75	100	125	150
β^s_d:	1·35	1·25	1·17	1·08	1·0

when $150 \le d \le 375$ (all mm):
$$\beta_d = 1\cdot23 - 5\cdot70 \times 10^{-8} d^2 \quad \text{for } t' - t_1 \le 365 \text{ days}$$
$$\beta_d = 1\cdot17 - 3\cdot19 \times 10^{-5} d^2 \quad \text{for } t' - t_1 \le 365 \text{ days}$$
when $d > 375$ mm:
$$\beta_d^s = 1\cdot2 e^{-0\cdot047 V/A}(1\cdot2 e^{-0\cdot12 V/A})$$
where V/A is the volume surface area ratio.

(iv) β_S^s = a coefficient for concrete consistency (III.29d)
$$\beta_S^s = 0\cdot89 + 8\cdot56 \times 10^{-5} S$$
where S is the slump of concrete.

(v) β_F^s = a coefficient for the content of fine aggregates (III.29e)
$$\beta_F^s = 0\cdot30 + 14 \times 10^{-4} F \qquad F \le 50\%$$
$$\beta_F^s = 0\cdot90 + 2 \times 10^{-3} F \qquad F > 50\%$$
where F is the fine aggregate content ($<4\cdot8$ mm).

(vi) β_{ce}^s = a coefficient to take into account the cement content (III.29f)
$$\beta_{ce}^s = 0\cdot75 + 2\cdot196 \times 10^{-7} \, ^\circ C$$

(vii) β_{AC}^s = a coefficient for estimating air content (III.29g)
$$\beta_{AC}^s = 0\cdot95 + 8 \times 10^{-3} \qquad A \ge 1\cdot0$$
where A is the air content by volume in %.

The development of shrinkage over time is also generally expressed as
$$\varepsilon_{st} = \frac{(t-t')}{35 + (t-t')} \varepsilon_{sh\infty} \qquad \text{(III.29h)}$$

III.3.1.3 The CEB–FIP 70 Method [220]
This code proposes prediction of shrinkage by
$$\varepsilon_{sh\infty} = \beta_H^{s'} \beta_{t'}^{s'} \beta_d^{s'} \beta_p^{s'} \beta_{td}^{s'} \qquad \text{(III.30)}$$

(i) $\beta_H^{s'}$ = a coefficient for relative humidity (III.30a)

H:	40	50	60	70	80	90	100
$\beta_H^s \times 10^{-6}$:	410	380	320	270	199	100	0

(ii) $\beta_{t'}^{s'}$ represents the age of concrete at loading (Table III.3) (III.30b)

(iii) $\beta_d^{s'}$ = a coefficient based on cross-sectional dimensions (III.30c)

d (cm):	100	200	300	400	500
$\beta_d^{s'}$:	1·10	0·80	0·65	0·55	0·45

Table III.3
$\beta_{t'}^s$ values for different water/cement ratios and unit weights

Water/cement ratio	Unit weight (kg/m³)			
	500	400	300	200
0·3	—	—	—	—
0·4	1·0	0·80	—	—
0·5	1·43	1·20	0·90	—
0·6	—	1·60	1·19	0·70
0·7	—	—	1·45	1·0
0·8	—	—	1·80	1·25

Table III.4
$\beta_{td}^{s'}$ values for different ages and values of d

d (mm)	Age (days)				
	1	10	100	1000	10 000
50	0·17	0·4	0·80	0·90	—
100	—	0·26	0·66	0·88	—
200	—	0·10	0·43	0·85	—
400	—	—	0·27	0·78	—
800	—	—	0·08	0·50	0·87

(iv) $\beta_p^{s'}$ = a coefficient based on longitudinal reinforcement (III.30d)
$(p = 100 A_{st}/A_c)$

p: 0   1   2   3   4   5
$\beta_p^{s'}$: 1·0 0·85 0·70 0·62 0·3 –

(v) $\beta_{td}^{s'}$ = a coefficient based on theoretical thickness versus the age of concrete at loading (Table III.4) (III.30e)

III.4 Temperature

Detailed discussions have been given earlier of the influence of temperature on the material properties of concrete. Bangash [221] has given a comprehensive analysis of movements caused by creep shrinkage and temperature. Temperature analysis using finite elements is given later in this book. Prior to such an analysis, it is essential to know the influence of various conditions on α_T, the thermal coefficient of expansion of concrete. The thermal coefficient of expansion varies with moisture content. It decreases with age, but it increases with temperature and richness of mix.

Hatt [229] carried out tests on oven-dried gravel concrete and evaluated coefficients of $7.2 \times 10^{-6}/°C$ at $15.6°C$ and $11.7 \times 10^{-6}/°C$ at $66°C$. Very little difference in these values was found for saturated and partially dried samples chosen for this concrete. Davis [225] carried out high- and low-temperature tests on 100 samples within the range of $-9°C$ to $+54°C$ and discovered a decrease in the coefficient of about 14%. Mitchell [226] and Walker et al. [230] reported some improvement of the thermal coefficient by using air-entraining admixtures. Berwanger [231] suggested that within a temperature range of $-40°C$ to $+27°C$ the coefficient reduces at an estimated rate of $0.02 \times 10^{-6}/°C$/day for specimens at between 28 to 200 days. After 200 days no specific changes occur in the magnitude of the thermal coefficient. Meyers [227] reported that the maximum coefficient of limestone concrete for partially dry specimens is 56% higher than for saturated specimens. Many others [231–234] have reported a 15% increase in the coefficient of the specimen in the partially dry state compared with in the saturated state. Emanuel and Hulsey [232] have reported, on the basis of extensive tests, an empirical formula which takes into consideration moisture content, age and mix type:

$$\alpha_T = f_T(f_M f_A B_P \alpha_s + B_{FA}\alpha_{FA} + B_{CA}\alpha_{CA}) \qquad (III.31)$$

where

f_T = a correction factor for temperature alterations (1·0 for a controlled environment; 0·86 for exposed conditions)

f_M = a correction factor for moisture

moisture content (%):	0	20	40	60	80	100
f_M:	1·2	1·2	1·5	1·8	1·6	1·0

f_A = a correction factor for age

moisture content (%):	0	20	40	60	80	90	100
f_A (0–6 months):	1·0	1·0	1·0	1·0	1·0	1·0	1·0
f_A (12–18 months):	1·0	1·0	0·99	0·90	0·97	1·0	—
f_A (old):	1·0	1·0	0·96	0·80	0·78	1·0	—

B_P = proportion by volume of paste

B_{FA} = proportion by volume of fine aggregate

B_{CA} = proportion by volume of coarse aggregate

α_s = thermal coefficient of expansion for saturated cement paste
  = $10.8 \times 10^{-6}/°C$ for evaluating B_P

α_{FA} = thermal coefficient of expansion for fine aggregate

α_{CA} = thermal coefficient of expansion for coarse aggregate

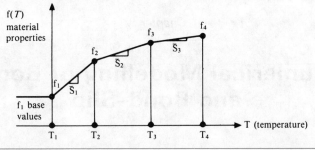

Slope	Break point	Function	Temperature
$\bar{S}_1 = (f_2 - f_1)/(T_2 - T_1)$	T_1	f_1	T_1
$\bar{S}_2 = (f_3 - f_2)/(T_3 - T_2)$	T_2	f_2	T_2
$\bar{S}_3 = (f_4 - f_3)/(T_4 - T_3)$	T_3	f_3	T_3
		f_4	T_4

Fig. III.7. Piecewise linear representation of temperature-dependent material properties.

The strains are given by

$$(d\varepsilon_T)_{ij} = (\alpha_T)_{ij} \, dT \qquad (III.32)$$

For isotropic cases eqn (III.32) is written as

$$(d\varepsilon_T)_{ii} = \alpha_T \, dT = \left[\bar{\alpha}_T + \frac{d\alpha_T}{dT}(T - T_0) \right] dT \qquad (III.33)$$

where T_0 is the initial temperature.

Terms inside the brackets of eqn (III.33) can be modelled using piecewise linear slopes and breaking points representing the temperature dependent material properties of concrete. This is shown in Fig. III.7. For details concerning heat conduction and thermal analysis of concrete, additional material can be found in Chapters VI and VII.

Chapter IV

Numerical Modelling of Bond and Bond–Slip

IV.1 INTRODUCTION

A great deal of work has been carried out on the bond between concrete and steel. It remains one of the most difficult topics in the study of concrete, and is still not properly understood. This is due to a number of theoretical and experimental problems encountered in the correct assessment of bond. Nevertheless it is understood that the bond between concrete and steel is due to the combination of adhesion, bearing action and friction. For a small slip there is an adhesion surface between the concrete and steel, and the resistance to slip is given by the shear strength of the fine particles of concrete filling the indentation of the surface of reinforcement. The value of the adhesion is zero when slip takes place. Bearing action is caused by the deformation of the reinforcing bars or cables in contact with concrete. Slip takes place when the concrete surrounding the reinforcement bars or cables is crushed. At the failure of the structure, bond generally depends on the friction at the failure surface. This friction is destroyed when large slip occurs. The bond–slip relationship is vital for bonded structures subjected to incremental and ultimate loads.

The variables which might affect bond between concrete and steel are:

(a) Concrete mix, temperature and humidity.
(b) The age and tensile/compressive strength at test.
(c) Bar or cable type, and the embedment length and depth.
(d) The loading speed and repeated loading.

The most commonly adopted methods of testing bond are the pull-out test and the tensile test. In the pull-out test, the steel in tension is in contact with concrete in compression. In this way meaningful results can be obtained for different types of bar or different concrete mixes and for the transfer zone of prestressing cables or tendons.

This chapter gives several numerical models for bond. A detailed bond–slip model is given which suits the finite element analysis presented in this text. The references given for bond [238–274] will help the reader understand details not covered in this book.

IV.2 SIMPLE MODELS FOR BOND BETWEEN STEEL AND CONCRETE

IV.2.1 BS 8110 Model for Bond

Two types of bond, namely anchorage and local bond are recognised. Both plain and deformed bars are considered. In cases of a simple pull-out, the anchorage bond stress is given by

$$\sigma_b = \sigma_s \phi / 4L \qquad (IV.1)$$

where σ_b = bond stress; σ_s = stress in the reinforcing bar; ϕ = the diameter of the bar; and L = the length of the embedded bar, or the development length.

The bond resistance at ultimate strength of the bar is computed as

$$L = 0.87 \sigma_{yt} \pi \phi^2 / 4 / \sigma_b \pi \phi L \qquad (IV.2)$$

where σ_{yt} is the yield strength of bar.

In the case of flexural and local bond, the code bases its conclusions on the ultimate shear. For a beam of uniform depth, the bond stress is given by

$$\sigma_b = V / d \sum O \qquad (IV.3)$$

where V = the ultimate shear; d = the depth of the beam; and $\sum O$ = the sum of the effective parameters of the bars in tension.

BS 8110 suggests that ultimate local bond must not exceed the following values when no slip is considered:

Plain bars	Grade 25 concrete	2·0	
	Grade 30 concrete	2·2	
Deformed bars	Grade 25 concrete	2·5	(IV.3a)
	Grade 30 concrete	2·8	

IV.2.2 Brice, Saliger and Wastlund Models

Figure IV.1 shows distributions of steel and concrete and bond stress distributions. It is assumed that two consecutive cracks occur and that the section in between has a constant bending moment. At each crack, the tensile stress in concrete $\sigma_t = 0$, and the maximum stress in the bars has a

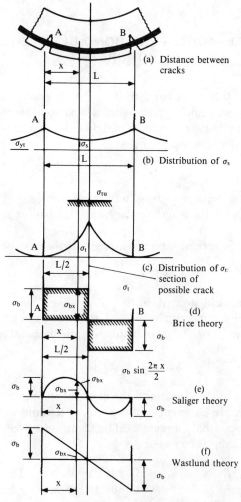

Fig. IV.I. Distribution of various stresses between cracks—based on different theories.

magnitude σ_{yt}. Away from the cracks, the bond between concrete and steel is introduced, whereby the bars transfer an increasing proportion of the total tensile force to the surrounding concrete. The maximum stress σ_{cu} in concrete reaches $x = L/2$ when the cracking takes place, if and when σ_t has a specific value σ_{tu}.

At a section x, the bond between the steel and concrete is given by

$$\sum O \int_0^x \sigma f(x/L) \, dx \qquad (IV.4)$$

where

$$\sigma_{bx} = \sigma_b f(x/L) \qquad (IV.4a)$$

$$\sigma_t, \sigma_s = \sigma_{yt} - (p/A)\sigma_b \int_0^x f(x/L) \, dx \qquad (IV.5)$$

$$M_x = \text{moment at } x = \sigma_x A_s z + \sigma_t z_c \qquad (IV.6)$$

From eqn (IV.5)

$$\sigma_t = (p\sigma_b z/z_c) \int_0^x f(x/L) \, dx \qquad (IV.7)$$

where z_c is the section modulus; and z is the lever arm.

At $L/2$, if $\sigma_t = \sigma_{tu}$ then the developing length L is given by

$$L = B_1 \frac{1}{p} \frac{z_c}{z} \frac{\sigma_{tu}}{\sigma_b} \qquad (IV.8)$$

where

$$\frac{1}{B_1} = \int_0^{1/2} f(x/L) \, d(x/L)$$

(a) When $B_1 = 2$ and $f(x/L) = 1$, the well-known Brice Theory [239] emerges

$$\sigma_{bx} = \sigma_b = \frac{2K_B}{1 + 3\dfrac{\Sigma\phi}{\Sigma d}} \qquad (IV.9)$$

where $K_B = 1\cdot 6$; $\phi =$ the diameter of the bars; and $d =$ the thickness of the concrete crossing the line of possible cracking, encountering the largest number of bars.

(b) When $B = \pi$ for $f(x/L) = 1$, Saliger's expression [240] is written as

$$\sigma_{bx} = \sigma_b \sin \frac{2\pi x}{L} \qquad (IV.10)$$

For flexure

$$f'_c/4\cdot 4 \text{ kg/cm}^2 \qquad (IV.11)$$

and for tension

$$\sigma_{tu} = f'_c/12$$

For deformed bars $\sigma_{ca} = \sigma_{tu}$ ranges between $f'_c/3$ and $f'_c/4$.
The corresponding crack width w is given by

$$w = \frac{\pi}{4} \frac{\sigma_{yt}}{E_c} \frac{\sigma_{tu}}{\sigma_b} \frac{\phi}{A_s} bd \qquad (IV.12)$$

where b = width and d = depth of the tension zone.

(c) When $\beta_1 = 4$ for $f(x/L) = 1$, it gives the equation of Wastlund [cited in Ref. 241]. The distribution of bond stress is given by

$$\sigma_{bx} = \left(1 - \frac{2x}{L}\right) K \sigma_{tu} \sqrt{\frac{w}{\phi}} \qquad (IV.13)$$

$$\sigma_b = K \sigma_{tu} \sqrt{\frac{w}{\phi}} \qquad (IV.13a)$$

where $\sigma_{tu} = f_{cu}/10$; K is for high tensile steels; and $\sqrt{\phi/w} \simeq 14$.

IV.3 SOMAYAJI AND SHAH [242] MODEL

This model considers [242, 243] the relationship between σ_{bx} and the second derivative of the local slip S_x. An exponential bond stress distribution function is assumed as well. The value of S_x

$$S_x = Ae^x + Be^{-x} + C\frac{x^2}{2} + Dx + E \qquad (IV.14)$$

where A–E are constants

$$A = (1 - e^{-L_t})\varepsilon_s^a/\bar{S}_D$$
$$B = (e^{L_t} - 1)\varepsilon_s^a/\bar{S}_D$$
$$C = (e^{-L_t} - e^{L_t})\varepsilon_s^a/\bar{S}_D$$
$$D = (e^{-L_t} + e^{L_t} - L_t e^{-L_t} + L_t e^{L_t} + 2)\varepsilon_s^a/\bar{S}_D$$
$$E = \varepsilon_s^a[-e^{L/2}(1 - e^{-L_t}) - e^{-L/2}(e^{L_t} - 1)L^2/8(e^{-L_t} - e^{L_t}) - L/2$$
$$\times (e^{-L_t} - e^{L_t} - L_t e^{-L_t} + L_t e^{L_t} + 2)]/\bar{S}_D$$
$$\bar{S}_D = (2e^{L_t} + 2e^{-L_t} + L_t e^{-L_t} - L_t e^{L_t} - 4)$$
$$\varepsilon_s^a = \text{applied strain}$$
$$L_t = \text{transfer length} \tag{IV.15}$$

Figure IV.2 shows typical distributions. The value of

$$P_{\text{tran}} = (P/1 + \bar{\alpha}_e p) \tag{IV.16}$$

where P is the applied load, $\bar{\alpha}_e$ the modular ratio and p the reinforcement ratio.

The transmission length given by Hoyer's method [245] is given by

$$L_t = \frac{\phi}{2\mu}(1 + v_c)\left(\frac{\bar{\alpha}_e}{v_s} - \frac{f_{pi}}{E_c}\right)\left(\frac{f_{pe}}{2f_{pi} - f_{pe}}\right) \tag{IV.17}$$

where μ is the coefficient of friction between steel and concrete; v_c, v_s = Poisson's ratio for concrete and steel, respectively; f_{pi} = initial prestress; and f_{pe} = effective prestress in steel.

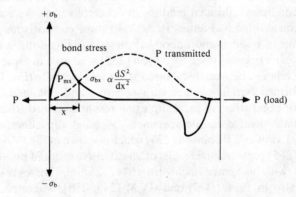

Fig. IV.2. Load and bond stress distribution.

IV.4 MARSHALL AND KRISHNAMURTHY MODEL

Marshall and Krishnamurthy [244] have predicted the transmission length for the prestressing cable as

$$L_t = \sqrt{f_{cu} \times 10^3/\bar{\beta}} \qquad (IV.18)$$

where $\bar{\beta}$ is constant depending upon the strand and wire

wire/strand (mm)	$\bar{\beta}$
2 mm wire	0·144
5 mm wire	0·0235
7 mm wire	0·0174
10 mm/7 wire strand	0·144
12·5 mm/19 wire strand	0·058
19 mm/7 wire strand	0·0235

C. G. Marshall [246] and W. T. Marshall [247] then propose a bond stress formula

$$\sigma_{bx} = \sigma_{b_{max}} e^{-4\psi x/\phi} \qquad (IV.19)$$

where σ_{bx} is the bond stress at a distance x from the free end; $\psi = $ constant ≈ 0.00725; $\sigma_{b_{max}} \not> 7.42 \, N/mm^2$ for $f_{cu} = 80 \, N/mm^2$; and ϕ is the diameter of wire.

IV.5 BOND AND BOND–SLIP MODELS

Bangash reported [248] the use of finite element analysis for bond and bond–slip conditions. Bonded tendons in concrete vessels were analysed under incremental loads. Figures IV.3–IV.4b show comparisons between the behaviour of bonded and unbonded prestressing tendons in concrete vessels. Similar work was done by Nilson [251] who used in finite element analysis the relation between local bond-stress and local slip (Fig. IV.5). He used a deformed bar in contact with concrete and obtained the data indirectly for bond stress to local slip using results obtained from several tests on axially loaded tensile specimens. A bond–slip curve was also obtained by Evans and Robinson [253] which is shown in Fig.IV.6. Stocker and Sozen [265] reported the results of an extensive series of pull-out tests on strands with an embedment length of 25 mm. These results are confirmed also in Figs. (IV.7) and (IV.8) [249, 250], where each curve represents the average results derived from many curves. From these

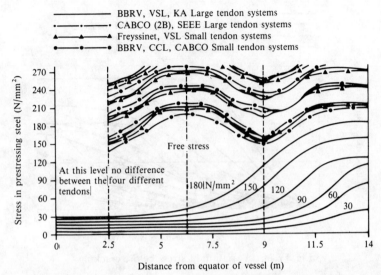

Fig. IV.3. Variation of steel stress with distance along tendon.

results, the relationship seems to indicate that bond stress decreases with increasing strand diameter and increases with increase in concrete strength. The unit bond force increases by approximately 10% for every 7.0 N/mm^2 increase in concrete strength. It is concluded that after a slip of 0.0025 mm, further values of slip increase rapidly and the bond strength continues to increase at a small rate. Nilson [251], using the finite element technique, analysed crack formation in a concentrically loaded tension member using the plane stress idealisation and the bond–slip curve for the deformed bars. The bond stress and steel stress distribution are shown in Fig. IV.5. Prior to cracking, both bond stress and steel stress decrease rapidly with distance from the member faces. As the load increases, bond deterioration occurs near the concrete faces and the maximum bond stress moves inward. Primary and secondary cracking produce destruction of bond. It was concluded that the distribution of longitudinal concrete stresses and the magnitude of the transverse stresses found were different from those given by elastic analysis. The difference is due to the gradual transfer of the tensile forces to concrete by bond. The finite element analysis carried out by Lutz and Gergely [266] concluded that with the assumed perfect bond after cracking, high interfacial tensile stresses existed between steel and concrete at the level of the crack. The separation in the case of plain bars would mean a total loss of bond. The crack width at the bar would be the same as the

120   *Concrete and Concrete Structures*

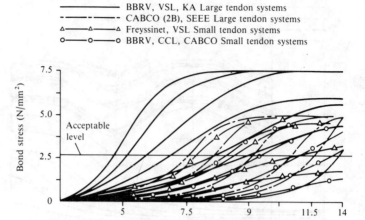

Fig. IV.4a.   Variation of stress before cracking.

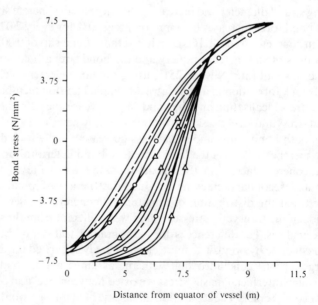

Fig. IV.4b.   Variation of stress after primary cracking.

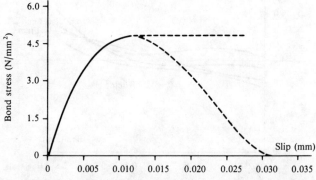

Fig. IV.5. Bond–slip curve for deformed bar [249].

width of the concrete member. In the case of deformed bars, due to prevention of the complete unloading of the concrete, a great deal of variation of crack width from the bar to the concrete surface existed.

The effects of repeated loading on bond have been reported in detail by Hanson and Kaar [254], Hanson and Hulsbos [263], and Edward and Picard [249]. Hanson and Hulsbos showed that the static bond strength decreased when preceded by repeated loads. When the ratio of the repeated load and static load is higher, the static bond strength (according to Hanson) became lower for the same number of load cycles. Hanson and Kaar [254] showed that under repeated loading, two beams with wires

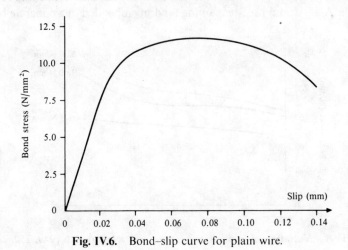

Fig. IV.6. Bond–slip curve for plain wire.

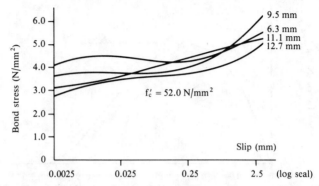

Fig. IV.7. Bond–slip curves for strands of different diameters [249, 250, 262].

failed in bond after 7200 cycles with an average bond stress of 85% of the average value determined from the static test. After 654 000 cycles, the average bond stress was 71% of the value from the static test.

Edwards and Picard [249] reported bond pull-out and tension tests on 12·7 mm wire strand. They used a 38 mm embedded length with three independent concrete covers (12·5 mm, 25 mm and 38 mm). Bond–slip curves showing elasto-plastic type of behaviour have been produced. These results indicate that with the increase in the concrete cover the average maximum bond strength decreases. Empirical equations for crack width and spacings are also given.

Morris [270, 271] carried out tests on unbonded tendons in containment vessels. A 54-strand post-tensioning tendon embedded in a concrete beam

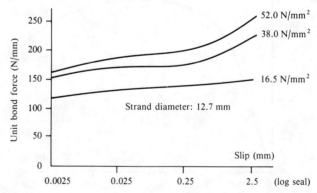

Fig. IV.8. Variation of bond strength with concrete strength [249, 250, 262].

was stressed to 76% of guaranteed ultimate tensile stress (g.u.t.s). After 31 days of grouting, the tendon was cut and the bond transfer length was found to be 3·1 mm to 3·7 mm. Naus [274] studied both grouted and nongrouted tendons in prestressed concrete pressure vessels. Flexural tests were performed on beams (3 m length × 0·15 m width × 0·31 m depth) prestressed with a 12·5 mm diameter seven-wire prestressed strand. The prestressing load was between 0·5 and 0·7 g.u.t.s with a loading of 0·074–74 kN/s. The results showed that grouted tendons had increased cracking and ultimate load.

Ngo and Scorcelis [272] used the finite element method on several singly reinforced concrete beams with different idealised cracking patterns. A bond linkage element (Fig. IV.9) is used for the bond between steel and concrete. The stiffness matrix for the linkage element is derived later on in this chapter and is similar to the one reported by Bangash [248].

Ahmed [262] developed the Ahmlink element which is a three-dimensional linkage element. He has tested his finite element analysis on a scale model of an octagonal slab, post-tensioned with 5 mm and 7 mm high-tensile wires placed in the two orthogonal directions. In addition, pull-out tests have been carried out on bonded prestressed concrete specimens in order to obtain local bond–slip data. This element is used along with other finite elements for the analysis of prestressed concrete vessels for a high-temperature reactor.

Figure IV.10 shows typical details of the pull-out specimens and the

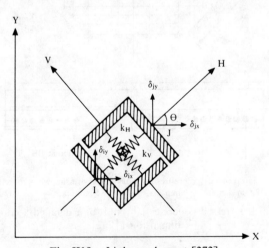

Fig. IV.9. Linkage element [272].

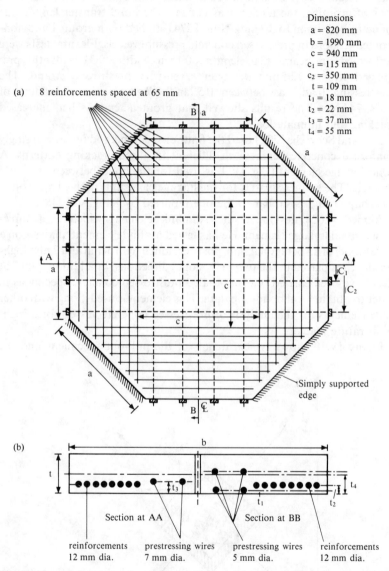

Fig. IV.10. (a) Prestressed concrete slab; and (b) a detailed section (see (a) for dimensions) [262].

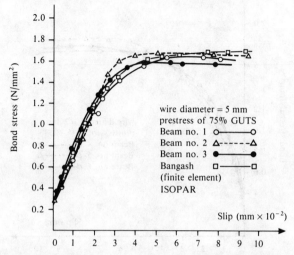

Fig. IV.11. Experimental bond stress–slip curve of 5 mm diameter prestressing wire.

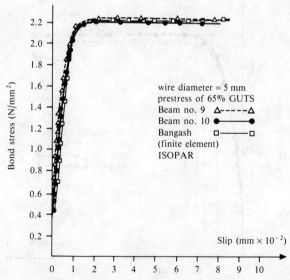

Fig. IV.12. Experimental bond stress–slip curve of 5 mm diameter prestressing wire [262].

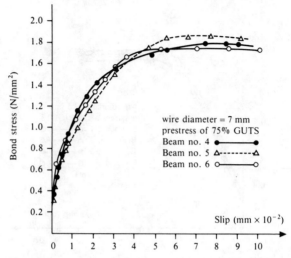

Fig. IV.13. Experimental bond stress–slip curve of 7 mm diameter prestressing wire.

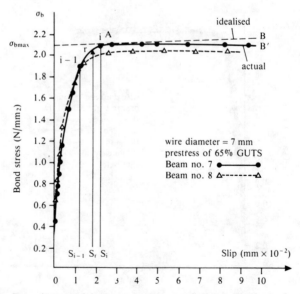

Fig. IV.14. Experimental bond stress–slip curve of 7 mm diameter prestressing wire [262].

concrete slab. Figures IV.11–IV.14 show bond stress–slip curves and prestressing wires.

IV.5.1 Bond Stress–Slip Constitutive Models
IV.5.1.1 Ngo–Scordelis Model

Figure IV.9 shows a two-dimensional linkage element to be incorporated into the finite element analysis. The springs are denoted by k_H and k_V respectively. The value of k_H is given by

$$k_H = d\sigma_b/dS \qquad (IV.20)$$

where k_H is the slope of the local bond slip curve; and $d\sigma_b$ = incremental local bond stress, dS = incremental local slip (both in the H direction).

For a bar surface area, the right-hand side of eqn (IV.20) is multiplied by $\pi\phi L/2$ which corresponds to one spring. The factor 2 for the bar surface representation indicates two-dimensional idealisation. In the axisymmetric case, when only one spring exists, this factor of 2 disappears. If the H and V axes have an angle θ with the x and y axes, the relationship between the spring deformation and the linkage force is given in a matrix form by

$$\begin{Bmatrix} F_H \\ F_V \end{Bmatrix} = \begin{bmatrix} k_H & 0 \\ 0 & k_V \end{bmatrix} \begin{Bmatrix} \delta_H \\ \delta_V \end{Bmatrix} \qquad (IV.21)$$

or the stress–strain relationship is written as

$$\begin{Bmatrix} \sigma_{bH} \\ \sigma_{bV} \end{Bmatrix} = \begin{bmatrix} k_H & 0 \\ 0 & k_V \end{bmatrix} \begin{Bmatrix} \varepsilon_{bH} \\ \varepsilon_{bV} \end{Bmatrix} \qquad (IV.22)$$

or

$$\{\sigma_b\} = [k_L]\{\varepsilon_b\} \qquad (IV.22a)$$

The strains and displacements are related by the displacement transformation matrix T'' (the H–V and x–y axes system)

$$\begin{Bmatrix} \varepsilon_{bH} \\ \varepsilon_{bV} \end{Bmatrix} = \begin{bmatrix} -C & -S & C & S \\ S & -C & -S & C \end{bmatrix} \begin{Bmatrix} \delta_{ix} \\ \delta_{iy} \\ \delta_{jx} \\ \delta_{jy} \end{Bmatrix} \qquad (IV.23)$$

$$C = \cos\theta \qquad S = \sin\theta \qquad (IV.23a)$$

or

$$\{\varepsilon\} = [T_b]\{\delta\} \qquad (IV.23b)$$

Bangash [221] included the nodal forces in the x–y system as

$$\begin{Bmatrix} F_{ix} \\ F_{iy} \\ F_{jx} \\ F_{jy} \end{Bmatrix} = [T_b]^{T''} \{F_b\} \qquad (\text{IV.24})$$

where

$$F_b = \begin{Bmatrix} F_H \\ F_V \end{Bmatrix}$$

The stiffness matrix $[K_L]_{xy}$ for the linkage element is thus given by:

$$[K_L]_{xy} = [T_b]^{T''}[k_L][T_b] \qquad (\text{IV.25})$$

Table IV.1 shows the matrix $(K_L)_{xy}$ for the linkage element in two dimensions.

Table IV.1
Two-dimensional linkage element matrix $(K_L)_{xy}$

$k_H C^2 + k_V S^2$	$k_H SC - k_V SC$	$-k_H C^2 - k_H S^2$	$-k_H SC + k_V SC$
$k_H SC - k_V SC$	$k_H S^2 + k_V C^2$	$-k_H SC + k_V SC$	$-k_H S^2 - k_V C^2$
$-k_H C^2 - k_V S^2$	$-k_H SC + k_V SC$	$k_H C^2 + k_V S^2$	$k_H SC - k_V SC$
$-k_H SC + k_V SC$	$-k_H S^2 - k_V C^2$	$k_H SC - k_V SC$	$k_H S^2 + k_V C^2$

where $S = \sin \beta$, $C = \cos \beta$. \qquad (IV.26)

When $\theta = 0$, i.e. the H and V axes coincide with x and y axes, the matrix $(K_L)_{xy}$ is given by

$$[K_L]_{xy} = \begin{bmatrix} k_H & 0 & -k_H & 0 \\ 0 & k_V & 0 & -k_V \\ -k_H & 0 & k_H & 0 \\ 0 & -k_V & 0 & k_V \end{bmatrix} \qquad (\text{IV.26a})$$

IV.5.1.2 Ahmed's Constitutive Model for Bond and Bond–slip
Ahmed used eqn (IV.20) in incremental form

$$k_H = \Delta\sigma_b / \Delta S \qquad (\text{IV.26b})$$

In order to model various types of interface characteristics, an experimental bond–slip curve idealised as shown in Fig. IV.5 is used. The nonlinear curve is idealised by a series of bond stress and slip points joined linearly. The slope (k_H) at point i of the bond–slip curve is given as

$$k_{Hi} = \frac{\sigma_{bi+1} - \sigma_{bi}}{S_{i+1} - S_i} = \Delta\sigma_{bi}/\Delta S_i \qquad \text{(IV.26c)}$$

and the other two spring coefficients in the vertical and lateral directions of the steel are taken as

$$\begin{aligned} k_{Vi} &= \alpha_m \times k_{Vi} \\ k_{Li} &= \alpha_m \times k_{Hi} \end{aligned} \qquad \text{(IV.26d)}$$

where α_m is a multiplying factor. Let I and J (Fig. IV.15) be line element nodes. x, y, z and x', y', z' are global and local co-ordinates. The direction cosines of the local axes (x', y', z') with respect to the global axes (x, y, z) are $(l, m, n), (p, q, \theta), (r, s, t)$. Let (I, J) be the line element nodes.

$$\begin{aligned} l &= (y_J - x_I)/L & m &= (y_J - y_I)/L \\ n &= (z_J - z_I)/L & p &= -m/\sqrt{1-n^2} \\ q &= l/\sqrt{1-n^2} & r &= -ln/\sqrt{1-n^2} \\ s &= -mn/\sqrt{1-n^2} & t &= \sqrt{1-n^2} \\ L &= \sqrt{(x_J - x_I)^2 + (y_J - y_I)^2 + (z_J - z_I)^2} \end{aligned} \qquad \text{(IV.27)}$$

Let $\Delta S_H, \Delta S_V$ and ΔS_L be the incremental slip in the horizontal, vertical and lateral directions of the steel element. The incremental relationship between the slip and the nodal displacements can be written as

$$\begin{Bmatrix} \Delta S_H \\ \Delta S_V \\ \Delta S_L \end{Bmatrix} = \begin{bmatrix} -l & -m & -n & l & m & n \\ -p & -q & -\theta & p & q & \theta \\ -r & -s & -t & r & s & t \end{bmatrix} \begin{Bmatrix} \delta_{ix} \\ \delta_{iy} \\ \delta_{iz} \\ \delta_{jx} \\ \delta_{jy} \\ \delta_{jz} \end{Bmatrix} \qquad \text{(IV.28)}$$

or

$$\{\Delta S\}_{H,V,L} = [T_b]_{l,m,n}^{r,s,t} \times \{\Delta U^e\} \qquad \text{(IV.28a)}$$

where (T_b) is the transformation matrix and $\{\Delta U^e\}$ are the global element displacements.

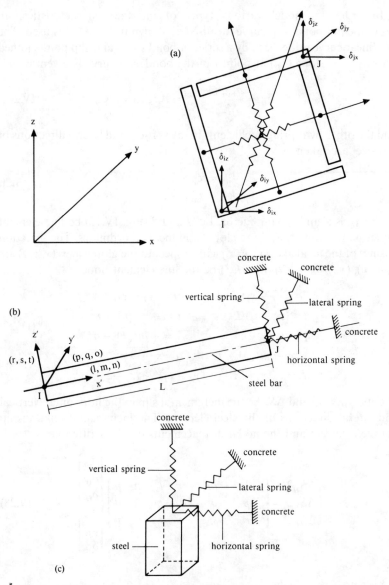

Fig. IV.15. Details of bond linkage element [262]. (a) Element; (b) direction cosines; (c) bond representation. xyz = global co-ordinate system; x', y', z' = local co-ordinate system; l, m, n, p, q, r, s, t = direction cosines; U, V, W = global displacement variables.

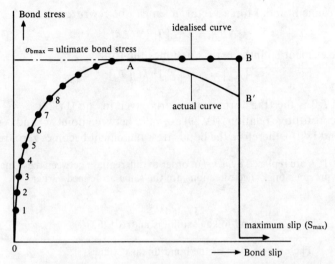

Fig. IV.16a. A typical idealised bond–slip curve.

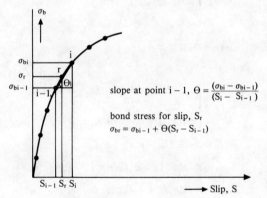

Fig. IV.16b. Linear interpolation of nonlinear bond–slip curve [262].

The local incremental bond stress and bond–slip may be written

$$\begin{Bmatrix} \Delta\sigma_{Hi} \\ \Delta\sigma_{Vi} \\ \Delta\sigma_{Li} \end{Bmatrix} = \begin{bmatrix} k_{Hi} & 0 & 0 \\ 0 & k_{Vi} & 0 \\ 0 & 0 & k_{Li} \end{bmatrix} \begin{Bmatrix} \Delta S_{Hi} \\ \Delta S_{Vi} \\ \Delta S_{Li} \end{Bmatrix} \quad \text{(IV.29)}$$

or

$$\Delta\sigma_{bi}^e = K_{bi}\,\Delta S_{Hi}^e \quad \text{(IV.29a)}$$

The element nodal force vector given can be rewritten as
$$\{\Delta P_i^e\} = \pi \, dL[T_b]^{T''}\{\Delta \sigma_{bi}^e\} \tag{IV.30}$$
and the element stiffness matrix assumes the form
$$[K_L]_{6\times 6} = \pi \, dL[T_b]^{T''} k_{bi}[T_b]^{T''} \tag{IV.31}$$
where $[T_b]$ is the transformation matrix given in eqn (IV.28).

The constitutive relations (IV.29) are valid between points O and A (Figs IV.15 and IV.16), therefore the bond stress increment becomes zero (region

Note: H, V, L are replaced by h, v, l in order to differentiate between these matrices and the previous ones. The directions are the same as defined earlier.

Table IV.2
Bond-linkage stiffness matrix [262]

The explicit form of the bond-linkage stiffness is:
$$K_b_{6\times 6} = \begin{bmatrix} K_{b11} & K_{b12} \\ K_{b21} & K_{b22} \end{bmatrix}$$

where

$$K_{b12} = K_{b21} = -K_{b11}$$

$$K_{b22} = K_{b11}$$

$$K_{b11}_{3\times 3} = \begin{bmatrix} K_{b11} & K_{b12} & K_{b13} \\ & K_{b22} & K_{b22} \\ \text{symmetry} & & K_{b33} \end{bmatrix}$$

$$K_{b11} = \pi dL(l^2 E_h + p^2 E_v + r^2 E_l)$$

$$K_{b22} = \pi dL(m^2 E_h + q^2 E_v + s^2 E_l)$$

$$K_{b33} = \pi dL(n^2 E_h + t^2 E_v)$$

$$K_{b12} = \pi dL(lm E_h + pq E_v + rs E_l)$$

$$K_{b13} = \pi dL(ln E_h + rt E_v)$$

$$K_{b23} = \pi dL(mn E_h + st E_l)$$

$$\left.\begin{array}{l} l, m, n \\ p, q, r \\ s, t \end{array}\right\} = \text{direction cosines}$$

$\pi dL = $ perimeter of the steel

AB), i.e. the slip occurs at constant bond stress. The bond is assumed to fail at point B (Fig. IV.16) when maximum allowable slip has been reached. At this point, the total bond stress is released. This creates a non-equilibrium state which is corrected by performing equilibrium iterations using the initial stress method.

Now a scheme is suggested to calculate correctly bond stress from a

Table IV.3
Stiffness matrix load vector for bond-linkage element [749]

$$[K_b]_{6\times 6} = \pi dL \begin{bmatrix} l^2k_h + p^2k_v + r^2k_l & lmk_h + pqk_v + rsk_l & lnk_h + rtk_v & -l^2k_h - p^2k_v - r^2k_l & -lmk_h - pqk_v - rsk_l & -lnk_h - rtk_v \\ & m^2k_h + q^2k_v + s^2k_l & mnk_h + stE_l & -lmk_h - pqk_v - rsk_l & -m^2k_h - q^2k_v - s^2k_l & -mnk_h - stk_l \\ & & h^2k_h + t^2k_v & -lnk_h - rtk_v & -mnk_h - stk_l & -h^2k_h - t^2k_v \\ & & & l^2k_h + p^2k_v + r^2k_l & lmk_h + pqk_v + rsk_l & lnk_h + rtk_v \\ & \text{symmetrical} & & & m^2k_h + q^2k_v + s^2k_l & mnk_h + stk_l \\ & & & & & n^2k_h + t^2k_v \end{bmatrix}$$

$$\Delta \mathbf{P}^e_{6 \times 1} = \pi dL \begin{Bmatrix} -l\Delta\sigma_h - p\Delta\sigma_v - r\Delta\sigma_l \\ -m\Delta\sigma_h - q\Delta\sigma_v - s\Delta\sigma_l \\ -n\Delta\sigma_h - t\Delta\sigma_l \\ l\Delta\sigma_h + p\Delta\sigma_v + r\Delta\sigma_l \\ m\Delta\sigma_h + q\Delta\sigma_v + s\Delta\sigma_l \\ n\Delta\sigma_h + t\Delta\sigma_l \end{Bmatrix}$$

πdL = perimeter of the steel

$\left.\begin{matrix} l, m, n, \\ p, q, r, \\ s, t \end{matrix}\right\}$ = direction cosines

specified bond–slip curve (Fig. IV.14). Let S_r be the total slip reached at any point in the calculation. This slip lies between the points $i-1$ and i on the specified curve. Bond stresses are calculated which are compatible with the total slip (S_r) by linear interpolation as

$$\sigma_{br} = \sigma_{bi-1} + \theta(S_r - S_{i-1}) \qquad \text{(IV.32)}$$

where $\theta =$ the slope of the specified curve at point $i-1$

$$\theta = \left(\frac{\sigma_{bi} - \sigma_{bi-1}}{S_i - S_{i-1}}\right) \qquad \text{(IV.33)}$$

Let the total bond stress calculated using the constitutive equation (IV.29) be σ_{bi}. The difference $(\sigma_{bi} - \sigma_{bv})$ is treated as initial stress and is corrected by performing equilibrium iterations.

Table IV.2 shows the bond linkage matrix K. The stresses in terms of nodal displacements can be written as

$$\{\Delta\sigma_b\} = \{K_b\}[T_b]\{\Delta U^e\} \qquad \text{(IV.34)}$$

A step-by-step procedure for computing bond stresses using three-dimensional finite elements and the above numerical model is given later in this book.

Bangash [749] carried out similar tests to Ahmed but with more tendons and a steel plate fastened by lugs to a concrete slab. Ahmed's linkage matrix was modified to include the bonds due to the lugs as well. The final matrix is given in Table IV.3.

Chapter V

Numerical Models for Fatigue of Concrete

V.1 INTRODUCTION

Substantial research has been carried out in recent years [276–285] on the fatigue of concrete structures and their components. The subject of fatigue is more important in concrete structures such as offshore gravity platforms, nuclear reactor vessels, heavy machine foundations, etc., where higher strength materials are used and at the same time the structures are expected to perform successfully under higher stress levels. Fatigue analysis is based on a period of time t_f, which is the planned life of the concrete structures. The value of t_f is taken to be 30 years plus.

Fatigue loading is generally a sequence of repetitive loads. This type of loading is distinctly divided between high-cycle low-amplitude and low-cycle high-amplitude. The distinction is based on an arbitrary number of cycles, normally within the range of 100–1800 cycles. Lower than 1800 cycles of high-amplitude fatigue loading can occur due to earthquakes or due to various sea states. For machine foundations the cycles could go up to 10^7.

A number of investigators [276–285] have reported that concrete specimens under cyclic compressive loadings receive progressive changes within their concrete matrix due to the growth of microcracking. Shah [275] reported that for all types of concrete in different environments and under controlled conditions, for a stress level equivalent to 50% of the compressive strength of concrete, the volume of concrete increases rather than decreases and microcracking occurs at the aggregate–paste interface. Reinforcement plays a great part in the deformation of concrete sections under repeated loading.

Holman [276] gives a relationship between strain ratio $\varepsilon_{max}/\varepsilon_0'$ and the cycle ratio N (number of cycles)/N_{TF} (total life in cycles). This relationship indicates that there is a gradual increase in strain between 10% and 80% of the N_{TF}; however, between 0 and 10% and over 80% the strain increment is fairly rapid. Wilson [285] concluded that the slow-cycle fatigue caused the bond cracks to expand, resulting in a decrease in the concrete durability. His results were based on first freezing concrete and then subjecting it to slow-cycle fatigue loading.

Some of the models for fatigue analysis which can easily be included in the global finite element analysis are given in the following paragraphs.

V.2 FATIGUE MODELS

V.2.1 Kakuta et al. Model [277]

Kakuta et al. [277] reported the following expression for the fatigue of concrete

$$\log N = 17\left[1 - \frac{(\sigma_{max} - \sigma_{min})/f_c'}{1 - (\sigma_{min}/f_c')}\right] \tag{V.1}$$

where N = the number of cycles; σ_{max} = the maximum stress of the cycle; σ_{min} = the minimum stress of the cycle; and f_c' = the cylindrical compressive strength of concrete.

V.2.2 Aas-Jacobson Model [278]

This model [278] is based on a linear relationship between $\log N$ and the cycles of stresses using a factor β_f based on statistical data

$$\log N = \frac{1}{\beta_f}\left[\frac{1 - (\sigma_{max}/f_c')}{1 - (\sigma_{max}/\sigma_{min})}\right] \tag{V.2}$$

A value of $\beta_f = 0.064$ was given by Aas-Jacobson.

V.2.3 Miner Model [279]

The main aspect of the Miner model [279], known as Miner's Rule, is the estimation of the cumulative effect of the stress history by adopting a simplified linear damage relationship. The cumulative usage factor $\bar{\eta}$ is given by

$$\bar{\eta} = \sum_{i=1}^{S} N_i/N_{fi} \tag{V.3}$$

where S = the number of stress blocks considered; N_i = the actual number of stress cycles for the stress block i; and N_{fi} = the permissible number of stress cycles if the stress block i is considered only, or the number of cycles that will cause failure at stress level i.

Stresses occur with variable amplitude in random order and they are substituted by a number of 'stress blocks'. Each stress block represents a number of constant-amplitude stress cycles.

Equation (V.3) is written for a fatigue damage as

$$\bar{\eta}_1 = \sum_{i=1}^{S} \frac{N_i}{K} S_i^{\bar{m}} \qquad (V.4)$$

where K is an empirical constant in the S–N equation; S = a fatigue stress cycle; N = the number of cycles to failure; $\bar{m}$ = an empirical constant in the S–N equation; and $S_i^{\bar{m}}$ = the expected value of $S^{\bar{m}}$ in the ith state. Failure occurs when $\bar{\eta} \geq 1$ (see Appendix IX for offshore platform fatigue results).

V.2.4 Waagaard Model [280]

Waagaard [280] uses a value of $\bar{\eta}$, the cumulative usage factor, of 0·2 rather than 1 for offshore structures. He represents the fatigue of concrete in compression in offshore structures as

$$\log N = 10\left[\left\{1\cdot 0 - \frac{\sigma_{max}}{\alpha_g(f'_c/\gamma)}\right\}\bigg/\left\{1\cdot 0 - \frac{\sigma_{min}}{\alpha_g(f'_c/\gamma)}\right\}\right] \qquad (V.5)$$

where α_g is the factor to account for flexural gradient, $1 \leq \alpha_g < 10$; γ is the partial safety factor for materials = 1·25; and σ_{min} is around $0\cdot 1 f'_c$.

V.2.5 JSCE Model [281]

The fatigue of concrete in compression is given by the Japan Society of Civil Engineers [281] as

$$\log N = 15\left[1 - \frac{(\sigma_{max} - \sigma_{min})}{(0\cdot 9 K' f'_c - \sigma_{min})}\right] \qquad (V.6)$$

where K' = a coefficient taken as equal to 0·85.

The characteristic fatigue strength for concrete F_{cr} is given by

$$F_{cr} = (0\cdot 9 K' f'_c - \sigma_{cp})\left(1 - \frac{\log N}{15}\right) \qquad (V.7)$$

where σ_{cp} = permanent compressive stress.

For deformed bars of diameter $\not> 32$ mm, the fatigue strength is given by the value of F_{sr}.

$$F_{sr} = \bar{C} \times 10^{-0.2(\log N - 6)} \quad \text{for } N < 6$$

or

$$= \bar{C} \times 10^{-0.1(\log N - 6)} \quad \text{for } N > 6 \quad \text{(V.8)}$$

where $\bar{C} = 160\,\text{Mps} - (\sigma_{tp}/3)$, σ_{tp} = permanent tensile stress in steel.

V.2.6 CEB–FIP Model [282]

According to CEB–FIP [282] recommendations, the stresses corresponding to fatigue strength must be determined by elastic analysis in which provision should be made for dynamic effects, creep and shrinkage, etc. The following conditions are checked:

(a)

$$\Delta\sigma = \sigma_{max} - \sigma_{min} \leq \Delta\sigma_{rep}/\gamma_{fat} \quad \text{(V.9)}$$

Both σ_{max} and σ_{min} have maximum values corresponding to 2×10^6 cycles. σ_{rep} = strength under repeated loads; $\gamma_{fat} = 1.25$.

The fatigue strength of concrete is the 50% of the fractile value obtained from the series of tests.

(b) Fatigue strength of reinforcement:

For 2×10^6 cycles, the characteristic strength of conventional steel and prestressing steel are 10% fractile and 50% fractile respectively.

For conventional steel

$$\sigma_{max} = 0.7\sigma_{yt}$$

For prestressing tendons

$$\sigma_{max} = 0.85\sigma_{yt}$$

The ratio

$$\Delta\sigma_{rep}/\gamma_{fat} = \Delta\sigma_{sf}/\gamma_{fat} \quad \text{(V.10)}$$

where $\gamma_{fat} = 1.15$ for conventional steel and $\gamma_{fat} = 1.25$ for prestressed anchorages.

The values of $\Delta\sigma_{sf}$ are given below:

Plain bars: 250 MPa
Deformed bars: 150 MPa
Prestressing tendons: 150–200 MPa

For welding zones a reduction of 0.4 of the above values is taken into consideration.

V.2.7 AASHTO Model [283]

The American Association of State Highway and Transportation Officials (AASHTO) [283] has recommended a limitation of stresses for fatigue for concrete, the maximum compressive stress must not exceed $0.5f'_c$. The stress range for the reinforcement is given by

$$\sigma_{sr} = 145 - 0.33\sigma_{min} + 54.9(r/h) \qquad (V.11)$$

where $r =$ the base radius of the rolled bar; $h =$ the height radius of the rolled bar; $\sigma_{min} =$ the minimum stress level; and $\sigma_{sr} =$ the stress range.

Where prestressing tendons are used the tendons shall offer 60–66% of the minimum tensile strength without failure under 5×10^5 cycles. The period of each cycle is counted from lower stress level to the upper stress level and back to the original lower stress level.

V.2.8 NCHR Model [284]

The National Cooperative Highway Research (NCHR) of the US Transportation Research Board studied bars from several manufacturers and as a result an empirical equation was developed [284] in order to predict the long-term limiting stress range below which fatigue damage will not occur. For hot-rolled bars, eqn (V.11) is validated. Above the fatigue limit of eqn (V.11), the safe fatigue life for all stress ranges is represented by eqn (V.12).

$$\log N = 6.1044 - 4.07 \times 10^{-5}\sigma_{rs} - 1.38 \times 10^{-5}\sigma_{min}$$
$$+ 0.71 \times 10^{-5}\sigma_{yt} - 0.0566 A_s + 0.3233 d \qquad (V.12)$$

where σ_{rs} and σ_{min} are in psi; $A_s =$ the area of the bar in (inches)^2; $d =$ the diameter of the bar in inches; and $\sigma_{yt} =$ ultimate steel strength.

Using SI units, eqn (V.12) is given by

$$\log N = 6.1044 - 590.8 \times 10^{-5}\sigma_{rs} - 199 \times 10^{-5}\sigma_{min}$$
$$+ 103 \times 10^{-5}\sigma_{yt} - 8.771 \times 10^{-5}A_s + 0.01265 d(r/h) \qquad (V.13)$$

where σ_{rs} and σ_{min} are in Mpa, A_s in mm^2, and d in mm.

V.2.9 Wilson Model [285]

Based on the ASTM freezing and thawing tests Wilson [285] concluded that the durability factor for concrete under cyclic loads can be calculated as

$$E_{DY} = n_i^2/n_0^2 \times 100 \qquad (V.14)$$

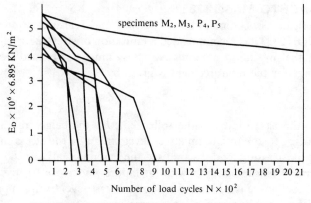

Test no: M_2, M_3, P_5	Cyclic load (% of ultimate)	Curing period (90, 28 days)
A_1	70	7 years
A_2	70	7 years
A_3	70	7 years
A_4	70	7 years
A_5	70	7 years
A_6	70	7 years
A_7	70	7 years
A_8	70	7 years

Fig. V.1. Dynamic modulus of elasticity versus number of load cycles (after Wilson [285]).

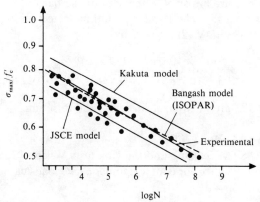

Fig. V.2. Comparison of two fatigue models with experimental tests.

where E_{DY} = the relative dynamic modulus of elasticity in percent after N cycles; n_0 = the fundamental transverse frequency at zero cycles; and n_i = the fundamental transverse frequency after N_i cycles.

The durability factor of the test specimens can be obtained by

$$DF = E_{DY} N_i / N_{fi} \qquad (V.15)$$

where N_{fi} = the specified number of cycles at which the dynamic load is terminated.

Several tests have been carried out for N_i and N_{fi} less or greater than 1800. The average durability factor for early freezing was 15·2, compared with 19·2 in a standard case without freezing. The average cycles to failure for freezing and non-freezing concrete were 851 and 1079. A slight improvement in the compressive strength was noted in early freezing concrete.

Figure V.1 shows some of these results. Figure V.2 shows a comparison of the Kakuta and JSCE models with the finite element experimental tests carried out on a prestressed concrete beam by the author.

PART II

This section contains a step-by-step finite element analysis for concrete structures in which provision is made for elastic–elasto-plastic static, dynamic, impact and blast effects. An attempt is made to integrate the elements of Part I and to present the simulation technique of concrete numerical models in a global finite element analysis.

Chapter VI

Finite Element Modelling of Concrete Structures

VI.1 INTRODUCTION

A step-by-step finite element analysis is developed in which a provision is made for elasto-plasticity, thermal creep, creep recovery, and cracking under static and dynamic conditions. Various concrete strength theories described in earlier chapters are included in the finite element analysis. These give facilities for the comparative study of results necessary for some sensitive structures. The analysis presented here also includes facilities for loading and unloading phenomena. Although greater emphasis is given to the use of isoparametric elements in the text, tables are provided which will assist in the automatic inclusion of other types of finite elements. Equations are set up in such a way that they can easily be replaced or modified to include other case studies. The initial stress or modified Newton–Raphson method is used for the major solution procedures. They, together with acceleration and convergence procedures can easily solve nonlinear, plasticity and cracking problems. Where dynamic problems exist, equations of motion are introduced in both linear and nonlinear cases. Expressions for displacements, velocities and accelerations are then obtained. Direct integration and Wilson-θ methods have been used throughout for the time dependent solutions of dynamically loaded structures. The main analysis of this chapter is supported by analyses given in various tables and appendices. For further study and additional information, the references cited in this text may be found useful.

VI.2 SOLID ISOPARAMETRIC ELEMENT REPRESENTING CONCRETE

Three-dimensional isoparametric elements are used and the functions relating to co-ordinate systems are expressed as follows:

$$X = \sum_{i=1}^{n} N_i(\xi,\eta,\zeta)X_i = N_1 x_1 + N_2 x_2 + \cdots = \{N\}^T\{X_n\}$$

$$Y = \sum_{i=1}^{n} N_i(\xi,\eta,\zeta)Y_i = N_1 y_1 + N_2 y_2 + \cdots \{N\}^T\{Y_n\} \qquad (VI.1)$$

$$Z = \sum_{i=1}^{n} N_i(\xi,\eta,\zeta)Z_i = N_1 z_1 + N_2 z_2 + \cdots = \{N\}^T\{Z_n\}$$

where $N_i(\xi,\eta,\zeta)$, $i = 1$ to n, are the interpolation functions in the curvilinear co-ordinates ξ, η, ζ, and X_i, Y_i and Z_i are the global X, Y, Z co-ordinates of mode i. The interpolation function N is also known as the shape function.

The terms $\{X_n\}$, $\{Y_n\}$ and $\{Z_n\}$ represent the nodal co-ordinates and $\{N\}^T$ is dependent on ξ, η, and ζ. Reference is made to Fig. VI.1 for important aspects of the mapping procedures. The most important aspect is to establish a one-to-one relationship between the derived and the parent element (Fig. VI.1). The necessary condition for a one-to-one relationship is the Jacobian determinant given in eqn (VI.2).

$$d[J] = \frac{\partial(X,Y,Z)}{\partial(\xi,\eta,\zeta)} = \begin{bmatrix} \dfrac{\partial X}{\partial \xi} & \dfrac{\partial X}{\partial \eta} & \dfrac{\partial X}{\partial \zeta} \\ \dfrac{\partial Y}{\partial \xi} & \dfrac{\partial Y}{\partial \eta} & \dfrac{\partial Y}{\partial \zeta} \\ \dfrac{\partial Z}{\partial \xi} & \dfrac{\partial Z}{\partial \eta} & \dfrac{\partial Z}{\partial \zeta} \end{bmatrix} \qquad (VI.2)$$

The Jacobian can also be written in a transposed form, i.e. columns replace rows and vice versa.

VI.2.1 The Shape Function
The relationship between displacements at any point in the local system (ξ, η, ζ) within the element and the nodal displacements is expressed in the

Finite Element Modelling of Concrete Structures

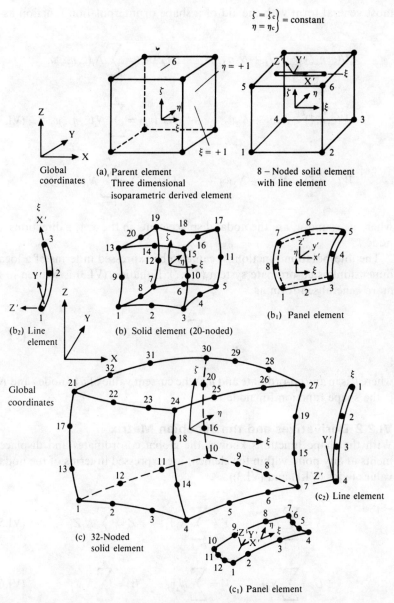

Note: To differentiate panel elements from others X', Y', Z' are replaced by x', y', z'.

Fig. VI.1. Finite elements.

most general form with the aid of a shape or interpolation function as

$$U(\xi,\eta,\zeta) = N_1 u_1 + N_2 u_2 + \cdots + (N)^T(u_n) = \sum_{i=1}^{n} N_i(\xi,\eta,\zeta) u_i$$

$$V(\xi,\eta,\zeta) = N_1 v_1 + N_2 v_2 + \cdots + (N)^T(v_n) = \sum_{i=1}^{n} N_i(\xi,\eta,\zeta) v_i \quad \text{(VI.3)}$$

$$W(\xi,\eta,\zeta) = N_1 w_1 + N_2 w_2 + \cdots + (N)^T(w_n) = \sum_{i=1}^{n} N_i(\xi,\eta,\zeta) w_i$$

where u_i, v_i and w_i are the nodal displacements in the x, y, z directions at node i.

The interpolation function N_i can also be expressed in terms of a local dimensionless co-ordinate system (ξ,η,ζ). Equation (VI.3) is written in a more generalised form as

$$\bar{u} = \sum_{i=1}^{n} N_i \bar{u}_i \quad \text{(VI.4)}$$

where $\bar{u}$ is in any co-ordinate and $\bar{u}_i$ is the current value of $\bar{u}$ at node i and N_i is the shape function for node i.

VI.2.2 Derivatives and the Jacobian Matrix

With the shape functions known, the global co-ordinates and displacements at any point within the element are expressed in terms of the nodal values (eqns (VI.1) and (VI.3))

$$X = \sum_{i=1}^{n} N_i X_i \quad Y = \sum_{i=1}^{n} N_i Y_i \quad Z = \sum_{i=1}^{n} N_i Z_i \quad \text{(VI.5)}$$

$$U = \sum_{i=1}^{n} N_i u_i \quad V = \sum_{i=1}^{n} N_i v_i \quad W = \sum_{i=1}^{n} N_i w_i \quad \text{(VI.6)}$$

where n is the number of nodes on element and X_i, Y_i, Z_i and u_i, v_i, w_i are the nodal co-ordinates and nodal displacements respectively.

The derivatives of shape functions with respect to global co-ordinates require the following transformation.

$$\begin{Bmatrix} \dfrac{\partial N_i}{\partial \xi} \\ \dfrac{\partial N_i}{\partial \eta} \\ \dfrac{\partial N_i}{\partial \zeta} \end{Bmatrix} = [J] \begin{Bmatrix} \dfrac{\partial N_i}{\partial X} \\ \dfrac{\partial N_i}{\partial Y} \\ \dfrac{\partial N_i}{\partial Z} \end{Bmatrix} \quad \text{(VI.7)}$$

where i is the current node number

$$\begin{Bmatrix} \dfrac{\partial N_i}{\partial X} \\ \dfrac{\partial N_i}{\partial Y} \\ \dfrac{\partial N_i}{\partial Z} \end{Bmatrix} = [J] \begin{Bmatrix} \dfrac{\partial N_i}{\partial \xi} \\ \dfrac{\partial N_i}{\partial \eta} \\ \dfrac{\partial N_i}{\partial \zeta} \end{Bmatrix} \quad \text{(VI.7a)}$$

where

$$[J] = \begin{bmatrix} \dfrac{\partial X}{\partial \xi} & \dfrac{\partial Y}{\partial \xi} & \dfrac{\partial Z}{\partial \xi} \\ \dfrac{\partial X}{\partial \eta} & \dfrac{\partial Y}{\partial \eta} & \dfrac{\partial Z}{\partial \eta} \\ \dfrac{\partial X}{\partial \zeta} & \dfrac{\partial Y}{\partial \zeta} & \dfrac{\partial Z}{\partial \zeta} \end{bmatrix} \quad \text{(VI.7b)}$$

where $[J]$ is a 3×3 Jacobian matrix. This matrix eventually plays a major role in the equilibrium equations:

$$\{F\} = [K]\{U\} \quad \text{(VI.8)}$$

where

$\{U\}$ = the nodal displacement vector

$$[K] = \sum_{i=1}^{n} [K_e] = \text{the total stiffness matrix}$$

$[K_e]$ = the element stiffness matrix

n = the number of elements

The determinant $d[J]$ as represented in eqn (VI.2) can be evaluated for a one-to-one relationship. Such derivatives of shape functions are given in Appendix I.

Displacements at nodes can now be expressed in the form of element displacement vectors

$$\{U\} = \begin{Bmatrix} \{u_1\}_1 \\ \{u_2\}_2 \\ \vdots \\ \{u_n\}_n \end{Bmatrix} \quad \text{(VI.9)}$$

The displacement field within each element can be expressed as

$$\{u\} = [N]\{u^e\} = \sum_{i=1}^{n} (N_i[I]\{u\}_i) \quad \text{(VI.10)}$$

where $\{u^e\}$ = the element nodal displacement vector; $\{u\}_i$ = the displacements at node i; $[N]$ = the element shape function matrix; N_i = the shape functions of node i.

VI.2.3 Determination of Strains

With the displacement known at all points within each element, the strains $\{\varepsilon\}$ can be expressed in the following form:

$$\{\varepsilon\} = \sum_{i=1}^{n} ([B_i]\{u\}_i = [B]\{u\} \quad \text{(VI.11)}$$

For the three-dimensional element the $[B]$ matrix of node i and $\{\varepsilon\}_{6 \times 1}$ are given below:

$$[B_i]_{6 \times 3} = \begin{bmatrix} \dfrac{\partial N_i}{\partial X} & 0 & 0 \\ 0 & \dfrac{\partial N_i}{\partial Y} & 0 \\ 0 & 0 & \dfrac{\partial N_i}{\partial Z} \\ \dfrac{\partial N_i}{\partial Y} & \dfrac{\partial N_i}{\partial X} & 0 \\ 0 & \dfrac{\partial N_i}{\partial Z} & \dfrac{\partial N_i}{\partial Y} \\ \dfrac{\partial N_i}{\partial Z} & 0 & \dfrac{\partial N_i}{\partial X} \end{bmatrix} \quad \text{(VI.12)}$$

$$\{\varepsilon\}_{6 \times 1} = [\varepsilon_X, \varepsilon_Y, \varepsilon_Z, \gamma_{XY}, \gamma_{YZ}, \gamma_{ZX}]^{T''} \quad \text{(VI.12a)}$$

Appendix II gives full details of the co-ordinate transformations between Cartesian and curvilinear axes. The dimension of $[B]$ for 8-, 20- and 32-noded elements is 6×24, 6×60 and 6×96 respectively.

VI.2.4 Determination of Stresses

The stresses $\{\sigma\}$ can be determined within each element from the strains as

$$\{\sigma\} = [D](\{\varepsilon\} - \{\varepsilon_0\}) + \{\sigma_o\} \tag{VI.13}$$

where $\{\sigma_0\}$ = initial stresses; $\{\varepsilon_0\}$ = initial strains; and $[D]$ = the material compliance matrix.

VI.2.5 Load Vectors and Material Stiffness Matrix

Appendix III gives the material matrices $[D]$ for both concrete and steel. Now virtual displacement $\{du^e\}$ is applied at the nodes; and the sum of work done, dW, by the stresses, and distributed body and surface forces over the element volume, vol, and surface, S, is given by

$$dW = [\{du^e\}^{T''}] \left(\int_{\text{vol}} [B]^{T''} \{\sigma\} \, dV - \int_S [N]^{T''} \{p\} \, dS - \int_{\text{vol}} [N]^{T''} \{G\} \, d\text{vol} \right) \tag{VI.14}$$

In order to maintain equilibrium within the element, a system of external nodal forces $\{p\}^e$ must be applied which will reduce virtual work (dW) to zero. Equation (VI.14) will take the form

$$\{du^e\}^{T''}\{p\}^e = \{du^e\}^{T''} \left(\int_{\text{vol}} [B]^{T''}\{\sigma\} \, d\text{vol} \right.$$

$$\left. - \int_S [N]^{T''}\{p\} \, dS - \int_{\text{vol}} [N]^{T''}\{G\} \, d\text{vol} \right) \tag{VI.14a}$$

where $\{p\}$ = the surface force per unit surface area; $\{G\}$ = the force per unit volume; and T'' = transpose of the matrix.

When the external work dW is related to the internal work dU

$$dW = dU \tag{VI.15}$$

The element stiffness matrix $[K]$

$$[K] = \int_{\text{vol}} B^{T''} DB \, d\text{vol} = \int_{-1}^{+1} \int_{-1}^{+1} \int_{-1}^{+1} B^{T''} DB \det J \, d\xi \, d\eta \, d\zeta \tag{VI.16}$$

The right-hand side of eqn (VI.16) is based on the Gaussian integration rule described in Appendix IV.

Equation (VI.14a) is valid for any virtual displacement $\{du^e\}$, thus it can be eliminated from both sides of eqns (VI.14a). Substituting eqns (VI.11) and (VI.13) into eqn (VI.14a) one obtains

$$\{P\}^e = \left(\int_{\text{vol}} [B]^{T''}[D][B]\,dV\right)\{u^e\} - \int_{\text{vol}} [B]^{T''}[D]\{\varepsilon_0\}\,dV$$

$$+ \int_{\text{vol}} [B]^{T''}\{\sigma_0\}\,dV - \int_S [N]^T\{p\}\,dS - \int_{\text{vol}} [N]^{T''}\{G\}\,dV \quad \text{(VI.17)}$$

Table (VI.1) gives miscellaneous loads and forces. Equation (VI.17) is the force–displacement relation with stiffness transformation. The terms in the eqn (VI.17) are defined as:

The element stiffness matrix

$$[K]^e = \int_{\text{vol}} [B]^{T''}[D][B]\,dV = \int_{-1}^{+1}\int_{-1}^{+1}\int_{-1}^{+1} B^{T''}DB \det J \,d\xi\,d\eta\,d\zeta$$
(VI.17a)

The nodal force due to body force

$$\{P_b\}^e = -\int_{\text{vol}} [N]^{T''}\{G\}\,dV = -\int_{-1}^{+1}\int_{-1}^{+1}\int_{-1}^{+1} N^{T''}G \det J\,d\xi\,d\eta\,d\zeta$$
(VI.17b)

The nodal force due to surface force

$$\{P_s\}^e = -\int_S [N]^{T''}\{p\}\,dS \quad \text{(VI.17c)}$$

The nodal force due to initial stress

$$\{P_{\sigma_0}\}^e = \int_{\text{vol}} [B]^{T''}\{\sigma_0\}\,dV = \int_{-1}^{+1}\int_{-1}^{+1}\int_{-1}^{+1} B^{T''}\sigma_0 \det J\,d\xi\,d\eta\,d\zeta$$
(VI.17d)

The nodal force due to initial strain

$$\{P_{\varepsilon_0}\}^e = \int_{\text{vol}} [B]^{T''}[D]\{\varepsilon_0\}\,dV = -\int_{-1}^{+1}\int_{-1}^{+1}\int_{-1}^{+1} B^{T''}D\varepsilon_0 \det J\,d\xi\,d\eta\,d\zeta$$
(VI.17e)

Equation (VI.17) can be rewritten as

$$\{F\}^e = [K]^e\{u\}^e + \{P_b\}^e + \{P_s\}^e + \{P_{\sigma_0}\}^e + \{P_{\varepsilon_0}\}^e \quad \text{(VI.17f)}$$

Table VI.1
Miscellaneous loads and forces [216]

Gravitational forces (surface forces)
Equivalent nodal force in the line of gravity Z-direction

$$\{P_s\}_i = \int_v [N^T]_i \begin{Bmatrix} 0 \\ 0 \\ -pg \end{Bmatrix} d\,\text{vol}$$

$$= \sum_{i=1}^{n}\sum_{i=1}^{n}\sum_{i=1}^{n} [N]_{i,jK}^T \begin{Bmatrix} 0 \\ 0 \\ -pg \end{Bmatrix} |J|_{i,j,K} W_1 W_j W_K$$

Body forces
Body force component per unit volume at (X, Y) point is

$$\begin{Bmatrix} f_x \\ f_y \\ f_z \end{Bmatrix} = \{\bar{f}\} = P_b \omega^2 \begin{Bmatrix} X \\ Y \\ 0 \end{Bmatrix}$$

$$0 \neq \int_v \{N\}^T \begin{Bmatrix} f_x \\ f_y \\ f_z \end{Bmatrix} d\,\text{vol}$$

in the case of isoparametric elements.

Concentrated loads
Concentrated loads away from the point

$$\{P_{\sigma 0}\} = N_i(\xi_1, \eta_1, \zeta_1) P$$
$$\xi_1 = \xi \quad \eta_1 = -\eta \quad \zeta = +1$$

Distributed loads

$$\{P_S\} = \int_{-1}^{+1}\int_{-1}^{+1}\int_{-1}^{+1} N_i^{T''}[p_x, p_y, p_z]^{T'} \begin{Bmatrix} \dfrac{\partial Y}{\partial \xi}\dfrac{\partial Z}{\partial \eta} - \dfrac{\partial Z}{\partial \xi}\dfrac{\partial Y}{\partial \eta} \\ \dfrac{\partial Z}{\partial \xi}\dfrac{\partial X}{\partial \eta} - \dfrac{\partial Z}{\partial \eta}\dfrac{\partial X}{\partial \xi} \\ \dfrac{\partial X}{\partial \xi}\dfrac{\partial Y}{\partial \eta} - \dfrac{\partial Y}{\partial \xi}\dfrac{\partial X}{\partial \eta} \end{Bmatrix} d\xi\, d\eta \quad \text{for } \zeta = \pm 1$$

similarly for $\xi = \pm 1$; $\eta = \pm 1$.

(continued)

Table VI.1—*contd.*

Thermal loads

$$\{P\}_T = \int B^T D\varepsilon_T \, dv$$

$$= \sum_{i=1}^{n}\sum_{j=1}^{n}\sum_{k=1}^{n} [B]_{i,jK}[D]\{\varepsilon_T\}|J|_{i,j,K} W_i W_j W_K$$

$$\{\varepsilon_T\} = [\alpha_T T, \alpha_T T, \alpha_T T \ \ 0 \ \ 0 \ \ 0]^T \qquad T = \sum_{i=1}^{n} N_i T_i$$

Creep loads

$$\{P\}_{ev} = \int B^T D\varepsilon_c \, dv$$

$$= \sum_{i=1}^{n}\sum_{j=1}^{n}\sum_{k=1}^{n} [B]_{i,jK}[D]\{\varepsilon_c\}_{i,jk} W_i W_j W_K \, d\xi \, d\eta \, d\zeta$$

The initial strain vector can be replaced, or the strains due to creep and shrinkage, $\{P^a\}^e$ or $\{P^s\}^e$ can be added to it. Hence these expressions can be added to eqn (VI.17f). In addition, the strain in eqn (VI.17e) can be the element swelling strain vector $\{P^{s'}\}^e$. In the case of dynamic loadings applied to structures, an acceleration load vector $\{P^a\}^e$ can be introduced into eqn (VI.17f) such that

$$\{P^a\}^e = [M]\{A_n\} = \sum_{i=1}^{n}[M_e] = P\sum_{i=1}^{n}[A_n]^{T''}[A_n]\, d[J]\, d\xi\, d\eta\, d\zeta \qquad \text{(VI.17g)}$$

where $[M]$ = the total mass matrix; $[M_e]$ = the element mass matrix; and $\{A_n\}$ = the nodal acceleration vector. If large displacements are considered, the nodal vector $\{P^*\}^e$ can also be incorporated into eqn (VI.17f).

Equation (VI.17f) is the force–displacement relation for each element. For the whole structure the stiffness matrix and load vectors are assembled according to the nodal incidences; overall equilibrium equations can be written in the following form, as before,

$$[K]\{U\} = \{F\} \qquad \text{(VI.18)}$$

Equation (VI.18) is solved for unknown displacements. The element strains can be obtained from nodal displacements. Linear material stresses are obtained from eqn (VI.13).

The same procedure is adopted for choosing other element types. Appendix I gives data for some well-known elements.

If boundary conditions are specified on $\{U\}$ to guarantee a unique solution, eqn (VI.17f) can be solved to obtain nodal point displacements at any node in the given structure. The equations with all degrees of freedom can be written as:

$$\begin{bmatrix} K & | & K_R \\ \hline K_R^{T''} & | & K_{RR} \end{bmatrix} \begin{Bmatrix} U \\ \hline U_R \end{Bmatrix} = \begin{Bmatrix} F \\ \hline F_R \end{Bmatrix} \quad \text{(VI.19)}$$

The subscript R represents reaction forces. The top half of the eqn (VI.19) is used to solve for $\{U\}$.

$$\{U\} = -[K]^{-1}[K_R]\{U_R\} + [K]^{-1}\{F\} \quad \text{(VI.20)}$$

The reaction forces $\{F_R\}$ are computed from the bottom half of the equation as

$$\{F_R\} = [K_R]^{T''}\{U\} + \{K_{RR}\}\{U_R\} \quad \text{(VI.21)}$$

Equation (VI.20) must be in equilibrium with eqn (VI.21).

VI.2.6 The Superelement and Substructuring

For large structures with complicated features, a substructure (superelement) may be adopted on the lines suggested in eqn (VI.19). This superelement may then be used as a reduced element from the collection of elements. If subscripts γ and γ' represent the retained and removed degrees of freedom of the equations partitioned into two groups, then the expressions in eqn (VI.19) can be written as

$$\begin{bmatrix} K_{\gamma\gamma} & | & K_{\gamma\gamma'} \\ \hline K_{\gamma'\gamma} & | & K_{\gamma'\gamma'} \end{bmatrix} \begin{Bmatrix} U_\gamma \\ \hline U_{\gamma'} \end{Bmatrix} = \begin{Bmatrix} F_\gamma \\ \hline F_{\gamma'} \end{Bmatrix} \quad \text{(VI.22)}$$

Equation (VI.22) when expanded assumes the following form:

$$\{F_\gamma\} = [K_{\gamma\gamma}]\{U_\gamma\} + [K_{\gamma\gamma'}]\{U_{\gamma'}\} \quad \text{(VI.22a)}$$
$$\{F_{\gamma'}\} = [K_{\gamma'\gamma}]\{U_\gamma\} + [K_{\gamma'\gamma}]\{U_{\gamma'}\} \quad \text{(VI.22b)}$$

When a dynamic analysis is carried out, the subscript γ (retained) represents the dynamic degrees of freedom.

When eqn (VI.22b) is solved, the value of $U_{\gamma'}$ is then written, similarly to eqn (VI.20),

$$\{U_{\gamma'}\} = [K_{\gamma'\gamma'}]^{-1}\{F_{\gamma'}\} - [K_{\gamma'\gamma'}]^{-1}[K_{\gamma'\gamma}]\{U_{\gamma}\} \tag{VI.23}$$

Substituting $\{U_{\gamma'}\}$ into eqn (VI.22a)

$$[[K_{\gamma\gamma}] - [K_{\gamma\gamma'}][K_{\gamma'\gamma'}]^{-1}[K_{\gamma'\gamma}]]\{U_{\gamma}\} = [\{F_{\gamma}\} - [K_{\gamma\gamma'}][K_{\gamma'\gamma'}]^{-1}\{F_{\gamma'}\}] \tag{VI.24}$$

or

$$[\bar{K}]\{\bar{U}\} = \{\bar{F}\} \tag{VI.24a}$$

where

$$[\bar{K}] = [K_{\gamma\gamma}] - [K_{\gamma\gamma'}][K_{\gamma'\gamma'}]^{-1}[K_{\gamma'\gamma}] \tag{VI.24b}$$

$$\{\bar{F}\} = \{F_{\gamma}\} - [K_{\gamma\gamma'}][K_{\gamma'\gamma'}]^{-1}\{F_{\gamma'}\} \tag{VI.24c}$$

$$\{\bar{U}\} = \{U_{\gamma}\} \tag{VI.24d}$$

and $[\bar{K}]$ and $\{\bar{F}\}$ are generally known as the substructure stiffness matrix and load vector, respectively.

In the above equations, the load vector for the substructure is taken as a total load vector. The same derivation may be applied to any number of independent load vectors. For example, one may wish to apply thermal, pressure, gravity and other loading conditions in varying proportions. Expanding the right-hand sides of eqns (VI.22a) and (VI.22b)

$$\{F_{\gamma}\} = \sum_{i=1}^{n} \{F_{\gamma i}\} \tag{VI.25}$$

$$\{F_{\gamma'}\} = \sum_{i=1}^{n} \{F_{\gamma' i}\} \tag{VI.26}$$

where $n =$ the number of independent load vectors.

Substituting into eqn (VI.24c)

$$\{\bar{F}\} = \sum_{i=1}^{n} \{F_{\gamma\gamma'}\} - [K_{\gamma\gamma'}][K_{\gamma'\gamma'}]^{-1} \sum_{i=1}^{n} \{F_{\gamma' i}\} \tag{VI.27}$$

The left-hand side of eqn (VI.27) is written as

$$\{\bar{F}\} = \sum_{i=1}^{n} \{\bar{F}_i\} \tag{VI.28}$$

Substituting eqn (VI.28) into eqn (VI.27), the following equation is achieved:

$$\{\bar{F}_i\} = \{F_{\gamma i}\} - [K_{\gamma\gamma'}][K_{\gamma\gamma'}]^{-1}\{F_{\gamma'i}\} \tag{VI.29}$$

VI.3 THE MEMBRANE ISOPARAMETRIC ELEMENTS

The membrane steel plane element acting with a concrete element to form a composite section is adopted for many concrete structures, such as bridge decks, concrete pressure and containment vessels, offshore gravity platforms, etc. They are treated as thin shell elements. These elements are capable of transmitting only the stresses in plane while assuming a constant strain along the thickness. These elements are compatible with the one face of the solid isoparametric elements representing concrete. The element local, global and curvilinear co-ordinate systems are shown in Figs VI.1a–c. Appendix II gives details of shape functions. Knowing these functions, and using eqns (VI.5) and (VI.6); the strain–displacement relation can be obtained. The thickness d of the element is interpolated as

$$d = \sum_{i=1}^{n} N_i d_i \tag{VI.30}$$

where $n =$ the number of nodes, and $d_i =$ the nodal thickness at node i.

As shown in Fig. VI.I, the local co-ordinate system x', y', z' with corresponding nodal freedom u', v', w' can be used to obtain the local strain field at that point.

$$\varepsilon_{x'} = \partial u'/\partial x' \qquad \varepsilon_{y'} = \frac{\partial v'}{\partial y'} \qquad \gamma_{x',y'} = \frac{\partial u'}{\partial y'} + \frac{\partial v'}{\partial x'} \tag{VI.31}$$

The x'-axis is tangential to the x_i-axis, the z'-axis is normal to the plane of the element, and the y'-axis is determined from the right-hand co-ordinate system. If $\hat{F}_\xi$ is the vector tangential to ξ-axis on the η-axes, the vector is written as

$$\hat{F}_\xi = \begin{Bmatrix} \dfrac{\partial x'}{\partial \xi} \\ \dfrac{\partial y'}{\partial \xi} \\ \dfrac{\partial z'}{\partial \xi} \end{Bmatrix} \tag{VI.32}$$

Similarly the $\hat{F}_\eta$ vector, in matrix form, is written as

$$\bar{F}_\eta = \begin{Bmatrix} \dfrac{\partial x'}{\partial \eta} \\ \dfrac{\partial y'}{\partial \eta} \\ \dfrac{\partial z'}{\partial \eta} \end{Bmatrix} \tag{VI.33}$$

The vector normal to these vectors will be shown as:

$$\hat{F}_{z'} = \hat{F}_\xi \times \hat{F}_\eta = \begin{bmatrix} \dfrac{\partial y'}{\partial \xi} & \dfrac{\partial z'}{\partial \eta} & -\dfrac{\partial z'}{\partial \xi} & \dfrac{\partial y'}{\partial \eta} \\ \dfrac{\partial z'}{\partial \xi} & \dfrac{\partial x'}{\partial \eta} & -\dfrac{\partial x'}{\partial \xi} & \dfrac{\partial z'}{\partial \eta} \\ \dfrac{\partial x'}{\partial \xi} & \dfrac{\partial y'}{\partial \eta} & -\dfrac{\partial y'}{\partial \xi} & \dfrac{\partial x'}{\partial \eta} \end{bmatrix} \tag{VI.34}$$

The above vectors are normalised in order to obtain direction cosines

$$\hat{\hat{F}}_z = \frac{\hat{F}_z}{|\hat{F}_z|} \tag{VI.35}$$

For the x'-axis

$$\hat{F}_{x'} = \frac{\hat{F}_\xi}{|\hat{F}_\xi|}$$

and for the y'-axis

$$\hat{\hat{F}}_y = \hat{\hat{F}}_z \times \hat{\hat{F}}_x$$

where

$$|\hat{F}_z| = \sqrt{\left(\frac{\partial x'}{\partial \xi}\frac{\partial z'}{\partial \eta} - \frac{\partial z'}{\partial \xi}\frac{\partial y'}{\partial \eta}\right)^2 + \left(\frac{\partial z'}{\partial \xi}\frac{\partial x'}{\partial \eta} - \frac{\partial x'}{\partial \xi}\frac{\partial z'}{\partial \eta}\right)^2}$$
$$+ \left(\frac{\partial x'}{\partial \xi}\frac{\partial y'}{\partial \eta} - \frac{\partial y'}{\partial \xi}\frac{\partial x'}{\partial \eta}\right)^2 \quad \text{(VI.35a)}$$

$$|\hat{F}_\xi| = \sqrt{\left(\frac{\partial x'}{\partial \xi}\right)^2 + \left(\frac{\partial y'}{\partial \xi}\right)^2 + \left(\frac{\partial z'}{\partial \xi}\right)^2}$$

The direction cosines in a matrix form for the local orthogonal Cartesian system assume the following form:

$$\{\hat{F}\}_{3\times 3} = \begin{Bmatrix} \hat{F}_{x'} \\ \hat{F}_{y'} \\ \hat{F}_{z'} \end{Bmatrix} = \begin{bmatrix} \hat{F}_{x',x} & \hat{F}_{y',x} & \hat{F}_{z',x} \\ \hat{F}_{x',y} & \hat{F}_{y',y} & \hat{F}_{z',y} \\ \hat{F}_{x',z} & \hat{F}_{z',z} & \hat{F}_{z',z} \end{bmatrix} \quad \text{(VI.36)}$$

VI.3.1 Derivatives

The local derivative of eqn (VI.31) must be obtained. To start with, sets of transformations are considered and the global derivatives are obtained using the method given in Appendix I.

$$\partial U_G = \begin{bmatrix} \frac{\partial u}{\partial x'} & \frac{\partial v}{\partial x'} & \frac{\partial w}{\partial x'} \\ \frac{\partial u}{\partial y'} & \frac{\partial v}{\partial y'} & \frac{\partial w}{\partial y'} \\ \frac{\partial u}{\partial z'} & \frac{\partial v}{\partial z'} & \frac{\partial w}{\partial z'} \end{bmatrix} = J^{-1} \begin{bmatrix} \frac{\partial u}{\partial \xi} & \frac{\partial v}{\partial \xi} & \frac{\partial w}{\partial \xi} \\ \frac{\partial u}{\partial \eta} & \frac{\partial v}{\partial \eta} & \frac{\partial w}{\partial \eta} \\ 0 & 0 & 0 \end{bmatrix} \quad \text{(VI.37)}$$

where

$$[J] = \begin{bmatrix} \frac{\partial x'}{\partial \xi} & \frac{\partial y'}{\partial \xi} & \frac{\partial z'}{\partial \xi} \\ \frac{\partial x'}{\partial \eta} & \frac{\partial y'}{\partial \eta} & \frac{\partial z'}{\partial \eta} \\ \hat{F}_{z',x}d & \hat{F}_{z',y}d & \hat{F}_{z',z}d \end{bmatrix} \quad \text{(VI.37a)}$$

The next step is to obtain the local derivatives and they are summarised below:

$$\partial U_L = \hat{F}^{T''} \partial \hat{U}_G \hat{F} = \sum_{i=1}^{n} \begin{bmatrix} \partial \hat{F}_1^i \partial u_1^i & \partial \hat{F}_1^i \partial u_2^i & \partial \hat{F}_1^i \partial u_3^i \\ \partial \hat{F}_2^i \partial u_1^i & \partial \hat{F}_1^i \partial u_2^i & \partial \hat{F}_2^i \partial u_3^i \\ \partial \hat{F}_3^i \partial u_1^i & \partial \hat{F}_3^i \partial u_2^i & \partial \hat{F}_3^i \partial u_3^i \end{bmatrix} \quad \text{(VI.38)}$$

where

$$\partial F_1^i = \hat{F}_{x',x}\frac{\partial N_i}{\partial x} + \hat{F}_{x',y}\frac{\partial N_i}{\partial y} + \hat{F}_{x',z}\frac{\partial N_i}{\partial z}$$

$$\partial F_2^i = \hat{F}_{y',x}\frac{\partial N_i}{\partial x} + \hat{F}_{y',y}\frac{\partial N_i}{\partial y} + \hat{F}_{y',z}\frac{\partial N_i}{\partial z}$$

$$\partial F_3^i = \hat{F}_{z',x}\frac{\partial N_i}{\partial x} + \hat{F}_{z',y}\frac{\partial N_i}{\partial y} + \hat{F}_{z',z}\frac{\partial N_i}{dz}$$

$$\partial u_1^i = \hat{F}_{x',x}u_i + \hat{F}_{x',y}v_i + \hat{F}_{x',z}w_i \quad \text{(VI.38a)}$$
$$\partial u_2^i = \hat{F}_{y',y}u_i + \hat{F}_{y',y}v_i + \hat{F}_{y',z}w_i$$
$$\partial u_3^i = \hat{F}_{z',z}u_i + \hat{F}_{z',y}v_i + \hat{F}_{z',z}w_i$$

Using eqns (VI.31) and (VI.38), the strain value is computed as

$$\{\varepsilon\} = \sum_{i=1}^{n} \begin{Bmatrix} \partial \hat{F}_1^i \partial u_1^i \\ \partial \hat{F}_2^i \partial u_2^i \\ \partial \hat{F}_2^i \partial u_1^i + \partial \hat{F}_1^i \partial u_2^i \end{Bmatrix} \quad \text{(VI.39)}$$

$$= [B]\{u^e\}$$

$$= \begin{Bmatrix} [B_1] \\ [B_2] \\ \vdots \\ [B_i] \\ \vdots \\ [B_n] \end{Bmatrix} \times \begin{Bmatrix} \{u_1\} \\ \{u_2\} \\ \{u_i\} \\ \{u_n\} \end{Bmatrix}$$

in which $[B_i]$ and $\{u_i\}$ are given as

$$[B]_{3\times 3} = \begin{bmatrix} \bar{F}_{x',x}\partial\hat{F}_1^i & \hat{F}_{x',y}\partial\hat{F}_1^i & \hat{F}_{x',z}\partial\hat{F}_1^i \\ \hat{F}_{y',x}\partial\hat{F}_2^i & \hat{F}_{y',y}\partial\hat{F}_2^i & \hat{F}_{y',z}\partial\hat{F}_2^i \\ (\hat{F}_{x',x}\partial\hat{F}_2^i + \hat{F}_{y',x}\partial F_1^i) & (\hat{F}_{x',y}\partial\hat{F}_2^i + \hat{F}_{y',y}\partial\hat{F}_1^i) & (\hat{F}_{x',z}\partial\hat{F}_2^i + \hat{F}_{y',z}\partial F_1^i) \end{bmatrix}$$
(VI.40)

$$\{u_i\}_{3\times 1} = \begin{Bmatrix} u_i \\ v_i \\ w^i \end{Bmatrix} \quad \text{(VI.40a)}$$

The local stresses at any point are written in usual manner in form of eqn (VI.13)

$$\{\sigma\} = [D](\{\varepsilon\} - \{\varepsilon_0\}) \quad \text{(VI.41)}$$

in which

$$\{\sigma'\} = [\sigma_{x'}, \sigma_{y'}, \tau_{x'y'}]^{T''} \qquad \{\varepsilon'_0\} = [\varepsilon_{x'0}, \varepsilon_{y'0}, 0]^{T''} \qquad \text{(VI.41a)}$$

For the plane stress case, the elastic material matrix is given by

$$[D] = \frac{E_s}{1 - v_s^2} \begin{bmatrix} 1 & v_s & 0 \\ v_s & 1 & 0 \\ 0 & 0 & \frac{1 - v_s}{2} \end{bmatrix} \qquad \text{(VI.42)}$$

where E_s and v_s are the modulus of elasticity and Poisson's ratio of steel. For $[D]$ in three-dimensional situations please refer to Appendix III. The element stiffness matrix for this element is given by

$$[K] = \int_{-1}^{+1} \int_{-1}^{+1} B^{T''} D B \det J \, d\xi \, d\eta \qquad \text{(VI.43)}$$

Again (as before), the three-dimensional situation can occur; in which case eqn VI.32 is modified. For further details about other elements, reference should be made to Appendix I.

VI.4 ISOPARAMETRIC LINE ELEMENTS

The method of derivation here for line elements is the same as for solid or membrane elements, the difference is only in the choice of displacement polynomials and the shape functions.

The line elements represent the conventional reinforcing bars or the prestressing tendons in concrete. The nodes of these elements depend upon the type of solid elements used for concrete. They are given below:

(a) A two-noded line element corresponds to an 8-noded solid element.
(b) A three-noded line element corresponds to a 20-noded solid element.
(c) A four-noded line element corresponds to a 30-noded solid element.

These line element nodes can be matched by placing them on top of the solid elements. Sometimes it is difficult to idealise them in this manner owing to the reinforcement layout. Such line elements can be placed in the body of the solid element. Where bond–slip analysis is carried out, these line elements can be placed on the nodes of the solid element or can be attached to spring elements as shown in Fig. VI.2.

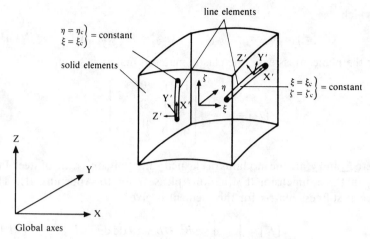

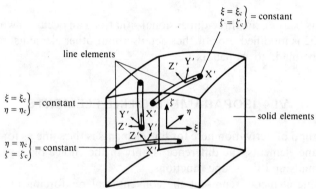

Fig. VI.2. Line elements in the body of a solid element.

IV.4.1 Two- Three- and Four-Noded Elements

The shape functions and derivatives for the isoparametric line elements are given below:

(a) Two-noded line element (Fig. VI.1a)

$$\begin{array}{ll} \textit{Shape functions} & \textit{Derivatives} \\ N_1 = \tfrac{1}{2}(1-\xi) & \dfrac{\partial N_1}{\partial \xi} = -\tfrac{1}{2} \\ N_2 = \tfrac{1}{2}(1+\xi) & \dfrac{\partial N_2}{\partial \xi} = \tfrac{1}{2} \end{array} \quad (\text{VI.44})$$

(b) *Three-noded line element (Figs VI. b_1 and b_2)*

Shape functions
Derivatives

$$N_1 = \tfrac{1}{2}(\xi - 1)\xi \qquad \frac{\partial N_1}{\partial \xi} = \xi - \tfrac{1}{2}$$

$$N_2 = 1 - \xi^2 \qquad \frac{\partial N_2}{\partial \xi} = 2\xi \qquad \text{(VI.45)}$$

$$N_3 = \tfrac{1}{2}(\xi + 1)\xi \qquad \frac{\partial N_3}{\partial \xi} = \xi + \tfrac{1}{2}$$

(c) *Four-noded line element (Figs VI. c_1 and c_2)*

Shape functions
Derivatives

$$N_1 = \tfrac{1}{3}(1 - \xi)(2\xi^2 - \tfrac{1}{2}) \qquad \frac{\partial N_1}{\partial \xi} = \tfrac{1}{3}(4\xi - 6\xi^2 + \tfrac{1}{2})$$

$$N_2 = \tfrac{4}{3}(\xi^2 - 1)(\xi - \tfrac{1}{2}) \qquad \frac{\partial N_2}{\partial \xi} = \frac{4}{3}(3\xi^2 - \xi - 1) \qquad \text{(VI.46)}$$

$$N_3 = \tfrac{4}{3}(1 - \xi^2)(\xi + \tfrac{1}{2}) \qquad \frac{\partial N_3}{\partial \xi} = \frac{4}{3}(1 - 3\xi^2 - \xi)$$

$$N_4 = \tfrac{1}{3}(1 + \xi)(2\xi^2 - \tfrac{1}{2}) \qquad \frac{\partial N_4}{\partial \xi} = \frac{1}{3}(4\xi + 6\xi^2 - \tfrac{1}{2})$$

VI.4.2 The Strain–Displacement Relation

At any point in the line element, the local Cartesian axis X' is tangential to the curvilinear axis. The local strain in the axial direction at any point can be written as

$$\varepsilon_{x'} = \partial U'/\partial X' \qquad \text{(VI.47)}$$

Using the displacement transformation, eqn (VI.47) is written as

$$\varepsilon_{x'} = \frac{1}{L}\left(l_1 \frac{\partial U}{\partial \xi} + m_1 \frac{\partial V}{\partial \xi} + n_1 \frac{\partial W}{\partial \xi}\right) \qquad \text{(VI.47a)}$$

where l_1, m_1, n_1 are the direction cosines of the X' axis and are written as

$$l_1 = \frac{\partial X}{\partial \xi}\bigg/L$$

$$m_1 = \frac{\partial Y}{\partial \xi}\bigg/L \qquad \text{(VI.47b)}$$

$$n_1 = \frac{\partial Z}{\partial \xi}\bigg/L$$

$$L = \sqrt{(\partial X/\partial \xi)^2 + (\partial Y/\partial \xi)^2 + (\partial Z/\partial \xi)^2} \qquad \text{(VI.47c)}$$

U, V, and W are the global nodal freedoms at any node, and U' is the local freedom in the X' direction. It is related according to

$$U' = l_1 U + m_1 V + n_1 W \tag{VI.48}$$

Equation (VI.47) in terms of the shape function derivatives is now written as

$$\varepsilon_{x'} = \frac{1}{L} \sum_{i=1}^{n} \left[l_1 \frac{\partial N_i}{\partial \xi}, m_1 \frac{\partial N_i}{\partial \xi}, n_1 \frac{\partial N_i}{\partial \xi} \right] \begin{Bmatrix} U_i \\ V_i \\ W_i \end{Bmatrix} \tag{VI.49}$$

where n is the number of nodes on the element.

Equation VI.49 is written also as

$$\{\varepsilon_{x'}\} = [B]\{U^e\} \tag{VI.50}$$

where

$$[B] = [[B_1], [B_2], [B_3] \ldots [B_i] \ldots \{B_n\}]^T \tag{VI.51}$$
$$\{U^e\} = [[U_1], [U_2], [U_3] \ldots [U_i] \ldots \{U_n\}]^T \tag{VI.52}$$

The stiffness matrix assumes the form

$$[K] = \int_{-1}^{+1} B^T E_s BA(\xi) L \, d\xi = \sum_{j=1}^{N} B_j^T E_s B_j L_j W_j A(\xi_j) \tag{VI.53}$$

where

$$A(\varepsilon)_j = \sum_{i=1}^{n} N_i A_j \tag{VI.53a}$$

where A = the cross-sectional area at node j; n = the number of nodes on the element; N = the number of integration points; and N_i = the shape function at node i.

The strain and stress are calculated as

$$\{\varepsilon_{x'}\} = [B]\{U^e\} \qquad \{\sigma_{x'}\} = [E_s]\{\varepsilon_{x'}\} \tag{VI.54}$$

where $\{U^e\}$ is the global nodal displacement of the element.

VI.4.3 Line Element in the Body of a Solid Element

The procedure for the strain–displacement and the stiffness matrix is the same as for the other line elements. The element must lie parallel to one of the curvilinear axes (ξ, η, ζ) of the solid element. The element may be anywhere in the solid element with maximum curvilinear co-ordinates $\xi = \pm 1$, $\eta = \pm 1$ and $\zeta = \pm 1$. The displacement U inside the element is written (Fig. VI.2) as

$$\{U\} = [\bar{N}]\{U^e\} = \sum_{i=1}^{n} [N_i][I]\{U_i\} \tag{VI.55}$$

$$[\bar{N}] = [N(\xi, \eta_c, \zeta_c)] \text{ (solid element)} \tag{VI.55a}$$

such that

$$\zeta = \zeta_c \quad \eta = \eta_c \text{ (constant)}$$

For the tangential directions of ξ and η

$$\hat{X}_\xi = \begin{Bmatrix} \dfrac{\partial X}{\partial \xi} \\ \dfrac{\partial Y}{\partial \xi} \\ \dfrac{\partial Z}{\partial \xi} \end{Bmatrix} \text{ (at } \eta = \eta_c, \zeta = \zeta_c) \quad \hat{X}_\eta = \begin{Bmatrix} \dfrac{\partial X}{\partial \eta} \\ \dfrac{\partial Y}{\partial \eta} \\ \dfrac{\partial Z}{\partial \eta} \end{Bmatrix} \text{ (at } \eta = \eta_c, \zeta = \zeta_c)$$

(VI.56)

A normal $\hat{Z}'$-axis vector, and $\hat{X}'$-axis and $\hat{Y}'$-axis vectors are defined as

$$\hat{Z}' = (\hat{X}_\xi \times \hat{X}_\eta)/|\hat{X}_\xi \times \hat{X}_\eta| \tag{VI.57}$$
$$\hat{X}' = \hat{X}_\xi/|\hat{X}_\xi|$$
$$\hat{Y}' = \hat{Z}' \times X'$$

$$\hat{F} = [X', Y', Z']^{T''} = \begin{bmatrix} l_1 & l_2 & l_3 \\ m_1 & m_2 & m_3 \\ n_1 & n_2 & n_3 \end{bmatrix}$$

Following eqn (VI.35) onward, the strains can be evaluated as

$$\varepsilon_{x'} = \frac{\partial U'}{\partial X'} = \sum_{i=1}^{n} \bar{a}(l_1 U_i + m_1 V_i + n_1 W_i) = [B]\{U^e\} \tag{VI.58}$$

$$\bar{a} = l_1 \frac{\partial N_i}{\partial x} + m_1 \frac{\partial N_i}{\partial y} + n_1 \frac{\partial N_i}{\partial z} \tag{VI.59}$$

Stresses are computed as

$$\{\sigma'_x\} = \{E_s\}\{\varepsilon'_x\} \tag{VI.60}$$

This analysis is needed when the prestressing tendons or conventional reinforcement cannot be modelled as line elements lying at the nodes of the solid elements. They can then be located at the body of the concrete element. This analysis will replace the ones adopted for the other line elements. For additional information, reference should be made to Appendix I.

VI.5 PLASTIC FLOW RULE AND STRESSES DURING ELASTOPLASTIC STRAINING

Many materials have been examined, including concrete. They behave elastically up to a certain stage of the loading beyond which plastic deformation takes place. During this plastic deformation the state of strain is not uniquely determined by the state of stress, as stated previously. In a uniaxial state of stress a simple rule is required to initiate yielding at any Gaussian point of the isoparametric element. In the multiaxial state of stress, there are an infinite number of possible combinations of stresses at which yielding starts. These can be examined by the flow rule. Morever, the flow rule supplements the elastic constitutive relationship; and the plastic strain increments are related to the plastic stress increments during the occurrence of plastic flow.

The yield criterion described by a failure surface in a multi-dimensional stress space is given by Bangash [350, 351]

$$L_F(\sigma^s_{ij}, \varepsilon^p, S_H) = 0 \tag{VI.61}$$

where

L_F = the yield function
σ^s_{ij} = the multi-dimensional stress state
ε^p = accumulated plastic strain
S_H = strain hardening or softening parameter

The general form of the yield surface given by eqn (VI.61) allows either isotropic or kinematic hardening of the material. For a given previous history, $L_F(\sigma_{ij})$ is always considered as a function of the current state of stress for which S_H is variable.

To give added generality, the plastic potential to which the normality principle is applicable is assumed as

$$L_Q(\sigma^s_{ij}, \varepsilon^p, SH_0) = 0 \tag{VI.62}$$

This allows non-associated plasticity to be dealt with and associated rules to be obtained as a special case by making

$$L_F = L_Q \tag{VI.63}$$

For a perfectly plastic material the yield surface of eqn (VI.61) remains constant. For a strain hardening material the yield surface must change with continued straining beyond the initial yield.

This phenomenon is included in eqn (VI.61) by allowing both L_F and S_H to be functions of the state of stress and the plastic deformation history. This means S_H will have a new value for every time dependent yielding. Further, if the material is unloaded and then loaded again, additional yielding cannot take place, unless the current value of S_H has been exceeded.

A unified approach for arriving at the incremental stress–strain equation based on eqn (VI.62) can be written in a combined tensor form in three dimensions as

$$f(\sigma, \varepsilon, \dot{\sigma}, \dot{\varepsilon}, S_H) = 0 \tag{VI.64}$$

where f is a definite representation of L_F and is a stress function. The only change is in the value of S_H, accounting for isotropic and anisotropic hardening, and allowing the function f to be dependent not only on the present state of stress or strain, but also on the hardening history according to previous states of stress and strain. The value of $(\sigma, \dot{\sigma})$ and $(\varepsilon, \dot{\varepsilon})$ must be in the plastic range, having total values of σ^p_{ij} and ε^p_{ij} respectively.

When eqn (VI.64) is satisfied then the total differential of f is written as

$$df = \frac{\partial f}{\partial \sigma P_{ij}} d\sigma^p_{ij} + \frac{\partial f}{\partial \varepsilon^p_{ij}} d\varepsilon^p + \frac{\partial f}{\partial S_H} dS_H \tag{VI.65}$$

The yield condition with ε^p_{ij} and S_H held constant can be interpreted as a yield surface in the multi-dimensional stress space, and is in conformity with eqn (VI.61). When $f < 0$ the condition indicates a purely elastic change towards the inside of the yield surface. In cases where plastic flow does not occur, the increments of plastic strain $d\varepsilon^p_{ij}$ and the change of hardening parameter dS_H will be automatically zero, and hence, in the case of the unloading, eqn (VI.65) is reduced to

$$df = \frac{\partial f}{\partial \sigma^p_{ij}} d\{\sigma^p_{ij}\} < 0 \tag{VI.66}$$

When $df = 0$, which is the case for neutral loading, no plastic strain changes occur and the hardening factor remains unchanged, then

$$df = \frac{\partial f}{\partial \sigma_{ij}} d\{\sigma_{ij}^p\} = 0 \qquad d\{\varepsilon_{ij}^p\} = 0 \qquad \text{(VI.66a)}$$

The quantity $d\sigma_{ij}$ is tangent to the surface for neutral loading. For the vector products to be zero, $\partial f/\partial \sigma_{ij}$ will be normal to the surface. When $d\sigma_{ij}$ is pointing to the outside of the surface the vector product will be positive; and this constitutes loading, with plastic flow taking place such that

$$df = \frac{\partial f}{\partial \sigma_{ij}} d\{\sigma_{ij}^p\} > 0 \qquad \text{(VI.67)}$$

The definition for the structural material stability postulates that during a load cycle that includes loading and unloading, the work performed has to be greater than zero, i.e.

$$d\varepsilon_{ij}^p \, d\sigma_{ij}^p \geq 0 \qquad \text{(VI.68)}$$

From eqns (VI.67) and (VI.68) for ideally plastic material the plastic strain increment $d\varepsilon_{ij}^p$ is proportional to the stress gradient of the yield surface

$$d\varepsilon_{ij}^p = \frac{\partial f}{\partial \sigma_{ij}^p} d\lambda \qquad \text{(VI.69)}$$

where $d\lambda = $ a constant of proportionality.

The normality rule given by eqn (VI.69) shows that the plastic strain increment has the same direction as the normal surface f in the stress space. At this stage both isotropic and kinematic hardening cases can be included in eqn (VI.69).

In order to perform the plastic analysis, the loading path is discretised into several linear load steps. The increments in stresses and strains for each step are then related by the rate constitutive laws mentioned earlier. The assumption is that the extension in the linearised step can be regarded as infinitesimal, and eqn (VI.69), which is true for all infinitesimal increments of stresses and strains, can be employed for small increments. Concrete structures for nuclear and offshore installations require small increments. The total strain increment $d\varepsilon_{ii}$ is the algebraic sum of the increments in elastic and plastic strains $d\bar{\varepsilon}_{ij}$ and $d\varepsilon_{ij}^p$ respectively. The equation for $d\varepsilon_{ij}'$ can be written as

$$d\varepsilon_{ij} = d\bar{\varepsilon}_{ij} + d\varepsilon_{ij}^p = d\varepsilon_{ij}^e + d\varepsilon_{ij}^T + d\varepsilon_{ij}^p \qquad \text{(VI.70)}$$

The stresses can be written as

$$d\sigma'_{ij} = [D] d\varepsilon_{kl} \tag{VI.71}$$

$$d\sigma_{ij} = [D] d\varepsilon_{kl} - d\lambda [D]^{-1} \frac{\partial f}{\partial \sigma_{kl}} \tag{VI.72}$$

The plastic strain relationship is given as

$$d\varepsilon = [D^*]^{-1} d\{\sigma\} \tag{VI.73}$$

The total strain increment $\{\Delta\varepsilon\}_{\text{TOT}}$ becomes

$$\{\Delta\varepsilon\}_{\text{TOT}} = [D^*]^{-1}\{\Delta\sigma\} + \frac{\partial f}{\partial \sigma} d\lambda \tag{VI.74}$$

When eqn (VI.65) is written in a compressed form as

$$df = \frac{\partial f}{\partial \{\sigma\}} d\{\sigma\} + \frac{\partial f}{\partial S_H} dS_H = 0 \tag{VI.75}$$

Replacing the second term for the hardening characteristics by $d\lambda$, the following equations are obtained

$$\left\{\frac{\partial f}{\partial \sigma}\right\}^T d\{\sigma\} - C d\lambda = 0 \tag{VI.76}$$

$$d\varepsilon_{\text{TOT}} = [D^*]^{-1} d\{\sigma\} + \left\{\frac{\partial f}{\partial \sigma}\right\} d\lambda \tag{VI.77}$$

$$0 = \left\{\frac{\partial f}{\partial \sigma}\right\}^T d\{\sigma\} - C d\lambda \tag{VI.78}$$

Multiplying eqn (VI.77) with $[D^*]$ throughout, eqn (VI.77) can be expressed as

$$d\{\sigma\} = [D^*]\{d\varepsilon\}_{\text{TOT}} - [D^*]\left\{\frac{\partial f}{\partial \sigma}\right\} d\lambda \tag{VI.79}$$

Equation (VI.79) is then substituted into eqn (VI.78), and eqn (VI.80) is obtained

$$0 = \frac{\partial f^T}{\partial \sigma}[D^*]\{d\varepsilon\}_{\text{TOT}} - \left\{\frac{\partial f}{\partial \sigma}\right\}^T [D^*]\left\{\frac{\partial f}{\partial \sigma}\right\} d\lambda - C d\lambda \tag{VI.80}$$

Equation (VI.70) is then combined with eqns (VI.69), (VI.77) and (VI.65) to give the final value of $d\lambda$

$$d\lambda = \frac{\left\{\dfrac{\partial f}{\partial \sigma}\right\}^T [D^*]\{d\varepsilon\}_{TOT}}{C + \left\{\dfrac{\partial f}{\partial \sigma}\right\}^T [D^*]\left\{\dfrac{\partial f}{\partial \sigma}\right\}} \qquad (VI.81)$$

The factor C in eqn (VI.81) is an unknown factor defining the state of elasto-plasticity. When $C = 0$, eqn (VI.81) gives a value of $d\lambda$ for a perfectly plastic situation with no hardening. The stress increments are given by:

$$\{d\sigma\}_{TOT} = [D^*] - \frac{[D^*]\left\{\dfrac{\partial f}{\partial \sigma}\right\}\left\{\dfrac{\partial f}{\partial \sigma}\right\}^T [D^*]}{C + \left\{\dfrac{\partial f}{\partial \sigma}\right\}^T [D^*]\left\{\dfrac{\partial f}{\partial \sigma}\right\}} \{d\varepsilon\}_{TOT} \qquad (VI.82)$$

for the elasto-plastic case,

$$\{d\sigma\}_{TOT} = [D^*] - [D_P]\{d_\varepsilon\}_{TOT} \qquad (VI.83)$$

$$[D_P] = \frac{[D^*]\left\{\dfrac{\partial f}{\partial \sigma}\right\}\left\{\dfrac{\partial f}{\partial \sigma}\right\}^T [D]}{C + \left\{\dfrac{\partial f}{\partial \sigma}\right\}^T [D]\left\{\dfrac{\partial f}{\partial \sigma}\right\}} \qquad (VI.84)$$

The true stress increment in eqn (VI.84) is the difference between the stress increment $[D_p]\{d\bar{\varepsilon}\}_{TOT}$ and the algebraic sum of stresses $[D]\{d\varepsilon\}$ due to elastic, creep, shrinkage, temperature, fatigue and other effects given in Part I.

The matrix $[D]$ has the flexibility to include any concrete failure models described elsewhere in this book.

The parameter C is given by Hill [336] as

$$C = 4/9\sigma_{eq}^2 S_H \qquad (VI.85)$$

where S_H is the slope of the curve and represents hardening.

Where the influence of the studs, lugs and any other type of anchorages is to be included in the finite element analysis, the steps given in Table VI.2 are considered together with solid, panel and line elements.

Table VI.2
Analytical formulation of the steel anchor s/studs

$$[K_{\text{TOT}}]\{\delta\}^* + \{F_\text{T}\} - \{R_\text{T}\} = 0$$

where

$$[K_{\text{TOT}}] = [K_l] + [K_a]$$

$$\{\delta\}^* = \begin{Bmatrix} \delta_{\text{un}} \\ \delta_b \end{Bmatrix} \quad \{F_\text{T}\} = \begin{Bmatrix} F_{\text{un}} \\ F_b \end{Bmatrix} \quad \{R_\text{T}\} = \begin{Bmatrix} R_{\text{un}} \\ R_b \end{Bmatrix}$$

$[K_{\text{TOT}}]$ = total stiffness matrix
$[K_e]$ = liner stiffness matrix
$[K_a]$ = a stud stiffness matrix
$\{F_\text{T}\}$ = total initial load vector
$\{R_\text{T}\}$ = total external load vector

Subscript

un = quantities corresponding to unknown displacement
b = quantities corresponding to restrained boundaries

$$[k_l]\{\delta_{\text{un}}\} + \{F_{\text{un}}\} = 0$$

$\{\varepsilon\} = [B]\{\delta\}$
$\{\sigma\} = [D](\{\varepsilon\} - \{\varepsilon_0\})$
$\{\bar{S}\}$ = anchor shear forces
    $= [K_a]\{\delta_{\text{un}}\}$

$$\{F_{\text{un}}\} = \int_v [B]^T[D]\{\varepsilon_0\}\,dv = \int_v [B]^T[D]\{\varepsilon_0\}\det[J]\,d\xi\,d\eta\,d\zeta$$

The plastic buckling matrix is given by

$$(K + \lambda K_G)F_\text{T} = 0$$

where K = the elasto-plastic stiffness matrix as a function of the current state of plastic deformation; and K_G = the initial stress geometric stiffness matrix.

The determinant

$$|K + \lambda_c K_G| = 0$$

The essential equation is characteristically triangularised for the ith loading step as

$$(K^i + \lambda_c K_G^i)F_\text{T}^i = 0 \qquad \lambda_c = 1 + E_{\text{ps}}$$

E_{ps} is an accuracy parameter

VI.5.1 Sample Cases
VI.5.1.1 Plastic Potential of the Same Form as the Yield Surface

$$f(J_2) = f(\sigma_{ij}) = 3I_2 = \tfrac{3}{2} S_{ij} S_{ij} \qquad \text{(VI.86)}$$

where

$$S_{ij} = \sigma_{ij} - \tfrac{1}{3}\sigma_{kk}\delta_{ij} \qquad \delta_{ij} = \text{the Kronecker delta}$$

Differentiating

$$\begin{aligned}
\frac{\partial f}{\partial \sigma_{ij}} &= \frac{3}{2}\partial(S_{kl}S_{kl})/\partial\sigma_{ij} \\
&= \frac{3}{2} S_{kl}\, \partial S_{mn}/\partial\sigma_{ij} = 3S_{ij}
\end{aligned} \qquad \text{(VI.87)}$$

The plastic strain increment is stated as

$$d\lambda \varepsilon_{ij}^{p} = \lambda S_{ij} \qquad \text{(VI.88)}$$

The equivalent plastic strain $\lambda \bar{\varepsilon}^p$ can be obtained as

$$d\lambda \bar{\varepsilon}^p = (\tfrac{2}{3} d\varepsilon_{ij}^p\, d\varepsilon_{ij}^p)^{1/2} \qquad \text{(VI.89)}$$

where

$$d\lambda = \frac{3}{2}\frac{d\bar{\varepsilon}^p}{\sigma_{eq}}$$

VI.5.1.2 Von Mises Yield Surface Associated with Isotropic Hardening

The yield function f is written as

$$f = \frac{T}{2}(\sigma_{ij}) - \tfrac{1}{3}\sigma_{eq}^2 \qquad \text{(VI.90)}$$

By differentiating, eqn (VI.90) becomes

$$df = (S_{ij}\, d\sigma_{ij}) - \tfrac{2}{3}\sigma_{eq}(\partial \sigma_{eq})/(\partial \bar{\varepsilon}^p)\{d\bar{\varepsilon}^p\} \qquad \text{(VI.91)}$$

Using eqn (VI.66) onwards, the values of $d\lambda$ and $d\sigma$ given in eqns (VI.81) and (VI.82) are given by

$$d\lambda = \left([D^*]_{i,j,k,l} - \frac{[D^*]_{mn} S_{pq} [D^*]_{pqkl}}{\sigma_{ij}[D^*]_{i,j,kl} S_{kl} + (\tfrac{2}{3}\sigma_{eq})^2 S_H} \right)\{d\varepsilon\}_{kl} \qquad \text{(VI.92)}$$

VI.6 DYNAMIC ANALYSIS

Dynamic analysis is a subject by itself. In this text a specific approach to the dynamic analysis of concrete structures is discussed. The equations of motion are discretised in time. The direct integration technique [285–363] and the Wilson-θ method are summarised. Where dynamic problems are tackled, the nonlinear equations of motion (coupled or uncoupled), have been solved using these methods.

VI.6.1 Nonlinear Transient Dynamic Analysis

The dynamic equilibrium at the nodes of a system of structural elements is formulated at a given time t as

$$[M]\{\ddot{U}_t\} + [C_t]\{\dot{U}_t\} + K'_t\{U_t\} = \{R_t\} \qquad \text{(VI.93)}$$

where $\{U_t\}$ and $\{R_t\}$ are the vectors of displacement and specified load, respectively. $[M]$ represents the mass matrix which is regarded as constant; $[C_t]$ and $[K'_t]$ are the damping and stiffness matrices, respectively. The subscript t is used for quantities at time t and a dot denotes a derivative with respect to time.

To formulate eqn (VI.93), discretisation with respect to time, using isoparametric finite elements, is performed. The simple applicable method is the numerical step-by-step integration of the coupled equations of motion, such as eqn VI.93. The response history is divided into time increments Δt, which are of equal length. The system is calculated for each Δt, with properties determined at the beginning of the interval. Only one matrix based on M, is excited. In addition, the direct integration technique allows a general damping matrix $[C_t]$ (which has to be specified explicitly), to be used without resorting to complex eigenvalues.

VI.6.1.2 Discretisation in the Time Domain

The equation of motion formulated at time $t = 0$, is written in the form

$$[M]\{\ddot{U}_0\} + [C_0]\{\dot{U}_0\} + [K_0]\{U_0\} = \{R_0\} \qquad \text{(VI.94)}$$

where the subscript zero has been introduced. At time t, all quantities are known. Equation (VI.93) is specified as

$$[M]\{\ddot{U}_t\} + ([C_0] + [\Delta C_{0 \to t}])\{\dot{U}_t\} + ([K'_0] + [\Delta K_{0 \to t}])\{U_t\} = \{R_t\} \qquad \text{(VI.95)}$$

or as

$$[M]\{\ddot{U}_t\} + [C_0]\{\dot{U}_t\} + [K_0]\{U_t\} = \{R_t\} + \{P_t\} = \{F\}_t \qquad \text{(VI.96)}$$

where the initial load $\{P_t\}$ is specified by

$$\{P_t\} = -[\Delta C_{0 \to t}]\{\dot{U}_t\} - \{\Delta K_{0 \to t}\}\{\delta_t\} \quad (VI.97)$$

To obtain the solution at time $t + \Delta t$, the equation is stated as

$$[M]\{\ddot{U}_{t+\Delta t}\} + [C_0]\{\dot{U}_{t+\Delta t}\} + [K_0]\{U_{t+\Delta t}\}$$
$$= \{R_{t+\Delta t}\} + \{P_t\} + \{\Delta P_{t \to t+\Delta t}\} \quad (VI.98)$$

$\{\Delta P_{t \to t+\Delta t}\}$ represents the influence of the nonlinearity during the time increment t and is determined by iteration and satisfied for $t + \tau$, where $\tau = \theta \Delta t$. ($\theta > 1 \cdot 37$ for an unconditionally stable method) applied to a linear problem. $[\Delta C_{0 \to t}]$ and $[\Delta K_{0 \to t}]$ represent the change of $[C]$ and $[K]$, respectively, from $t = 0$ to t.

To obtain the solution at time $t + \Delta t$, eqn (VI.98) can be written as

$$[M]\{\ddot{U}_{t+\Delta t}\} + [C_0]\{\dot{U}_{t+\Delta t}\} + [K_0]\{U_{t+\Delta t}\}$$
$$= \{R_{t+\Delta t}\} + \{F_t\} + \{\Delta F_{t \to t+\Delta t}\} \quad (VI.99)$$

$\{\Delta P_{t \to t+\Delta t}\}$ represents the influence of the nonlinearity during the time increment t and is determined by iteration

$$\{\Delta P_{t \to t+\Delta t}\} = -[\Delta C_{0 \to t}]\{\Delta \dot{U}_{t \to t+\Delta t}\} - [\Delta C_{t \to t+\Delta t}](\{\dot{U}\} + \{\Delta \dot{U}_{t \to t+\Delta t}\})$$
$$- [\Delta K_{0 \to t}]\{\Delta U_{t \to t+\Delta t}\} - [\Delta K_{t \to t+\Delta t}](\{U_t\} + \{\Delta U_{t \to t+\Delta t}\}) \quad (VI.100)$$

$(\Delta P_{t \to t+\Delta t})$ is calculated using the initial stress approach.

A modified Newton–Raphson or initial stress approach is adopted for solving these nonlinear equations. A step-by-step integration method is given in Table VI.3. Using these methods along with acceleration and convergence procedures described in this chapter and Appendix V can allow successful solution of finite element based problems.

VI.6.2 Reduced Linear Transient Dynamic Analysis

This is a reduced form of nonlinear transient dynamic analysis. This analysis is carried out faster than the nonlinear analysis since the matrix in eqn (IV.93) requires to be inverted once, and the analysis is reduced to a series of matrix multiplications and essential degrees of freedom (dynamic or master of freedoms) to characterise the response of the system. The analysis generally has restrictions such as constant $[M]$, $[C_t]$, $[K_t]$ and time interval for all iterations and nodal forces applied at dynamic or master degrees of freedom.

Equations (I) and (J) of Table VI.3 are written in a different form by considering the above-mentioned restraints.

VI.6.2.1 Quadratic Integration

$$\left(\frac{1}{\Delta t^2}[M]_R + \frac{3}{2\Delta t}[\hat{C}_t]_R + [K_t]_R\right)\{U_t\}_R = \{F(t)\}_R$$

$$+ [M]_R \frac{1}{\Delta t^2}(2\{U_{t-1}\}_R - \{U_{t-2}\}_R)$$

$$+ \frac{1}{\Delta t}(2\{U_{t-1}\}_R - \tfrac{1}{2}\{U_{t-2}\}_R) \tag{VI.101}$$

The symbol R represents reduced matrices and vectors.

VI.6.2.2 Cubic Integration

$$\left(\frac{2}{\Delta t^2}[M]_R + \frac{11}{6\Delta t}[C_t]_R + [K_t]_R\right)\{U_t\}_R = \{F(t)\}$$

$$+ [M]_R \frac{1}{\Delta t^2}(5\{U_{t-1}\}_R - 4\{U_{t-2}\}_R + \{U_{t-3}\})$$

$$+ [C_t]_R \frac{1}{\Delta t^2}(3\{U_{t-1}\}_R - \tfrac{3}{2}\{U_{t-2}\}_R + \tfrac{1}{3}\{U_{t-3}\}_R) \tag{VI.102}$$

VI.6.3 Mode Frequency Analysis

The equation of motion for an undamped structure with no applied forces is written as

$$[M]\{\ddot{U}_t\} + [K'_t]\{U_t\} = \{0\} \tag{VI.103}$$

$[K'_t]$ the structure stiffness matrix, may include stress-stiffening effects.

The system of equations is initially condensed down to those involved with the master (dynamic) degrees of freedom.

The number of dynamic degrees of freedom should at least be equal to two times the selected frequencies. The reduced form of eqn (VI.103) can be written as

$$[M]_R\{\ddot{U}_t\}_R + [K'_t]_R\{U\}_R = \{0\} \tag{VI.104}$$

For a linear system, free vibrations of harmonic type are written as

$$\{U_t\}_R = \{\psi_i\}_R \cos \omega_i t \tag{VI.105}$$

where $\{\psi_i\}_R$ = the eigenvector representing the shape of the ith frequency; ω_i = the ith frequency (radians/unit time); and t = time.

Table VI.3
Summary of step-by-step integration method

Initialisation
(1) Effective stiffness matrix $[K_0^*] = (6/\tau^2)[M] + \dfrac{3}{\tau}[C_0] + [K_0]$ (A)
(2) Triangularise $[K_0^*]$

For each time step:
Calculation of displacement $\{U_{t+\tau}\}$
(1) Constant part of the effective load vector

$$\{R_{t+\tau}^*\} = \{R_t\} + \theta(\{R_{t+\Delta t}\} - \{R_t\}) + \{F_t\} + [M]$$
$$\times \left(\left(\dfrac{6}{\tau^2}\right)\{U_t\} + \dfrac{6}{\tau}\{\dot{U}_t\} + 2\{\ddot{U}_t\}\right) + [C_0]\left(\dfrac{3}{\tau}\{U_t\} + 2\{\dot{U}_t\} + \dfrac{\tau}{2}\{\ddot{U}_t\}\right) \quad \text{(B)}$$

(2) Initialisation $i = 0$, $\{\Delta P_{t \to t+\tau}^i\} = 0$
(3) Iteration

(a) $i \to i+1$
(b) Effective load vector $\{R_{t+\tau \text{tot}}^*\} = \{R_{t+\tau}^*\} + \{\Delta P_{t \to t+\tau}^{i-1}\}$ (C)
(c) Displacement $\{U_{t+\tau}^i\} [K_0^*]\{U_{t+\tau}^i\} = \{R_{t+\tau_{\text{tot}}}^{*i}\}$ (D)
(d) Velocity $\{\dot{U}_{t+\tau}^i\} + (3/\tau)(\{U_{t+\tau}^i\} - \{U_t\}) - 2\{\dot{U}_t\} - \dfrac{\tau}{2}\{\ddot{U}_t\}$
(e) Change of initial load vector caused by the nonlinear behaviour of the material $\{\Delta P_{t \to t+\tau}^i\}$

$$\{\Delta P_{t \to t+\tau}^i\} = -[\Delta C_{0 \to t}](\{\dot{U}_{t+\tau}^i\} - \{\dot{U}_t\}) - [\Delta C_{t \to t+\tau}^i]\{\dot{U}_{t+\tau}^i\}$$
$$[\Delta K_{0 \to t}](\{U_{t+\tau}^i\} - \{U_t\}) - [\Delta K_{t \to t+\Delta t}^i]\{U_{t+\tau}^i\} \quad \text{(E)}$$

In fact, $\{\Delta P_{t \to t+\tau}^i\}$ is calculated using the initial-stress method.

(f) Iteration convergence

$$\|\{\Delta P_{t \to t+\tau}^i\} - \{\Delta P_{t \to t+\tau}^{i-1}\}\| / \|\{\Delta P_{t \to t+\tau}^i\}\| < \text{tol} \quad \text{(F)}$$

or analogously, on stresses.
Note: $\{P\}$ could be any value of $\{F\}$.

Calculation of velocity, acceleration
Calculate new acceleration $\{\ddot{U}_{t+\Delta t}\}$, velocity $\{\dot{U}_{t+\Delta t}\}$, displacement $\{U_{t+\Delta t}\}$ and initial load $\{P_{t+\Delta t}\}$:

$$\{\ddot{U}_{t+\Delta t}\} = (6/\theta\tau^2)(\{U_{t+\tau}\} - \{U_t\}) - (6/\tau\theta)\{\dot{U}_t\} + \left(1 - \dfrac{3}{\theta}\right)\{\ddot{U}_t\}$$

$$\{\dot{U}_{t+\Delta t}\} = \{\dot{U}_t\} + \dfrac{\tau}{2\theta}\{\ddot{U}_t\} + \{\ddot{U}_{t+\Delta t}\}$$

Table VI.3—contd.

$$\{U_{t+\Delta t}\} = \{U_t\} + \frac{\tau}{0}\{\dot{U}_t\} + (\tau^2/60^2)(2\{\ddot{U}_t\} + \{\ddot{U}_{t+\Delta t}\})$$

$$\{P_{t+\Delta t}\} = \{P_t\} + \{\Delta P^i_{t \to t+\tau}\} \tag{G}$$

Calculation by quadratic integration

When the velocity varies linearly and the acceleration is constant across the time interval, appropriate substitutions are made into eqn (VI.93) giving

$$[f_1[M] + f_2[C_t] + [K'_t]]\{U_t\} = \{F(t)\} + \{f_3([C_t], [M], U_t, U_{t2},...)\} \tag{H}$$

where f_1, f_2, f_3 = functions of time.

This results in an implicit time integration procedure. The only unknown is $\{U_t\}$ at each time point and this is calculated in the same way as in static analysis. Equation (H) is then written as:

$$\left(\frac{2}{\Delta t_0 \Delta t_{01}}[M] + \frac{2\Delta t_0 + \Delta t_1}{\Delta t_0 \Delta t_{01}}[C] + [K'_t]\right)\{U_t\}$$

$$= \{F(t)\} + [M]\left(\frac{2}{\Delta t_0 \Delta t_1}\{U_{t-1}\} - \frac{2}{\Delta t_1 \Delta t_{01}}\{U_{t-2}\}\right)$$

$$+ [C_t]\left(\frac{\Delta t_{01}}{\Delta t_0 \Delta t_1}\{U_{t-1}\} - \frac{\Delta t_0}{\Delta t_{01} \Delta t_1}\{U_{t-2}\}\right) \tag{I}$$

where

$\Delta t_0 = t_0 - t_1$ t_0 = time of current iteration
$\Delta t_1 = t_1 - t_2$ t_1 = time of previous iteration
$\Delta t_2 = t_2 - t_3$ t_2 = time before previous iteration
                  t_3 = time before t_2
$\Delta t_2 = \Delta t_0 + \Delta t_1 = t_0 - t_2$

Calculation by cubic integration

Equation (H) becomes cubic and hence is written as

$$(a_1[M] + a_2[C_t] + [K'_t])\{U_t\} = \{F(t)\} + [M](a_3\{U_{t-1}\} - a_4\{U_{t-2}\} + a_5\{U_{t-3}\})$$
$$+ [C](a_6\{U_{t-1}\} - a_7\{U_{t-2}\} + a_8\{U_{t-3}\}) \tag{J}$$

where a_1 to a_8 are functions of the time increments; these functions are derived by inverting a four by four matrix.

For clear-cut solutions, the size of the time step between adjacent iterations should not be more than a factor of 10 in nonlinear cases and should not be reduced by more than a factor of 2 where plasticity exists.

Equation (VI.103) assumes the form

$$(-\omega_i^2[M]_R + [K''_t]_R\{\psi_i\}_R = \{0\} \qquad (VI.106)$$

which is an eigenvalue problem with n values of ω^2 and n eigenvectors $\{\psi_i\}_R$ which satisfy eqn (VI.106), where n is the number of dynamic degrees of freedom. Using standard iteration procedures, eqn (IV.106) will yield a complete set of eigenvalues and eigenvectors.

Each eigenvector, $\{\psi_i\}_R$ is then normalised such that:

$$\{\psi_i\}_R^{T''}[M]_R\{\psi_i\}_R = 1 \qquad (VI.107)$$

These n eigenvectors are now expanded to the full set of structure modal displacement degrees of freedom. Using Section VI.2.6

$$\{\psi_{\gamma'i}\}_R = [K_{\gamma'\gamma'}]^{-1}[K_{\gamma'\gamma}]\{\psi_i\}_R \qquad (VI.108)$$

where $\{\psi_i\}_R =$ the slave degree of freedom vector of mode i; and $[K_{\gamma'\gamma'}]$, $[K_{\gamma'\gamma}] =$ submatrix parts as shown in eqn (VI.22) onwards.

The above dynamic analysis approach is generally adopted for structures subjected to normal dynamic loads, wind, wave and seismic loads. The above analysis, with modifications, is also applied to missile and aircraft explosions/impact problems.

VI.6.4 Spectrum Analysis

Spectrum analysis is an extension of the mode frequency analysis, with both base and force excitation options. The base excitation option is generally suitable for seismic and wave applications. A direction vector and a response spectrum table will be needed in addition to the data and parameters required for the reduced modal analysis. The response spectrum table generally includes displacements, velocities and accelerations. The force excitation is, in general, used for wind and space structures and missile/aircraft impact. It requires a force distribution and an amplitude multiplier table in addition to the data and parameters needed for the reduced modal analysis. A study of the mass distribution is made. Generally the masses are kept close to the reaction points on the finite element mesh rather than the (master) degrees of freedom. It is important to calculate the participation factors in relation to a given excitation direction. The base and forced excitations are given below:

$$\tilde{\gamma}_i = \{\psi_i\}_R^{T''}[M]\{\tilde{b}\} \quad \text{for the base excitation} \qquad (VI.109)$$

$$\tilde{\gamma}_i = \{\psi_i\}_R^{T''}\{F_t\} \quad \text{for the force excitation} \qquad (VI.110)$$

where $\{\bar{b}\}$ = the unit vector of the excitation direction; and $\{F_t\}$ = an input force vector, such as eqn (VI.93).

The values of $\{\psi\}_R$ are normalised, and the reduced displacement vector is calculated from the eigenvector by using a mode coefficient $\tilde{M}$.

$$\{\tilde{U}\}_i = [\tilde{M}_i]\{\psi\}_i \tag{VI.111}$$

where $\{\tilde{U}\}_i$ = the reduced displacement vector; and $[\tilde{M}_i]$ = the mode coefficient and where (a) for velocity spectra

$$[\tilde{M}_i] = \frac{[V_{si}]\{\tilde{\gamma}_i\}}{\omega_i} \tag{VI.112}$$

(V_{si} = spectral velocity for the ith mode); (b) for force spectra

$$[\tilde{M}_i] = \frac{[\bar{J}_{si}]\{\tilde{\gamma}_i\}}{\omega_i^2} \tag{VI.113}$$

($\bar{J}_{si}$ = spectral force for the ith mode); (c)

$$[\tilde{M}_i] = \frac{[a_{si}]\{\tilde{\gamma}_i\}}{\omega_i^2} \tag{VI.114}$$

(a_{si} = spectral acceleration for the ith mode); (d)

$$[\tilde{M}_i] = \frac{[U_{si}]\{\tilde{\gamma}_i\}}{\omega_i^2} \tag{VI.115}$$

(U_{si} = spectral displacement for the ith mode).

$\{U\}_i$ may be expanded to compute all the displacements, as was done in eqns (VI.19) onwards.

$$\{U_{\gamma'}\}_i = [K_{\gamma'\gamma'}]^{-1}[K_{\gamma'\gamma}]\{U_i\}_R \tag{VI.116}$$

where $\{U_{\gamma'}\}_i$ = the slave degree of freedom vector of mode i; and $[K_{\gamma'\gamma'}]$, $[K_{\gamma'\gamma}]$ = submatrix parts.

Sometimes an equivalent mass M_i^e is needed for the ith mode since it may not be a function of excitation direction. This M_i^e is computed as

$$[M_i^e] = 1/\{\psi_i\}_R^{T''}\{\psi_i\}_R \tag{VI.117}$$

This is derived from the definition of the diagonal matrix of equivalent masses $[M^e]$

$$[\psi]_R^{T''}[M_e][\psi]_R = [I] \tag{VI.117a}$$

where $[I]$ = the identity matrix; and $[\psi]_R$ = a square matrix containing all mode shape vectors.

Where damping is included, the damping ratio D_{Ri} for the data input, including damping C_e, is given for a matrix of coupling coefficients as

$$D_{Ri} = C_e \omega_{i/2} \qquad (VI.118)$$

where ω_i is the undamped natural frequency of the ith mode.

In between the modes i and j, a modified damping ratio D'_{Ri} is needed to take into account the concrete structures subjected to wave and seismic effects.

$$D'_{Ri} = D_{Ri} + 2/t_e \omega_i \qquad (VI.119)$$

where t_e is the duration.

VI.6.5 Impact/Explosion

Structural response of concrete structures subjected to relatively fast loading rates, such as those from missile and aircraft impact/explosion, bombs and nuclear detonations, etc., can be influenced by the effect of strain rate on the material properties. These material changes lead to an instantaneous change in the strength of materials such as concrete. Section VI.5 is invoked along with the above equations of motion. The normality rules and the proportionality factor $d\lambda$ are used as given in Section VI.5.

The $d\lambda$ values are substituted into eqn (VI.72) to give an expression for the stress state of the form

$$d\dot{\sigma}_{ij} = [D]\{d\dot{\varepsilon}_{kl}\} + \tilde{\gamma}_{ij}\{\ddot{\varepsilon}\}^p \qquad (VI.120)$$

where, using Von Mises criterion,

$$\tilde{\gamma}_{ij} = \frac{[D^*]\{S_{mn}\}\{S_{kl}\}\{\bar{D}\}}{\alpha^*} \qquad (VI.121)$$

$$\alpha^* = \tfrac{4}{9}\sigma_{eq}^2 \frac{\partial \sigma_{eq}}{\partial \varepsilon^p} + S_{ij}[D]\{S_{kl}\}$$

$$[D^*] = D_{ijmn} \qquad (VI.122)$$

$$[\bar{D}] = D_{pqkl}$$

the term $\gamma_{ij}\dot{\varepsilon}^p$ can be implemented. Where deformation rates do not change too rapidly the term $\gamma_{ij}\dot{\varepsilon}^p$ is neglected. The rest of the procedure is the same as for general dynamic analysis.

VI.7 BUCKLING ANALYSIS

The eigenvalue buckling (bifurcation) analysis is vital where a steel liner is fully or partially anchored to concrete, such as in concrete bridges, tunnels, pressure and containment vessels, cells of offshore platforms, underground shelters, structures for hydroelectricity and irrigation, etc.

Buckling of the liner or its embedded anchors in concrete is possible. The buckling matrix is developed so that at appropriate stages the liner and its anchor system is checked against buckling. The equation below gives the eigenvalue buckling by bifurcation:

$$([K]^e + \lambda_{ei}[K^s])\{\psi\}_i = \{0\} \quad \text{(VI.123)}$$

where

$[K]^e$ = the elastic stiffness matrix
$[K^s]$ = the stress stiffness matrix
λ_{ei} = the ith eigenvalue (used to multiply the loads which generated $[K^s]$)
$\{\psi\}_i$ = the ith eigenvector of displacements

The first step in the solution is to reduce eqn (VI.123) to its static or dynamic buckling (master) degrees of freedom. The elastic stiffness matrix $[K]^e$ is reduced and the stress stiffness matrix $[K^s]$ is reduced in a manner identical to that by which the mass matrix is reduced. Hence eqn (VI.123) becomes:

$$([K]_R + \lambda_{ei}[K^s])\{\psi_i\}_R = 0 \quad \text{(VI.124)}$$

where $[K]_R$ can be replaced by $[K'_t]_R$ when dynamic problems are involved.

When a geometric stiffness matrix is included, the plastic buckling matrix is given as

$$([K^p]_R + \lambda_{ci}[K]_G)\{\psi_i\}_R = 0 \quad \text{(VI.125)}$$

where $\lambda_c = 1 + E_{ps}$, and E_{ps} = accuracy parameters.

Table VI.2 gives a brief breakdown of the buckling phenomenon. $\{\psi_i\}_R$ is related to the total initial load vector $\{F_T\}$ of the liner and the studs. Equations (VI.124) and (VI.125) are further condensed by again reducing out those rows and columns of the matrices that have a positive value on the main diagonal of the stress or geometric matrix. The matrices are inverted such that an accuracy is achieved having the lowest eigenvalue. A standard iteration procedure is adopted for the solution of these matrices. The eigenvectors are then normalised such that each has the largest value of 1·0.

Chapter VII

Material Modelling Simulation for Finite Element Formulation

VII.1 INTRODUCTION

This chapter deals with the step-by-step numerical modelling of various concrete strengths and failure theories as applied to finite elements. Theoretical expressions are arranged in such a way that they can easily be incorporated into the finite element analysis given in Chapter VI. The simulation of the material modelling gives a unified analysis for the strength and failure conditions of concrete structures. The theoretical models given in this chapter cover One- to Five-Parameter Models: Bulk and Shear Moduli Models, Endochronic Theory, Temperature, Heat Conduction, Shrinkage, Creep, Bond, Fatigue and Cracking.

VII.2 UNIAXIAL STRESS–STRAIN CURVE

Figure VII.1 shows an idealised stress–strain curve in compression and in tension. Curve B is obtained in a number of ways, as described in Chapter 1; from A to B' eqns (I.1), (I.2), (I.5), (I.7) and (I.8) can be chosen depending upon the choice of the individual concrete strength theory. The stress–strain values are incorporated into eqn (VI.71). Using incremental techniques described in Chapter VI, a series of points can be chosen on the curve from A to B' and then, invoking eqn (VI.71), they can be included in the finite element cycles. Where softening is included (B' to B^{IV}), eqn (I.5) is used. If B' to B^{IV} is treated as a straight line, then the value of σ^0 shall not be less than $0.15\sigma_{cu}$. In the case of cyclic loads, the value of σ^0 shall be limited

Material Modelling Simulation for Finite Element Formulation

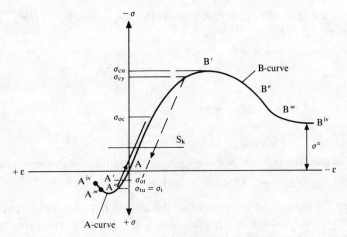

Fig. VII.1(a). Typical stress–strain curve.

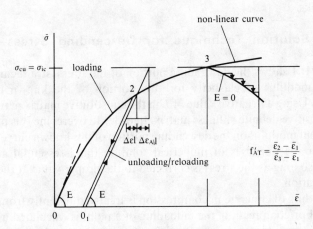

Fig. VII.1(b). Solution technique for stress–strain curve (loading–unloading scheme).

to $0.2f_c''$. The B'–B^{IV} curve can also be evaluated from the following equation:

$$\frac{\sigma}{\sigma_{cu}} = \frac{2(\varepsilon/\varepsilon_{cu})}{1 + (\varepsilon/\varepsilon_{cu})^2} \tag{VII.1}$$

Curve A represents the tension cut-off scheme.

Bicanic and Zienkiewicz [297] define the descending curve for concrete softening, i.e. B'B^{IV}:

$$\sigma^0 = c'/\gamma^* \qquad (VII.2)$$

where

$$c' = \frac{d\sigma'}{dt'} \frac{1}{E_0} + \gamma^* \sigma'$$

$$\log \gamma = a_0 + a_1 \log \dot{\varepsilon}_c \quad \text{or} \quad \gamma^* = 10^{a_0}(\dot{\varepsilon}_c)^{a_1}$$

$\dot{\varepsilon}_c$ = strain rate $\quad \sigma' = \sigma - \sigma_{0c} \quad \sigma_{0c}$ = initial stress of concrete

$\qquad a_0 = -2.9299 \qquad a_1 = 0.76 \qquad 0 \le \dot{\varepsilon}_c \le 10^{-5}$

Figure VII.1 shows an idealisation of the tension part denoted by curve A. The maximum value is not greater than $0.55\sqrt{f'_c}$, and beyond this tensile failure is assumed to occur. In finite element analysis, this limitation is easily achieved in standard computer output, giving normal and principal stresses and cracks.

VII.2.1 Solution Technique for Descending Stress–Strain Curve

Figure VII.1 shows the descending portion of the stress–strain curve. The tangent modulus of elasticity for this portion of the known curve is negative. Using a negative value of E in the constitutive matrix may lead to a non-positive definite stiffness matrix. In order to overcome this problem, the tangent modulus on the descending portion of this known curve is set to zero. For the purposes of numerical solution this is essential and permissible so long as the stress is corrected to the proper value at the end of each iteration.

Unloading of concrete in compression is treated differently from that in tension. For compression, the unloading of a point is calculated using the maximum equivalent strain in the finite element analysis. If the equivalent strain ($\sqrt{\frac{2}{3}\varepsilon_{ij}\varepsilon_{ij}}$) at any point is less than that of the equivalent strain in the previous load increment, the point is maintained. The unloading is treated as elastic, as shown in Fig. VII.2. The equivalent strains of all points from where unloading started can now be stored. For example, unloading in Fig. VII.2 starts at point 2 and follows the linear line 2–0$_1$, and the equivalent strain of point 2 is stored. Reloading of this point will follow the unloading line 0$_1$–2 until the equivalent strain of point 2 is reached. After that, the nonlinear curve is followed. Now upon reloading at point 1 (on the same curve) the path 1–2–3 is followed. Path 1–2–3 of the strain increment covers

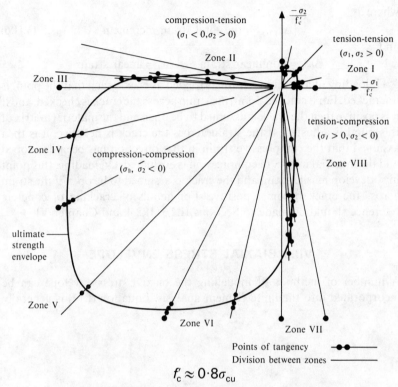

Fig. VII.2. Simulation of a typical biaxial state curve.

both linear and nonlinear parts. The linear and nonlinear strains are calculated as described below.

If F_{AT} is a fraction of a linear strain increment, its value is computed as

$$F_{AT} = (\bar{\varepsilon}_2 - \bar{\varepsilon}_1)/(\bar{\varepsilon}_3 - \bar{\varepsilon}_1)$$

where $\bar{\varepsilon}_1$, $\bar{\varepsilon}_2$ and $\bar{\varepsilon}_3$ are the equivalent strains at points 1, 2 and 3, respectively:

$$\bar{\varepsilon}_1 = \sqrt{\tfrac{2}{3}\varepsilon_1^{T''}\varepsilon_1} \qquad \bar{\varepsilon}_2 = \sqrt{\tfrac{2}{3}\varepsilon_2^{T''}\varepsilon_2} \qquad \bar{\varepsilon}_3 = \sqrt{\tfrac{2}{3}\varepsilon_3^{T''}\varepsilon_3} \qquad \text{(VII.3)}$$

The linear part of the strain increment is given as

$$\{\Delta\varepsilon_l\} = F_{AT}\{\Delta\varepsilon\}_{TOT} = F_{AT}\{\Delta\varepsilon_l + \Delta\varepsilon_{nl}\} \qquad \text{(VII.4)}$$

The nonlinear part of the strain increment is then computed as

$$\{\Delta\varepsilon_{nl}\} = (1 - F_{AT})\{\Delta\varepsilon\}_{TOT} \qquad \text{(VII.5)}$$

where

$$\{\Delta\varepsilon\}_{TOT} = \text{the total strain increment} \qquad (VII.6)$$

and

$\Delta\varepsilon_{nl}$ = nonlinear strain $\Delta\varepsilon_l$ = linear strain

Unloading of a point in tension is treated as elastic such that the point is uncracked. For a cracked point, the strain across the crack is checked, and if it is positive then the crack is assumed to be open and the material matrix of this point is adjusted. If the strain across the crack is negative, it is then assumed that the compressive strain in that direction has been developed and the material matrix in compression is evaluated. On loading, this point may develop tensile strain and the crack is assumed to be open if the strain across the crack becomes positive. For details of cracking of concrete reference should be made to Sections II.2.3, II.2.4 and Chapter VI.

VII.3 BIAXIAL STRESS ENVELOPE

A number of methods of modelling the biaxial stress envelope can be incorporated into the finite element analysis. Equation (I.13) is generally

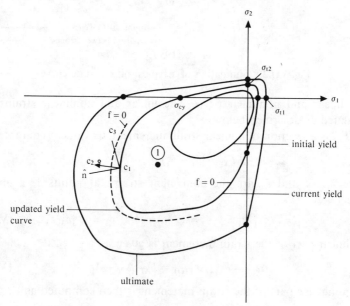

Fig. VII.3. Yield curves and points in stress space.

used. The curve can be defined in the biaxial principal stress space as shown in Fig. VII.3 in terms of ultimate strength in tension σ_{tu} and compression σ_{cu}, respectively, as obtained from the uniaxial stress–strain curve shown in Fig. VII.1. If f is the yield function, various regions of the stress space termed 'zones' can then be given analytical expressions. The peak values and limitations given in Section I.3 are marked. In the finite element analysis, curves for tension–tension regions, compression–tension regions and compression–compression regions are represented by their respective equations in Section I.3. In the compression–compression region under directions 1 and 2, the mathematical relations are given for the biaxial stress envelope given in Fig. VII.3. The initial yield curve and current yield curve can easily be plotted within the ultimate strength envelope, as indicated in Fig. VII.3. Three different yield strengths are required in order to define the current yield curve which, in turn, requires three different strain hardening measurements. Assuming that increments in plastic strain are normal to the yield curve and the stress point is located on the current yield curve, then

$$\frac{\partial \varepsilon^p}{\partial t} = \text{vector of the principal plastic strain rate}$$

$$= \dot{\varepsilon}^p$$

$$= \left[\frac{\partial f'}{\partial \sigma_1}, \frac{\partial f'}{\partial \sigma_2}\right]^t d\lambda \qquad (\text{VII.7})$$

where f' is $f(\sigma_1, \sigma_2, \sigma_{cy}, \sigma_{t1}, \sigma_{t2})$; $d\lambda$ is constant of proportionality; and t is time. For any plastic strain, the rate of equivalent plastic strain $\partial \varepsilon^p_{eq}/\partial t$ can be determined on the lines suggested in eqn (VI.71).

This equivalent plastic strain rate is now decomposed arbitrarily into proportions which contribute to hardening of any yield strength. If three decomposition parameters are A_1, A_2 and A_3, the equivalent plastic strains are written as

$$\varepsilon^p_c = \sum_0^t A_1 \frac{\partial \varepsilon^p_{eq}}{\partial t} dt \qquad \varepsilon^p_{t1} = \sum_0^t A_2 \frac{\partial \varepsilon^p_{eq}}{\partial t} dt \qquad \varepsilon^p_{t2} = \sum_0^t A_3 \frac{\partial \varepsilon^p_{eq}}{\partial t} dt \qquad (\text{VII.8})$$

where ε^p_c, ε^p_{t1} and ε^p_{t2} are equivalent compressive plastic strain and tensile strains. Table VII.1 gives the decomposition parameters adopted by Elwi and Murray [5]. Assuming $\{\sigma\}_1$, $\{\varepsilon\}_1$, $\{\varepsilon^p\}_1$, A_1, A_2 and A_3, a new value of $\{\varepsilon\}$ is imposed such that

$$\{\Delta \varepsilon\} = \{\varepsilon\} - \{\varepsilon\}_1 \qquad (\text{VII.9})$$

Table VII.1
Yield function and decomposition parameters (Elwi & Murray [5])

Zone	f	Limits $\dfrac{\sigma_1}{f_c} = \dfrac{\sigma}{\sigma_{cy}}$	Limits $\dfrac{\sigma_2}{f_c} = \dfrac{\sigma}{\sigma_{cy}}$	A_1	A_2	A_3
1	$\sqrt{\sigma_{t1}\sigma_{t2}}\, f_1\!\left(\dfrac{\sigma_1}{\sigma_{t1}}, \dfrac{\sigma_2}{\sigma_{t2}}\right)$	≥ 0	≥ 0	0	$^a\!\left(\dfrac{b_1}{b}\right)^2$	$^a\!\left(\dfrac{b_2}{b}\right)^2$
2	$\sigma_2 + \sigma_{t2} f_2\!\left(\dfrac{\sigma_1}{\sigma_{cy}}\right)$	$\left(-\dfrac{\sigma_2}{2\sigma_{t2}}, 0\right)$	> 0	$-\dfrac{\sigma_1}{\sigma_{cy}}$	0	$1 - A_1$
3	$\sigma_2 + \sigma_{t2} f_3\!\left(\dfrac{\sigma_1}{\sigma_{cy}}\right)$	$< -\dfrac{\sigma_2}{2\sigma_{t2}}$	> 0	$-\dfrac{\sigma_1}{\sigma_{cy}}$	0	$1 - A_1$
4	$\sigma_{cy} f_4\!\left(\dfrac{\sigma_2}{\sigma_{cy}}\right) - \sigma_1$	$< \dfrac{1{\cdot}58491\,\sigma_2}{\sigma_{cy}}$	≤ 0	1	0	0
5	$\sigma_{cy} f_5\!\left(\dfrac{\sigma_1}{\sigma_{cy}}, \dfrac{\sigma_2}{\sigma_{cy}}\right) - \sigma_{cy}$	$\left(\dfrac{1{\cdot}58491\,\sigma_2}{\sigma_{cy}}, \dfrac{\sigma_2}{1{\cdot}58491\,\sigma_{cy}}\right)$	≤ 0	1	0	0
6	$\sigma_{cy} f_4\!\left(\dfrac{\sigma_1}{\sigma_{cy}}\right) - \sigma_2$	$\left(\dfrac{\sigma_2}{1{\cdot}58491\,\sigma_{cy}}, 0\right)$	≤ 0	1	0	0
7	$\sigma_1 + \sigma_{t1} f_3\!\left(\dfrac{\sigma_2}{\sigma_{cy}}\right)$	> 0	$< -\dfrac{\sigma_1}{2\sigma_{t1}}$	$-\dfrac{\sigma_2}{\sigma_{cy}}$	$1 - A_1$	0
8	$\sigma_1 + \sigma_{t1} f_2\!\left(\dfrac{\sigma_2}{\sigma_{cy}}\right)$	> 0	$\left(-\dfrac{\sigma_1}{2\sigma_{t1}}, 0\right)$	$-\dfrac{\sigma_2}{\sigma_{cy}}$	$1 - A_1$	0

$^a\, b = \left|\dfrac{\partial f}{\partial \sigma}\right| = $ the magnitude of $\dfrac{\partial f}{\partial \sigma}$; b_1, b_2 are the components of b.

Table VII.2
Step-by-step simulation of the biaxial envelope using the Kupfer model

---

The Kupfer model is chosen such that I_1 and I_2 are the first and second stress invariants. The yield criterion is written as

$$f(I_1, I_2) = \sqrt{K'_1 I_1 + 3K'_2 I_2} = \sigma_{eq} \tag{A}$$

where K'_1 and K'_2 are material parameters.

Kupfer's results for

$$\sigma_1/\sigma_2 = \infty \quad \text{(uniaxial compression)}$$

and

$$\sigma_1/\sigma_2 = 1 \quad \text{(equal biaxial compression)}$$

are

$$K'_1 = 0{\cdot}356\sigma_{eq} \qquad K'_2 = 1{\cdot}356\sigma_{eq} \tag{B}$$

For other values of σ_1/σ_2 reference should be made to Section I.3.

The strain increment $\Delta\varepsilon^p$ for the flow rule is given by

$$\Delta\varepsilon^p = d\lambda \frac{\partial f(\sigma)}{\partial \sigma} \tag{C}$$

where $d\lambda$ is a factor for the plastic strain increment from eqn (VI.81) and $\partial f/\partial \sigma$ is a vector normal to the current load surface.

The parameter or stress function, in this case from eqn (C), is written as

$$f(\sigma) = \left[2 \frac{K'_1}{2\sigma_{eq}} \sigma_{eq} I_1 + 3K'_2 I_2 \right]^{1/2} = \frac{K_1}{2\sigma_{eq}} I_1 + \left[\left(\frac{K_1}{2\sigma_{eq}}\right)^2 + 3K_2 I_2 \right]^{1/2} \tag{D}$$

Flow vectors

$$\frac{\partial f(\sigma)}{\partial \sigma} = \left[\underset{①}{\frac{\partial f}{\partial \sigma_x}}, \underset{②}{\frac{\partial f}{\partial \sigma_y}}, \underset{③}{\frac{\partial f}{\partial \tau_{xy}}}, \underset{④}{\frac{\partial f}{\partial \sigma_z}}, \underset{⑤}{\frac{\partial f}{\partial \tau_{yz}}}, \underset{⑥}{\frac{\partial f}{\partial \tau_{zx}}} \right] \tag{E}$$

where

$$\frac{\partial f(\sigma_1)}{\partial \sigma_1} = \frac{K'_1}{2\sigma_{eq}} + \left[2\left\{ \left(\frac{K'_1}{2\sigma_{eq}}\right)^2 + K_2 \right\} \sigma_x + \left\{ 2\left(\frac{K'_1}{2\sigma_{eq}}\right)^2 - K'_2(\sigma_y + \sigma_z) \right\} \right] \Big/ K^*$$

$$= L_1 + [L_2 \sigma_x + L_3(\sigma_y + \sigma_z)]/K^*$$

$$\frac{\partial f(\sigma_2)}{\partial \sigma_2} = L'_1 + [L'_2 \sigma_y + L'_3(\sigma_z + \sigma_x)]/K^*$$

$$\frac{\partial f(\sigma_3)}{\partial \sigma_3} = L''_1 + [L''_2 \sigma_z + L''_3(\sigma_x + \sigma_y)]/K^*$$

(continued)

Table VII.2—contd.

$$\frac{\partial f(\sigma_4)}{\partial \sigma_4} = 6K_2'\tau_{xy}/K^*$$

$$\frac{\partial f(\sigma_5)}{\partial \sigma_5} = 6K_2'\tau_{yz}/K^*$$

$$\frac{\partial f(\sigma_6)}{\partial \sigma_6} = 6K_2'\tau_{zx}/K^*$$

$$K^* = 2[L_2'(\sigma_x^2 + \sigma_y^2 + \sigma_z^2) + L_3'(\sigma_x\sigma_y + \sigma_y\sigma_z + \sigma_z\sigma_x) + 3K_2'(\tau_{xy}^2 + \tau_{yz}^2 + \tau_{zx}^2)]^{1/2}$$

$\{d\sigma\}$ = the elasto-plastic incremental stress = $\{D^*\}\{d\varepsilon\}$ \hfill (F)

The elasto-plastic material matrix $\{D^*\}$ is given in Chapter VI.

The rest of the procedure for crushing, compression–tension combinations, and cracking are given in earlier chapters and Section VII.11. For example, for crushing

$$3J_2 = \varepsilon_{cu}^2 \hspace{2cm} (G)$$

where J_2 is the deviatoric strain invariant.

For compressive behaviour of cracked concrete and where hardening S_H is involved, the value of S_H given below is substituted in the elasto-plastic material matrix $\{D^*\}_{ep}$:

$$S_H = E\left[\left\{1 - 20 \text{ to } 50 \times \frac{\varepsilon_t}{\varepsilon_p}\right\}^2 - 1\right]$$

where E = Young's modulus; ε_t = tensile strain; and ε_p = plastic strain.

Next $\sigma_{cy}(A_1)$, $\sigma_{t1}(A_2)$ and $\sigma_{t2}(A_3)$ are computed. It is then easy to compute (Fig. VII.4) the corresponding values of $\{\Delta\sigma\}$ and $\{\sigma\}_2$:

$$\{\Delta\sigma\} = [D]\{\Delta\varepsilon\} \quad \text{and} \quad \{\sigma\}_2 = \{\sigma\}_1 + \{\Delta\sigma\} \hspace{1cm} (VII.10)$$

If

$$f[\{\sigma\}_2, \{\sigma_{cy}\}, \{\sigma_{t1}\}, \{\sigma_{t2}\}] \le 0 \hspace{1cm} (VII.11)$$

then

$$\{\sigma\}_1 = \{\sigma\}_2 \quad \{\varepsilon\}_1 = \{\varepsilon\}$$

When $f[\{\sigma\}_1, \{\sigma_{cy}\}, \{\sigma_{t1}\}, \{\sigma_{t2}\}] \ge 0$ it is necessary to take into consideration a factor which can bring the stress point to the yield surface. If that factor is F_{AT} as shown in Fig. VII.1, then

$$\{\sigma\} = \{\sigma\}_1 + F_{AT}[D]\{\Delta\varepsilon\} \quad \{\Delta\varepsilon\} = (1 - F_{AT})\{\Delta\varepsilon\} \quad \text{(VII.12)}$$

The third case is when $\{\sigma\} = \{\sigma\}_1$ hardening parameters and decomposition parameters are updated, and the following value of $\{\Delta\sigma\}$ is computed:

$$\{\Delta\sigma\} = [D]_{ep}\{\Delta\varepsilon\} \quad \text{(VII.13)}$$

The rest of the solution procedure given in Sections II.2.3, II.2.4 and Appendix V is adopted for strength and crack evaluations. Table VII.2 gives a second version of the biaxial stress envelope and the tension curve in the finite element analysis. Loading surfaces for concrete constitutive models are given in Chapters I and II.

VII.4 SIMULATION PROCEDURES FOR ONE- TO FIVE-PARAMETER MODELS

VII.4.1 One-Parameter Model

Equations (II.1)–(II.4) give various cases of the one-parameter model. During loading it is necessary that the yield function of any one of these equations must be satisfied such that the stress state is on the yield surface. The simplest form of the isotropic-hardening Von Mises model is $f = J_2 = \frac{1}{2}S_{ij}S_{ij}$, where f is the loading function. If λ_s is a scalar function such that $\lambda_s = \lambda_s(J_2)$, then

$$d\varepsilon_{ij}^p = \lambda(J_2)\frac{\partial f}{\partial \sigma_{ij}} dJ_2 = \lambda_s(J_2)S_{ij} dJ_2 \quad \text{(VII.14)}$$

When both sides of eqn (VII.14) are squared, it yields

$$d\varepsilon_{ij}^p d\varepsilon_{ij}^p = \lambda_s^2(J_2)2J_2(dJ_2)^2 \quad \text{(VII.15)}$$

The value of $\int \sqrt{d\varepsilon_{ij}^p d\varepsilon_{ij}^p}$ is a function of $J_2 = \frac{1}{3}\sigma_{eq}^2$.

For a one-parameter model

$$\sigma_{eq} = \sqrt{3J_2} \quad \text{(VII.16)}$$

where

$$J_2 = \frac{1}{2}(S_x^2 + S_y^2 + S_z^2) + \tau_{xy}^2 + \tau_{yz}^2 + \tau_{zx}^2 \quad \text{(VII.17)}$$
$$S_x = \sigma_x - \sigma_0 \quad S_y = \sigma_y - \sigma_0 \quad S_z = \sigma_z - \sigma_0 \quad \text{(VII.17a)}$$
$$\sigma_0 = \text{mean stress} = \frac{1}{3}(\sigma_x + \sigma_y + \sigma_z)$$

The equivalent plastic strain ε_p is defined as

$$\varepsilon_p = \int \sqrt{\{d\varepsilon_{ij}^p\}^{T''}\{d\varepsilon_{ij}^p\}} \tag{VII.18}$$

$$d\varepsilon_p = d\lambda \sqrt{\left\{\frac{\partial f}{\partial \sigma_{ij}}\right\}^{T''}\left\{\frac{\partial f}{\partial \sigma_{ij}}\right\}} \tag{VII.18a}$$

where $d\lambda$ is a proportionality factor and f is a stress function of the given flow rule.

Knowing $d\lambda$, and upon solving for $d\lambda$, the following equation yields

$$\{d\varepsilon_{ij}^p\} = \{d\varepsilon_p\} \frac{\{\partial f/\partial \sigma_0\}}{\sqrt{\{\partial f/\partial \sigma_{ij}\}^{T''}\{\partial f/\partial \sigma_{ij}\}}} \tag{VII.19}$$

The stress increment can now be written as

$$\{d\sigma\} = [D](\{d\varepsilon\} - \{d\varepsilon^p\}) \tag{VII.20}$$

where $\{d\varepsilon\}$ and $\{d\varepsilon^p\}$ are the strain vectors and are related as

$$\{d\varepsilon^p\} = [C]\{d\varepsilon\} \tag{VII.21}$$

The matrix $[C]$ transforms the total strain increments into plastic strain increments. Substituting eqn (VII.21) into eqn (VII.20), the well-known equation expressed in the finite element analysis is derived as

$$\{d\sigma\} = [D]_{ep}\{d\varepsilon\} \tag{VII.22}$$

The plastic material matrix $[D_p]$ is derived in the way shown in Chapter VI.

VII.4.2 Two-Parameter Model

Equations (II.5)–(II.10) give various expressions for the two-parameter model. All these equations can easily be linked to the associated flow rule. For example, eqn (II.10) is written as

$$f(\sigma) = f(I, \sqrt{J_2}) = K \tag{VII.23}$$

$$\frac{\partial f}{\partial \sigma_{ij}} = \frac{\partial f}{\partial I_1}\frac{\partial I_1}{\partial \sigma_{ij}} + \frac{\partial f}{\partial \sqrt{J_2}}\frac{\partial \sqrt{J_2}}{\partial \sigma_{ij}} \tag{VII.24}$$

$$= \frac{\partial f}{\partial I}\delta_{ij} + \tfrac{1}{2}\sqrt{J_2}\frac{\partial f}{\partial \sqrt{J_2}} S_{ij} \tag{VII.24a}$$

where δ_{ij} is the Kronecker delta and S_{ij} are the principal stress deviators.

This value is substituted in an equivalent plastic strain equation, such as is given in eqn (VII.18a). The procedure given from eqn (VII.19) onwards is now adopted for the stress and strain increments.

When eqn (II.10) is written in its original form

$$f = \alpha I_1 + \sqrt{J_2} - K = 0 \qquad (\text{VII.25})$$

the plastic stress–strain relation is written after computing $\partial f/\partial \sigma_{ij}$. The final expression is given as

$$\frac{\partial f}{\partial \sigma_{ij}} = \alpha \delta_{ij} + \tfrac{1}{2}\sqrt{J_2} S_{ij} \qquad (\text{VII.25a})$$

The plastic strain $d\varepsilon^p_{ij} = d\lambda(\partial f/\partial \sigma_{ij})$ is then evaluated, and this is followed by the rest of the procedure given earlier.

Now the expression given in eqn (II.5) for the Mohr–Coulomb criterion can be rewritten for the elasto-plastic formulation to suit the finite element analysis. Chen [151] proposes the form of yield function f for the Mohr–Coulomb criterion by rewriting eqn (II.5) as

$$f(\sigma_{ij}) = I_1 \sin \phi + \frac{3 \sin \theta (1 - \sin \phi) + \sqrt{3} \cos \theta (3 + \sin \phi)}{2} \sqrt{J_2}$$

$$- 3c \cos \phi = 0 \qquad (\text{VII.26})$$

The chain rule given in eqn (VII.24) is extended to cater for the third parameter θ. The gradient $\partial f/\partial \sigma_{ij}$ is written as

$$\frac{\partial f}{\partial \sigma_{ij}} = \frac{\partial f}{\partial I_1}\frac{\partial I_1}{\partial \sigma_{ij}} + \frac{\partial f}{\partial J_2}\frac{\partial J_2}{\partial \sigma_{ij}} + \frac{\partial f}{\partial J_3}\frac{\partial J_3}{\partial \sigma_{ij}} \qquad (\text{VII.27})$$

Partial differentiation of the above terms is given below:

$$\frac{\partial f}{\partial I_1} = \sin \phi \qquad \frac{\partial \theta}{\partial J_2} = \frac{3\sqrt{3}}{4 \sin 3\theta} \frac{J_3}{J_2^{5/2}} \qquad \frac{\partial \theta}{\partial J_3} = \frac{-\sqrt{3}}{2 \sin \theta J_2^{3/2}} \qquad (\text{VII.28})$$

$$\frac{\partial f}{\partial J_2} = \frac{3 \sin \theta (1 - \sin \phi) + \sqrt{3}(3 + \sin \phi) \cos \theta}{4 J_2}$$

$$+ \frac{3\sqrt{3} J_3}{8 J_2^2 \sin 3\theta} [3(1 - \sin \phi) \cos \theta - \sqrt{3}(3 + \sin \phi) \sin \theta] \qquad (\text{VII.29})$$

$$\frac{\partial f}{\partial J_3} = -\frac{\sqrt{3} [3(1 - \sin \phi) \cos \theta - \sqrt{3}(3 + \sin \phi) \sin \theta]}{4 J_2 \sin 3\theta} \qquad (\text{VII.30})$$

The value of $\partial f/\partial \sigma_{ij}$ is calculated and the procedure given earlier in Section

VII.4.1 is followed for the stress–strain calculation. Crushing and cracking can then be linked to this model as stated in other cases elsewhere in the text.

VII.4.3 Three-Parameter Model

The three-parameter model of William and Wanke has already been discussed in Section II.1.3. In the case of elasto-plastic formulation of the three-parameter model, eqn (II.13) is chosen; this predicts failure or yielding provided the average stresses σ_a, τ_a and the angle of similarity θ satisfy this equation. The gradient direction is again evaluated using the chain rule of differentiation:

$$\frac{\partial f}{\partial \sigma_{ij}} = \frac{\partial f}{\partial \sigma_a}\frac{\partial \sigma_a}{\partial \sigma_{ij}} + \frac{\partial f}{\partial \tau_a}\frac{\partial \tau_a}{\partial \sigma_{ij}} + \frac{\partial f}{\partial \theta}\frac{\partial \theta}{\partial \sigma_{ij}} \qquad \text{(VII.31)}$$

The terms for σ_a and τ_a are given in eqn (II.13a). Various terms on the right-hand side of eqn (VII.31) are evaluated on the lines suggested in eqn (VII.27):

$$\frac{\partial f}{\partial \sigma_a}\frac{\partial \sigma_a}{\partial \sigma_{ij}} = \frac{1}{zf_c''}\tfrac{1}{3}\delta_{ij} \qquad \text{(VII.31a)}$$

$$\frac{\partial f}{\partial \tau_a}\frac{\partial \tau_a}{\partial \sigma_{ij}} = \frac{1}{\rho(\theta)f_c''}\frac{1}{5\tau_a}S_{ij} \qquad \text{(VII.31b)}$$

$$\frac{\partial f}{\partial \theta}\frac{\partial \theta}{\partial \sigma_{ij}} = \frac{\partial f}{\partial \rho}\frac{\partial \rho}{\partial \theta}\frac{\partial \theta}{\partial \sigma_{ij}} \qquad \text{(VII.31c)}$$

The values of $\rho(\theta)$ are given in eqn (II.12a). In eqn (VII.31c), the terms are calculated as

$$\frac{\partial f}{\partial \theta} = -\frac{1}{\rho^2}\frac{\tau_a}{f_c''} \qquad \text{(VII.31d)}$$

From eqn (II.12a)

$$\frac{\partial \rho}{\partial \theta} = \frac{V\dfrac{dU}{d\theta} - U\dfrac{dV}{d\theta}}{V^2} \qquad \text{(VII.31e)}$$

where

$$U = 2\rho_c(\rho_c^2 - \rho_t^2)\cos\theta + \rho_c(2\rho_t - \rho_c)[\alpha_1'' + 5\rho_t^2 - 4\rho_t\rho_c]^{1/2}$$
$$\alpha_1'' = 4(\rho_c^2 - \rho_t^2)\cos^2\theta \qquad \text{(VII.31f)}$$

$$\frac{dU}{d\theta} = -2\rho_c(\rho_c^2 - \rho_t^2)\sin\theta\frac{4\rho_c(\rho_c - 2\rho_t)(\rho_c^2 - \rho_t^2)\sin\theta\cos\theta}{[\alpha' + 5\rho_t^2 - 4\rho_t\rho_c]^{1/2}} \qquad \text{(VII.31g)}$$

$$V = \alpha_1'' + (\rho_c - 2\rho_t)^2 \quad \frac{dV}{d\theta} = -8(\rho_c^2 - \rho_t^2)\sin\theta\cos\theta \quad \text{(VII.31h)}$$

The term $\partial\theta/\partial\sigma_{ij}$ is computed by differentiating the angle θ given in eqn (II.12b).
Again the angle $\cos\theta$ is denoted by similar terms given in eqn (VII.31e). Equation (II.12b) is written as

$$\theta = \cos^{-1}\left(\frac{U}{V}\right)$$

$$\frac{\partial\theta}{\partial\sigma_{ij}} = \frac{\partial\cos^{-1}}{\partial\sigma_{ij}}\left(\frac{U}{V}\right) = \frac{1}{\sin\theta}\left[\frac{V\dfrac{\partial U}{\partial\sigma_{ij}} - U\dfrac{\partial V}{\partial\sigma_{ij}}}{V^2}\right] \quad \text{(VII.31i)}$$

$$U = 2\sigma_1 - \sigma_2 - \sigma_3 + \tau_{xy} = 0 + \tau_{yz} = 0 + \tau_{xz} = 0 \quad \text{(VII.31j)}$$

$$\frac{\partial U}{\partial\sigma_{ij}} = [2, -1, -1, 0, 0, 0] \quad \text{(VII.31k)}$$

$$V = \sqrt{2}\left[(\sigma_1 - \sigma_2)^2 + (\sigma_2 - \sigma_3)^2 + (\sigma_3 - \sigma_1)^2\right]^{1/2}$$

$$\frac{\partial V}{\partial\sigma_{ij}} = \frac{\sqrt{6}}{\sqrt{5}} \frac{\sqrt{15\tfrac{1}{3}}[2\sigma_1 - \sigma_2 - \sigma_3, 2\sigma_2 - \sigma_3 - \sigma_1, 2\sigma_3 - \sigma_1 - \sigma_2, 0, 0, 0]}{\sqrt{(\sigma_1 - \sigma_2)^2 + (\sigma_2 - \sigma_3)^2 + (\sigma_3 - \sigma_1)^2}}$$

$$= \frac{5[2\sigma_1 - \sigma_2 - \sigma_3, 2\sigma_2 - \sigma_3 - \sigma_1, 2\sigma_3 - \sigma_1 - \sigma_2, 0, 0, 0]}{\sqrt{(\sigma_1 - \sigma_2)^2 + (\sigma_2 - \sigma_3)^2 + (\sigma_3 - \sigma_1)^2}} \quad \text{(VII.31l)}$$

Equations (VII.31a)–(VII.31l) are substituted into eqn (VII.31) for evaluation of the gradient $\partial f/\partial\sigma_i$. The procedure given in Section VII.4.1 onward is adopted together with the finite element technique given in Chapter VI for crushing and cracking.

VII.4.4 Four-Parameter Model

An analytical failure model in which four parameters are included has been developed by Ottoson [24, 25]. This failure surface is described in Section II.1.4 and is defined by eqn (II.14). The same chain rule of partial differentiation is applied. The gradient $\{\partial f/\partial\sigma\}$ is given by Bangash [351, 749]:

$$C = \frac{\partial f}{\partial(\sigma)} = \frac{\partial f}{\partial I_1}\frac{\partial I_1}{\partial\sigma} + \frac{\partial f}{\partial J_2}\frac{\partial J_2}{\partial\sigma} + \frac{\partial f}{\partial\cos 3\theta}\frac{\partial\cos 3\theta}{\partial\sigma}$$

$$= a_1 c_1 + a_2 c_2 + a_3 c_3 \quad \text{(VII.32)}$$

$$a_1 = \frac{\partial f}{\partial I_1} = \frac{B}{f_c'}$$

$$a_2 = \frac{\partial f}{\partial J_2} = \frac{\partial}{\partial J_2}\frac{AJ_2}{f_c'^2} + \frac{\partial}{\partial J_2}\frac{\lambda J_2^{1/2}}{f_c'^2} = \frac{A}{f_c'^2} + \frac{\partial}{\partial J_2}\frac{\lambda J_2^{1/2}}{f_c'^2}$$

$$\frac{\partial}{\partial J_2}\frac{\lambda J_2^{1/2}}{f_c'^2} = \frac{\lambda}{f_c'^2}\frac{1}{2}J^{-1/2} + \frac{J^{1/2}}{f_c'}\frac{\partial(\lambda)}{\partial J_2} \qquad (\text{VII.33})$$

$$\frac{\partial(\lambda)}{\partial J_2} = \frac{\partial}{\partial J_2}\left\{K_1 \cos\left[\tfrac{1}{3}\cos^{-1}\left(K\frac{3\sqrt{3}}{2}\frac{\sqrt{3}}{J_2^{3/2}}\right)\right]\right\} \qquad (\text{VII.34})$$

where
$$\cos 3\theta \geq 0$$

$$= -K_1 \sin(P)\frac{1}{3}\frac{(-1)}{\sqrt{1-t^2}} K_2\left(\frac{3}{2}\right)\sqrt{3}J_3\left(-\frac{3}{2}\right)J_2^{-6/2}$$

$$\equiv -K_1 \sin(P)\frac{1}{3J}\left(K_2\frac{3\sqrt{3}}{2}\frac{\sqrt{3}}{J_2^{3/2}}\right)\frac{1}{\sqrt{1-t^2}}\left(+\frac{3}{2}\right)$$

$$\equiv -K_1\frac{t}{\sqrt{1-t^2}}\sin(P)\frac{1}{2J} = -\frac{K_1}{2J}\frac{t}{\sqrt{1-t^2}}\sin(P) \qquad t = K_2\frac{3\sqrt{3}}{2}\frac{J_3}{J_2^{3/2}}$$

$$\frac{\partial(\lambda)}{\partial J_2} = \frac{\partial}{\partial J_2}\left\{K_1\cos\left[\frac{\pi}{3} - \frac{1}{3}\cos^{-1}\left(-K_2\frac{3\sqrt{3}}{2}\frac{J_3}{J_2^{3/2}}\right)\right]\right\} \qquad \cos\theta \leq 0$$

$$= K_1(-\sin P)\left[-\frac{1}{3}\frac{(-1)}{\sqrt{1-t^2}}\left(-K_2\frac{3\sqrt{3}}{2}J_3\right)\left(-\frac{3}{2}\right)J^{-6/2}\right]$$

$$= \frac{K_1}{T}(-\sin P)\left[-\frac{1}{3}\frac{t}{\sqrt{1-t^2}}\left(-\frac{3}{2}\right)\right] = \frac{-K}{2J}(\sin P)\frac{t}{\sqrt{1-t^2}} \qquad (\text{VII.35})$$

From this equation
$$a_2 = \frac{A}{f_c'^2} + \frac{1}{2J_2^{1/2}f_c'}\left\{\lambda - K_1\sin\left[\frac{1}{3}\cos^{-1}\left(K_2\frac{3\sqrt{3}}{2}\frac{J_3}{J_2^{3/2}}\right)\right]\frac{t}{\sqrt{1-t^2}}\right\}; \cos 3\theta \geq 0$$

$$= \frac{A}{f_c'^2} - \frac{1}{2J_2^{1/2}f_c'}\left\{\lambda - K_1\sin\left[\frac{\pi}{3} - \frac{1}{3}\cos^{-1}\left(-K_2\frac{3\sqrt{3}}{2}\frac{\sqrt{3}}{J_2^{3/2}}\right)\frac{t}{\sqrt{1-t^2}}\right]\right\}$$

$$\qquad (\text{VII.36})$$

$$a_3 = \frac{\partial}{\partial \cos 3\theta}\left(\lambda\frac{J_2^{1/2}}{f_c'}\right) \qquad \cos 3\theta \geq 0 \qquad (\text{VII.37})$$

$$= \frac{J_2^{1/2}}{f_c'} \frac{1}{3} K_1 \sin\left[\frac{1}{3}\cos^{-1}(K_2\cos 3\theta)\right] \frac{K_2}{\sqrt{1-t^2}}$$

$$= \frac{1}{3} \frac{K_1 K_2 J_2^{1/2}}{f_c'\sqrt{1-t^2}} \sin\left[\frac{1}{3}\cos^{-1}(K_2\cos 3\theta)\right] \qquad \cos 3\theta \geq 0 \qquad \text{(VII.38)}$$

$$a_3 = \frac{K_1 K_2 J_2^{1/2}}{3\sqrt{1-t^2}f_c'} \sin\left[\frac{\pi}{3} - \frac{1}{3}\cos^{-1}(-K_2\cos 3\theta)\right] \qquad \cos 3\theta \leq 0$$

Where $t = K_2 \cos 3\theta$

$$J_2 = [\tfrac{1}{2}(S_x^2 + S_y^2 + S_z^2) + \tau_{xy}^2 + \tau_{yz}^2 + \tau_{zx}^2]$$

VII. 4.4.1 The Values of C_1, C_2, C_3

$$C_1 = \frac{\partial I_1}{\partial \{\sigma\}} = \begin{Bmatrix} 1 \\ 1 \\ 1 \\ 0 \\ 0 \\ 0 \end{Bmatrix}$$

$$C_2 = \frac{\partial J_2}{\partial \{\sigma\}} = \frac{\partial J_2}{\partial S_x}\frac{\partial S_x}{\partial \{\sigma\}} + \frac{\partial J_2}{\partial S_y}\frac{\partial S_y}{\partial \{\sigma\}} + \frac{\partial J_2}{\partial S_z}\frac{\partial S_z}{\partial \{\sigma\}}$$

$$+ \frac{\partial J_2}{\partial \tau_{xy}}\frac{\partial \tau_{xy}}{\partial \{\sigma\}} + \frac{\partial J_2}{\partial \tau_{yz}}\frac{\partial \tau_{yz}}{\partial \{\sigma\}} + \frac{\partial J_2}{\partial \tau_{zx}}\frac{\partial \tau_{zx}}{\partial \{\sigma\}}$$

$$= S_x \begin{Bmatrix} \tfrac{2}{3} \\ -\tfrac{1}{3} \\ \tfrac{1}{3} \\ 0 \\ 0 \\ 0 \end{Bmatrix} + S_y \begin{Bmatrix} -\tfrac{1}{3} \\ \tfrac{2}{3} \\ -\tfrac{1}{3} \\ 0 \\ 0 \\ 0 \end{Bmatrix} + S_z \begin{Bmatrix} -\tfrac{1}{3} \\ -\tfrac{1}{3} \\ \tfrac{2}{3} \\ 0 \\ 0 \\ 0 \end{Bmatrix} + 2 \begin{Bmatrix} 0 \\ 0 \\ 0 \\ \tau_{xy} \\ \tau_{yz} \\ \tau_{zx} \end{Bmatrix} \qquad \text{(VII.39)}$$

or

$$C_2 = \begin{Bmatrix} \tfrac{1}{3}(2S_x - S_y - S_z) \\ \tfrac{1}{3}(2S_y - S_x - S_z) \\ \tfrac{1}{3}(2S_z - S_x - S_y) \\ 2\tau_{xy} \\ 2\tau_{yz} \\ 2\tau_{zx} \end{Bmatrix} = \begin{Bmatrix} S_x \\ S_y \\ S_z \\ 2\tau_{xy} \\ 2\tau_{yz} \\ 2\tau_{zx} \end{Bmatrix} \qquad \text{(VII.40)}$$

$$\cos 3\theta = \frac{3\sqrt{3}}{2} \times \frac{J_3}{J_2^{3/2}} \qquad (VII.41)$$

$$C_3 = \frac{\partial \cos 3\theta}{\partial J_3} = \frac{\partial \cos 3\theta}{\partial J_3}\frac{\partial J_3}{\partial \{\sigma\}} + \frac{\partial \cos 3\theta}{\partial J_2}\frac{\partial J_2}{\partial \{\sigma\}} \qquad (VII.42)$$

From eqn (VII.42)

$$\frac{\partial \cos 3\theta}{\partial J_3} = \frac{3\sqrt{3}}{2J_2^{3/2}} \qquad \frac{\partial \cos 3\theta}{2J_2} = \left(\frac{3\sqrt{3}}{2}J_3\right)\left(\frac{-\frac{3}{2}}{J_2^{5/2}}\right) = -\frac{9}{4}\sqrt{3}\frac{J_3}{J_2^{5/2}} \qquad (VII.43)$$

Hence

$$C_3 = \frac{3}{2}\frac{\sqrt{3}}{J_2^{3/2}}\frac{\partial J_3}{\partial \{\sigma\}} - \frac{9}{4}\sqrt{3}\frac{J_3}{J_2^{5/2}}\frac{\partial J_2}{\partial \{\sigma\}} = \frac{3}{2}\frac{\sqrt{3}}{J_2^{3/2}}\left[\frac{\partial J_3}{\partial \{\sigma\}} - \frac{3}{2}\frac{J_3}{J_2}\frac{\partial J_2}{\partial \{\sigma\}}\right] \qquad (VII.44)$$

Now

$$J_3 = [S_x S_y S_z + 2\tau_{xy}\tau_{yz}\tau_{zx} - S_x\tau_{yz}^2 - S_y\tau_{xz}^2 - S_z\tau_{xy}^2]$$

$$\frac{\partial J_3}{\partial \{\sigma\}} = \frac{\partial J_3}{\partial S_x}\frac{\partial S_x}{\partial \{\sigma\}} + \frac{\partial J_3}{\partial S_y}\frac{\partial S_y}{\partial \{\sigma\}} + \frac{\partial J_3}{\partial S_z}\frac{\partial S_z}{\partial \{\sigma\}}$$

$$+ \frac{\partial J_3}{\partial \tau_{xy}}\frac{\partial \tau_{xy}}{\partial \{\sigma\}} + \frac{\partial J_3}{\partial \tau_{yz}}\frac{\partial \tau_{yz}}{\partial \{\sigma\}} + \frac{\partial \sqrt{3}}{\partial \tau_{zx}}\frac{\partial \tau_{zx}}{\partial \{\sigma\}}$$

$$\frac{\partial J_3}{\partial S_x} = S_y S_z - \tau_{yz}^2 \qquad \frac{\partial J_3}{\partial S_y} = S_x S_z - \tau_{xz}^2 \qquad \frac{\partial J_3}{\partial S_z} = S_x S_y - \tau_{xy}^2 \qquad (VII.45)$$

$$S_x = \tfrac{1}{3}(2\sigma_x - \sigma_y - \sigma_z) \qquad (VII.46)$$

$$\frac{\partial S_x}{\partial \{\sigma\}} = \tfrac{1}{3}\begin{Bmatrix}2\\-1\\-1\\0\\0\\0\end{Bmatrix} \qquad \frac{\partial S_y}{\partial \{\sigma\}} = \tfrac{1}{3}\begin{Bmatrix}-1\\2\\-1\\0\\0\\0\end{Bmatrix} \qquad \frac{\partial S_y}{\partial \{\sigma\}} = \tfrac{1}{3}\begin{Bmatrix}-1\\-1\\2\\0\\0\\0\end{Bmatrix} \qquad (VII.47)$$

$$\frac{\partial J_3}{\partial \tau_{xy}} = 2\tau_{yz}\tau_{zx} - 2S_z\tau_{xy} \qquad \frac{\partial J_3}{\partial \tau_{yz}} = 2\tau_{xy}\tau_{zx} - 2S_x\tau_{yz} \qquad \frac{\partial J_3}{\partial \tau_{xz}} = 2\tau_{xy}\tau_{yz} - 2S_y\tau_{xz}$$
$$(VII.48)$$

$$\frac{\partial \tau_{xy}}{\partial \{\sigma\}} = \begin{Bmatrix}0\\0\\0\\1\\0\\0\end{Bmatrix} \qquad \frac{\partial \tau_{yz}}{\partial \{\sigma\}} = \begin{Bmatrix}0\\0\\0\\0\\1\\0\end{Bmatrix} \qquad \frac{\partial \tau_{zx}}{\partial \{\sigma\}} = \begin{Bmatrix}0\\0\\0\\0\\0\\1\end{Bmatrix} \qquad (VII.49)$$

Material Modelling Simulation for Finite Element Formulation

$$\frac{\partial J_3}{\partial \{\sigma\}} = \begin{Bmatrix} \frac{1}{3}[2(S_y S_z - \tau_{yz}^2) - (S_x S_z - \tau_{xz}^2) - (S_x S_y - \tau_{xy}^2)] \\ \frac{1}{3}[-(S_y S_z - \tau_{yz}^2) + 2(S_x S_z - \tau_{xz}^2) - (S_x S_y - \tau_{xy}^2)] \\ \frac{1}{3}[-(S_y S_z - \tau_{yz}^2) - 2(S_x S_z - \tau_{xz}^2) + 2(S_x S_y - \tau_{xy}^2)] \\ 2(\tau_{yz}\tau_{zx} - S_z \tau_{xy}) \\ 2(\tau_{xy}\tau_{zx} - S_x \tau_{yz}) \\ 2(\tau_{xy}\tau_{yz} - S_y \tau_{xz}) \end{Bmatrix} \quad \text{(VII.50)}$$

Equation (VII.50) is further simplified as

$$\frac{\partial J_3}{\partial \{\sigma\}} = \begin{Bmatrix} \frac{1}{3}[2S_y S_z - S_x S_z - S_x S_y - 2\tau_{yz}^2 + \tau_{xz}^2 + \tau_{xy}^2] \\ \frac{1}{3}[2S_x S_z - S_y S_z - S_x S_y - 2\tau_{xz}^2 + \tau_{yz}^2 + \tau_{xy}^2] \\ \frac{1}{3}[2S_x S_y - S_y S_z - S_x S_z - 2\tau_{xy}^2 + \tau_{yz}^2 + \tau_{zx}^2] \\ 2(\tau_{yz}\tau_{zx} - S_z \tau_{xy}) \\ 2(\tau_{xy}\tau_{zx} - S_x \tau_{yz}) \\ 2(\tau_{xy}\tau_{yz} - S_y \tau_{xz}) \end{Bmatrix} \quad \text{(VII.50a)}$$

From the flow rule of the normality principle, the following relationship exists between the plastic strain increment and the plastic stress increment:

$$d\{\varepsilon\} = \lambda \frac{\partial f}{\partial \{\sigma\}} \quad \text{(VII.51)}$$

This equation can be interpreted as requiring the normality of the plastic strain increment vector to yield the surface in the hyper space of n stress dimensions. As before $d\lambda$ is the proportionality constant.

For stress increments of infinitesimal size, the change of strain can be divided into elastic and plastic parts, thus (as before)

$$d\{\varepsilon\} = d\{\varepsilon\}_e + d\{\varepsilon\}_p \quad \text{(VII.52)}$$

The elastic increment of stress and strain is related to an isotropic material property matrix $[D]$ by

$$d\{\varepsilon\}_e = [D]^{-1} d\{\sigma\} \quad \text{(VII.53)}$$

From eqns (VII.51)–(VII.53) the following equation is established:

$$d\{\varepsilon\} = [D]^{-1} d\{\sigma\} + d\lambda \left\{ \frac{\partial f}{\partial \sigma} \right\} \quad \text{(VII.54)}$$

The function stresses, on differentiation, can be written as

$$df = \frac{\partial f}{\partial \sigma_1} d\sigma_1 + \frac{\partial f}{\partial \sigma_2} d\sigma_2 + \cdots \qquad 0 = \left\{ \frac{\partial f}{\partial \sigma} \right\}^{T''} d\{\sigma\} \quad \text{(VII.55)}$$

Equation (VII.55) together with eqn (VII.56) can be written in matrix form as

$$d\{\varepsilon\} = [D]_{ep}^{-1} d\{\sigma\} \qquad (VII.56)$$

or

$$\begin{Bmatrix} d\varepsilon_x \\ d\varepsilon_y \\ d\varepsilon_z \\ d\gamma_{xy} \\ d\gamma_{yz} \\ d\gamma_{zx} \\ \hline 0 \end{Bmatrix} = \left[\begin{array}{c|c} [D]^{-1} & \begin{array}{c} \frac{\partial f}{\partial \sigma_x} \\ \frac{\partial f}{\partial \sigma_y} \\ \frac{\partial f}{\partial \sigma_z} \\ \frac{\partial f}{\partial \tau_{xy}} \\ \frac{\partial f}{\partial \tau_{yz}} \\ \frac{\partial f}{\partial \tau_{zx}} \end{array} \\ \hline \frac{\partial f}{\partial \sigma_x} \; \frac{\partial f}{\partial \sigma_y} \; \frac{\partial f}{\partial \sigma_z} \; \frac{\partial f}{\partial \tau_{xy}} \; \frac{\partial f}{\partial \tau_{yz}} \; \frac{\partial f}{\partial \tau_{zx}} & 0 \end{array} \right] \begin{Bmatrix} d\sigma_x \\ d\sigma_y \\ d\sigma_z \\ d\tau_{xy} \\ d\tau_{yz} \\ d\tau_{zx} \\ \hline d\lambda \end{Bmatrix} \qquad (VII.57)$$

Inversion of the above matrix $[D]^{-1}$ will give stresses

$$d\{\sigma\} = [D]_{ep} d\{\varepsilon\} \qquad (VII.58)$$

The explicit form of the elasto-plastic material matrix $[D]_{ep}$ is given by

$$[D]_{ep} = [D] - [D]\left\{\frac{\partial f}{\partial \sigma}\right\}\left\{\frac{\partial f}{\partial \sigma}\right\}^T [D] \left[\left\{\frac{\partial f}{\partial \sigma}\right\}^T [D] \left\{\frac{\partial f}{\partial \sigma}\right\}\right]^{-1} \qquad (VII.59)$$

The value of $\{\partial f/\partial \sigma\}$ has been evaluated above.
The rest of the procedure using finite elements is described in Chapter VI.

VII.4.5 Five-Parameter Model
A full description of this model is given in Section II.1.5. Equation (II.17) predicts failure. The same failure surface in terms of tensile and compressive meridians is given by eqn (II.18). The average normal stress σ_a

is given by eqn (II.19) and in brief is

$$\rho(\sigma_a, \theta) = \frac{(c+t)}{v} \tag{VII.60}$$

$$c = 2(\rho_c^3 - \rho_c\rho_t^2)\cos\theta$$
$$t = (2\rho_t\rho_c - \rho_c^2)[\alpha_1'' + 5\rho_t^2 - 4\rho_t\rho_c]^{1/2}$$
$$v = \alpha_1'' + (\rho_c - 2\rho_t)^2$$

The chain rule of differentiation is used in order to evaluate the gradient $\partial f/\partial \sigma_{ij}$:

$$\frac{\partial f}{\partial \sigma_{ij}} = \frac{\partial f}{\partial \rho}\frac{\partial \rho}{\partial \sigma_a}\frac{\partial \sigma_a}{\partial \sigma_{ij}} + \frac{\partial f}{\partial \rho}\frac{\partial \rho}{\partial \theta}\frac{\partial \theta}{\partial \sigma_{ij}} + \frac{\partial f}{\partial \tau_a}\frac{\partial \tau_a}{\partial \sigma_{ij}} \tag{VII.61}$$

Values for the terms in eqn (VII.61) are given below:

$$\frac{\partial f}{\partial \rho} = \frac{-\tau_a}{f_c'}\frac{1}{\rho^2} \tag{VII.61a}$$

where

$$\frac{\partial \rho}{\partial \sigma_a} = \frac{1}{v}\left[\left(\frac{\partial c}{\partial \sigma_a}\right) + \left(\frac{\partial t}{\partial \sigma_a}\right)\right] - (c+t)\frac{\partial v}{\partial \sigma_a} \tag{VII.61b}$$

$$\frac{\partial c}{\partial \sigma_a} = 2\cos\theta\left[(3\rho_c^2 - \rho_t^2)\frac{d\rho_c}{d\sigma_a} - 2\rho_c\rho_t\frac{d\rho_t}{d\sigma_a}\right] \tag{VII.61c}$$

$$\frac{\partial t}{\partial \sigma_a} = \left[2\rho_c\frac{d\rho_t}{d\sigma_a} + 2(\rho_t - \rho_c)\frac{d\rho_c}{d\sigma_a}\right][\alpha_1'' + 5\rho_t^2 - 4\rho_c\rho_t]^{1/2}$$
$$+ \frac{(2\rho_c\rho_t - \rho_c^2)\left[(+10\rho_t - 8\rho_t\cos^2\theta - 4\rho_c)\frac{d\rho_t}{d\sigma_a} + (8\rho_c\cos^2\theta - 4\rho_t)\frac{d\rho_c}{d\sigma_a}\right]}{2[\alpha_1'' + 5\rho_t^2 - 4\rho_c\rho_t]^{1/2}}$$
$$\tag{VII.61d}$$

$$\frac{\partial v}{\partial \sigma_a} = (8\rho_t\sin^2\theta - 4\rho_c)\frac{d\rho_t}{d\sigma_a} + (8\rho_c\cos^2\theta + 2\rho_c - 4\rho_t)\frac{d\rho_c}{d\sigma_a} \tag{VII.61e}$$

From eqn (II.20) the vectors ρ_c and ρ_t are evaluated.

In eqn (VII.61d) $d\rho_c/d\sigma_a$ is taken from eqn (II.20) as

$$\frac{d\rho_c}{d\sigma_a} = \sqrt{5}b_1 + \frac{2\sqrt{5}b_2\sigma_a}{f_c'} \tag{VII.61f}$$

Similarly

$$\frac{d\rho_t}{d\sigma_a} = \sqrt{5}\,a_1 + \frac{2\sqrt{5}\,a_2\sigma_a}{f'_c}$$

The value of the term $\partial\sigma_a/\partial\sigma_{ij}$ is already known and is equal to $\tfrac{1}{3}S_{ij}$. Once the gradient is known the elasto-plastic formulation described in this chapter and in Chapter VI can be adopted.

VII.5 Creep Model

Experiments have been carried out under various stress combinations for volumetric and deviatoric creep strain behaviour. The stress–strain relationship is given in Part I for creep under multiaxial conditions. The normal modulus of elasticity, E, is replaced by inverse of the uniaxial creep compliance, $J(t' - \tau')$. The elastic Poisson ratio is replaced by the creep Poisson ratio. Hence the multiaxial creep strain as given in Part I may be written as

$$\varepsilon_{ij}^c = J(t'' - \tau')\{(1 + v_c)\sigma_{ij} - v_c^*\delta_{ij}\sigma_{kk}\} \qquad (\text{VII.62})$$

where v_c^* is the creep Poisson ratio along any principal axis. In incremental form over a pseudo-time $\Delta t''$, this may be written as

$$\Delta\varepsilon_{ij}^c = \Delta J(t'' - \tau')\{(1 + v_c)\sigma_{ij} - v_c^*\delta_{ij}\sigma_{kk}\} \qquad (\text{VII.63})$$

where $\Delta J(t'' - \tau')$ is the uniaxial creep compliance over an increment of $\Delta t''$; v_c^* is the creep Poisson ratio; and δ_{ij} is the Kronecker delta. Equation (VII.63) can be written by rearranging it in matrix form as

$$\begin{Bmatrix}\Delta\varepsilon_x^c \\ \Delta\varepsilon_y^c \\ \Delta\varepsilon_z^c \\ \Delta\gamma_{xy}^c \\ \Delta\gamma_{yz}^c \\ \Delta\gamma_{zx}^c\end{Bmatrix} = \Delta J(t'' - \tau')\begin{bmatrix} 1 & -v_c^* & -v_c^* & 0 & 0 & 0 \\ -v_c^* & 1 & -v_c^* & 0 & 0 & 0 \\ -v_c^* & -v_c^* & 1 & 0 & 0 & 0 \\ 0 & 0 & 0 & 2(1 + v_c^*) & 0 & 0 \\ 0 & 0 & 0 & 0 & 2(1 + v_c^*) & 0 \\ 0 & 0 & 0 & 0 & 0 & 2(1 + v_c^*)\end{bmatrix}\begin{Bmatrix}\sigma_x \\ \sigma_y \\ \sigma_z \\ \tau_{xy} \\ \tau_{yz} \\ \tau_{zx}\end{Bmatrix}$$

(VII.64)

$$\Delta\varepsilon^c = \Delta J(t'' - \tau')[D]\{\sigma\} \qquad (\text{VII.64a})$$

where $[D]$ is the material compliance matrix with a constant value of v_c^*. For variable Poisson ratio value and Young's modulus along different axes, reference should be made to matrices $[D]$ in Appendix I with the relevant constituents modified.

The creep strain (or increment of strain) in finite element analysis is treated as 'initial strain'. The kinematically equivalent loads arising from creep strains have been discussed elsewhere. This equivalent load may be written using the creep strain increment as

$$\Delta F^c = \int_{vol} B^{T''} D \, \Delta \varepsilon^c \, d\,vol \quad \text{(VII.65)}$$

or
$$= \int_{-1}^{+1} \int_{-1}^{+1} \int_{-1}^{+1} B^{T''} D \, \Delta \varepsilon^c \det J \, d\xi \, d\eta \, d\zeta \quad \text{(VII.65a)}$$

or
$$= \sum_{i=1}^{nx} \sum_{j=1}^{ny} \sum_{k=1}^{nz} W_i W_j W_k \bar{g} \det J$$

where nx, ny, nz are Gauss points; W_i, W_j, W_k are the weighting coefficients; and $\det J$ is the determinant of the Jacobian:

$$\bar{g} = B^{T''} D \, \Delta \varepsilon_c \quad \text{(VII.65b)}$$

where B = the strain–displacement matrix; D = the elastic or nonlinear material compliance matrix; and $\Delta \varepsilon_c$ = the creep strain increment.

The following steps are performed to include the effects of creep:

1. At time $t' = 0$, elastic or linear analysis is carried out. If thermal loads are considered, calculate the kinematically equivalent load as

$$\{\Delta F\}_i = \int_V B^{T''} D \, \Delta \varepsilon_c \, d\,vol \quad \text{(VII.65c)}$$

   Calculate stresses, strains and other parameters as before.

2. Apply a small time increment, $\Delta t''$, during which the stresses are assumed to remain constant. The creep strain increment for concrete is calculated using eqn (VII.64) of the form

$$\Delta \varepsilon_c = f(\sigma, \Delta t'', t', T) \quad \text{(VII.65d)}$$

3. Convert the creep strain increment into the kinematically equivalent load using eqn (VII.65).

4. The rest of the procedure is given in this chapter and in Chapter VI for crushing, cracking, plasticity and solution procedures.

VII.6 Thermal Stress–Strain Model

The thermal strain in finite element analysis is treated as 'initial strain' similar to the creep strain. The kinematically equivalent load due to temperature rise, $\Delta T'$, has a similar form:

$$\{\Delta F\}^{T''} = \int [B]^{T''} [D] \{\Delta \varepsilon_T\} \, d\,vol$$

$$= \int_{-1}^{+1} \int_{-1}^{+1} \int_{-1}^{+1} [B]^{T''} [D] \{\Delta \varepsilon_T\} \, d|J| \, d\xi \, d\eta \, d\zeta \quad \text{(VII.66)}$$

where $\{\Delta F\}$ is the thermal strain increment of the material. For concrete, as a three-dimensional material, the thermal strain increment is written as

$$\begin{Bmatrix} \Delta\varepsilon_x \\ \Delta\varepsilon_y \\ \Delta\varepsilon_z \\ \Delta\gamma_{xy} \\ \Delta\gamma_{yz} \\ \Delta\gamma_{zx} \end{Bmatrix}^{T''} = \begin{Bmatrix} \alpha_T \Delta T' \\ \alpha_T \Delta T' \\ \alpha_T \Delta T' \\ 0 \\ 0 \\ 0 \end{Bmatrix} \tag{VII.67}$$

For a plane stress condition

$$\begin{Bmatrix} \Delta\varepsilon_x \\ \Delta\varepsilon_y \\ \Delta\gamma_{xy} \end{Bmatrix} = \begin{Bmatrix} \alpha_T \Delta T' \\ \alpha_T \Delta T' \\ 0 \end{Bmatrix} \tag{VII.68}$$

where α_T = the coefficient of thermal expansion for the materials.

Tendons and reinforcing bars are treated as uniaxial material, for which the thermal strains are

$$\{\Delta\varepsilon_x\} = \{\alpha_T \Delta T'\} \tag{VII.69}$$

The rest of the procedure given in Chapter VI is followed, as mentioned in Section VII.4.4.

VII.7 NONLINEAR BOND LINKAGE ELEMENT [262]

The Ahmlink element procedure described in Part I can easily be included in the finite element analysis by adopting the following steps.

(a) Calculate the incremental slip from the nodal displacements:

$$\{\Delta S_i\} = [T'']\{\Delta U_i\} \tag{VII.70}$$

where $[T'']$ is the transformation matrix and $\{\Delta U_i\}$ are the element nodal displacements.

The total slip at iteration i is calculated as

$$\{S_i\} = \{S_{i-1}\} + \{\Delta S_i\} \tag{VII.71}$$

(b) Calculate the incremental stress based on the bond stress at iteration $i-1$:

$$\{\Delta\sigma_{b_i}\} = [E_b^t]\{\sigma_{b_{i-1}}\}\{\Delta S_i\} \tag{VII.72}$$

Total stress is equal to

$$\{\sigma_{b_i}\}_{\text{TOT}} = \{\sigma_{b_{i-1}}\} + \{\Delta\sigma_{b_i}\} \tag{VII.73}$$

(c) Check the state of the bond, i.e. whether the bond is broken or not, and calculate the stress accordingly.

— If $|S_i| > S_{max}$, set flag $I_{flat} = 1$, i.e. bond is broken. At this point, the bond stress is instantaneously dropped to zero, i.e. $\{\sigma_{b_i}\} = 0 \cdot 0$, where $\{S_{max}\}$ is the maximum slip allowed.

— If $|S_i| < S_{max}$, calculate the bond stress which is compatible with the slip, S_i. This is obtained by linear interpolation of a nonlinear bond–slip curve. The curve is simulated by multi-linear lines. Figure VII.4 gives the scheme for the linear interpolation. Let $\{\sigma_b^r\}$ be the bond stress compatible with the slip S_i. The difference between $\{\sigma_{b_i}\}$ and $\{\sigma_b^r\}$ is treated as initial stress and this may be converted into nodal loads on the structure under consideration:

$$\{\Delta \sigma_D\} = \{\sigma_{b_i}\} - \{\sigma_b^r\} \tag{VII.74}$$

The correct stress is written as

$$\{\sigma_{b_i}\}_{cor} = \{\sigma_{b_i}\} - \{\Delta \sigma_D\} \tag{VII.75}$$

(d) Total internal equivalent loads and residuals are calculated as

$$\{F\}_{internal} = \pi \, dL[T]^{T''} \{\sigma_{b_i}\} \qquad \{R\} = \{F\}_{external} - \{F\}_{internal} \tag{VII.76}$$

The rest of the procedure is the same as for the other above cases.

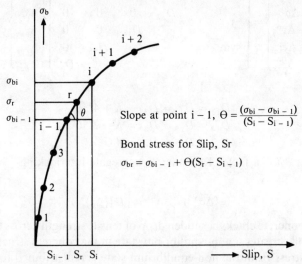

Fig. VII.4. Linear interpretation of nonlinear bond–slip curve.

VII.8 BULK AND SHEAR MODULI MODEL

This has been given in detail in Section II.1.7. The material compliance matrix $[D]$ is given in eqn (II.55). Using various parameters given in eqns (II.56)–(II.62), the stress–strain equation is written as

$$\{\Delta\sigma\} = [D]\{\Delta\varepsilon\} \tag{VII.77}$$

The rest of the procedure for the finite element analysis is given in detail in this chapter as well as in Chapter VI.

VII.9 ENDOCHRONIC CRACKING MODEL

The basic constitutive equations of the endochronic model are given in Section II.1.8, in which deviatoric stress–strain increments are formulated together with stress invariants. The inelastic dilatancy using shear and bulk moduli is included in the derivation of various expressions. Based on the cracking criteria given in Chapter VI, the endochronic constitutive relations can be extended for cracking. Based on the material matrix given in eqn (II.55), the left-hand side of the same equation is modified as

$$\begin{Bmatrix} \Delta\sigma_x + \Delta\sigma_x^p \\ \Delta\sigma_y + \Delta\sigma_y^p \\ \Delta\sigma_z + \Delta\sigma_z^p \\ \Delta\tau_{xy} + \Delta\tau_{xy}^p \\ \Delta\tau_{yz} + \Delta\tau_{yz}^p \\ \Delta\tau_{zx} + \Delta\tau_{zx}^p \end{Bmatrix} = \begin{bmatrix} D_{11} & D_{12} & D_{13} & 0 & 0 & 0 \\ & D_{22} & D_{23} & 0 & 0 & 0 \\ & & D_{33} & 0 & 0 & 0 \\ & & & D_{44} & 0 & 0 \\ \text{sym.} & & & & D_{55} & 0 \\ & & & & & D_{66} \end{bmatrix} \begin{Bmatrix} \Delta\varepsilon_x \\ \Delta\varepsilon_y \\ \Delta\varepsilon_z \\ \Delta\gamma_{xy} \\ \Delta\gamma_{yz} \\ \Delta\gamma_{zx} \end{Bmatrix} \tag{VII.78}$$

where

$$D_{11} = D_{22} = D_{33} = K + 4/3G$$
$$D_{12} = D_{13} = D_{23} = K - 2/3G \tag{VII.79}$$
$$D_{44} = D_{55} = D_{66} = \beta'G$$

(β' = a factor for concrete aggregate interlocking)

or

$$\{\Delta\sigma\} + \{\Delta\sigma^p\} = [D]\{\Delta\varepsilon\} \tag{VII.80}$$

When concrete cracks, a sudden drop of tensile strength across the crack occurs. This creates a non-equilibrium state in the concrete structure. The residual stress from the non-equilibrium state is redistributed to another part of the structure. The material matrix $[D]$ is modified to include the reduction of the stiffness across the crack. The material constitutive relations in the crack co-ordinate system are written as

$$\{\Delta\sigma^*\} = \{\Delta\sigma^{p*}\} = [D^*]\{\Delta\varepsilon^*\} \tag{VII.81}$$

Assuming concrete has a crack normal to any axis, x, then

$$\Delta\sigma_x^* = 0$$

The procedure described in Section II.2 is adopted. The value of $\Delta\sigma_x^p$ is given by

$$\Delta\sigma_x^{p*} = D_{11}\Delta\varepsilon_x^* + D_{12}\Delta\varepsilon_y^* + D_{13}\Delta\varepsilon_z^* = \Delta\sigma_x^{p*}$$

$$\Delta\varepsilon_x^* = \frac{\Delta\sigma_x^{p*}}{D_{11}} - \frac{D_{12}}{D_{11}}\Delta\varepsilon_y^* - \frac{D_{13}}{D_{11}}\Delta\varepsilon_y^* \tag{VII.82}$$

Using eqn (VII.82), the stresses represented by eqn (VII.81) are written in a local crack co-ordinate system as

$$\Delta\sigma_y^* + \Delta\sigma_y^{p*} - \frac{D_{12}}{D_{11}}\Delta\sigma_x^{p*} = \left(D_{22} - \frac{D_{12}D_{21}}{D_{11}}\right)\Delta\varepsilon_y^* + \left(D_{23} - \frac{D_{12}D_{13}}{D_{11}}\right)\Delta\varepsilon_z^*$$

$$\Delta\sigma_z^* + \Delta\sigma_z^{p*} - \frac{D_{13}}{D_{11}}\Delta\sigma_x^{p*} = \left(D_{23} - \frac{D_{12}D_{13}}{D_{11}}\right)\Delta\varepsilon_y^* + \left(D_{33} - \frac{D_{13}D_{31}}{D_{11}}\right)\Delta\varepsilon_z^*$$

$$\Delta\tau_{xy}^* + \Delta\tau_{xy}^{p*} = \beta' D_{44}\gamma_{xy}^* \tag{VII.83}$$

$$\Delta\tau_{yz}^* + \Delta\tau_{yz}^{p*} = D_{55}\Delta\gamma_{yz}^*$$

$$\Delta\tau_{zx}^* + \Delta\tau_{zx}^{p*} = \beta' D_{66}\Delta\gamma_{zx}^*$$

Cracks in other directions are dealt with by modifying the matrix $[D]$ using the same principles. Where cracks are open in all three directions, the value of $[D]$ has all elements equal to zero such that

$$[D] = [0] \tag{VII.84}$$

Other details of crack initiation, closure and reopening, with and without the inclusion of concrete aggregate interlocking, are given in Section II.2 and in Appendix II.

VII.10 HEAT-CONDUCTION MODEL

Steady state and transient heat-conduction models are developed and problems are solved using finite elements. The equations of heat conduction generally include the effects of surface areas with convection and radiation boundary conditions.

VII.10.1 Steady State
The basic thermal equilibrium equation is written as

$$[\bar{K}]\{T\} = \{Q\} \qquad \text{(VII.85)}$$

where $[K]$ = the thermal conductivity matrix; $\{Q\}$ = the heat flow vector; and $\{T\}$ = the vector of the nodal point temperatures.

VII.10.2 Transient
The basic thermal diffusion equation is written as

$$[C]_s\{\dot{T}\} + [K]\{T\} = \{Q\} \qquad \text{(VII.86)}$$

where $[C]_s$ is the specific heat matrix.

For large and complicated structures a substructuring system on the lines suggested in Section VI.2.6 is adopted. In order to distinguish the dynamic transient equations (eqns VI.23 and VI.24) from these, the governing equation for the thermal substructure is written as

$$[C]_s\{\dot{T}\} + [K]_c\{T\} = \{Q\} \qquad \text{(VII.86a)}$$

where

$$[K]_c = [K_{\gamma\gamma}] - [K_{\gamma\gamma'}][K_{\gamma'\gamma'}]^{-1}[K_{\gamma'\gamma}]$$
$$[C]_s = [C_{\gamma\gamma}]_s - [K_{\gamma\gamma'}][K_{\gamma'\gamma'}]^{-1}[C_{\gamma'\gamma}]_s - [C_{\gamma'\gamma}]_s[K_{\gamma'\gamma'}]^{-1}[K_{\gamma'\gamma}]$$
$$\quad + [K_{\gamma\gamma'}][K_{\gamma'\gamma'}]^{-1}[C_{\gamma'\gamma'}]_s[K_{\gamma'\gamma'}]^{-1}[K_{\gamma'\gamma}]$$
$$\{Q\} = \{Q_\gamma\} - [K_{\gamma\gamma'}][K_{\gamma'\gamma'}]^{-1}\{Q_{\gamma'}\}$$
$$\{T\} = \{T_\gamma\}$$

Hence

$$\{T_{\gamma'}\} = [K_{\gamma'\gamma'}]^{-1}\{Q_{\gamma'}\} - [K_{\gamma'\gamma'}]^{-1}[K_{\gamma'\gamma}]\{T_\gamma\} \qquad \text{(VII.86b)}$$

In the above equations the convection surfaces are not included.

The temperature at point X, Y, Z is written as

$$T(X, Y, Z) = \{N\}^{T''}\{T\} \qquad \text{(VII.87)}$$

where $\{N\}$ = the vector of shape functions and $\{T\}$ = the nodal temperature vector.

$\{S\}^T$, the vector of temperature, is related to $\{T\}$ using the definition of $\{S\}^T$ and eqn (VII.87) to give

$$\{S\}^T = [B]\{T\} \qquad \text{(VII.88)}$$

where

$$[B] = \begin{bmatrix} \left\{\dfrac{\partial N}{\partial X}\right\}^{T''} \\ \left\{\dfrac{\partial N}{\partial Y}\right\}^{T''} \\ \left\{\dfrac{\partial N}{\partial Z}\right\}^{T''} \end{bmatrix} \qquad \{S\}^T = \begin{bmatrix} \left(\dfrac{\partial T}{\partial X}\right) \\ \left(\dfrac{\partial T}{\partial Y}\right) \\ \left(\dfrac{\partial T}{\partial Z}\right) \end{bmatrix}_{x,y,z}$$

In thermal terms, the virtual work within one element, i.e. internal work, is written as

$$\Delta U = \int \{\Delta S^T\}^{T''} \{Q\}_{X,Y,Z} \, \mathrm{d\,vol} \qquad (\text{VII.89})$$

where

$$\{Q\} = \begin{Bmatrix} Q_X \\ Q_Y \\ Q_Z \end{Bmatrix} = \text{the vector of heat flow}$$

Q_X = heat flow in the x direction per unit area

Q_Y = heat flow in the y direction per unit area

Q_Z = heat flow in the z direction per unit area

The heat flows are related to the temperature gradients using the following relation:

$$\{Q\} = [K^c]\{S\}^T \qquad (\text{VII.90})$$

where the conductivity matrix $[K^c]$ is given by

$$[K^c] = \begin{bmatrix} K_{XX} & 0 & 0 \\ 0 & K_{YY} & 0 \\ 0 & 0 & K_{ZZ} \end{bmatrix}_c$$

The internal work is written again as

$$\Delta U = \{\partial T\}^{T''} \int_{\text{vol}} [B]^{T''}[D][B] \, \mathrm{d\,vol} \, \{T\} \qquad (\text{VII.91})$$

The virtual internal work associated with convection surfaces is given by

$$\Delta U = \int_A \partial \Delta T \hat{Q} \, \mathrm{d}A \qquad (\text{VII.92})$$

where $\wedge$ = the direction normal to the surface:

$$\partial \Delta T = \partial T(X, Y, Z)|_s \qquad \text{(VII.92a)}$$

The heat flow $\hat{Q}$ over the unit area is given by

$$\hat{Q} = e_f \Delta T \qquad \text{(VII.93)}$$

where e_f = the film coefficient for heat transfer of the surface and $\partial T(X, Y, Z)|_s$ is the temperature function evaluated at the convection surface. In finite element form the above equations are then written as

$$\Delta U = \{\partial T_e\}^{T''} c_f \int_{\text{area}} \{N_s\}^{T''} \, d(\text{area}) \{T_e\} - \{\partial T_e\}^{T''} c_f T_B \int_{\text{area}} \{N_s\} \, d(\text{area})$$
$$\text{(VII.93a)}$$

where $\{N_s\}$ are the shape functions evaluated at the convection surface and T_B is the temperature of the coolant.

Similarly the virtual work associated with specific heat, i.e. thermal damping, is written as

$$\Delta U = \int_V \partial T(X, Y, Z) H \, d\text{vol} \qquad \text{(VII.94)}$$

where

$H = D_c C_p \dfrac{\partial T(X, Y, Z)}{\partial t}$ = total heat change per unit volume per unit time

D_c = density

C_p = specific heat

When D_c, C_p and T do not vary over the element, eqn (VII.94) is written

$$\partial U = \{\partial T\}^{T''} D_c C_p \int_V \{N\}\{N\}^{T''} \, d\text{vol} \frac{\partial}{\partial t} \{T\} \qquad \text{(VII.94a)}$$

If H_r is the heat generation rate per unit volume, the expression for ∂U is written as

$$\partial U = \{\partial T\}^{T''} H_r \int_V \{N\} \, d\text{vol} \qquad \text{(VII.94b)}$$

When all the above effects are combined, the following expression is obtained:

$$[[K] + [K^c]]\{T\} + [C]_s \frac{\partial}{\partial t}\{T_e\} = \{Q^c\} + \{Q^h\} + \{Q\} \qquad \text{(VII.95)}$$

where

- $[K]$ is the material matrix $= \int_V B^{T''} DB \, d\,\text{vol}$
- $[K^c]$ is the conductivity matrix $= e_f \int_A \{N_s\}\{N_s\}^{T''} dA$
- $[C]_s$ is the specific heat matrix $= \rho C_p \int_V \{N\}\{N\}^{T''} dV$
- $\{Q^c\}$ is a total element matrix heat flow matrix per unit area
  $= e_f T_B \int_A \{N_s\} dA$
- $\{Q^h\}$ is a total element matrix heat generated matrix per unit volume
  $= H_r \int_V \{N\} dV$
- t is time

For the nonlinear transient analysis a time integration scheme is adopted on the lines suggested in the dynamic analysis given in Chapter VI.

The usual heat-conduction matrices and thermal load vectors are given below:

$$[K_e] = \begin{bmatrix} K_T + S_1 & -K_T & 0 & 0 & 0 \\ -K_T & 2K_T & -K_T & 0 & 0 \\ 0 & -K_T & 2K_T & -K_T & 0 \\ 0 & 0 & -K_T & 2K_T & -K_T \\ 0 & 0 & 0 & -K_T & K_T + S_2 \end{bmatrix} \quad \text{(VII.96)}$$

where

$$K_T = \frac{4(A)K_{XX}^c}{d}$$

$S_i = C_{i,f}(A)$

$K_{XX}^c =$ thermal conductivity

$\bar{t} =$ thickness

$f =$ film coefficient

$A =$ area

The thermal damping (specific heat) matrix is

$$[C_e]_s = \begin{bmatrix} 2B_t & B_t & 0 & 0 & 0 \\ B_t & 4B_t & B_t & 0 & 0 \\ 0 & B_t & 4B_t & B_t & 0 \\ 0 & 0 & B_t & 4B_t & B_t \\ 0 & 0 & 0 & B_t & 2B_t \end{bmatrix} \quad \text{(VII.96a)}$$

where

$$B = \frac{(A)C_d dD_c}{24}$$

D_c = density
C_p = specific heat (°C)
A = area

The thermal load vector is

$$\{Q\} = \begin{Bmatrix} k_t + \bar{g}_1 \\ 2k_t \\ 2k_t \\ 2k_t \\ k_t + \bar{g}_2 \end{Bmatrix} \qquad (\text{VII.97})$$

where

$$k_t = \frac{(A)H_r t}{8}$$

H_r = heat generation rate
$\bar{g}_i = S_i T_b$
T_b = bulk temperature
A = area

VII.11 MISCELLANEOUS ORTHOTROPIC CONSTITUTIVE MODELS

VII.11.1 Liu, Nilson and Slate [70]

An introduction is given in Section I.3 to this orthotropic constitutive model for biaxial compression of concrete subjected to proportional loading. Liu et al. [70] developed an expression similar to that of Saenz [4] (eqn (I.8)) using specific boundary conditions for biaxial loading and this is given as

$$\sigma_{cu} = \sigma_i = E_0 \varepsilon_{ie}/(1-v)\bar{\alpha}\left[1 + \left(\frac{1}{1-v\bar{\alpha}}\frac{E_0}{E_{sec}} - 2\right)\frac{\varepsilon_{ie}}{\varepsilon_1} + \left(\frac{\varepsilon_{ie}}{\varepsilon_1}\right)^2\right] \qquad (\text{VII.98})$$

where $E_{sec} = \sigma_p/\varepsilon_p$; σ_p and ε_p = the maximum principal stress and corresponding strain; and $\bar{\alpha}$ = the principal stress ratio (eqn (I.18), $\bar{\alpha} = \sigma_2/\sigma_1$).

If $\bar{\alpha}$ is kept at zero, eqn (VII.98) reduces to the uniaxial stress–strain curve suggested by Saenz [4]. This model is extended to include compression–tension and tension–tension cases. They are given below in conjunction with eqns (I.15)–(I.19).

Compression–tension

$$\sigma_1/f_c' = \frac{1}{\bar{\alpha} f_c'/\sigma_t} \qquad (VII.99)$$

Tension–tension

$$\sigma_1/f_c' = \frac{1}{f_c'/\sigma_t} \quad \text{for } 1 \geq \bar{\alpha} \geq 0 \qquad (VII.100)$$

Using the experimental data of Tasuji and co-workers [73, 74], the above equations are extended to cover all states of stress. The values of ε_1 in the biaxial state at maximum stress in the major and minor principal directions are given below:

	Major direction	Minor direction
eqn (VII.99):	$\varepsilon_1 = (360 + 0.59\sigma_1) \times 10^{-6}$	$\varepsilon_1 = (1180 - 2.58\sigma_1) \times 10^{-6}$
eqn (VII.100):	$\varepsilon_1 = 150 \times 10^{-6}$	$\varepsilon_1 = (-26 + 52.2\sigma_1) \times 10^{-6}$

Using the finite element method for the plain stress conditions, the incremental stress–strain curve with the above boundary limits is written as

$$\begin{Bmatrix} \Delta\sigma_1 \\ \Delta\sigma_2 \\ \Delta\tau_{12} \end{Bmatrix} = \begin{bmatrix} \dfrac{E'_{01}}{L_{11}} & \dfrac{vE'_{02}}{L_{11}} & 0 \\ \dfrac{vE'_{02}}{L_{11}} & \dfrac{E'_{02}}{L_{11}} & 0 \\ 0 & 0 & \dfrac{E'_{01}E'_{02}}{E'_{01}E'_{02} + L_{22}} \end{bmatrix} \times \begin{Bmatrix} \Delta\varepsilon_1 \\ \Delta\varepsilon_2 \\ \Delta\gamma_{12} \end{Bmatrix} \qquad (VII.101)$$

where

$$L_{11} = 1 - v^2 \frac{E_{02}}{E_{01}} \qquad L_{22} = \frac{2vE'_{02}}{1 - v^2 E'_{02}/E'_{01}}$$

where E'_{01} and E'_{02} are the tangent moduli of elasticity as a function of the state of stress and strain in the principal stress directions 1 and 2, which are also the principal axes of orthotropy. Equation (VII.98) is considered for the evaluation of the E'_0 values. These values are given in Chapter I for various conditions. Liu et al. [70] obtain the value of E'_0 by differentiating eqn (VII.98) with respect to strain and taking into consideration the microcrack confining effect:

$$E'_{01} = \frac{E_0\left[1 - \left(\frac{\varepsilon_{ie}}{\varepsilon_1}\right)^2\right]}{\left[1 + \left(\frac{1}{1-v\bar{\alpha}}\frac{E_0}{E_{sec}} - 2\right)\frac{\varepsilon_{ie}}{\varepsilon_1} + \left(\frac{\varepsilon_{ie}}{\varepsilon_1}\right)^2\right]} \qquad (VII.102)$$

The value of $i = 1, 2, 3$ for the respective orthotropic axes. The shear modulus G (eqn (VII.101)) has been obtained using the assumption that no interaction exists between the principal normal stress and shear strains caused by the stress-induced anisotropy. In the finite element analysis the value of G is taken from the data given in Chapter I and the normal stresses are obtained from eqn (VII.101). The rest of the method for computing stresses, strains and cracks is given in Chapter VI. The program ISOPAR (described elsewhere) or any other finite element package can be extended to include the above model for the stress–strain relationships of a concrete structure subjected to biaxial loading.

VII.11.2 Darwin and Pecknold [7a,b]

The constitutive model proposed by Darwin and Pecknold [7a,b] has already been introduced in Chapter I. As mentioned earlier, concrete is treated as an orthotropic nonlinear elastic material. The stress–strain relations for use in the finite element analysis are written in incremental form as

$$\begin{Bmatrix} \Delta\sigma_1 \\ \Delta\sigma_2 \\ \Delta\tau_{12} \end{Bmatrix} = \frac{1}{1-v^2} \begin{bmatrix} E'_{01} & v\sqrt{E'_{01}E'_{02}} & 0 \\ & E'_{02} & 0 \\ \text{sym.} & & \frac{1}{4}(E'_{01} + E'_{02} - 2vE'_{01}E'_{02}) \end{bmatrix} \begin{Bmatrix} \Delta\varepsilon_1 \\ \Delta\varepsilon_2 \\ \Delta\gamma_{12} \end{Bmatrix}$$

(VII.103)

The tangent moduli E'_{01} and E'_{02} are evaluated using Saenz's equation (I.8). The failure envelope given by Kupfer and Gerstle [31] is used as described in Section I.3 for computing $\sigma_i = \sigma_{cu}$ and ε'_{i0}. The Poisson ratio given in eqn (I.57) is used for the uniaxial compression and tension–compression states. For various limiting cases, namely compression–compression and compression–tension, reference should be made to Section I.3. In this model the biaxial tension stress state is given by choosing the strain as

$$\varepsilon_{it} = \sigma_t/E_0 \qquad (VII.104)$$

It is estimated that a constant Poisson ratio of 0·2 gives a good account of this model for the stress levels between 75 and 80% of the value of f'_c. Beyond this stress range, the Poisson ratio increases sharply. As explained in the finite element formulation in Chapter VI, the equivalent uniaxial strains are accumulated during a typical loading history in order to determine the tangent moduli E'_{01} and E'_{02} from the stress–strain curves. This work has been extended as described in Chapter I to include cyclic loading capability. The stress–strain relationship given in eqn (VII.103) replaces other relations in the finite element analysis and the rest of Chapter VI is then adhered to for the analysis of concrete structures.

VII.11.3 Elwi and Murray [5]

Elwi and Murray [5] have developed a three-dimensional constitutive model for concrete using the hypoelastic concept given in Chapter II, Section II.1.6.1. Here they have extended it to include orthotropicity. In addition, the model incorporates the equivalent uniaxial strain concept of Darwin and Pecknold [7a,b], the nonlinearity of Saenz [4], and the failure surface of Willam and Wanke [130], which is described in Chapter II, Section II.1.3. The incremental stress–strain relations referred to the orthotropic principal axes 1, 2 and 3 are written in the following form:

$$\{d\sigma\} = [D]\{d\varepsilon\} \quad \text{(VII.105)}$$

in which the tangent material stiffness matrix $[D]$ is given by

$$[D] = \frac{1}{\bar{v}}\begin{bmatrix} D_{11} & D_{12} & D_{13} & 0 \\ & D_{22} & D_{23} & 0 \\ & & D_{33} & 0 \\ & & & D_{44} \end{bmatrix} \quad \text{(VII.106)}$$

The tangent material stiffness matrix given in eqn (VII.106) can easily be compared with similar compliance matrices for concrete given in Chapters II and VI and in Appendix III. Where other cases are included, they can replace eqn (VII.106) in the finite element analysis. The elements of the matrix $[D]$ in eqn (VII.106) are defined below:

$$D_{11} = E'_{01}(1 - v_{32}^2) \qquad D_{12} = \sqrt{E'_{01}E'_{02}}\,[v_{13}v_{32} + v_{12}]$$
$$D_{13} = \sqrt{E'_{01}E'_{03}}\,[v_{12}v_{32} + v_{13}] \qquad D_{22} = E'_{20}(1 - v_{13}^2)$$
$$D_{23} = \sqrt{E'_{02}E'_{03}}\,[v_{12}v_{13} + v_{32}] \qquad D_{33} = E'_{03}(1 - v_{12}^2)$$
$$D_{44} = \bar{v}G_{12} \quad \text{(VII.106a)}$$
$$\bar{v} = 1 - v_{12}^2 - v_{23}^2 - v_{13}^2 - 2v_{12}v_{23}v_{13}$$
$$v_{12}^2 = v_{12}v_{21} \qquad v_{23}^2 = v_{23}v_{32} \qquad v_{13}^2 = v_{13}v_{31}$$
$$G_{12} = \frac{1}{4\bar{v}}[E'_{01} + E'_{02} - 2v_{12}\sqrt{E'_{01}E'_{02}} - (E'_{01}v_{23} + E'_{02}v_{31})^2]$$

Equation (VII.105) is then written in the form

$$\begin{Bmatrix} d\sigma_1 \\ d\sigma_2 \\ d\sigma_3 \\ d\tau_{12} \end{Bmatrix} = \begin{bmatrix} E'_{01}B_{11} & E'_{02}B_{12} & E'_{01}B_{13} & 0 \\ E'_{02}B_{21} & E'_{02}B_{22} & E'_{02}B_{23} & 0 \\ E'_{03}B_{31} & E'_{03}B_{32} & E'_{03}B_{33} & 0 \\ 0 & 0 & 0 & G_{12} \end{bmatrix} \begin{Bmatrix} d\varepsilon_1 \\ d\varepsilon_2 \\ d\varepsilon_3 \\ d\gamma_{12} \end{Bmatrix} \quad \text{(VII.107)}$$

Using the equivalent uniaxial strain ε_{ie} ($i = 1, 2, 3$), the right-hand side strain matrix is then

$$d\varepsilon = B_{i1}\,d\varepsilon_1 + B_{i2}\,d\varepsilon_2 + B_{i3}\,d\varepsilon_3 \quad \text{(VII.108)}$$

In order to be truly three-dimensional, additional values of G_{23} and G_{31} are included in $[D]$ of eqns (VII.106) and (VII.107) as given in Appendix III. Throughout the finite element analysis the above material matrix $[D]$ is adopted.

It is interesting to note that the biaxial stress–strain relationship (eqn (I.8)) suggested by Saenz [4] is generalised for assessing tensile and compressive response. Equation (I.8a) represents the equivalent uniaxial strain. The incremental moduli in eqn (VII.107) can then be determined by differentiating eqn (I.8a) which is given in eqn (I.66). The Poisson ratio v_i used is given in eqn (I.66). The maximum value of v_i is 0·5 corresponding to zero incremental volume change. The five-parameter model can be used in conjunction with the above equations to obtain the three values of σ_i for each stress ratio. The values of ε_{ie} corresponding to σ_i values are determined by assuming a surface in the equivalent uniaxial strain space that has the same form as the ultimate strength surface. The rest of the finite element procedure given in Chapter VI can then easily be adopted for analysing concrete structures. In this regard reference should be made to the finite element formulation given in Chapter VI and the associated appendices.

VII.11.4 Strain Hardening Models
VII.11.4.1 Chen and Chen [148]
This constitutive model for concrete has been developed using the principle of work hardening plasticity; concrete is modelled as a continuous, isotropic and linearly elasto-plastic strain hardening fraculure material. The transition from the elastic to plastic stage is represented by a discontinuous point on an initially discontinuous surface. Unloading and loading in the plastic range are assumed to follow the initial elastic modulus. At a limiting tensile stress or strain, a crack is assumed to occur in a plane normal to the direction of the stress or strain; under compression

the concrete becomes perfectly plastic at the ultimate stress. The surface fails or fractures when crushing occurs at a limiting crushing strain or stress. The failure surface is simulated using biaxial test data for two different conditions:

Compression–compression

$$F_u(\sigma_{ij}) = J_2 + \tfrac{1}{3}A_u I_1 - \tau_u^2 = 0 \qquad \text{(VII.109)}$$

Tension–tension or tension–compression

$$F_u(\sigma_{ij}) = J_2 - \tfrac{1}{6}I_1^2 + \tfrac{1}{3}A_u I_1 - \tau_u^2 = 0 \qquad \text{(VII.110)}$$

where A_u and τ_u are material constants generally determined from tests.

Using a combination of isotropic and kinematic hardening rules in which the loading surface experiences translation and uniform expansion, eqns (VII.109) and (VII.110) are rewritten as

Compression–compression

$$F(\sigma_{ij}) = \frac{J_2 + (\beta''/3)I_1}{1 - (\alpha/3)I_1} - \tau^2(\varepsilon_p) = 0 \qquad \text{(VII.111)}$$

Tension–tension and tension–compression

$$F(\sigma_{ij}) = \frac{J_2 - (1/6)I_1^2 + (\beta''/3)I_1}{1 - (\alpha''/3)I_1} - \tau^2(\varepsilon_p) = 0 \qquad \text{(VII.112)}$$

where α'' and β'' are constants; $\tau = \tau_u$ for the failure surface; $\tau = \tau_0$ for the yield surface; and $\tau = f(\varepsilon_p) = $ a function of plastic strain.

The finite element method described in Chapter VI is used for the stress–strain relations with a particular associated flow rule.

VII.11.4.2 Buyukozturk [71]

The strain hardening model proposed by Buyukozturk [71] predicts yield and failure of concrete under multiaxial stress states using the Mohr–Coulomb internal friction theory. The failure law is based on biaxial experimental data. For biaxial compression and compression–tension states, the following expression is proposed:

$$f(I_1, J_2) = 3J_2 + \sqrt{3}\,\beta'' \tau I_1 + \alpha'' I_1^2 - \tau^2 = 0 \qquad \text{(VII.113)}$$

where α'', β'' and τ are experimental constants. The experimental constants are evaluated, and upon insertion eqn (VII.113) is rewritten as

$$3J_2 + f'_c I_1 + \frac{I_1^2}{5} = \frac{f'^2_c}{9} \qquad \text{(VII.114)}$$

The initial yield surface to a stress of one-third of the ultimate strength $(f'_c/3)$ is given by

$$3J_2 + f'_c/3I_1 + \frac{I_1^2}{5} = \frac{(f'_c/3)^2}{9} \qquad \text{(VII.115)}$$

This now defines the motion of the subsequent yield surfaces during plastic deformation. The flow rules described in Chapter VI can now be invoked to obtain the incremental stress–strain relations. The rest of the procedure in the finite element formulation is then followed for the analysis of concrete structures.

PART III

This part considers the application of the analyses to a variety of concrete structures, ranging from beams, slabs and concrete-filled columns to containment vessels, offshore platforms, LNG tanks, etc. It is supported by relevant appendices (I–X).

PART III

Chapter VIII

Application to Engineering Problems

VIII.1 INTRODUCTION

Various analyses and numerical models given in earlier chapters are now applied to selected engineering problems. As far as possible the analytical results produced for various case studies have been compared with those available from experimental results and other published data. For all experiments concrete cubes and cylinders were tested under uniaxial, biaxial and triaxial loading conditions. Appendix VI gives a summary of the experimental and theoretical results.

VIII.2 CASE STUDY No. 1—REINFORCED AND PRESTRESSED CONCRETE BEAMS

VIII.2.1 Reinforced Concrete Beams

A model reinforced concrete beam with reinforcement details is shown in Fig. VIII.1. Cube crushing tests showed that the concrete mix is of grade 30 ($f_{cu} = \sigma_{cu} = 30\,\text{N/mm}^2$). Separate tensile tests were carried out on the reinforcements. The characteristic strength of a steel bar was determined as $410\,\text{N/mm}^2$. The beam was placed on simple supports as shown in Fig. VIII.2. Dimensions and the positions of studs on strain gauges were recorded. The load was applied in increments to failure. The first load recorded was due to self-weight of the beam and the equipment. The second was of 10 kN, and all other loads were in increments of 5 kN. At the end of each load increment, the inclinometer, demec and strain gauge readings were recorded. Figures VIII.3–VIII.6 show the values of strains for various depths at all four stud columns shown in Fig. VIII.2. Figure VIII.7 shows

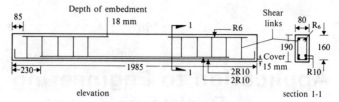

Fig. VIII.1. Reinforcement details.

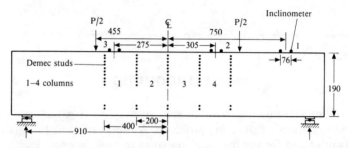

Fig. VIII.2. Position of supports, loads, and demec and inclinometer studs.

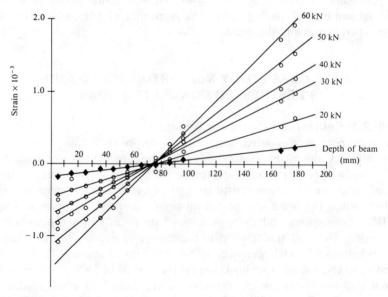

Fig. VIII.3. Strains for column 1. (——, Experimental; —O—, theoretical; —◆—, additional theoretical, 10 kN.)

Application to Engineering Problems

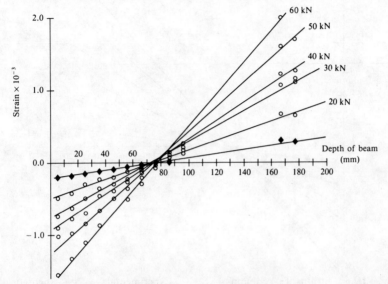

Fig. VIII.4. Strains for column 2. (———, Experimental; —O—, theoretical; —◆—, additional theoretical, 10 kN.)

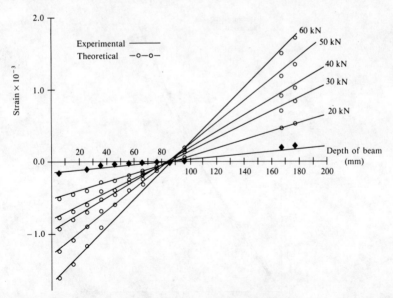

Fig. VIII.5. Strains for column 3. (———, Experimental; —O—, theoretical; —◆—, additional theoretical, 10 kN.)

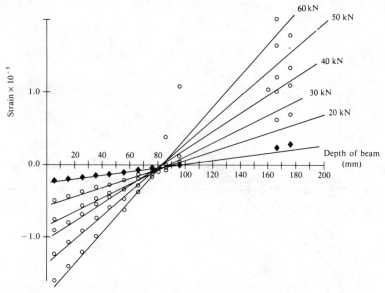

Fig. VIII.6. Strains for column 4. (———, Experimental; —○—, theoretical; —◆—, additional theoretical, 10 kN.)

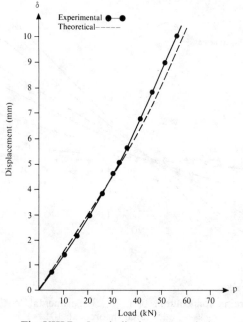

Fig. VIII.7. Load–displacement relation.

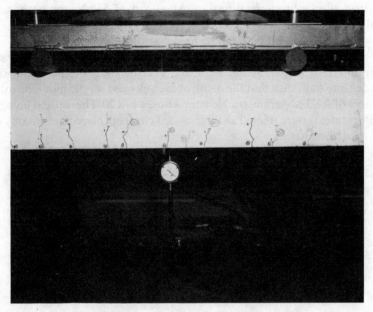

Fig. VIII.8a. Cracking of a beam (at 40 kN).

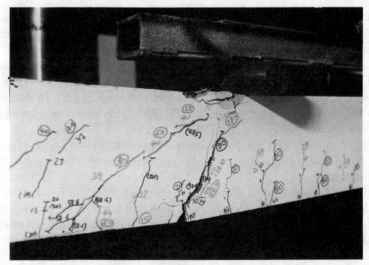

Fig. VIII.8b. Cracking and crushing (at 55 kN).

the load–displacement relationship. At a load of 55 kN, the beam was finally crushed.

Figures VIII.8a and b show the final cracking stages.

Figure VIII.9 shows the finite element mesh of the same concrete beam with 20-noded isoparametric elements. The thickness of the beam was divided into half, such that the depth of each element was 95 mm. The total number of solid isoparametric elements chosen was 20. The top and bottom reinforcements were treated as three-noded elements placed in the body of the solid elements.

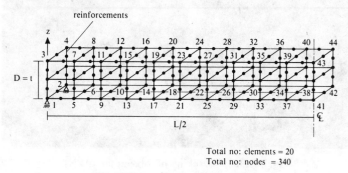

Total no: elements = 20
Total no: nodes = 340

Fig. VIII.9. Finite element mesh.

A three-dimensional finite element analysis, as given in Chapters VI and VII, was carried out using a five-parameter model and the cracking criteria given in Sections II.1.5 and VII.4.5. Both E and v values were assumed to change along the three principal axes with initial values of $200 \, GN/m^2$ and 0·15, respectively. Concrete aggregate interlocking factor $\beta' = 0.5$ was considered in the final evaluation of the material matrix $[D]$. In the final element analysis using the program ISOPAR, the load was applied in increments of 5 kN.

Figures VIII.3–VIII.6 show the comparative values of strain for various depths. These results are in good agreement with those from the experiment.

The load–displacement relationship obtained from the finite element analysis again gives a good correlation with that given by the experiment. These results are plotted in Fig. VIII.7. The failure load predicted by the finite element analysis was about 7% higher than that given by the experiment. Figure VIII.10 shows a comparative study of crack patterns from both theory and experiment. The finite element analysis predicted the

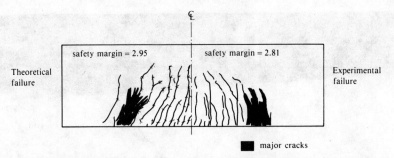

Fig. VIII.10. Comparative results of theoretical and experimental failure.

maximum stress in the reinforcement as 395 N/mm² in the bottom bars. One of the bottom bars in the experiment was severely damaged with a V-notch. The safety margins against the design load predicted by the analysis and the experiment were found to be 2·95 and 2·81, respectively.

VIII.2.2 Prestressed Concrete Beams

In this case the size of the beam is the same as in the case of the reinforced concrete beam given in Fig. VIII.1. This beam was recast with certain minor changes, as shown in Fig. VIII.11. These changes are the span of 1800 mm between the supports and the effective depth of 127 mm due to two straight ducts for prestressing tendons.

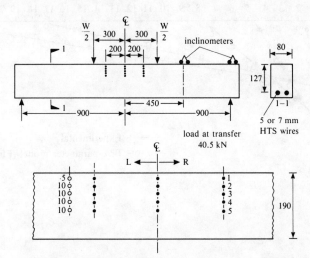

Fig. VIII.11. Details of the prestressed concrete beam.

Fig. VIII.12. Collapse mode of the beam.

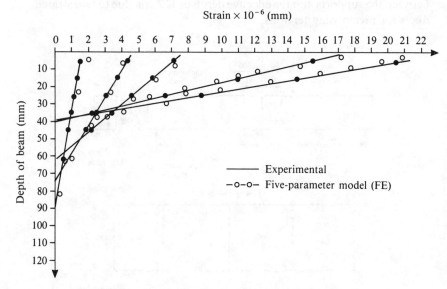

Fig. VIII.13. Strain distribution along the depth of the beam.

The beam was pretensioned by means of a hydraulic jack. The prestressing force at transfer per tendon was 40·5 kN. Grade 30 concrete was selected for this beam. First, ignoring applied load, displacements, rotations and strains were measured due to dead loads of the equipment and the self-weight of the beam. Loads at two intervals of 5 kN were applied. Afterwards the beam was subjected to loads of five increments each of 1 kN. The last four loads prior to failure of the beam were in increments of 0·5 kN. The beam finally failed at about 17·5 kN. Figure VIII.12 shows the collapse mode of the beam. Figure VIII.13 shows the strain distribution along the depth of the beam plotted for 5, 10, 12, 15 and 17·5 kN loads. Figure VIII.14a shows the final load–displacement curve, given by a solid line.

In the finite element analysis, the same mesh as given in Fig. VIII.10 was adopted. This time the line elements were placed at 127 mm from the top of the beam. As the tendons were grouped, the Ahmlink element (representing bond) was used along with 20-noded solid and three-noded line elements. The material properties were the same except for the stress–strain relation of the prestressing tendons. Only the strain distribution results using the five-parameter model are plotted along with the experimental results in Fig. VIII.13. The results are in good agreement. The load–displacement relation using endochronic, four- and five-parameter models are plotted in Fig. VIII.14a. The endochronic theory gave close results above the 12 kN load,

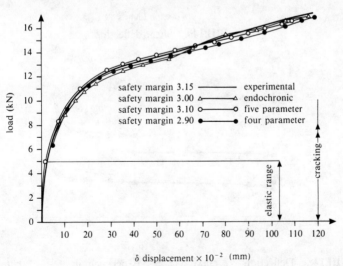

Fig. VIII.14a. Load–displacement relation.

otherwise it showed a lower bound effect. The five-parameter model bore close resemblance to the experimental behaviour of this beam. The four-parameter model, as expected, gave results which indicated a lower bound phenomenon as soon as the beam deformation becomes nonlinear. In all cases, as is evident from Fig. VIII.14a, the collapse load was slightly lower than that obtained from the experiment. The difference ranges between 1 and 5%. This difference, although acceptable, could have been due to non-optimisation of the mesh, assumptions made in failure models, and the assumptions for the boundary conditions of the experimental model.

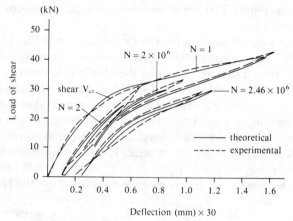

Fig. VIII.14b. Shear deflection.

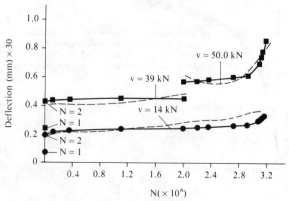

Fig. VIII.14c. Deflection–N curve. (■–■, Experimental; –––, theoretical (ISOPAR).)

Material properties at the time of experiment might be slightly different from those considered in the finite element analysis.

The idealised stress–strain curve for tendons given in Appendix IX had been adopted for this particular problem. The program ISOPAR has the flexibility to include other types of stress–strain curves and material properties. For this particular problem concrete aggregate interlocking was ignored. The crack pattern given in Fig. VIII.11 is identical for this beam as well; only the major crack is obtained at a much lower load.

Next the same beam was studied in order to evaluate the repeated-load shear strength. The beam was subjected to a symmetrical two-point loading and was loaded statically in increments to approximately 80% of the flexural capacity determined above. This was sufficient to cause significant inclined cracking. After unloading (the procedure described in Chapters VI and VII) the beam was then subjected to additional cycles of statically applied loads (4 cycles/s). The repeated load was stopped at 2×10^6 cycles of its design repeated-loading. The load increments adopted were equal to 5% of the flexural capacity of the beam. The maximum crack width computed was 10 mm. Figure VIII.14b shows the load–deflection curve. Figure VIII.14c shows the deflection–cycle curve.

VIII.3 CASE STUDY No. 2—REINFORCED CONCRETE SLABS

VIII.3.1 Reinforced Concrete Rectangular Slabs Subject to Concentrated Loads

The dimensions of a reinforced concrete slab made of grade 30 concrete are shown in Fig. VIII.15a. The slab is 610 mm × 910 mm × 52 mm thick. It is reinforced at the bottom only with a light steel mesh reinforcement (6 mm diameter at 80 mm c/c). Based on several tensile tests, the yield strength of the mesh is around 413 N/mm^2. This reinforced concrete slab, as shown in Fig. VIII.15b, was simply supported at the corners. The increments of the point-load were applied centrally. Four deflection gauges were located 150 mm from the corners. Table VIII.1 shows the proofing ring calibration. The slab was subjected to incremental loadings. At a certain stage of loading between 10 and 11 kN, the tensile stress in the concrete reached the modulus of rupture and cracking started in the zones of maximum tensile stress. The deflections recorded immediately after this stage were nonlinear. Excessive cracks were formed around 16 kN. As shown in Fig. VIII.16 the major cracks were formed in the central area of the slab and the corners of the

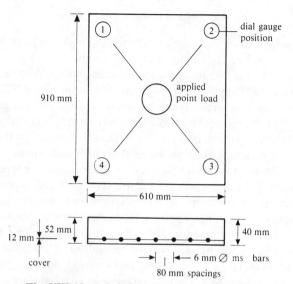

Fig. VIII.15a. Reinforced concrete slab model.

Fig. VIII.15b. A reinforced concrete slab on simple supports with a proofing ring and dial gauges.

Fig. VIII.16. Major cracks in the model slab.

Table VIII.1
Proofing ring calibration

Load (kN)	Division	Proofing ring calibration
10	91	
20	182	
30	270	
40	360	
50	450	0·110 67 kN
60	540	per division
70	635	
80	720	
90	820	
100	910	

newly developed crack zone of a square shape (240 mm square) were linked to the corners of the slab by the radial cracks. The preferential cracks around these major cracks started developing and propagating in large numbers at a load of 14·94 kN. The slab failed completely at a load of 18·59 kN. It was a plug type of failure. The reinforcement mesh was examined. At the level of major cracks the mesh was either yielded or completely ruptured. The load–displacement relationship for this slab is given in Figs VIII.17a and VIII.17b together with the relationship obtained using finite elements. The concrete modelling was based on the five-parameter concept.

Three-dimensional isoparametric 20-noded solid elements together with three-noded line elements placed in the body of the element were considered for this slab. The finite element mesh scheme for this slab is given in Fig. VIII.18. Three failure models, namely endochronic, five-parameter and hypoelastic were used. The load–displacement relation was plotted for each of these failure models. The collapse loads obtained from the finite element analysis and with the use of the four- and five-parameter models were 17·5 and 16·80 kN, respectively. The collapse load from the endochronic theory was 19·5 kN. The three failure models associated with the finite element analysis gave identical crack patterns. There were changes in the direction of such cracks but these were confined to only 10 small peripheral cracks. In all cases, the crack patterns obtained from the finite element analysis were close to the ones shown in Fig. VIII.19. Comparative study of the theoretical and experimental crack patterns given in Figs VIII.16 and VIII.19 shows marked similarities in the initiation of major cracks. No attempts have been made gradually to remove the load and to

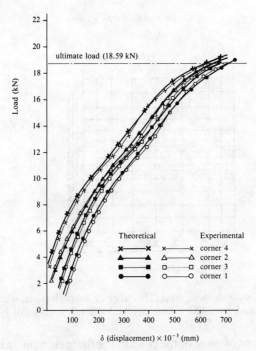

Fig. VIII.17a. Load versus deflection at corners. (Note: The slab has four corners where displacements are evaluated.)

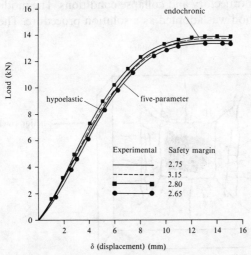

Fig. VIII.17b. Load–displacement relationship for a square slab under patch loads.

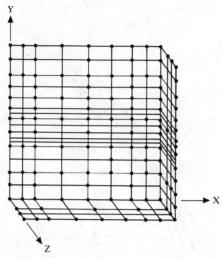

Fig. VIII.18. Finite element mesh. (Number of solid elements = 384; number of nodes = 7400; Gauss points $2 \times 2 \times 2$; tol > 0·001.)

assess the closure of some of the cracks. The finite element analysis showed stresses in the reinforcement well above 20% of the yield stress at the time of collapse. The loads were applied in increments of 2 kN up to 10 kN, and afterwards small increments of 0·5 kN were considered in order to assess the final stress trajectory and collapse conditions. The modified Newton–Raphson method was adopted as a solution procedure. The convergence

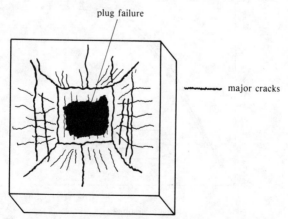

Fig. VIII.19. Crack pattern from finite element analysis.

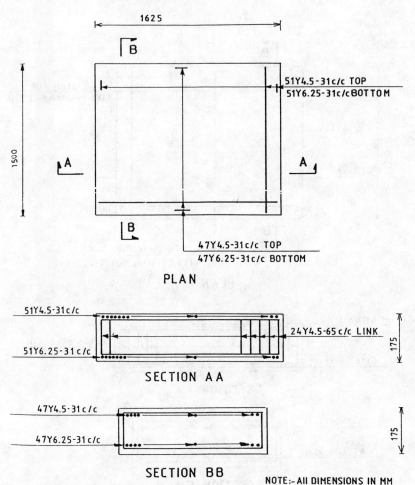

Fig. VIII.20. Detail of target slab B16 (courtesy of UK Atomic Energy Authority, Winfrith).

criterion described in Chapters VI and VII has been adopted so as to converge the solutions of the nonlinear equations.

VIII.3.2 Reinforced Micro-Concrete Square Slabs Subject to Patch Loads

An identical test was carried out with a micro-concrete slab subjected to four patched loads. The slab was 510 mm square and 25 mm thick

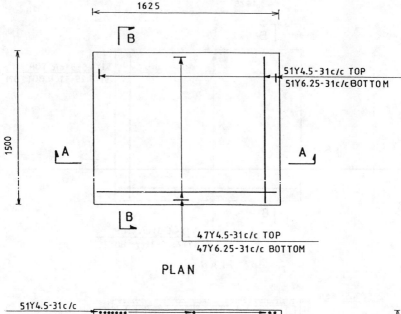

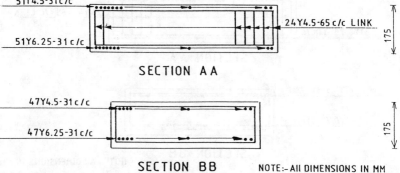

Fig. VIII.21. Detail of target slab B16 (courtesy of UK Atomic Energy Authority, Winfrith).

reinforced with 1·55 mm mesh with 13 mm spacings. Figure VIII.17b shows a comparative study of the load–displacement relationship using the finite element mesh scheme of Fig. VIII.18.

VIII.3.3 Reinforced Concrete Slabs Under Impact Loads

Two other reinforced concrete slabs were tested under impact loads by the UK Atomic Energy Authority at Winfrith [769–772]. Figures VIII.20 and

Table VIII.2
Data for slabs subject to impact loads

Test No. M126
Missile: collapsing steel tube type $11\frac{1}{4}$
Impact velocity: 237 m/s
Target: type $16\frac{1}{4}$ 1·625 m × 1·5 m × 0·175 m
   bending reinforcement   ⌀4·5 mm impact face, ⌀6·25 mm rear face on 31-mm pitch
   shear reinforcement   ⌀4·5 mm 0·37% of plan area in central region

Concrete strength measurements:

Age (days)	28	
152 mm cubes	34·4 MPa std dev.	2·0 MPa
75 mm cubes	32·0	2·6 MPa
75 mm × 75 mm × 300 mm beams	3·0	0·4 MPa
Brazilian	2·5	0·3 MPa
⌀152 mm × 300 mm cylinders	33·7	2·0 MPa
Panel age at date of test	28 days	

Test No. M289
Missile: type $11\frac{1}{4}$, mass 15·6 kg
Impact velocity: 240 m/s
Target: B26($\frac{1}{4}$)   1500 mm × 1625 mm × 150 mm
   bending reinforcement   ⌀4·5/⌀6·25 mm at 37/28 mm front/rear
   shear reinforcement   ⌀5·0 mm at 62 mm

Concrete strength measurements:

Age (days)	35
152 mm cubes	45·1 + 1·6 MPa
75 mm cubes	61·3 + 2·9 MPa
75 mm × 75 mm × 300 mm beams	3·8 + 0·4 MPa
Brazilian	3·8 + 0·9 MPa
⌀152 mm × 300 mm cylinders	37·0 + 4·7 MPa
Panel age at date of test	36 days

VIII.21 show the reinforcement layout for the slabs. Table VIII.2 shows the data from the impact test. Figure VIII.22 shows a typical soft missile used as an impactor, and Fig. VIII.23 shows a load–time function.

A three-dimensional finite element analysis has been carried out using the program ISOPAR.

Three-dimensional 20-noded isoparametric elements have been used to represent the concrete. Each edge of this isoparametric element forms a parabola, so that eight nodes define the corner of the element and a further 12 nodes define the position of the 'mid-point' of each edge. This element

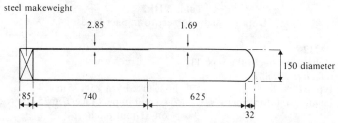

Fig. VIII.22. Dimensions of soft missiles, type 12, used for tests M126 and M289 (all dimensions in mm) (courtesy of UK Atomic Energy Authority, Winfrith).

has 27 Gaussian integration points located in three layers within the element. There are three global co-ordinates in the x-, y- and z-directions, and there are three global degrees of freedom, U, V and W. There are six components of stress and strain.

The reinforcement is represented by the three-dimensional line element. This element is an isoparametric, three-dimensional empty block which contains reinforcing bars running in patterns designed by the user through the user subroutine, LINE. These elements are used in conjunction with 20-noded elements, which represent the concrete. The combination of these

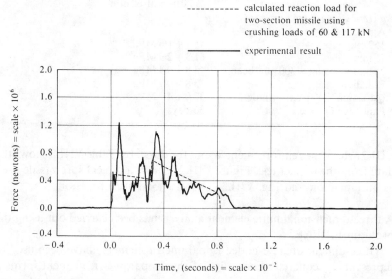

Fig. VIII.23. Load–time function for type 11 missile load at 236 m/s impact (courtesy of UK Atomic Energy Authority, Winfrith).

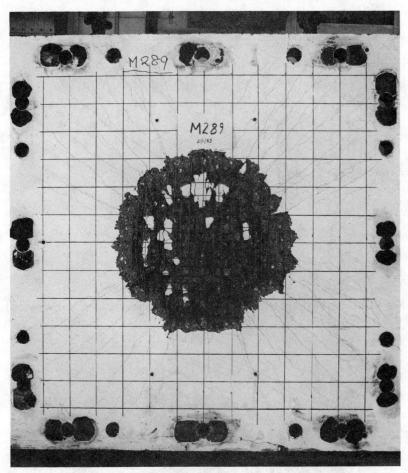

Fig. VIII.24. Bottom face of target B26 after impact test.

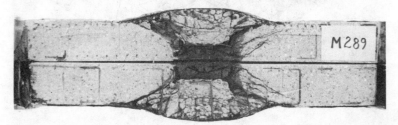

Fig. VIII.25. Cross-section through target B26 after impact test.

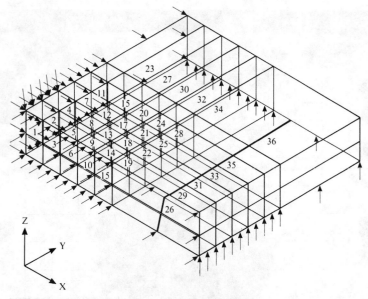

Fig. VIII.26a. Model of one-quarter of the slab showing support condition load application.

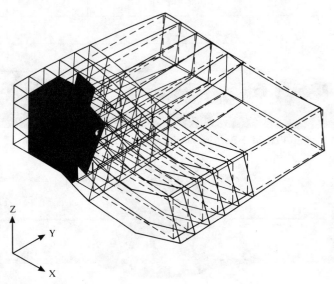

Fig. VIII.26b. Mode 2 of target B16—plugged-out zone in three dimensions.

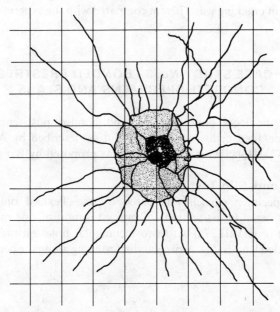

Fig. VIII.26c. Final element results for shear cone and cracking.

two elements approximates the reinforced concrete behaviour, since, in the isoparametric elements, each edge of this element forms a parabola. The line elements are integrated using a numerical scheme based on Gauss quadrature. Each layer contains nine integration points. In each of these elements there are five layers and a total of 45 integration points. The Newmark direct integration method ($\alpha = \frac{1}{2}$; $\beta = \frac{1}{4}$) was applied. The force was applied keeping in mind the size of the time step, the numerical damping factor and the total transient time. Nodal displacements, velocities, accelerations, principal stresses and strains, plastic strains and crack strains were computed. Concrete modelling adopted was based on Ottoson's four-parameter model and the cracking criteria adopted in this book. The total number of elements, 509, was chosen using a close mesh at the centre. Figures VIII.24 and VIII.25 show the results from the experiment. Figures VIII.26a, VIII.26b and VIII.26c summarise the results obtained from the program ISOPAR. The maximum penetration found on the face at which the impact occurred was 17·5 mm compared with the experimental value of 20 mm. On the rear face the diameter of the concrete cover scabbed was 700 mm compared with the experimental value of 650.

The spread of cracking was 370 mm compared with the experimental value of 300 mm.

VIII.4 CASE STUDY No. 3—BONDED PRESTRESSED CONCRETE SPECIMENS AND SLABS

Studies of the local bond strength and bonded post-tensioned slab specimens detailed below were carried out as described by Ahmad and Bangash [873], together with modifications proposed by Bangash [749].

VIII.4.1 Bond–Slip Test Specimen

The test specimen chosen was based on the 'flexural only case'. A rectangular beam (cross-section 152 mm × 152 mm) of one metre length was chosen (Fig. VIII.27a) with two rectangular holes on each side of a 38 mm bonded length (L_b). A circular duct running along the beam through

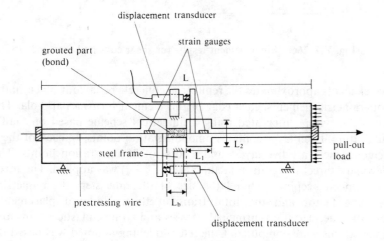

dimensions:
L = 1000.0 mm
L_b = 38.0 mm
L_1 = 76.0 mm
L_2 = 38.0 mm

Material data
$E_s = 200 \times 10^3$ N/mm^2
$E_c = 27.5 \times 10^3$ N/mm^2
f_{cu} = 45 N/mm^2
$v = 0.17$

σ_y (conventional steel) = 297 N/mm^2
σ_y (high tensile wire) = 1340 N/mm^2

Fig. VIII.27a. Details of prestressed concrete pull-out specimen [262].

the centre of the cross-section was cast void for later insertion of the prestressing wire. This duct between rectangular holes in the central part of the specimen was grouted using cement grout after post-tensioning.

Two strain gauges (gauge factor 2·5) were mounted onto the wires 50 mm from the centre of the wires prior to the insertion of the wires in the ducts. The strains on the wires were recorded using a Peekel strain indicator (Fig. VIII.27b). Four high-precision displacement transducers were placed at

Fig. VIII.27b. General views of the test specimen with transducers and load cells (after [873] and [262], with permission).

both ends of the bonded length (with two at each end). The slip of the wires was measured using these transducers with a measurement accuracy of 10^{-5} mm. Slip was directly recorded from the transducer measuring unit. The load was applied using a hydraulic jack and the load increment was recorded using a load cell. The load was applied in increments of 0·25 kN. To study the influence of various parameters on the bond between the prestressing wire and the grout, two main parameters were investigated: the magnitude of the prestress force and the wire diameter. Wires of 5 and 7 mm diameter were prestressed to 65 and 75% of the guaranteed ultimate tensile strength (GUTS). Ten beam specimens were tested. Table VIII.3 shows the prestressing loads and wire diameters for these beams. The slip was obtained by taking an average value of the transducer readings. Table VIII.4 gives the results of bond test.

Table VIII.3
Parameters of bond–slip specimen

Beam number	Diameters (mm)	Amount of prestress (% GUTS)
1	5	75
2	5	75
3	5	75
4	7	75
5	7	75
6	7	75
7	7	65
8	7	65
9	5	65
10	5	65

The finite element analysis was carried out using three-dimensional isoparametric elements representing concrete and line elements for tendons with three nodes in conjunction with an Ahmlink element. The mesh elements were similar to those shown in Fig. VIII.9 but with reduced sizes. The load was applied incrementally; stresses, strains and cracks were produced at the onset of the nonlinear behaviour of the specimen. Bond stress–slip relationships were computed, for all the specimens, using the program N-SARVE [262]. The analytical results are plotted in Figs VIII.28–VIII.31. In some cases the results obtained from the analysis disagreed with those from the experiments. This is because of the

Table VIII.4
Bond test results

Beam number	Diameter (mm)	Prestress force (% GUTS)	σ_{bmax} (N/mm^2)	Average σ_{bmax} (N/mm^2)	Slip (mm)	Average slip (mm)	E_h (N/mm^2/mm)	Average E_h (N/mm^2/mm)
1		75	1·6		0·044		380	
2	5	75	1·64	1·64	0·052	0·048	540	473·3
3		75	1·68		0·05		500	
9		65	2·2	2·21	0·019	0·0205	509	526·0
10		65	2·22		0·022		543	
4		75	1·84		0·055		510	
5	7	75	1·81	1·8	0·065	0·057	520	523·3
6		75	1·75		0·052		540	
7		65	2·1	2·05	0·026	0·025	466	483·0
8		65	2·0		0·024		500	

σ_{bmax} = ultimate (maximum) bond stress.
E_h = initial slope or slip modulus (average value of E_h = 501·4 N/mm^2/mm).

complicated nature of the steel–concrete interface factors, such as local shrinkage and settlement of grout, which were not taken into the finite element analysis carried out by the program ISOPAR.

Figure VIII.31 shows the results for beams 7 and 8 in which 7 mm diameter wire was used with a prestress of 65% GUTS. In this case, as shown in Fig. VIII.31, a relatively large scatter of results was obtained for maximum bond stress. An average value of maximum bond stress was found to be 2·30 N/mm^2 (2·05 N/mm^2 from experiment) at an average slip of 0·025 mm when bond stress stops increasing. In this case curves indicate less nonlinearity, and maximum bond stress is relatively larger than that for beams 3, 4 and 5. The maximum slip at which the bond stress stops increasing is less here than for beams 3, 4 and 5. Figure VIII.29 shows the results for beams 9 and 10 in which 5 mm diameter wire with a prestress of 65% GUTS was used. The average maximum bond stress is found to be 2·40 N/mm^2 (test 2·21 N/mm^2). The slip at which this maximum bond stress was obtained is 0·0205 mm. Comparing these results with those of beams 1, 2 and 3 it can be seen that the maximum bond stress is larger with a smaller slip.

From the above results it may be concluded that the maximum bond stress decreases with the increase of prestressing force in the wire and at a lower slip value. The effect on maximum bond stress due to the variation of steel wire diameters is negligible. For any definite conclusions about large

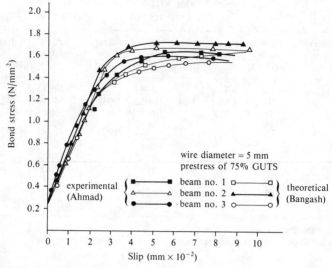

Fig. VIII.28. Experimental bond stress–slip curve of 5 mm diameter prestressing wire.

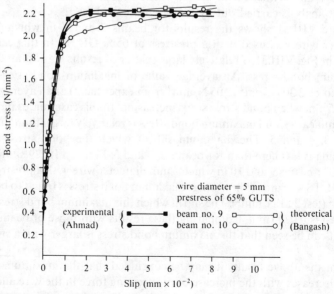

Fig. VIII.29. Experimental bond stress–slip curve of 5 mm diameter prestressing wire.

Application to Engineering Problems

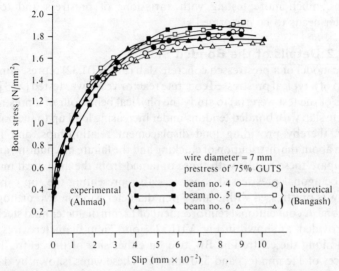

Fig. VIII.30. Experimental bond stress–slip curve of 7 mm diameter prestressing wire.

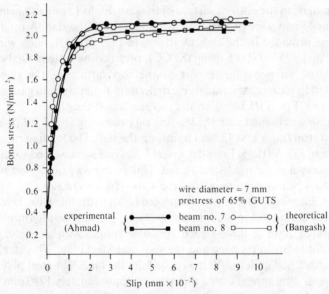

Fig. VIII.31. Experimental bond stress–slip curve of 7 mm diameter prestressing wire.

tendons, much more testing with variations of prestress and tendon diameter needs to be performed.

VIII.4.2 Details of the Bonded Slab

A scale model of a prestressed concrete slab (Fig. VIII.32) representing the top cap of a typical prestressed concrete reactor vessel was tested [873]. The objects of the test were: (a) to study the physical behaviour of a prestressed concrete slab with bonded tendons under increasing load up to the point of failure, thereby providing load–displacement relationships, and information about the distribution of cracking and the failure mechanism; and (b) to compare these results with those obtained from the analytical models.

An octagonal prestressed concrete slab representing the top cap of a vessel was prestressed with 5 and 7 mm diameter wires in two orthogonal directions. A conventional reinforcement of 12 mm diameter mild steel bars was provided, as shown in Fig. VIII.32. Four 7 mm diameter wires were placed along the centre line BB, two on either side of the centre line at distances of 115 mm (c_1) and 350 mm (c_2). These wires (shown by dashed lines) were at a height of 36·5 mm from the bottom surface of the slab. Eight 5 mm diameter wires were placed along the centre line AA, four on either side of the centre line at distances of 115 and 350 mm, and at heights of 18 and 55 mm from the bottom surface of the slab. Eight 12 mm diameter bars spaced at 65 mm were provided in one quadrant of the slab. A total of 32 bars were provided in the slab, as shown in Fig. VIII.33). Each wire was stressed up to 75% GUTS using the CCL prestressing system. Dial gauges were placed on one-quarter part of the top surface of the slab (Fig. VIII.34(a)) to measure the transverse deflection. Rosette strain gauges were also placed (Fig. VIII.34(b)) on the top surface of the slab.

The slab was loaded with the loading rig producing four load points on the main steel frame and 12 load points on the slab. The loading rig system is shown in Fig. VIII.35. Four deep steel I sections were welded together at 90° to make a main rigid steel frame. This frame was supported by two strong steel beams running across the slab, which were supported by the columns. Each rigid frame was connected to an hydraulic jack. Hydraulic jacks were connected to a common hydraulic pump. Each hydraulic ram was lowered onto a three-legged spreader frame, with the load being transferred between each leg and the slab by a steel plate. This steel plate had a rubber pad underneath it to distribute the load. The steel plates and rubber pads were spread over a diameter of approximately 1780 mm. Load cells (100 kN) were used on each hydraulic jack for accurate measurement of load. The surface strains of the concrete were measured using rosette

Application to Engineering Problems

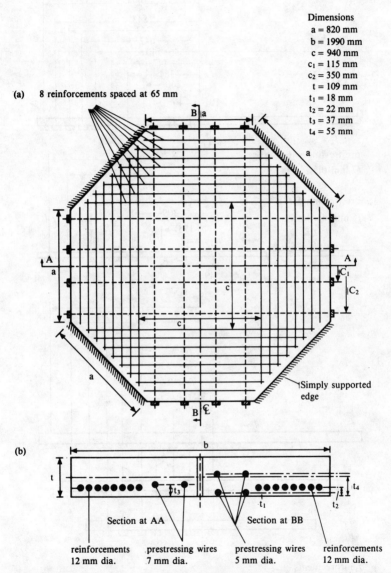

Fig. VIII.32. (a) Prestressed concrete slab and (b) a detailed section (see (a) for dimensions) [262].

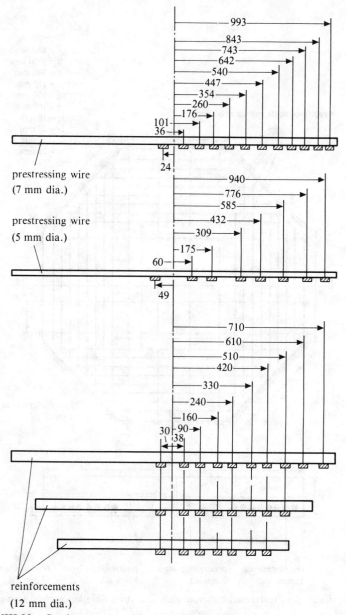

Fig. VIII.33. Strain gauge positions on prestressing wires and reinforcements [262] (all dimensions in mm).

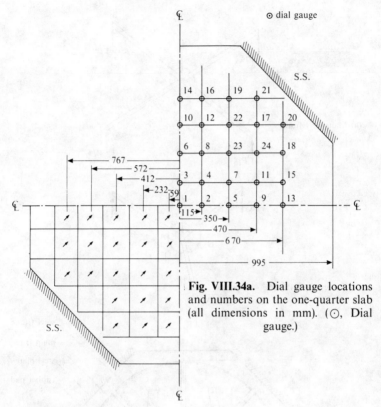

Fig. VIII.34a. Dial gauge locations and numbers on the one-quarter slab (all dimensions in mm). (⊙, Dial gauge.)

Fig. VIII.34b. Rosette strain gauge locations on the one-quarter slab [873, 262] (all dimensions in mm). (↗, Rosette strain gauge.)

strain gauges (Fig. VIII.34). The strain gauges and the rosette strain gauges were connected to a data logger (compulog) to process the results. For each load increment, the strains were recorded by compulog, and these were also printed on a teletype.

The vertical deflections were measured on the top surface of the slab using dial gauges which were capable of measuring with an accuracy of 2×10^{-3} mm. The dial gauges were placed on only one-quarter of the slab due to symmetry and were clamped using the rigid steel beam running across the slab. Figure VIII.36 shows the model slab ready for testing.

The load was applied in small increments using a hydraulic pump which

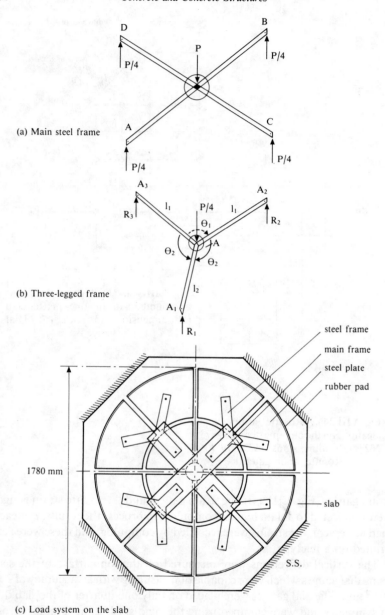

(a) Main steel frame

(b) Three-legged frame

(c) Load system on the slab

Fig. VIII.35. Load system of the rig, [873] and after Ahmad [262]. ($l_1 = 284$ mm; $l_2 = 204$ mm; $\theta_1 = 141.5$; $\theta_2 = 109.25$.)

had four hydraulic jacks connected to it as stated earlier. At each load increment, the readings of the dial gauges and the strain gauges were recorded. In the early stages of the loading larger load increments of 20 kN (0·032 N/mm^2) were applied. In the later stages of loading, smaller load increments of 8 kN (0·012 86 N/mm^2) were applied. The bottom part of the slab was painted with whitewash in order to show the cracks. The part of the slab which was vulnerable to cracking was divided into small, 50 mm

Fig. VIII.36. Model slab of octagonal shape with instrumentation, [873] and after Ahmad [262].

square, divisions in order to accurately estimate the positions and propagation of cracks. The slab was examined for cracks at each load increment. As cracks appeared, they were marked with a dark pencil so that they were visible in a mirror placed underneath the slab. A camera was mounted at the corner of the slab to take photographs of the cracks reflected in the mirror.

The total load up to the point of failure of the slab was applied in 53 increments. The results obtained are plotted in the form of load–displacement curves. The experimental load–displacement curve at the centre of the slab for the entire load history is shown in Fig. VIII.37. The deflected shapes of the slab along two centre lines (i.e. AA and BB) for

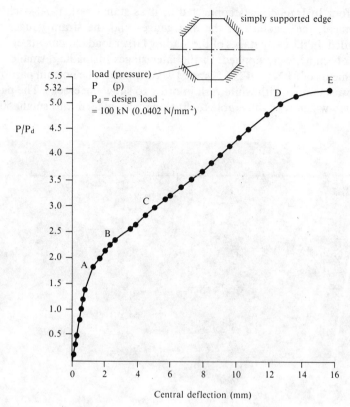

Fig. VIII.37. Experimental load–displacement curve of the octagonal prestressed concrete slab, [873] and after Ahmad [262].

various loads are shown in Figs VIII.38 and VIII.39. Photographs of cracks at the bottom surface of the slab were taken at various stages of loading. The collapse mode of Fig. VIII.40 showed crack patterns at failure. Just before failure the cracks at the bottom surface had developed, as shown in Fig. VIII.41.

The slab behaved linearly up to a load of 160 kN. The initiation of the first set of cracks occurred at a load of 180 kN. The vulnerable load stages are 200–250 kN. Figure VIII.41 indicates the position and distribution of cracking at the surface of the slab. The slab finally failed at a load of 544 kN, when deflections could not be measured as they were beyond the limit of the measuring gauges. The safety margin against the design load was 3·95.

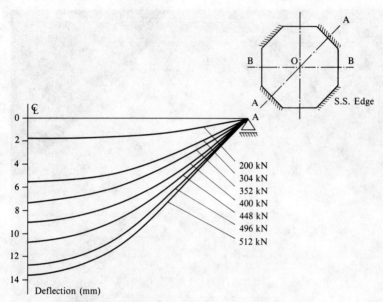

Fig. VIII.38. Experimental deflected shapes along AA.

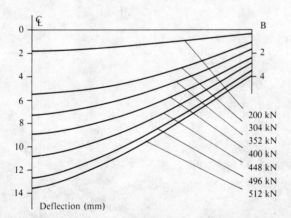

Fig. VIII.39. Experimental deflected shapes along BB, [873] and after Ahmad [262].

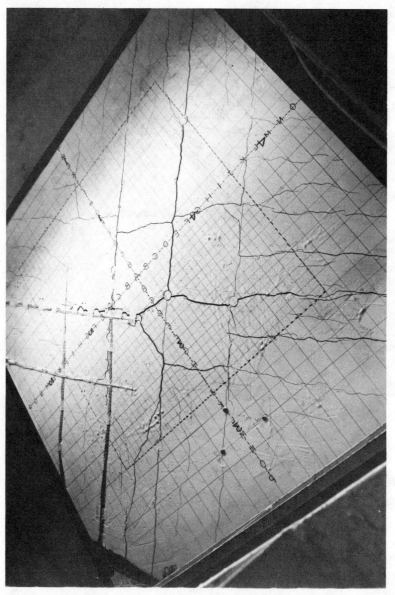

Fig. VIII.40. A 'post-mortem' of the octagonal slab.

Application to Engineering Problems

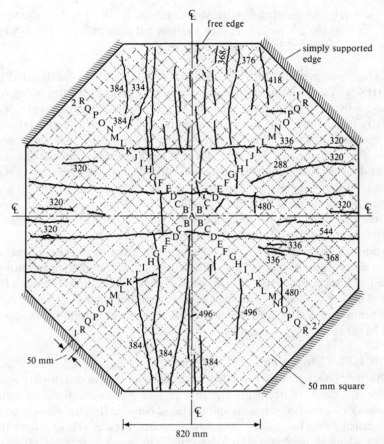

Fig. VIII.41. Experimental cracks at the bottom surface of the slab just before failure.

Ahmad and Bangash [873, 749] carried out finite element analyses for the following three cases:

Case I   Bonded Prestressed Concrete Slab—the interface behaviour between the steel and concrete was modelled using the linkage (Ahmlink) element.

Case II Perfectly Bonded Prestressed Concrete Slab—the steel element was placed on one side of the solid element and a perfect (rigid) bond assumed between the two.

Case III   Unbonded Prestressed Concrete Slab—the steel elements were not included in the analysis but their prestress forces were included in the analysis of the slab.

Due to symmetry, only one-quarter of the slab was analysed (Fig. VIII.42a). This slab was restrained from moving in the y direction along AB (i.e. $v = 0$) and in the x direction along AC (i.e. $u = 0$). The centre point A was restrained in both the x and y directions (i.e. $u = 0, v = 0$). The edge DE was supported in the z direction (i.e. $w = 0$). Figure VIII.36 shows these boundary conditions. The finite element mesh of one-quarter of the slab was dimensioned, as shown in Fig. VIII.42b. The choice of this mesh was made so that the prestressing wires and reinforcements lay on the sides of the solid elements. The nodes of solid elements and line elements were connected either by Ahmlink elements for the bonded slab or rigidly interconnected for the perfectly bonded slab. Where a large amount of ordinary reinforcement is closely spaced in the slab, they were modelled as line elements in the body of the solid elements. Such elements are shown as dashed lines in Fig. VIII.42b. Two solid elements were chosen through the thickness of the slab. The line elements were placed at an effective depth of 72·5 mm (Fig. VIII.42b). The details of finite element meshes for the three different cases are given below.

VIII.4.2.1 Case I: Bonded Slab (Ahmlink)

The slab concrete was modelled using eight-noded isoparametric solid elements, the reinforcements and prestressing wires were modelled using two-noded axial line elements and the bond between the steel and concrete modelled using two-noded Ahmlink elements. Figure VIII.43 shows the finite element mesh for this case only. The mesh comprises 60 solid isoparametric elements, 48 line elements, 28 Ahmlink elements and 40 body axial line elements, which are shown as dashed lines in Fig. VIII.43. The nodes for steel and concrete are represented by different numbers, although they occupy the same points in space. The Ahmlink elements (joining steel and concrete) are located at nodal points along the steel–concrete interfaces. Figure VIII.43 shows the bond of Ahmlink elements connected to solid elements. Here solid element nodes 26, 29, 47 and 50 are connected to the line element nodes 130, 131, 137 and 138 through the Ahmlink elements. Node numbers 26 and 130, 29 and 131, 47 and 137, and 50 and 138 have the same co-ordinates in space. There are a total of 150 nodes and 450 degrees of freedom for this case.

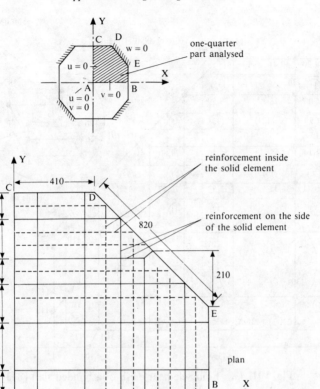

Fig. VIII.42a. Octagonal prestressed concrete slab (all dimensions in mm).

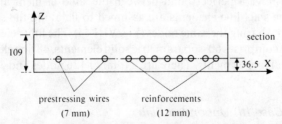

Fig. VIII.42b. Main dimensions of finite element mesh of the slab [873] (all dimensions in mm).

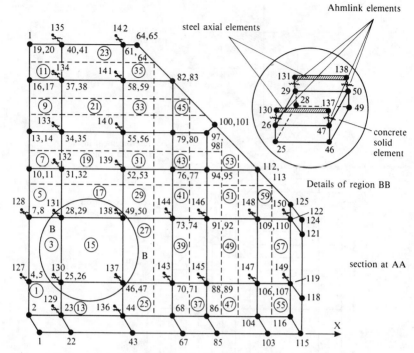

Fig. VIII.43. Finite element mesh of a bonded slab [873].

VIII.4.2.2 Case II: Perfectly Bonded Slab
The finite element mesh for this case is shown in Fig. VIII.44. The mesh is similar to the bonded slab case except that the Ahmlink elements are replaced by traditional line elements rigidly placed on solid elements. It is assumed to have a perfect bond between the solid element and the line element. The steel line elements are assumed to lie along the sides of the solid element sharing the same nodes (Fig. VIII.44). The finite element mesh for this case comprises 60 isoparametric solid elements, 48 line elements and 40 body axial line elements. There are a total of 126 nodes and 378 degrees of freedom.

VIII.4.2.3 Case III: Unbonded Slab
In this case the line elements representing prestressing steel are not included in this analysis since they are unbonded. The finite element mesh (Fig. VIII.44) in this case is identical to the case of perfectly bonded slab except

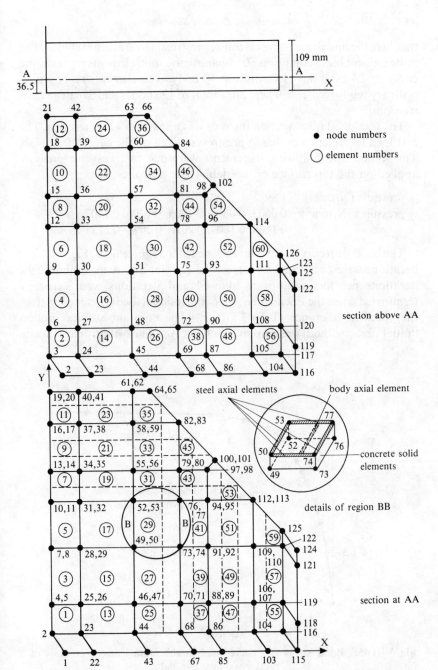

Fig. VIII.44. Finite element mesh of a perfectly bonded slab [873].

that here the line elements representing prestressing wires are excluded. The finite element mesh comprises 60 isoparametric solid elements representing concrete, 24 axial line elements and 40 body axial elements representing ordinary reinforcements. There are a total of 126 nodes and 378 degrees of freedom.

The total load (pressure) on the slab was applied in 11 increments. The first load increment was due to prestress applied on the edges of the slab. The second and following increments were due to transverse pressure applied on the top surface of the slab. These pressures were:

prestress (N/mm^2): 1·89
pressures (N/mm^2): 0·0603, 0·0804, 0·1005, 0·1206, 0·1407,
0·1608, 0·1809, 0·201, 0·2074, 0·2138

The bond–slip curve used is given for each load increment (Fig. VIII.45); iterations were performed to correct the equilibrium. A maximum of 12 iterations per increment were allowed and iterations were generally terminated when the Euclidian norm of the residual loads became less than the specified tolerance (Tol = 3×10^{-2}). The constant stiffness option (initial stress method) was employed for all cases. The orthotropic concrete

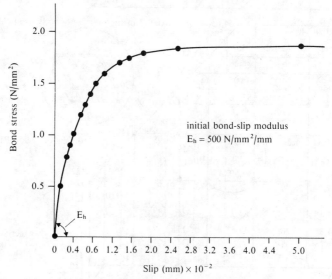

Fig. VIII.45. Bond stress–slip curve used in the finite element analysis of a prestressed concrete slab.

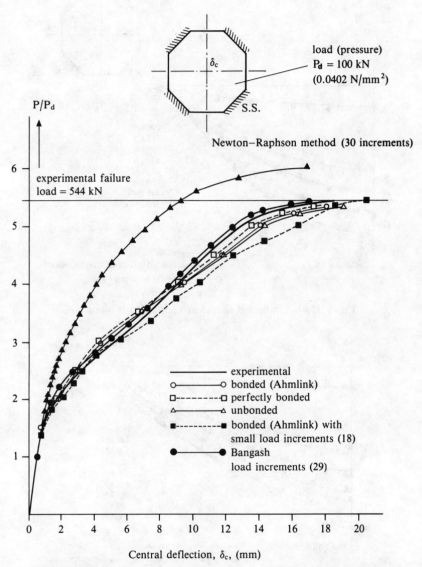

Fig. VIII.46. Load–displacement curves of prestressed concrete slabs.

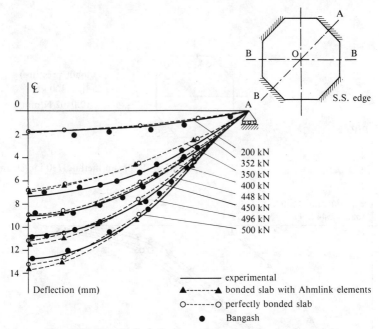

Fig. VIII.47. Deflected shapes at various loads along AA.

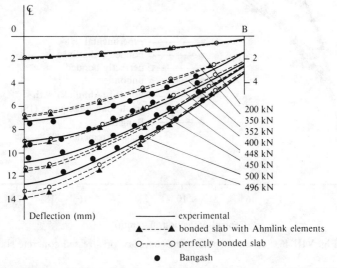

Fig. VIII.48. Deflected shapes at various loads along BB [749, 873].

model in compression, the tension cut-off model for cracking of concrete and the elasto-plastic model with strain hardening for steel were adopted throughout.

The results obtained from these analyses are plotted in the form of load–displacement curves, crack patterns and deflected shapes, as shown in Figs VIII.46–VIII.51. Cracking of concrete has a particular strong influence on the behaviour of the slab. Figure VIII.46 shows the load–central deflection history of the slab. The experimental curve is also shown for comparison. It is seen that the predicted results for three cases compare favourably with the experimental results. These curves show that they are in very good

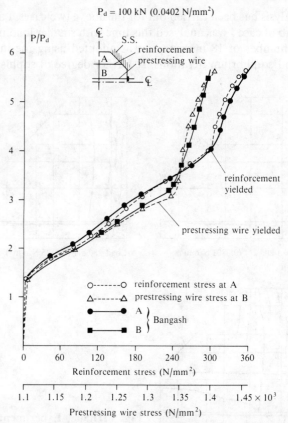

Fig. VIII.49. Variation of steel stress with load in a prestressed concrete slab [749, 873].

agreement in the early stages of loading (between 180 kN (0·0723 N/mm²) and 280 kN (0·112 N/mm²)). At this stage most of the cracking takes place.

The difference in the ultimate strength of the two cases is about 1·4%. In addition, the third case of the unbonded slab has also been analysed for comparison. The computed ultimate load is reasonably close to the other two analyses, but the computed displacements at the ultimate load are much too large. The reason is explained earlier. Figures VIII.47 and VIII.48 illustrate the deflected shapes along the two centre lines of the slab. The computed and the experimental deflected shapes are in close agreement. Figure VIII.49 shows the maximum steel stress (reinforcement and prestressing tendon) as a function of applied load as computed for the analysis.

A re-analysis has been carried out for the above two cases, I and II. The bonded slab of case I was analysed this time with smaller load increments. The total number of 18 increments was adopted using the initial stress method and equilibrium iterations. A further degree of sophistication in

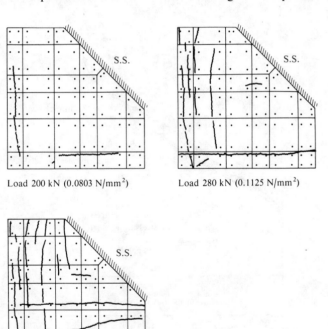

Fig. VIII.50. Experimental cracks at the bottom surface of the slab [873].

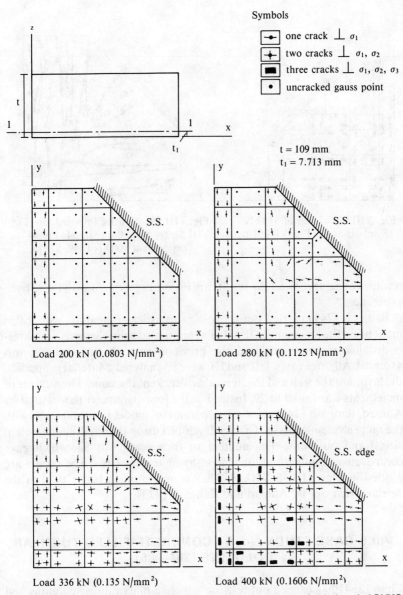

Fig. VIII.51. Crack patterns of the slab at 1-1; slab type—perfectly bonded [873].

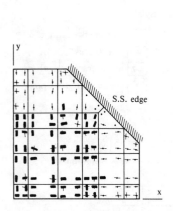 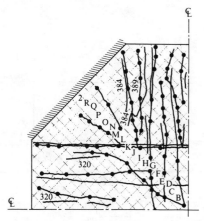

Fig. VIII.52. Load of 544 kN. (Symbols as for Fig. VIII.51.)

Fig. VIII.53. Cracks imposed on Fig. VIII.52. (—●—, Bangash 644 kN; ——, experiment 400 kN [849, 873].)

results was achieved and the results are in good agreement with those from experiment.

Bangash [749] carried out a further sophisticated analysis with a fine mesh using the program ISOPAR. Twenty-noded solid elements and three-noded line elements replaced the previous ones. A perfect bond was assumed. All three cases, I, II and III, were re-analysed. Material properties, the layout of the slab and the iterations were kept the same. The number of increments was raised to 29. Instead of the four-parameter model used by Ahmad, Bangash [749] used a five-parameter model in conjunction with the finite element analysis. Cracks developed using ISOPAR match well in size(s) and positions. The number of increments was 53, which gave comparative results on an increment-by-increment basis. The results are plotted in Figs VIII.46–VIII.53. They compare much better than in the work carried out by Ahmad and Bangash [873].

VIII.5 CASE STUDY No. 4—CONCRETE-FILLED TUBULAR COLUMNS [364–386]

Chan [364] has carried out tests on concrete-filled tubular columns and compared their results with those from the known empirical formulae summarised in Table VIII.5 [364–376]. For the concrete mix design a

Table VIII.5
Empirical column formulae

1. *ACI–NBC allowable load formulae* [374]

$$p = 0.25 f_c\left(1 - 0.000\,025\frac{L^2}{r_c^2}\right)A_c + f_s A_s$$

where the allowable steel stress f_s is given by

$$f_s = \left[20\,000 - 70\frac{KL}{r_s}\right]\frac{\sigma_y}{33\,000} \qquad f_s \not\leq \frac{145\,000\,000}{(\bar{K}L/r_s)^2}$$

where $\bar{K}$ = the length fixity factor for columns, and r_c and r_s are radii of gyration. Safety margins are 2·18–4·32.

2. *Knowles and Park formulae* [365, 366]
The ultimate load of concrete-filled tubular columns is given by the recommended equation (all values in imperial units):

$$P_u = 2qf_c' A_c(\sqrt{q^2+1} - q) + f_y A_s\left[1 - \frac{f_y(\bar{K}L/T_s)^2}{4\pi^2 E_s}\right]$$

when

$$\frac{\bar{K}S}{r_s} < \sqrt{\frac{2\pi^2 E_s}{f_y}}$$

$$P_u = 2qf_c' A_c(\sqrt{q^2+1} - q) + \frac{A_s \pi^2 E_s}{(\bar{K}L/r_s)^2}$$

when

$$\frac{\bar{K}S}{r_s} > \sqrt{\frac{2\pi^2 E_s}{f_y}}$$

where
P_u = ultimate load of composite column = $162\cdot 8 D_c^{1.5}/\sqrt{f_c}\bar{K}L/r_c$
$f_{cu} = 0\cdot 85 f_c$ (lb/in²)
f_y = yield strength of steel
D_c = unit weight of concrete (lb/ft³)
$\bar{K}$ = effective length factor
f_c = compressive strength of concrete (lb/in²)

For SI units the following conversion is adopted:

$$1\,\text{lb/in}^2 = 6\cdot 845\,\text{kN/m}^2 \qquad 1\,\text{lb/ft}^3 = 0\cdot 27\,\text{kN/m}^3$$

Safety margins are 1·40–3·91.

continued

Table VIII.5—contd

3. Furlong, Kloppel and Godr formulae [371–374]
The strength interaction formula is (all values are in imperial units)

$$\left(\frac{P}{P_0}\right)^2 + \left(\frac{M}{M_0}\right)^2 \leq 1\cdot 0$$

in which for circular tubes

$$P_0 = A_s f_y + A_c f_c \sqrt{\frac{f_y}{0\cdot 0018 E_s}}$$

$$M_0 = \frac{f_y}{6}(D_0^3 - D_i^3)$$

$$M_0 = \frac{f_y}{6} D_0^3 \{1 - (1-p)^{3/2}\}$$

and for square tubes

$$P_0 = A_g p f_y + (1-p) f_c \sqrt{\frac{f_y}{0\cdot 0018 E_s}}$$

$$M_0 = \frac{f_y}{4}(D_0^3 - D_i^3)$$

$$M_0 = \frac{f_y}{4} D_0^3 \{1 - (1-p)^{3/2}\}$$

where

P = ultimate axial load when a finite moment also exists
P_0 = axial load capacity in the absence of moment
M = ultimate flexural load when a finite axial load also exists
M_0 = flexural capacity in the absence of axial load
D_o = outside diameter of steel tubes
D_i = inside diameter of steel tubes
p = ratio between steel area and gross area

Safety margins are 1·16–4·41.

---

course aggregate of uncrushed irregular gravel of 38 mm maximum size was suggested. The characteristic strength obtained of the concrete was 61·47 N/mm² with an average value for E_c of 57·14 kN/mm². A tensile test specimen of the steel tube was taken by cutting the tube longitudinally into pieces of about 20 mm width. A universal testing machine was used and the extension of the coupon was measured using an extensometer over a gauge

Application to Engineering Problems 273

Fig. VIII.54. Top: arrangement of the apparatus in the Losenhausen compressive machine for strain and deflection measurements. Bottom: failure of 2·3 m length of the column [364].

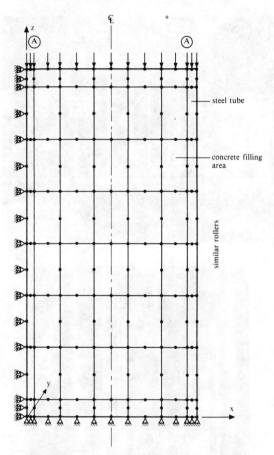

Fig. VIII.55a. Finite element mesh scheme for a concrete-filled tubular elevation (20-noded solid isoparametric column element).

length of 50 mm. A complete stress–strain curve was obtained, from which the yield stress of 305 kN/mm² corresponded to the design yield strain of 2000×10^{-6}. A short length of tube about 250 mm long was tested in compression to failure. A 2000 kN Losenhausen (Fig. VIII.54) compression machine was used. The longitudinal strain at the centre was measured using two electrical resistance gauges positioned opposite to each other. Strain readings were taken at suitable increments until the load reached 100 kN. The procedure was repeated three times to get the average elastic modulus of steel. The column lengths chosen were 0·25, 0·75, 1·3, 2·1, 2·3, 2·5, 2·7 and

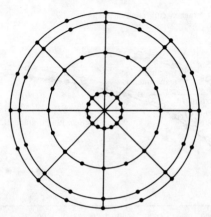

Fig. VIII.55b. Finite element mesh scheme for a concrete tubular column plan.

2·9 metres. The load–strain curves were produced and the possible failure mode was established.

A three-dimensional finite element scheme devised by the author for tubular columns is shown in Figs VIII.55a and VIII.55b. A Von Mises failure criterion was chosen for the steel tube together with the bulk and shear moduli model for the concrete described in Chapters VI and VII. In order to assess the frictional effect between the concrete and the steel two

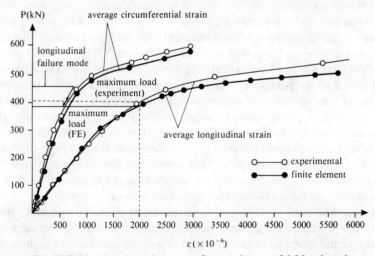

Fig. VIII.56. Load–strain curves for specimens of 0·25 m length.

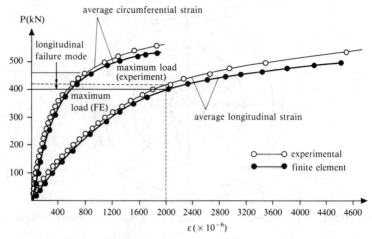

Fig. VIII.57. Load–strain curves for specimens of 0·75 m length.

cases were examined, namely the application at nodes and loads caused by friction and the introduction of a special element known as the gap element, the details of which are given in Appendix I. Owing to low frictional forces, the results in both cases differed by between 4 and 6%. Integration order ($3 \times 3 \times 3$) was used for computing stresses, strains, plasticity of the steel and cracking of the concrete. After 100 kN of load, an incremental load of 10 kN was adopted. At the end of each iteration the results were examined. Acceleration and convergence procedures discussed in Chapter VI were adopted. All long columns failed in flexure and the short columns failed by yielding of steel and cracking of concrete. Figures VIII.56–VIII.63 give a comparative study of analytical and experimental results. These results indicate that the parabolic decrement hypothesis gives a good representation of the redistribution of longitudinal and circumferential stresses in the column in the nonlinear range. In the finite element analysis the boundary conditions (Figs VIII.55a and VIII.55b) chosen for the column were very close to those imposed in the experiment. Results from the finite element analysis are around 10–12% lower than those obtained from the experiment. The difference may be ascribed to the slight end fixity not considered in the analysis; and the exact boundary conditions at the bottom could not be established from the report produced by Chan [364]. It is interesting to note from the experiment that no slip between the concrete and steel had occurred. Hence it was not necessary to include the bond stress–slip analysis discussed earlier. Using the Von Mises criterion

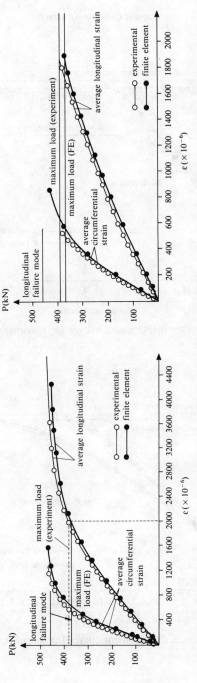

Fig. VIII.58. Load–strain curves for specimens of 1·3 m length.

Fig. VIII.59. Load–strain curves for specimens of 3·1 m length.

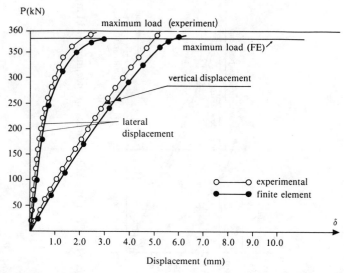

Fig. VIII.60. Load versus displacement curves for specimens of 2·1 m length.

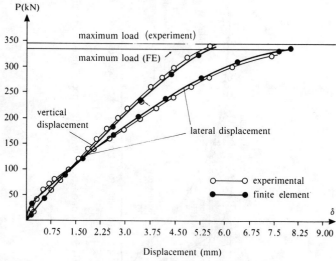

Fig. VIII.61. Load versus displacement curves for specimens of 2·5 m length.

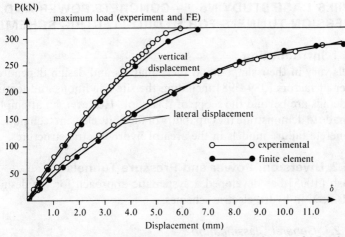

Fig. VIII.62. Load versus displacement curves for specimens of 2·7 m length.

for the steel tube, the results from the finite element analysis showed that during the elastic stage the deviation of strains from the average strain was generally in the range of 15%, corresponding to 20% evaluated from the experiments. The finite element analysis also indicated the early initiation of plastic zones along the longitudinal axis. The concrete corresponding to such zones had crushed when reaching a strain value of 0·0035. The safety margins from both analyses differ by 12–15%.

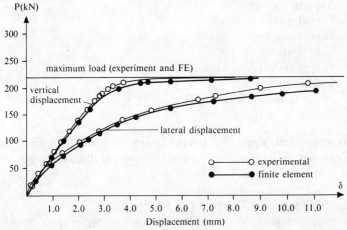

Fig. VIII.63. Load versus displacement curves for specimens of 2·9 m length.

VIII.6 CASE STUDY No. 5—CONCRETE POWER AND DIVERSION TUNNELS FOR HYDRO-ELECTRIC SCHEMES

VIII.6.1 Introduction
Tunnels vary in their shapes, and their analysis and design depend on a number of factors [389–398] including the site conditions and loadings. Such details are beyond the scope of this book. However, an attempt has been made to demonstrate the capability of the finite element technique and the concrete failure models in the area of hydro-electric structures.

VIII.6.2 Diversion, Power and Pressure Tunnels
Bangash [1005] has developed a systematic approach for the design of tunnels used in hydro-electric schemes.

VIII.6.2.1 General Classification
The tunnels used in hydro-electric schemes are generally classified as:

(a) Tunnels to divert flow during construction which are eventually closed using a combination of gates and concrete plugs
(b) Tunnels to divert water from one scheme to another
(c) Tunnels to be converted to power or pressure tunnels

VIII.6.2.2 Components of Tunnels
The following are the general components of a tunnel:

(a) Approach channel
(b) Tunnel intake structure
(c) Tunnel proper
(d) Gate shaft
(e) Tunnel transition and plug
(f) Slide gate chamber
(g) Portal structure and stop logs
(h) Discharge channel

This section only deals with the portal structure and the tunnel proper. Loadings pertaining to other components are not discussed; they can be included in certain cases.

VIII.6.2.3 Shape and Thickness of Lining
Many shapes based on hydraulic and structural criteria have been investigated and compared. Only circular tunnels that have transitioned

from rectangular openings at the intake level have become popular. The thicknesses of single concrete linings of tunnels and also of steel linings for concrete tunnels depend on the following well-known factors:

(a) Size and shape of the tunnel and the character of the rock formation around it
(b) External and internal loading conditions, including stability of the steel lining
(c) Stress distribution before and after rock mass excavation
(d) Boundary conditions, including types of supports and spaces for concreting requirements

Prior to an accurate estimate of a designed thickness of lining, a rule of thumb is available to give an initial thickness for the analysis. Both the US Bureau of Reclamation and the US Corps of Engineers generally adopt thicknesses of the order of 19–25·4 mm of concrete lining per $\frac{1}{3}$ m of finished diameter of tunnel. Regarding steel liners, the thickness based on an unstiffened shell is generally 6 mm per 344·75 kN/m^2 internal pressure based on a 3 m internal diameter. Table VIII.6 gives a detailed description of various parameters for recent designs of diversion, power and pressure tunnels.

VIII.6.2.4 Hydraulic and Structural Criteria

Generally the design of a tunnel is governed by two independent criteria, namely hydraulic criteria and structural criteria. The final design is always a compromise between the salient features of both these criteria. Again both these criteria are specifically geared to the individual project or scheme and the problems surrounding each one of them. The following criteria based on existing available statistics give reasonable guidance for tunnel design.

(a) Hydraulic criteria. It is essential to have a complete record of the following major items:

(i) A complete history of the catchment area, and specifically the hydrology of the site and operational conditions.
(ii) Establishment of upstream and downstream elevation at the design discharge and normal water surface and an estimate of hydraulic pressures.
(iii) Design discharge through the tunnels, the size of tunnels, etc.
(iv) The closure water level of the bulkhead and plug completion if it is a diversion tunnel; the minimum velocity during the diversion period.

Table VIII.6
Existing tunnel parameters [1005]

Type[a]	Scheme	Rock type	E (tons/ft²) Young's modulus	f_c (tons/ft²) compressive strength	Average inside diameter (ft/in)	Average lining (concrete) thickness (in)	Steel lining (in)	Pressure inside tunnel (psi)	Steel liner stress (tons/in²)	Concrete strength (lb/in²)	Safety factor
D	Fort Peck (USA)	Clay-shale	750–10000	5–50	24–8	36	—	—	—	3000	3·21
P	Garrison (USA)	Clay-shale containing lignite coal beds	800–4000	5–75–150	29–0	36	—	—	—	—	
P	Oahe (USA)	Shale	800–4000	5–75–150	26–0	33	—	—	—	3000	3·95
D			800–4000	5–75–150	22–0	30	—	—	—		
P			2000–10000	5–150	24–0	30	—	—	—	3000	4·5
D			2000–10000	5–150	19–9	28	—	—	—		
D	Akosombo (Ghana)	Quartzite	3000–5000	720–1370	30–0	24 u/s[b] 12 d/s	—	—	—	3000	2·75
P	Bandama (Ivory Coast)	Greenstone	2850	1000–3600 (500 design)	23–0 (7 m)	24 and 23	$\frac{9}{16}$–$1\frac{7}{8}$	410	20000 permissible 60000 yield	3000	2·75
			570				$1\frac{1}{2}$				
D	Rama-ganga (India)	Clay-shale interbedded with sand rock	3000–7000 7850–520000	1·83–6·44	30–0	30	—	—	—	3000	3·3
P			3000–7000 7850–520000		31–0	30	—	—	—	3000	

	Name	Rock								
D	Mangla (Pakistan)	Clay and sandstone bedrock	6 200–14 400	26·5–41·5	26 and 30 u/s 30 and 31 d/s	36	—	—	—	3 750	3·85
P	Burrendong (Australia)	Weathered greywack and slate	3 000–4 000	1 200	27-0	24	—	—	—	3 000	6·9
P	Yamuna (India)	Quartzite and slate and phyllites	Average 3 000	1 000	18-0	18	$\tfrac{3}{8}$–$2\tfrac{1}{8}$	—	—	3 000	5·1
D	Roselend (France)	—	—	—	12-6	20 (average)	$\tfrac{1}{2}$ with 4·7 × 4·7 × $\tfrac{1}{2}$ angle iron stiffeners	230	—	3 000	3·93
P	Bersimis No. 1 No. 2 (Canada)	—	—	—	5 No. 10-0 5 No. 17-0 reduced to 12-0	20 (average)	$1\tfrac{1}{16}$–$2\tfrac{1}{16}$ $1\tfrac{1}{16}$–$2\tfrac{1}{16}$	381 162	—	3 000	5·2
P	Warsak (Pakistan)	Metamorphis schists and gneisses with phyllite	700 (average)	80	39-0 and 6 No. 18-0	18 (average)	$\tfrac{5}{8}$–$1\tfrac{5}{16}$	100 (average)	—	3 000	4·0
P	Tumut No. 1	Granite gneiss with biotite granite	700–1 425	12	12-0	12	$\tfrac{1}{2}$–$1\tfrac{5}{8}$	325	Ultimate tensile strength 26–32 yield 15	3 000	3·5
	No. 2				11-6	12	$\tfrac{1}{2}$–$1\tfrac{5}{8}$	320	Ultimate tensile strength 26–32 yield 15	3 000	

^{a} D = diversion tunnel; P = power tunnel or pressure tunnels.
^{b} u/s = upstream; d/s = downstream.
(1 ton/m² ≈ 10 kN/m², 1 inch = 25·4 mm.)

(v) For power and pressure tunnels an accurate estimate of the surge tank diameter and the area, including load demand and rejection.

(b) Structural criteria (external loading or pressure). For design computation, empirical values are given by Bangash, based on Terzaghi [389], assuming the rock load acts as a uniformly distributed load over a tunnel diameter. These loads are based on assumptions that the rock is associated with some roof yielding due to failing of the interlocking action. In loose and fracture rock the sides also yield, thereby developing an active pressure state along the shear planes at an angle of $45° \pm \alpha$, where α denotes the angle of internal friction. If B and H represent the width and height of the tunnel, the following values from Terzaghi give a fair estimate of the extent of rock load on tunnels.

Rock load acting at depth of more than $1.5(B+H)$

(i) Moderately blocky and seamy rock: $0.25B$–$0.35(B+H)$
(ii) Very blocky and seamy: 0.35–$1.1(B+H)$
(iii) Completely crushed but chemically intact: $1.1(B+H)$
(iv) Squeezing rock, moderate depth: 1.1–$2.1(B+H)$
(v) Squeezing rock, great depth: 2.1–$4.5(B+H)$
(vi) Swelling rock: up to 85 m

(c) Load combination. Lateral restraint offered by the surrounding rock is extremely important and must be investigated while designing the tunnel lining. Various theories exist to take into account the pre-existing stress in rock and the redistribution of stresses in the surrounding rocks after tunnelling.

In addition, tunnels and their components are designed for the following loading cases, separately and in combination:

(1) Transition and the portion of the tunnel upstream from the plug:
External pressure from rock + hydrostatic pressure at normal water surface.
(2) Tunnel downstream from the plug:
External pressure from rock + weight of lining + zero hydrostatic pressure.
(3) Rock load + weight of lining:
Rock load with submerged weight of the rock below water table + weight of lining + minimum hydrostatic head + empty tunnel.

(4) Rock load + submerged weight of the rock below water table + weight of lining + contained water with internal pressure + zero external hydrostatic load.
(5) Stop logs:
Closure at specific level + hydrostatic pressure at normal water surface + any external loading.
(6) Tunnel plug:
Hydrostatic pressure at design elevation.
(7) Pier walls:
Stop log + wheelgate closed + maximum external pressure.

In the absence of suitable data, where applicable the tunnel components are designed to take into account the following average loads:

(i) Gate shaft house: $1.676 \, kN/m^2$
(ii) Hoist deck: $9.576 \, kN/m^2$
(iii) Inspection deck: $12.768 \, kN/m^2$
(iv) Gate support deck: $12.768 \, kN/m^2$
(v) Cut and cover section: $12.768 \, kN/m^2$
(vi) Fixed wheel gate storage area: surcharge equivalent of 1 m minimum of fill
(vii) Wind pressure: $0.96 \, kN/m^2$ on the vertical projection of exposed area
(viii) 10% seismic value of dead and permanent live loads applied at the centre of the mass in any horizontal direction
(ix) Hydrostatic + full uplift pressures arising from upstream or tail water conditions

(d) Materials and stresses and other elastic constants. Materials such as concrete and steel have been discussed in this text in detail.

VIII.6.2.5 Data for Diversion Tunnels

A typical diversion scheme, similar to the Akosombo scheme [390] with altered parameters, is presented as a design example. It is considered to be the most economic layout of a diversion tunnel in rock. An interesting feature is the introduction of a single pier housing slide gate chambers and a fixed wheel gate shaft with stop logs covering rectangular openings. The pier is placed symmetrically about the centre line of the tunnel. Figures VIII.64 and VIII.65 show a plan and sectional elevation of this tunnel. Figures VIII.66–VIII.70 give a full account of the way the 12·192 m transition is made from a rectangular portal to the beginning of a circular

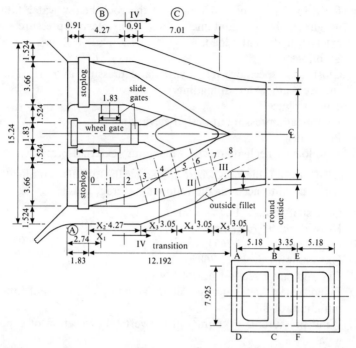

Fig. VIII.64. Plan and section IV-IV of the tunnel. (Note: All dimensions in metres.)

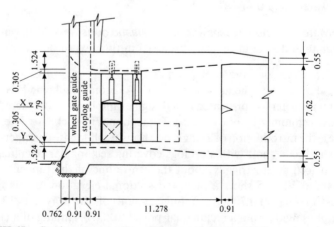

Fig. VIII.65. Section at centre line of the tunnel. (Note: All dimensions in metres.)

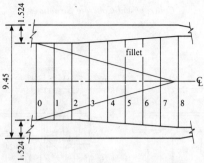

Fig. VIII.66. Profile at centre line of tunnel. (Note: All dimensions in metres.)

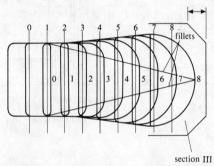

Fig. VIII.67. Sections along centre line of tunnel.

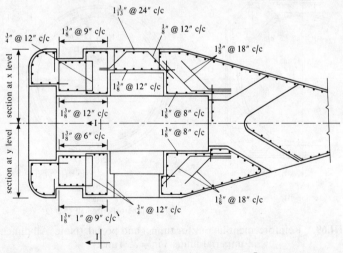

Fig. VIII.68. Reinforcement plan. All vertical bars $= \frac{7}{8}$ in × 12 in (c/c) and all horizontal bars $= \frac{3}{4}$ in × 12 in (c/c), unless otherwise noted. (Note: All dimensions in imperial units; 1 in = 25·4 mm.)

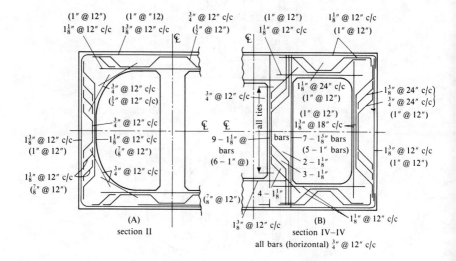
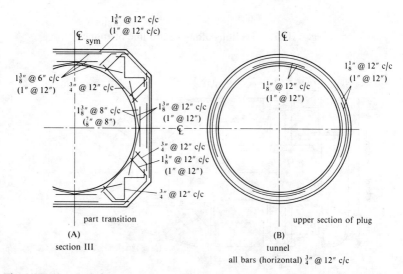

Fig. VIII.69. Reinforcement layout for tunnel and portal. (Note: All dimensions in imperial units; 1 in = 25·4 mm.)

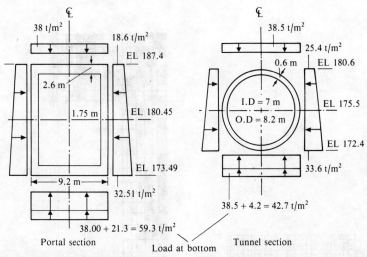

Fig. VIII.70. Bandama tunnel: portal and tunnel sections. (Note: Loads in original MKS units; 1 ton = 9·964 kN.)

tunnel. It is assumed that after the diversion of flow the tunnel is finally closed with a permanent concrete plug.

Tunnel diameter = 7·62 m
Area of tunnel section = 45·54 m^2
For reinforcement (plan and sections) see Figs VIII.68 and VIII.69
External loading intensity and water elevation
Maximum water surface elevation = 206·00
All other elevations are shown in Fig. VIII.70

Vertical:
Invert head of water = 32·5 m
Road load on top = 54·8 kN/m^2
Assumed load at top = 378·6 kN/m^2
Self load of section = 196 kN/m^2
Intensity of loading due to self load = 21·22 kN/m^2
  therefore load at bottom = 590·86 kN/m^2

Lateral:
At top = 185·33 kN/m^2
At bottom = 323·93 kN/m^2

Internal loading:
= 254·58 kN/m^2

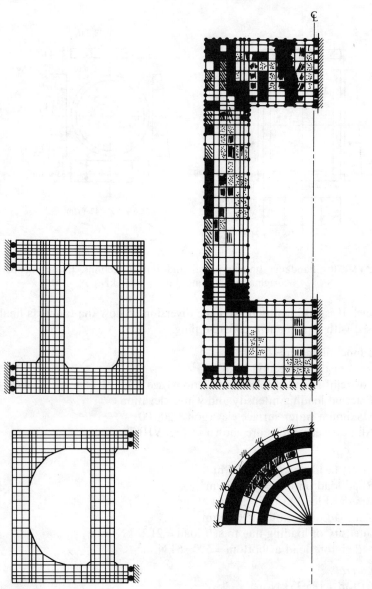

Fig. VIII.71. Finite element mesh.

Fig. VIII.72. Finite element mesh and cracking. (■, Cracks in three dimensions; ⦀, radial cracks; ▨, circumferential cracks; ▦, heavy principal stresses.)

Tunnel section (Fig. VIII.69)
External loading: maximum water surface elevation = 206·00

Vertical:
Invert head of water = 334·79 kN/m^2
Rock load on top = 488·2 kN/m^2
  therefore load at top = 383·6 kN/m^2
Self load of section = 340·77 kN/m^2
Intensity of loading due to self load = 41·85 kN/m^2
  therefore load at bottom = 425·46 kN/m^2

Lateral:
Hydrostatic head at top = 253·086 kN/m^2
Hydrostatic head at bottom = 334·79 kN/m^2

Internal loading:
Water head = 293·94 kN/m^2
Shear in rock as for plug key τ = 206·85 kN/m^2

VIII.6.2.6 Finite Element Analysis

Three-dimensional finite element analyses using eight-noded solid isoparametric elements and two-noded line elements were used. Where the steel lining interfaced with concrete, displacements at the nodes of the two meshes were matched. When lugs were provided the finite element analysis covered the lugs by invoking Table VI.2. External and internal loads were transformed into nodal loads. The tunnel concrete was analysed using the finite element mesh scheme given in Fig. VIII.71. The existing layout of the reinforcement designed by conventional analysis was used. The line elements and their positions in the solid elements were as in the original layout. Stresses, strains and principal stresses were computed. Afterwards inelastic analysis was carried out as described in Chapter VI using the endochronic cracking models formulated in Chapters VI and VII. Twenty increments each of 10 kN were considered. The early cracking in concrete started at about 1·65 times the load. The concrete section of the tunnel crushed at about 2·75 times the design load.

Figure VIII.72 shows a 'post mortem' of the tunnel at about 2·75 times the indicated design loads shown in Fig. VIII.70. Figure VIII.73 shows reinforcement for maximum principal stresses at 2·5 times the design loads.

Table VIII.6 shows the parameters and the safety margins for other types of tunnels using finite element analysis inclusive of the concrete failure

292   Concrete and Concrete Structures

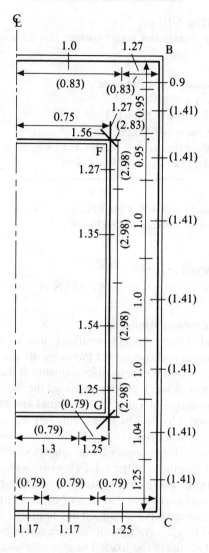

reinforcement comparison

── reinforcements in sq. in. per ft.

Fig. VIII.73. Tunnel reinforcement. (Note: Reinforcements in imperial units; 1 in = 25·4 mm.)

models (endochronic) and the cracking equations given in Chapters VI and VII.

VIII.7 CASE STUDY No. 6—PRESTRESSED CONCRETE NUCLEAR REACTOR VESSELS

VIII.7.1 General

The use of nuclear energy in the production of electric power involves substantial structural systems comprised of pressure vessels to house the reactor. Most of the light water and pressurised water reactors in operation at present are sheltered by two major structural units: the primary container, which is a steel pressure vessel for holding the coolant systems under pressure, and a secondary container providing a second protective shield. An alternative to the use of liquid coolants is a pressurised gas system. The coolant in this case is either carbon dioxide or helium. The current British and the French reactors, using these coolants, are called Advanced Gas-Cooled Reactors (AGR) and High-Temperature Gas-Cooled Reactors (HTR or HTGCR). Efficient operation of these nuclear power plants requires a large electrical output. This requirement calls for a large nuclear core space and much larger primary pressure vessels and supporting structures of sizeable dimensions. The inherent difficulties that would be involved in the fabrication and transportation of the pressure vessel steel units of the size necessary for such a reactor have led to the need for a more versatile type of structure. The vessel must, undoubtedly, be built on site and meet all the serviceability and safety requirements. The solution adopted has been the use of prestressed concrete reactor vessels (PCRV).

One incentive for adopting PCRV is the economic advantages to be expected. These advantages arise, for example, from:

(a) the ability to contain large reactors with high pressures and temperatures;
(b) simplification of plant layout;
(c) the fact that a highly developed steel fabrication industry is not necessary.

Attractive features from a safety point of view include:

(a) physical isolation of the steel prestressing tendons and reinforcement from sources of heat and radiation and from the primary coolant;
(b) the high degree of redundancy in the prestressing systems.

The shapes used, so far, for PCRVs have varied from a cylinder bounded by two inverted, non-prestressed, hemispherical domes, torospherical domes to spherical shells. The current trend in PCRV configuration is the use of thick-walled cylinders, the ends of which are closed by flat slabs known as caps. The boilers and circulators are either housed within the main cavity or within the thickness of the walls—the latter is known as a *multi-cavity type vessel*. This study is concerned with such vessels.

VIII.7.2 Historical Development

The first application of prestressed concrete as a reactor vessel was in the Marcoule G-2 and G-3 installations in France. These are horizontally placed concrete cylinders with concave domes as caps. The vessel main cylinder is wrapped with prestressing cables describing an arc of 270° which are finally anchored to prestressed concrete foundations under the vessel. The domes are not prestressed, resulting in a substantial loss of vessel volume. The stand pipe and the control rod areas are badly situated for the charge machine. The French have recognised these limitations of the Marcoule vessels. However, in the meantime the requirements for high-quality steel and more rigorous inspections for both British and the French reactor vessels has reawakened interest in concrete vessels. Consequently, in planning the construction of the EDF-3 reactor, the French designers decided to re-examine the application of prestressed concrete to reactor vessels. The design selected provides a vessel with a cylindrical shape and with prestressing cables placed in layers parallel and perpendicular to the long axis of the vessel. Supplemental prestressing steel is also provided at the corners where the stress concentrations are high.

The British interest in concrete pressure vessels began with some experimental work done by the General Electric Company and the Simon-Carves group on cylindrical vessels prestressed circumferentially by wire winding. Based on these tests and other tests carried out by the United Kingdom Atomic Energy Authority (UKAEA), Waters and Barrett [399] published design studies for cylindrical and spherical vessels containing advanced gas-cooled reactors. They carried out a comparative study of steel and concrete pressure vessels and concluded that there were significant economic and technical attractions to adopting such vessels. They then presented a detailed design philosophy in which particular attention was paid to the short- and long-term effects of tendon loads on concrete, the premature failure of the liner, and the optimum choice of the conventional reinforcement as an anti-crack steel. Apart from highlighting design and constructional problems, they suggested a list of research programmes. At a time when no suitable code of practice and no completely reliable mathematical tools existed for analysing and designing such structures, their recommendations prompted many UK firms with interests in nuclear technology to carry out research and development in this field. The first order placed by the Central Electricity Generating Board (CEGB) was for the Oldbury Power Station. The Nuclear Power Group (TNPG) proposed a vessel design that was a hybrid of the French Marcoule and EDF-3 concepts. The vessel is of cylindrical shape with flat top and bottom caps, and it utilises alternate layers of helical cables which wrap the vessel in reverse directions. These cables are anchored in groups in specially provided top and bottom galleries. Both longitudinal and lateral prestressing loads are produced from this single prestressing system.

By the time of completion of this vessel a great deal of knowledge had been gained from both the British and the French experience in this field. Another concept was under consideration at the CEGB, namely a spherical vessel containing the entire reactor internals. Owing to immediate requirements for more electrical output, the CEGB gave the go-ahead to a group member of the British Nuclear Design Company [412, 456] to design and construct such a vessel for the Wylfa power plant. This vessel has a spherical inner surface and the prestressing is provided by a six-sided polyhedron made up of cables anchored to buttresses along the vessel circumference. Each tendon is laid perpendicular to the preceding one. Since the prestressing tendons in the Wylfa Nuclear Power Station are nearly straight, the possibility of large frictional effects is remote. The prestressing loads are uniform and this is partly due to its spherical shape. Compared with cylindrical vessels, the Wylfa spherical vessel effectively utilises the entire prestressing system, and the peak stressing is considerably less.

In the meantime, a number of variations on the French EDF-4 (now Bugey) reactor were under consideration. It was decided to have a vertical cylindrical pressure envelope divided into two chambers: the upper one supports the reactor core whilst the lower one contains the boilers or heat exchangers. These affect the size of the internal cavity and the prestressing system. Complete data for these vessels was published by Marsh and Melese [400], based on detailed literature surveys and the preliminary designs carried out by the Frankline Institute [401].

Both the Frankline Institute [401] and the Oak Ridge Laboratory [886] in the USA made a start on the research and development required for the Fort St Vrain high temperature gas-cooled reactor. The choice was made of a cylinder with flat top and bottom caps. Both in the wall and in the cap the prestressing was by means of cables of variable curvature.

In the meantime, in the UK a great deal of experience was gained from the Oldbury and Wylfa vessels. Problems associated with the plant layout and the construction of the Wylfa vessel were such that the British designers had to rethink the use of cylindrical vessels. The CEGB finally approved the cylindrical vessel of the Dungeness B Power Station using large tendons in the vertical and in the circumferential directions. The circumferential tendons are anchored on specially formed buttresses. This time, using a minimum number of large tendons, a greater concentration of forces has been achieved for the limited available space in the vessel.

However, the Oldbury contractors carried out optimisation studies on the Hinkley B and Hunterston B power stations. Slight changes in the vessel parameters were inevitable. This time true helices were adopted for the prestressing tendons rather than the 'barrelised' type adopted for the Oldbury vessel.

At this stage, the concept of multi-cavity vessels was under active consideration. Boilers and circulators occupied a large space within the main cavity. The dimensions of the main cavity were reduced by sheltering the boilers and circulators within the thickness of the wall. Burrow reported [456] on providing the wire-winding system for the circumferential prestressing loads on the vessel, thus leaving the main vessel thicknesses for the large capacity longitudinal tendons only. Such a vessel was accepted for the AGR system adopted for both the Hartlepool and Heysham power stations. In the last few years substantial progress has been made in the development of such vessels and much experimental data has been produced from both model tests and on-the-spot measurement of the prototype vessel.

VIII.7.3 Problems Associated with Vessels

Although the multi-cavity vessel has much to offer, it is by no means an easy structure to model. Many designers and structural analysts have tried to tackle the various problems associated with these structures. In many cases they have claimed some achievements but for some reason they have not made them public. However, most are united on the design criteria for such vessels and these are:

(a) the vessel should be designed for elastic response to all possible combinations of loads during operation;
(b) it must show a progressive mode of failure under increasing gas pressures with large deformation to warn against impending failure; and
(c) the vessel must have an acceptable safety margin against failure.

In order to meet the above criteria satisfactorily, one has first to validate the performance of the individual components forming such structures. The major components are:

(a) concrete,
(b) prestressing tendons,
(c) the liner and other penetrations, and
(d) bonded primary and secondary reinforcements.

The problems associated with each one of these are enormous. For example, a certain amount of evidence must be available to demonstrate that for short- and long-term loadings (under multiaxial compressive stresses) concrete can safely withstand higher compressive stresses than are generally acceptable under uniaxial loading. For all service load conditions, including start-up and shut-down, both initially and at the end of the vessel's life, the stress–strain characteristics for the concrete should take account of the age, temperature and time under load. It is considered also that limited cracking may be accepted provided due regard is paid to any significant redistribution of the stresses which may arise, due to lack of integrity and leak tightness of the liner. Where local concentrations of stress occur, due to the presence of embedments or other discontinuities in the vessel geometry, these should be assessed individually. In such cases, due regard should be paid to the effects of increased creep rates or tensile cracking on the distribution of stresses in the vessel concrete and the influence which they may have on the strains in the vessel liner.

Stresses and strains, deflections and cracking in the vessel should be analysed for all relevant combinations of mechanical and thermal loads which can arise under normal service and ultimate conditions. In such cases the prestressing forces play a great part. The tendon forces adopted in each analysis should include allowances for the most severe effects of friction and loss of prestress. A proper method of analysis together with short- and long-term experimental tests are required to design tendon systems for cyclic loads produced by gas pressures. These requirements are translated in terms of range of stress or strain cycles in the steel liner, the concrete, or the other components in the nuclear islands. Beyond the elastic range, the method of assessing the ultimate behaviour of prestressing tendons and their anchorages is very important for predicting a safety margin for the vessel. This, of course, depends on the time at which the incident occurs to the vessel and the rate of increased gas pressure and whether or not the tendons are grouted.

The liner undoubtedly represents a vital safety element of the vessel. Indeed, it serves the fundamental purpose of forming an impervious barrier to the cooling gas. The reactor's operability depends on its integrity. The vessel will collapse, not at the ultimate collapse pressure coincident with the failure of the prestressing tendons but at the liner failure pressure which might be lower. This would require knowledge of the topology and magnitude of the cracking on the internal face of the concrete in contact with the liner. It would be necessary, therefore, to have a clear idea of the cracking mode of the vessel for increasing pressures up to the collapse pressure. Moreover, the liner integrity is also dependent upon the performance of the liner anchorages and cooling pipes, buckling stability and fatigue in general areas, the insulation and temperature distribution.

The choice and the distribution of the conventional bonded steel reinforcement in the vessel main areas and around the penetrations are extremely important should initial cracking occur due to moisture migration and unpredictable shrinkage and creep in concrete. The bonded steel reinforcement is needed in areas where extreme stresses under serviceability conditions cannot be avoided. Above all, the progressive failure of vessels depends on the amount and the distribution of such reinforcements.

Now turning to the vessel behaviour under extreme loads, it is necessary to provide some means of limiting the effect on the vessel of an excessive rise in internal operating pressure. The magnitude and rate of a postulated pressure rise can be determined only by reference to the characteristics of the particular reactor contained. It is common practice to provide automatic venting devices, such as safety valves, for this purpose. An alternative is to design the vessel so that it is self-venting by partial structural failure. It is generally recognised that this alternative cannot be relied upon at present.

It is customary and advantageous to make provision to verify the state of the vessel. This may be done by installed instrumentation and/or by periodic in-service inspection. Such measures, which are taken in a manner appropriate to the particular situation, serve to verify the vessel integrity and to confirm the design criteria. However, this does not rule out the main problem of numerically assessing such design criteria. Under any circumstances the design philosophy is based on the recognition of two or more modes in the vessel response to increasing pressure, and this cannot be met by simple experimental models.

The design objective is to ensure that a particular response to the imposed load can be achieved in each mode and that this behaviour is consistent with the appropriate operational and predetermined fault conditions.

Over a range of pressures and temperatures, including normal operating conditions, the vessel will respond to short-term variations in pressure in an elastic manner. This facilitates machine analysis of stress and strain in the vessel. Long-term stresses and strains are affected by shrinkage and creep of the concrete, relaxation of tendons and possibly fatigue. In this range of response the effects of short- and long-term behaviour can be combined to demonstrate that stresses and strains are limited to acceptable values.

Beyond the elastic range the response becomes increasingly inelastic and nonlinear. The vessel would not be expected to enter this phase except under the most severe overpressure fault conditions. In this phase the vessel is stable but may experience permanent damage. It is in this phase of vessel response that some limit states occur.

The ultimate load condition, in which the vessel is incapable of sustaining any further increase of internal pressure, is a further limit state. Evaluation of the ultimate load provides a measure of the factor of safety above design conditions.

VIII.7.4 Vessel Layouts and Finite Element Mesh Schemes
Typical layouts of the Dungeness B, Oldbury and HTGCR vessels are given in Figs VIII.74–VIII.76 respectively. Figures VIII.77–VIII.80 show the finite element mesh layout of the Dungeness B, Oldbury, Hartlepool and HTGCR vessels respectively.

The mesh layouts take into consideration the vessel penetrations, prestressing configurations and other features such as the existence of buttresses, galleries and the liner lugs and cooling pipes.

VIII.7.5 Design Analysis
The objective of this analysis is to demonstrate that reactor vessels will meet both the serviceability and ultimate limit state conditions, taking into consideration the gas increasing pressure.

VIII.7.5.1 Service Conditions
Under service conditions it is important to know the technique and precision with which the vessel geometry can be reproduced by analytical means. A number of key loading cases are given below.

(a) Initial prestressing loads: Analysis includes losses due to friction and elastic shortening and creep; this is to see that no abnormal stresses develop in the vessel concrete.

(b) Proof test: In this analysis the vessel is tested at the commissioning stage for changes in strains and deflection due to proof pressure loading.
(c) Normal operating conditions: This is intended to observe the vessel operating normally under extreme loads caused by the combination of prestress, design pressure and temperature gradient together with the effects of applicable environmental loadings. This can be at an early-life or late-life operating condition.
(d) Start-up and shut-down conditions: These are transient and quasi-transient conditions normally imposed on the steady conditions of normal operation. Together they reproduce the extremes of the early-life shut-down/start-up cycle. The late-life conditions are also examined by changing the material properties and temperature transient conditions.
(e) Deviations from operating conditions: A number of inaccuracies arise from the simple treatment of loading cases; these mainly concern the effect of concrete creep on stresses, unloading and cooling effects on the vessel, and the allowance for low-probability loadings arising from accident or fault conditions. Analyses are required of these uncertainties.

VIII.7.5.2 Ultimate Conditions
A validity analysis is required to check the mode of failure assumed or calculated or derived from the experimental tests. This analysis should cover cases such as the intact and ruptured liner, failure of local areas, the possibility of shear failure in top caps or at cavity/wall ligaments and cracks in hot spots (areas around cooling pipes).

VIII.7.5.3 Methods of Analysis
Many methods are available to designers. The most common [399–429, 750, 751] are based on the finite element, finite difference or dynamic relaxation, lumped parameter, and limit state methods. This book gives earlier the step-by-step approach of the finite element method. In some service and fault conditions it is necessary to consider the influence of external hazards and environmental conditions. Major external hazards are seismic disturbances, wind, missiles and aircraft crashes. These hazards are considered for the case of containment vessels.

VIII.7.5.4 Model Testing
The purpose of a model test is to verify the methods of analysis and to

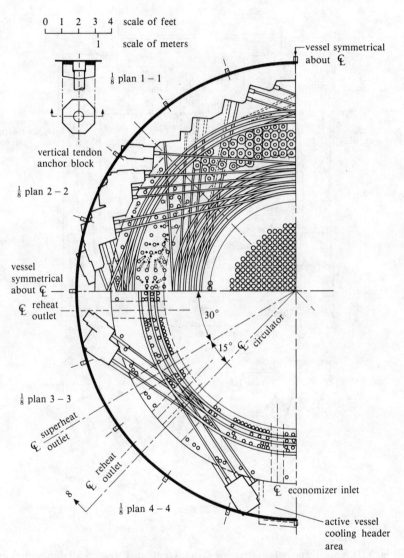

Fig. VIII.74. Dungeness B vessel prestressing layout (plan).

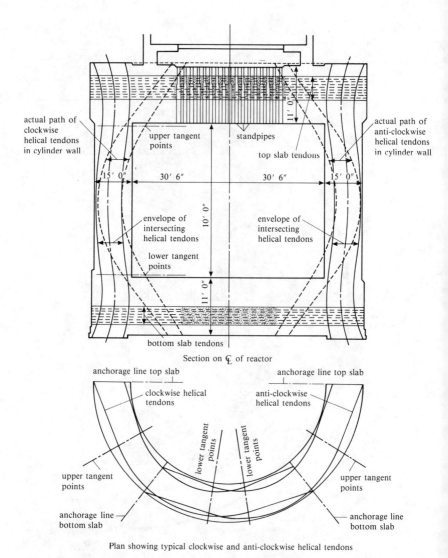

Fig. VIII.75a. Principal prestressing systems (patented). (Note: All dimensions in imperial units; 1 ft = 0·3048 m, 1 in = 25·4 mm.) [14, 415, 418]

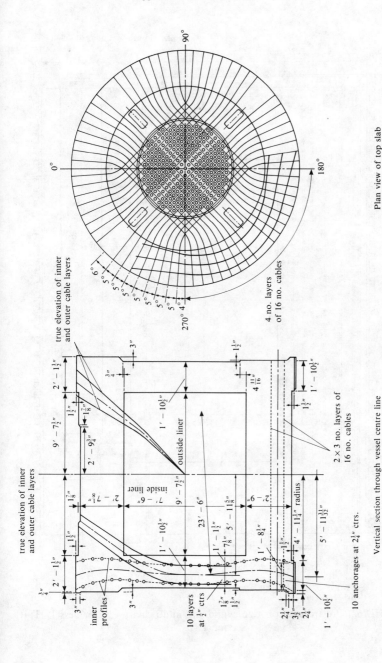

Fig. VIII.75b. Model for Oldbury nuclear power station—layout of prestressing cables and cable profiles. (Note: All dimensions in imperial units; 1 ft = 0·3048 m, 1 in = 25·4 mm.) [14, 415, 418]

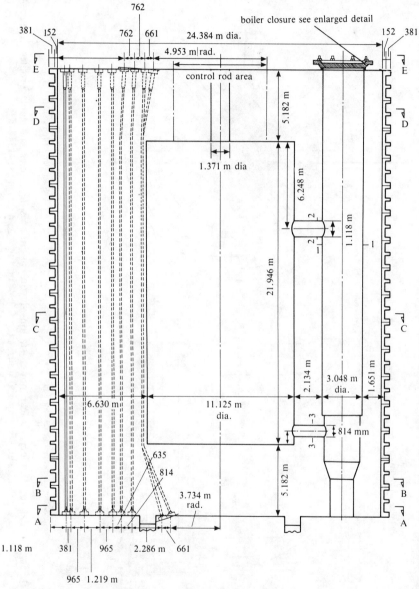

Fig. VIII.76. Prestressed concrete cylindrical multi-cavity reactor vessel, HTR, section F–F.

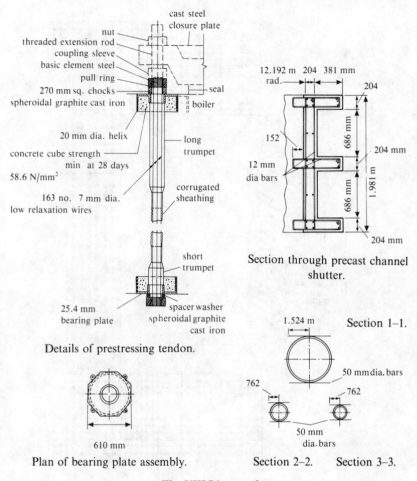

Fig. VIII.76—contd.

provide a visible physical demonstration of the adequacy of the design requirements. As described earlier, model tests have been carried out to verify the service and ultimate state behaviour of vessels [418–429], and to assess prestressing, reinforcement and liner requirements [418–429]. The scale chosen for a specific model depends on the objective of the test and on the reduction in size of important vessel components. For a complete vessel a suitable scale varies from 1:25 to 1:10, and scales of 1:25 and 1:50 have been used for the investigation of the top slab behaviour. A large number of

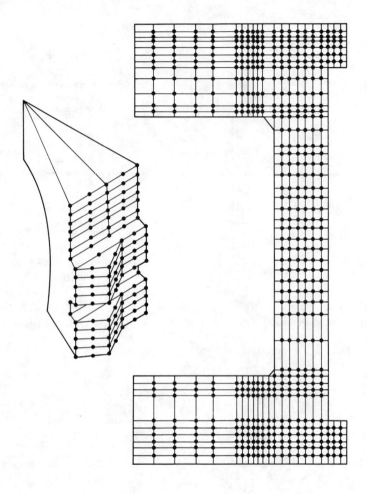

no. of nodes = 2500
no. of elements = 850

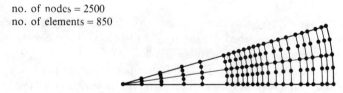

Fig. VIII.77. Finite element idealisation of Dungeness B vessel. (Note: For clarity certain nodes are not shown.)

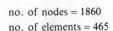

no. of nodes = 1860
no. of elements = 465

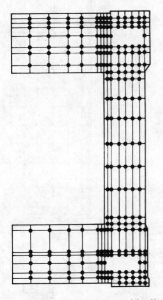

Fig. VIII.78. Finite element idealisation of Oldbury vessel.

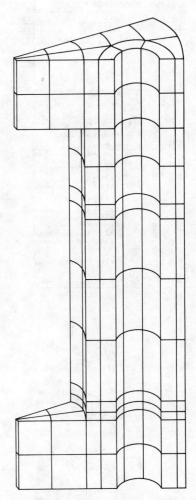

Fig. VIII.79. Prestressed concrete cylindrical multi-cavity reactor vessel (HTR)—3-D finite element.

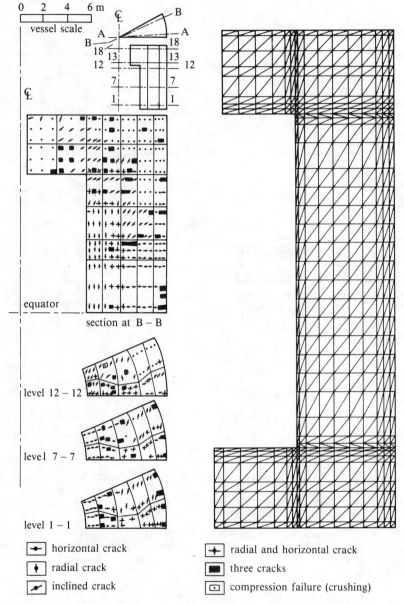

Fig. VIII.80. Prestressed concrete cylindrical 2-D finite element pressure vessel.

models for shear modes of failure have been tested [418–423] with varying geometries, materials, prestressing loads, reinforcement and penetrations. Some pneumatic tests have been carried out using gas as the pressurising medium [418–423]. Most of these tests produced progressive failures with sufficient means available for detecting the elastic limit, the onset of cracking and the failure mode. Many individual tests have been carried out to assess the integrity of the liners, penetrations, closures and thermal protection systems [461–483]. The finite element analysis and constitutive model discussed earlier are applied in Section VIII.7.5.5 to validate the proposed analysis.

VIII.7.5.5 Analysis of Results

(a) *Dungeness B vessel.* Figures VIII.81 and VIII.82 show gas pressure versus vessel deflection for the top cap. Here only 12 increments of pressure are plotted. For various pressures the trend in the incremental deflection is

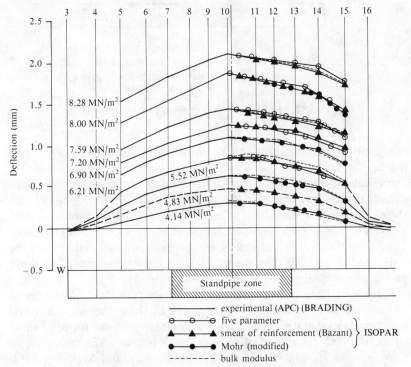

Fig. VIII.81. Deflection versus pressure: 4·14–8·28 MN/m².

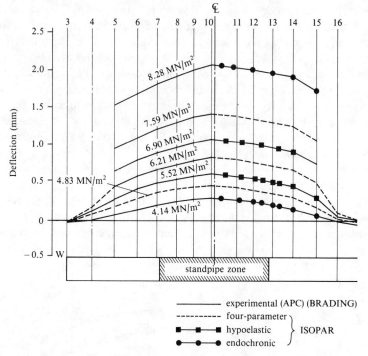

Fig. VIII.82. Deflection versus pressure: 4·14–8·28 MN/m².

fairly regular and no abrupt change in the deflections between any two increments has been found. Finite element analysis using three concrete numerical models, namely the four-parameter, hypoelastic and endochronic models, has been carried out on the model as well. Figure VIII.83 shows the pressure–deflection relation for two areas of the barrel wall. These results from the analysis are summarised in Figs VIII.81 to VIII.83b. The final curves are the idealised curves for pressure versus deflection of the wall and the top cap. The shapes of these curves are similar to those produced by Brading and McKillen [418, paper 3].

Figures VIII.83a and VIII.84 show the various principal stresses under normal operation of the vessel. Figure VIII.85 shows the maximum and minimum principal stresses at 2·5 times the design pressure. Based on these stress contours, assessment of various cracks is made. The results are in good agreement with those provided by Brading and McKillen [418, paper 3] and as shown in Fig. VIII.86a. Figure VIII.86b shows a failure mode of the top cap along the thickness; this was provided by ENL of Italy [423].

Since the Dungeness B vessel is stressed by circumferential tendons as well, it is essential to give the stress pattern under the anchorages. Figures VIII.87a and VIII.87b show maximum and minimum principal stresses, for normal operation, in zones where buttresses are located.

For ease of discussion in the following paragraphs M denotes cracks derived from the published model test results and FE denotes cracks in various zones predicted by the finite element and the limit state analyses.

Various stress trajectories at the design gas pressure ($P_{GD} = 3 \cdot 3 \, \text{MN/m}^2$) and at 2·5 times the design gas pressures ($8 \cdot 25 \, \text{MN/m}^2$) have already been discussed. At pressure $P_G = 4 \, \text{MN/m}^2$ ($1 \cdot 21 P_{GD}$) both FE and M indicate minor local cracking in the haunch zone initiated in both the radial and circumferential directions. In the top cap radial cracks appear from FE analysis. Model tests [418–423] indicate that initial local cracking at the top cap in the radial direction has occurred at a pressure of $3 \cdot 46 \, \text{MN/m}^2$. This value has been raised to $3 \cdot 8 \, \text{MN/m}^2$ in an isolated top cap test [418–423].

When the pressure is raised to $4 \cdot 7 \, \text{MN/m}^2$ ($1 \cdot 42 P_{GD}$), no change is found in the radial crack, but in both the radial and circumferential directions a narrow band indicating a plastic zone has appeared. A slight increment in the length of crack is noticed in the haunch zone. Here both FE and M agree with each other. At the same pressure at the haunch near the bottom cap, a localised plastic zone has developed. However, at the same pressure the M analysis indicates a plastic hinge together with a crack at the bottom haunch. The 1/25th scale model test [418–423] has not indicated this feature, probably because of the difficulty of seeing and measuring such a crack. However, the tests show extreme stress concentration around this pressure in these local areas.

Raising the pressure to $5 \cdot 1 \, \text{MN/m}^2$ ($1 \cdot 545 P_{GD}$) the plastic zone has been increased only in depth. FE suggests that plasticity is followed by minor flexural cracks and some premature shear cracks initiating within this zone. Radial-cum-circumferential cracks have also been predicted by model tests.

At this pressure the top haunch crack indicated has been extended and so has the bottom haunch crack under this loading. These crack positions are in full agreement with those from the model tests M. Due to high stress concentrations produced in the top haunch by the model tests, there might be evidence of some kind of cracking but this has not been confirmed. Both FE and M predict surface cracking at the equator of the vessel. Clearly this

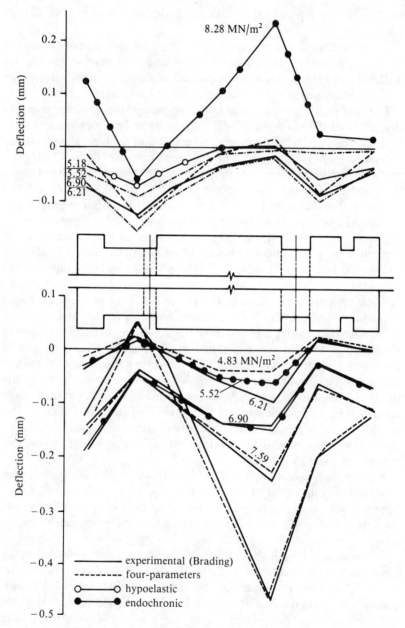

Fig. VIII.83a. Deflected shape—sides of model, ultimate condition.

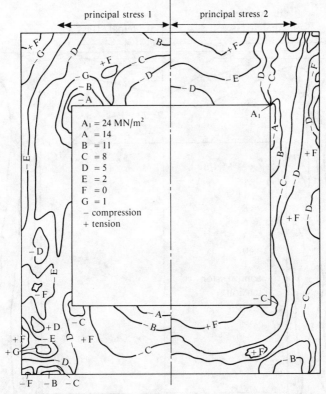

Fig. VIII.83b. Principal stresses 1 and 2—normal operation.

again has not been revealed in published test data. At a pressure of 5·8 MN/m² (1·76P_{GD}), FE data indicate no change in the top cap plastic zone. It is clearly established that the localised shear-compression cracks predicted in the top cap by the model tests M have not changed much. There is an overall improvement in the plastic zones in the haunch areas.

However, there is no change of the loading in the surface crack at the equator except that FE indicates a slight extension of the plasticity zone at and around the equator with a possibility of isolated flexural cracks. Just opposite to the bottom haunch crack at the outside face of the vessel, a plastic zone has developed. Simultaneously, another plastic zone appears in areas along the centre line of the bottom cap. They have both been predicted by FE. No evidence is available from model tests.

When the pressure is raised to 6·5 MN/m² (1·97P_{GD}) it becomes evident

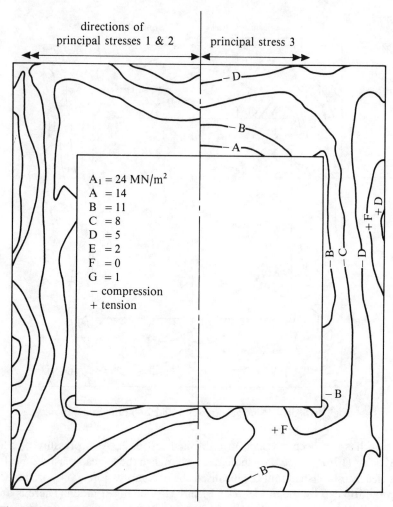

Fig. VIII.84. Principal stress 3 and directions of the major principal stresses—normal operation.

that the top cap plastic zone has now been slightly modified by the FE data. In addition, both flexural and shear cracks in isolated pockets within this plasticity zone are visible.

The plastic zones at the haunches and areas close to them predicted by FE are now suddenly increased. These zones have practically the same size as predicted by M earlier.

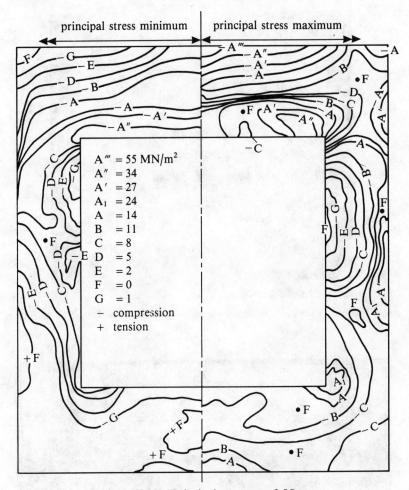

Fig. VIII.85. Principal stresses at $2.5 P_{GD}$.

However, there is a change in the plastic zone in the bottom cap; but at the equator FE and M show flexural cracks, although the positions indicated by both vary. The experimental tests [418–423] show only surface flexural cracks half-way between those predicted by the finite element analysis.

When the pressure is increased to $7.2\,\text{MN/m}^2$ ($2.14 P_{GD}$) the plastic zone has been extended far beyond that predicted in the earlier results. Nevertheless, within this zone the crack sizes predicted by FE have been

Fig. VIII.86a. Flexural failure of caps (courtesy of Atomic Power Construction and ICE [418]).

Fig. VIII.86b. Plug-cum-flexure failure (courtesy of Professor Scotto, ENL, Italy).

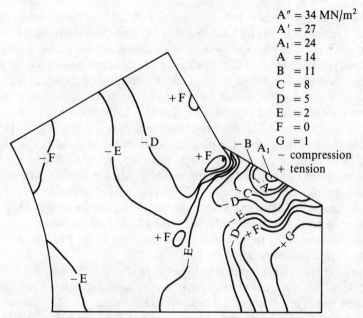

Fig. VIII.87a. Maximum principal stresses—normal operation.

$A'' = 34 \text{ MN/m}^2$
$A' = 27$
$A_1 = 24$
$A = 14$
$B = 11$
$C = 8$
$D = 5$
$E = 2$
$F = 0$
$G = 1$
− compression
+ tension

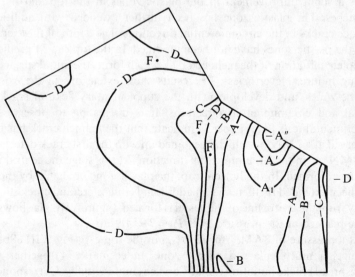

Fig. VIII.87b. Minimum principal stresses—normal operation.

stabilised and no further change has been observed. The shear compression crack and the inclined crack predicted by FE have been enlarged. This has been observed using M and the results are superimposed on the emerging inclined crack. There is a further extension of plasticity zones in the top and bottom haunches predicted by FE. From FE results and change in the wall cracks in all directions this is negligible, as there is no substantial improvement on the previous results. However, at the haunch in the bottom cap a two-pronged crack has been observed. This is due to the fact that results from M indicate crack positions both in the cap and in the barrel wall. The bottom cap plastic zone near the centre line predicted by FE has been extended radially and circumferentially with a possibility of a minor flexural crack near the centre line. There is a marked improvement in the plastic zone at the equator and near the outside face of the wall. Additional superficial cracks at the outside face have also been predicted within this zone. It is interesting to note that the plastic zone under FE is almost the same as that predicted by M.

Now, with a further increase in pressure reaching 7.9 MN/m^2 ($2.39 P_{GD}$), additional flexural cracks have been observed using FE. Some cracks have disappeared in minor areas and new cracks have emerged. The crack at the top haunch is two-pronged and jointed together at the corner. From M cracks have been predicted in the top cap and the inside zone in the wall. FE gives a slight improvement in the plastic zones in the top and bottom haunches. The plastic zone has been further extended with additional surface cracks in the circumferential directions. The depth of these cracks and the plastic zones have not been modified. In the top cap M predicts a complete initiation of the inclined crack which forms the initial surface for a plug failure. Nevertheless, FE results show extensions in the existing plastic zones, and development in the supplementary flexural cracks in radial and circumferential directions. It is interesting to observe that experimental tests [418, paper 3] indicate that the haunch crack has been observed at 7.7 MN/m^2 which has joined up with the first crack detected at 5.18 MN/m^2 in the circumferential direction. At this stage the central plug has been formed. In the bottom cap, the plastic zone predicted by the FE has been extended further, and an additional circumferential crack slightly away from the centre line of the vessel is formed. Figure VIII.88a shows the stage reached at the pressure level $P_G = 7.9 \text{ MN/m}^2$.

At a pressure of 8.6 MN/m^2 ($2.6 P_{GD}$) in the top cap (Fig. VIII.88b) FE predicts a much enlarged plastic zone under mark 74 together with widespread cracks initiating radially and circumferentially. Corresponding to this limited state analysis (LS) shows cracking (mark 60) with a visible

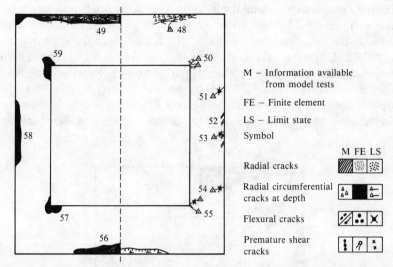

Fig. VIII.88a. Dungeness B prestressed concrete reactor vessel ($P_G = 7 \cdot 9 \, \text{MN/m}^2$).

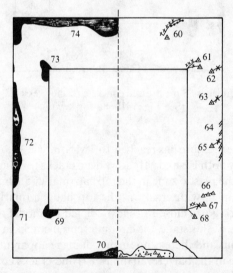

Fig. VIII.88b. Dungeness B prestressed concrete reactor vessel ($P_G = 8 \cdot 6 \, \text{MN/m}^2$). (See Fig. VIII.88a for key.)

line for a plug failure. Along the barrel wall an additional crack (mark 62) has been predicted by M. Cracks under marks 61, 63, 65, 67 and 68 have been slightly modified and extended. The results from the experimental tests [418–423] show a crack above the duct (mark 66). This is in line with the plastic zone and cracks developed (mark 72) by FE. Results from LS do not predict a crack at this stage. No change has been envisaged by FE in the top and bottom haunch plastic zones (mark 73 and 69) and the bottom cap plastic zone, and cracking radially and circumferentially under mark 70.

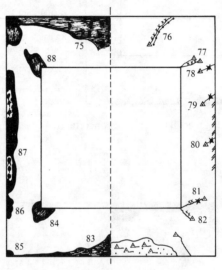

Fig. VIII.89. Dungeness B prestressed concrete reactor vessel ($P_G = 10.7 \text{ MN/m}^2$). (See Fig. VIII.88a for key.)

Now when the pressure has reached 10.7 MN/m^2 ($\approx 3.25 P_{GD}$) the top cap (Fig. VIII.89) (by both FE and M) shows deep cracks for plug failure (marks 75 and 76) and the plastic zone in the top cap (mark 75) almost reaches the plastic zone coupled with cracks at the top haunch (mark 88). The plug failure line (mark 76) is close to that predicted by model tests [418–423]. Apart from that, at this stage, the circumferential cracks in the top cap are still discontinuous and there is no further damage apparent to the top cap. The plastic zone in the barrel wall (mark 87) and cracks contained therein have enlarged their boundaries. The trend is identical to that exhibited by LS over the area shown under marks 77 to 82. In both cases cracks have not gone so deep despite excessive deflections. The bottom haunch (mark 84) and the bottom cap (marks 83 and 85) have their plastic zones extended and visible cracks have been predicted by FE. Corresponding to this the M shows an inclined crack, indicating the first complete initiation of the plug failure. The corner crack 71 (Fig. VIII.89) discussed earlier has been stabilised but an additional plastic zone has been discovered under mark 86.

At this stage it is fairly evident that both caps are more vulnerable to failure than the barrel wall.

When the pressure is raised to $12{\cdot}8\,\mathrm{MN/m^2}$ ($3{\cdot}88P_{\mathrm{GD}}$) new cracks have been predicted by LS and now the total number of cracks is shown under marks 90 to 98 in Fig. VIII.90. Both M and FE have predicted the extended failure zone in the top cap (marks 89 and 105). It is interesting to observe that both LS and FE crack zones (marks 102 to 104) have been extended further along the height of the vessel but have not made any further advance within the thickness of the wall. On the other hand, both top and bottom haunch cracks (marks 101 and 104) have made a significant advance. The bottom cap plastic zone (mark 101) has reached the plastic zone of the wall (mark 102). The continuity has not yet been broken. On the other hand, the top haunch plastic zone (mark 104) has reached the wall

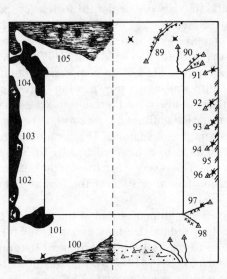

Fig. VIII.90. Dungeness B prestressed concrete reactor vessel ($P_{\mathrm{G}} = 128\,\mathrm{MN/m^2}$). (See Fig. VIII.88a for key.)

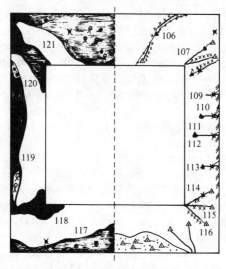

Fig. VIII.91. Dungeness B prestressed concrete reactor vessel ($P_G = 13 \cdot 5 \, \text{MN/m}^2$). (See Fig. VIII.88a for key.)

under its own influence. However, there is still a continuity. The bottom cap plastic zone (mark 100) has been enlarged with a few additional cracks. Here LS gives two closed cracks in the bottom cap but the depth of these cracks remains practically the same.

Now, raising the pressure to $13 \cdot 5 \, \text{MN/m}^2$ ($4 \cdot 1 P_{GD}$) there is a complete plug failure in the top cap, and a shear–compression failure at the junction of the top cap and the wall, as are shown in Fig. VIII.91 under marks 106, 107, 120 and 121. The study of the LS results indicates that haunch cracks (mark 107) have made a significant advance within the thickness of the wall causing a shear–compression failure. The remaining cracks in the wall (marks 107 to 115) have, likewise, predicted by FE under mark 119, not made much progress. This is very much true for the bottom cap, except that a shear–compression failure exists near the bottom haunch under mark 118.

With an assumed 2% ultimate strain in the tendons, the possibility of failure of some longitudinal tendons and a group of circumferential tendons lying close to marks 106, 107, 120 and 121 cannot be ruled out. From the experimental tests [418–423] it is evident that the central core (plug) has come out at a pressure of $13 \cdot 8 \, \text{MN/m}^2$ ($1850 \, \text{lb/in}^2$) and two tendons in the model have failed at a pressure of $10 \cdot 2 \, \text{MN/m}^2$.

It is concluded that the failure of this vessel has occurred owing to the complete removal of the central core or plug in the top cap at a pressure around four times the design pressure. The FE and the M analyses are in good agreement with each other and with most of the published data from the experimental tests. Throughout the vessel there is no premature failure of the liner.

(b) Oldbury vessel. For the model test of the Oldbury vessel, gas pressure versus deflections is plotted in Figs VIII.92 and VIII.93. The results from finite element analysis have been plotted for comparison in the same figures. Because of great variations in tendon layout, the pressure versus deflection curves are given for various positions or locations. Therefore each curve takes into account the contribution it has received from the exact position of tendons, conventional steel and the liner in the original layout. In order to test these results, some of them are plotted in Fig. VIII.94 along with those from the model tests (four tests) carried out by Eddie *et al.* and others [418–423]. Both the horizontal and vertical deflections are in good agreement. However, in Fig. VIII.95 although the computed axial strains are in good agreement with those measured on the prototype [418–423], the hoop or circumferential strains are wide apart. The only explanation that can be offered is the difference between the exact chosen tendon layout of the vessel parameters and that of the prototype. The percentage difference is still within 5–8%. A further comparison is given in Fig. VIII.95(a–c) which is based on the monitoring of the vessel's performance for a period of 10 years.

Figures VIII.96 and VIII.97 show plots for various principal stresses. It is interesting to note that in Fig. VIII.98 the effect of removal of cracked concrete at 1·81 times the design pressure is shown. Figures VIII.99 and VIII.100 show the principal stresses at 2·5 times design pressure. No cracks have been observed for a pressure range of 2·66–3·75 MN/m². At a pressure of 3·99 MN/m² ($1·455 P_{GD}$) a plastic zone in the top cap centre has been predicted by FE. Both FE and M show a plastic zone and a minor crack at the top haunch. Radial hair cracks together with a plastic zone have also been predicted by FE. However, M assumes a crack width at the equator and finally proves that it is the only crack that exists in the wall at this pressure.

At a pressure of 4·665 MN/m² ($1·75 P_{GD}$), a wider and deeper plastic zone appeared (mark 6) in the top cap. It is interesting to note that the top haunch plastic zone has also been increased. In addition, a minor plastic zone has been predicted by FE at the bottom cap haunch. Moreover, at the

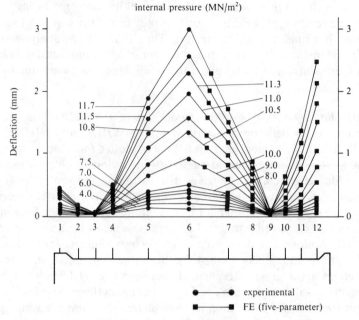

Fig. VIII.92. Horizontal deflection of wall (ultimate pressure)—Oldbury vessel. [418; experimental]

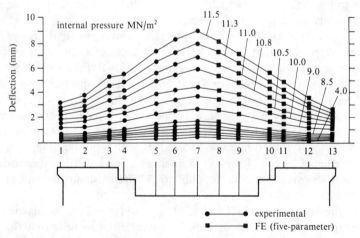

Fig. VIII.93. Vertical deflection of top cap (ultimate pressure)—Oldbury vessel. [418; experimental]

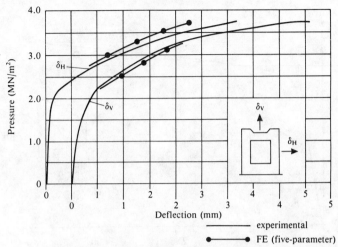

Fig. VIII.94. Pressure–deflection curve of Oldbury vessel.
[418; experimental]

upper part of the side wall a wider radial crack has occurred and this crack has also been confirmed by the model tests, but at a pressure of 4·825 MN/m² (1·815P_{GD}). This must have been produced even earlier in the model, since such a crack cannot suddenly appear at such a marginal pressure difference of 0·065P_{GD}. The results from FE clearly indicate that there is a plastic zone appearing at the equator.

Raising the pressure to 5·32 MN/m² (1·995P_{GD}) FE indicates that the vessel top cap plastic zone has been abruptly increased and additional flexural cracks have emerged. The plastic zone in this case has reached the areas surrounding the central hole in the stand pipe area and the prestressing galleries. Model tests [418–423] show similar effects in the top cap. The results from M give two surface flexural cracks and a shear–compression crack initiating from the gallery zone. Both top and bottom haunch plastic zones of this vessel predicted by FE are fairly large. Here there is some disagreement between FE and M on the type of crack. The top haunch M results show both flexural and premature shear cracks. On the other hand, the results from FE clearly indicate the possibility of cracks caused by the principal tensile stresses caused by shear. However, there is an agreement between the two analyses on the initiation of flexural cracks in the bottom cap haunch. The magnitude of these cracks is not greater than 0·007 mm. Practically there is no change in the flexural crack. The bottom cap has an enlarged plastic zone.

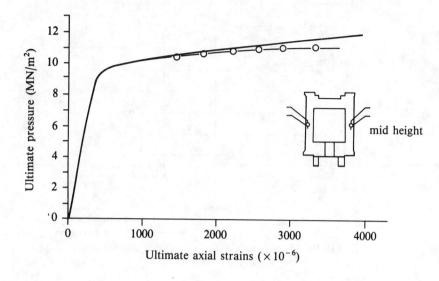

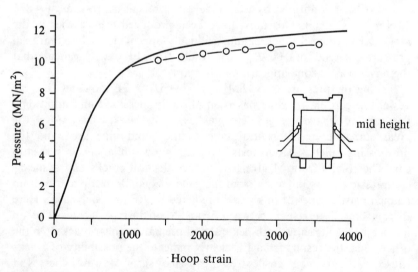

Fig. VIII.95. Hoop and axial strains at wall mid-height—Oldbury reactor vessel (conditions at ultimate pressure). (——, Finite element (five-parameter); —○—, experimental.)

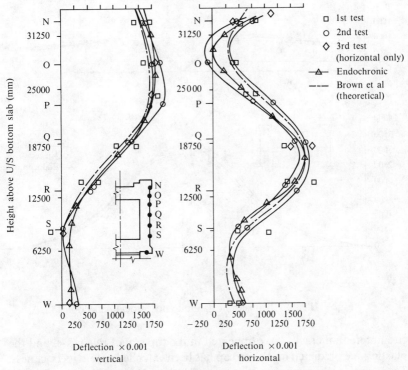

Fig. VIII.95a. Wall horizontal and vertical deflections.

Reports [418–424] suggest that a horizontal-cum-radial crack has appeared above the circulator penetration, but the published data [418–428] have not confirmed it. In the barrel wall M shows flexural cracks in both radial and circumferential directions. Examination shows that cracks in radial directions do exist at pressures between 4·664 and 5·32 MN/m². They have been superimposed for comparison. The results obtained from FE show plastic zones and surface cracks. Obviously there is a slight discrepancy. It seems that FE does not agree with M up to a pressure of 5·32 MN/m² for the creation of plastic zones or surface cracks. The difference is in the modelling of the layouts. No test data are available for this vessel or any other cylindrical vessels [418–428] for such cracks appearing at 1·995 times the design pressure. However, there seems to be some agreement in this area at pressures above 5·32 MN/m².

For a pressure of 5·985 MN/m² ($2·26P_{GD}$) additional radial and

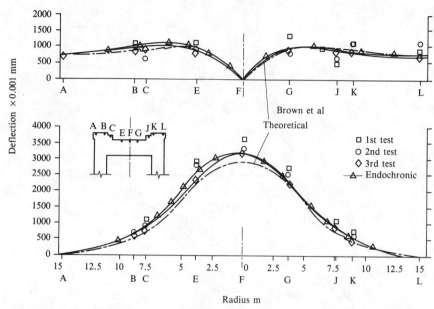

Fig. VIII.95b. Top slab horizontal and vertical deflections.

circumferential cracks have appeared in the top cap of this vessel and the plastic zone predicted in the top cap has been extended within its thickness. There is virtually no change in the overall length of this plastic zone. Results from M predict two flexural cracks, one within and the other just outside the stand pipe perimeter. In addition, the shear–compression failure line has been initiated. At the outside of the top prestressing gallery, a circumferential crack has been predicted in a localised zone. However, in the top cap haunch (mark 27) FE predicts a much wider plastic zone. Here both FE and M are in agreement with the model test results. At the upper part of the vessel side wall, both FE and the model tests are now in agreement.

In the bottom cap haunch the plastic zone predicted by FE is now slightly changed in shape but still no crack has appeared. In the bottom cap itself, the plastic zone predicted by FE now extends from the centre line of the vessel to the inside edge of the lower prestressing gallery. Some flexural cracks are visible in the bottom cap of a magnitude not greater than 0·009 mm. Here both M and FE partially agree with each other. No information from model tests is available to support this. However, in critical regions of the wall cracks have appeared and here both FE and M

Application to Engineering Problems 329

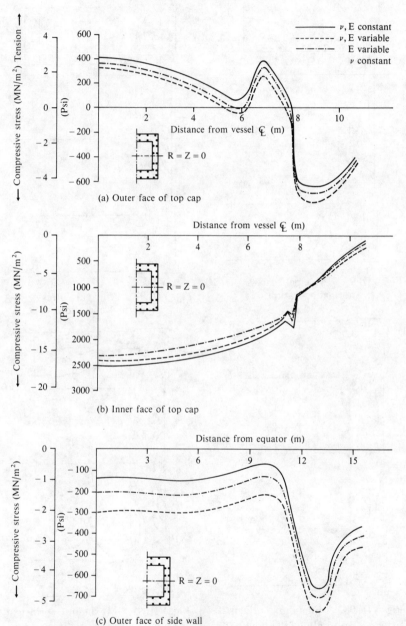

Fig. VIII.95c. Circumferential stresses—operating conditions only.

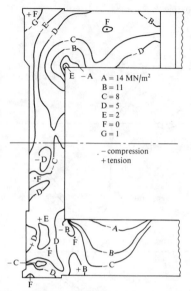

Fig. VIII.96. Principal stress 1—normal operation.

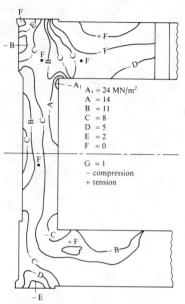

Fig. VIII.97. Principal stress 2—normal operation.

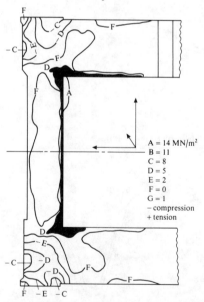

Fig. VIII.98. Minimum principal stress—pressurisation after long-term shut-down showing effect of removal of cracked concrete (A).

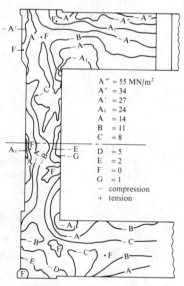

Fig. VIII.99. Maximum principal stresses at $2 \cdot 5 P_{GD}$.

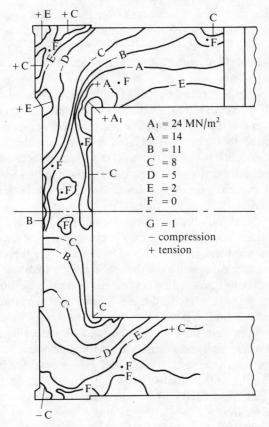

Fig. VIII.100. Minimum principal stresses at $2 \cdot 5 P_{GD}$.

are in full agreement. Again within this area model cracks measured from test data [418–430] have been shown side-by-side with the computed cracks. Finite element analysis predicts radial cracks, and model tests [418–430] show that these cracks are justifiable, although the FE 'spread zone' is twice that of the model test data.

At a pressure of $6 \cdot 65 \text{ MN/m}^2$ ($2 \cdot 5 P_{GD}$) additional cracks have appeared and the plastic zones have been increased along the thickness of the vessels. Here only lengths and widths of cracks in the radial and circumferential directions have been extended.

It is interesting to note that in the top cap there is a straight competition between the flexural and premature shear cracks. Superimposed on these are the reproduction of cracks obtained from the cracked model [418–431]. Both crack pattern analyses from the model in the top cap are very similar. One novel feature is the re-emergence of the radial crack predicted by the model test [418–431] reaching the circular penetration. This has also been predicted by FE.

When the pressure is raised to 7·98 MN/m² ($3·0P_{GD}$) model test results in the top cap of the vessel have shown zones of considerable cracking. In the actual prototype vessel, FE predicts a crack zone, but not as much as is given by the model test. On the other hand, in the top cap the flexural failure zone advances beyond the plug failure. At this stage it is necessary to show a cracked (before failure model) model, which is given in Fig. VIII.101. Apart from a slight modification of the plastic zones, the cracking behaviour is fairly consistent.

As shown in Fig. VIII.101 for a pressure of 8·645 MN/m² ($3·25P_{GD}$), the top cap plasticity zone predicted by FE comes very close to the model prediction, as is evident from marks 65 and 68. In the top cap, the plug failure line is catching up with the flexural cracks, and at one place it is a pure shear failure line proceeding beyond the flexural zone. Both haunches (marks 67 and 76) have not gained much. The bottom cap (mark 77) plastic zone and cracks have increased in size and in number. The sizes of these cracks is no more than 0·9 mm, but the tendons have also reached 3% strain, which is very high and almost reaches the yield point. The barrel wall is developing cracks and plastic zones under pure flexure. The biggest crack in the barrel wall is at the equator. Most of the top gallery zone in the top cap has cracked. Model tests predicted a portion of gallery cracking at a pressure of 7·98 MN/m².

When the pressure is raised to 9·975 MN/m² ($3·75P_{GD}$), cracks have changed their positions (Fig. VIII.102), but the failure patterns given by FE and M are consistent. The plastic zones near the top gallery (mark 79) have been extended. It seems that the top haunch plastic zone is very close to the one approaching from the top gallery. Both bottom cap (marks 82 and 83) and the barrel wall plastic zones predicted by FE have been very slightly modified. Tendons passing through these cracks have almost reached 3·8% strain, and their failure has been predicted. The top cap is broken into two sectors. The longitudinal wall cracks have joined the circumferential cracks of the wall and the caps.

At a pressure of 11·305 MN/m² ($4·24P_{GD}$) the plastic zone at the top haunch has reached the gallery zone (mark 85), as shown in Fig. VIII.103. Both in the wall and in the caps the cracks have advanced further. Model test results are in agreement with those predicted by FE and M. However, at

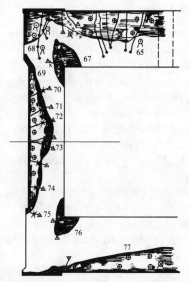

Fig. VIII.101. Oldbury prestressed concrete reactor pressure vessel ($P_G = 8\cdot645$ MN/m^2). (See Fig. VIII.88a for key.)

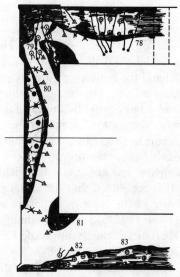

Fig. VIII.102. Oldbury prestressed concrete reactor pressure vessel ($P_G = 9\cdot975$ MN/m^2). (See Fig. VIII.88a for key.)

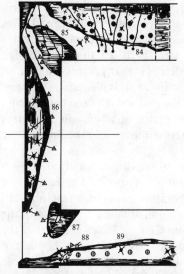

Fig. VIII.103. Oldbury prestressed concrete reactor pressure vessel ($P_G = 11\cdot505$ MN/m^2). (See Fig. VIII.88a for key.)

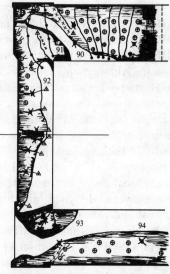

Fig. VIII.104. Oldbury prestressed concrete reactor pressure vessel ($P_G = 11\cdot97$ MN/m^2). (See Fig. VIII.88a for key.)

a pressure of 11·97 MN/m² (4·5P_{GD}) the vessel top cap and top corner (Fig. VIII.103) have been separated. The bottom cap haunch plastic zone has reached the wall plastic zone but the cracks (mark 93) have not joined. The wall and the bottom cap are broken into several pieces but have not been separated yet. The top cap failure is predominantly flexural, although shear–compression failure causing the removal of the plug is just one step away. The tendons in most places have reached 4·25% strain. The crack sizes are not exceeding permissible values, and premature liner failure is not possible provided no initial defects exist. The conventional steel at the haunches and at equator has yielded.

It is concluded that the vessel predominant failure is governed by the failure of the top cap in flexure and the tendons at a pressure of 9·975 MN/m² (3·75P_{GD}). The remaining failure mechanisms (Figs VIII.103 and VIII.104) are purely an academic exercise for investigating any reserve strength left in the vessel.

This failure mechanism is different from that predicted for the Dungeness B vessel, although they both have an identical shape and reactor system layout but with different prestressing system and gas design pressure.

(c) High-temperature gas-cooled reactor vessel (HTGCR). Figures VIII.105 and VIII.106 show gas pressure versus deflection, and Figs VIII.107 and VIII.108 show principal stresses at normal operation. Figures VIII.109 and VIII.110 show principal stresses at 2·5 times the design pressure. Although this vessel layout is different in many respects to that of the Hartlepool reactor, the stress flow in many areas is identical. The stress concentrations around the boilers and at the junction between the caps and the walls are similar in both cases. However, the quantities do differ. It is because of these and some dissimilar stress contours that the final failure modes in both vessels differ.

Figures VIII.79 and VIII.80 show three-dimensional and two-dimensional finite element mesh generation schemes of a prestressed concrete vessel designed for a high-temperature gas-cooled reactor. Both internal and external loads have been computed in accordance with the method given in Chapter VI. The program CREEP, which is a part of the program ISOPAR, gives results for when the vessel undergoes the effect of creep. First a normal operation condition (prestress + pressure + temperature) is considered. This vessel is analysed with and without creep. The axial, radial and circumferential stresses have been computed. The influence of creep when the Young's modulus is constant or variable is

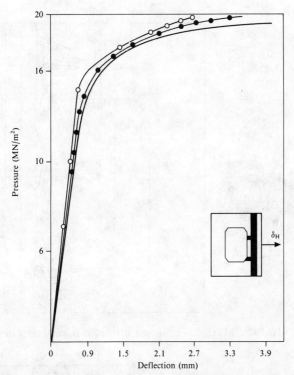

Fig. VIII.105. Deflection curve of wall of HTGCR reactor vessel. (———, Experimental; —○—, FE (four-parameter); —●—, FE (five-parameter).)

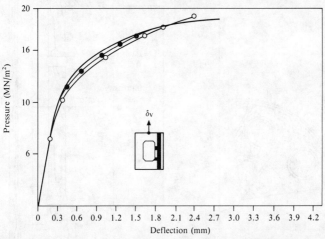

Fig. VIII.106. Pressure–deflection curve for top cap of HTGCR reactor vessel. (———, Experimental; —○—, FE (four-parameter); —●—, FE (five-parameter).)

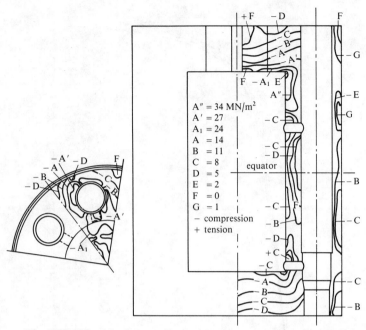

Fig. VIII.107. Maximum principal stresses at normal operation.

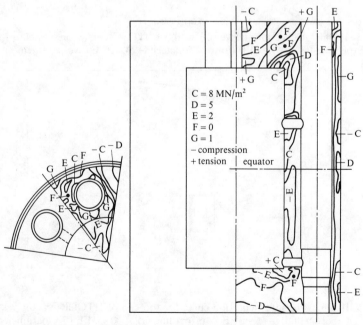

Fig. VIII.108. Minimum principal stresses at normal operation.

Application to Engineering Problems

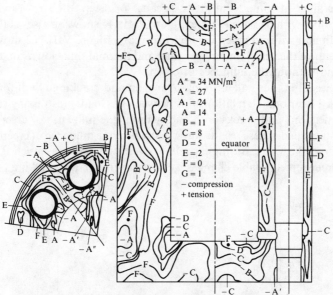

Fig. VIII.109. Maximum principal stresses at $2{\cdot}5P_{GD}$.

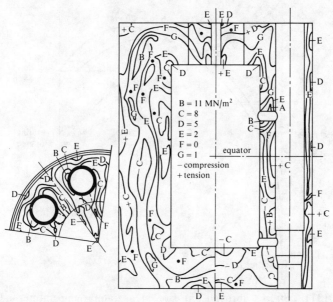

Fig. VIII.110. Minimum principal stresses at $2{\cdot}5P_{GD}$.

considerable. In some local areas, because of creep, the stresses change significantly with passage of time. Figure VIII.111 shows stresses at the top cap with and without creep for up to 20 years. Figure VIII.112 shows the behaviour of the vessel under operating and thermal shut-down conditions with and without the influence of creep.

The vessel is then taken to elasto-plastic and cracking conditions. As discussed later on, the relationships between gas load, prestressing strains, and deflections and cracks with and without the influence of creep have been established. These results show that where the influence of creep is not considered, the load-carrying capacity is overestimated by as much as 25%. With creep, considerable changes were discovered in the cracking pattern

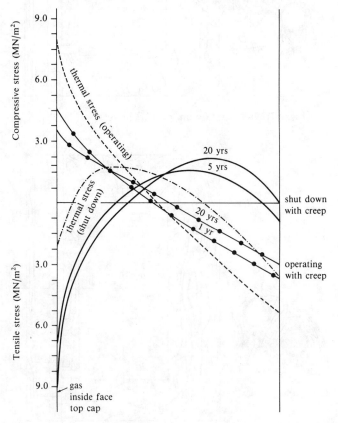

Fig. VIII.111. Stresses through top cap—at operating temperature and shut-down.

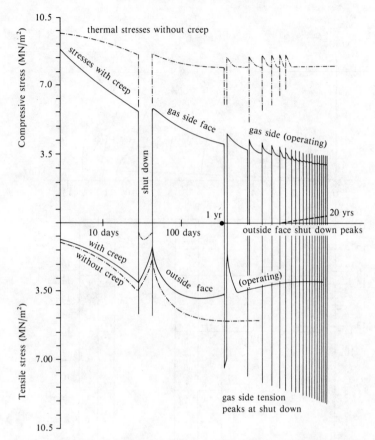

Fig. VIII.112. Stresses in top cap showing influence of shut-down.

of the vessel and the overstress conditions in local areas (stand pipes and boiler/circulator penetrations).

Ahmad [262] has idealised the same vessel, as shown in Fig. VIII.113. He has used the vessel bonded and unbonded, and produced deformations (Figs VIII.113–VIII.116) for normal operation and for 40 years of creep. He has also produced a graph (Fig. VIII.117) showing deformation for this vessel at various pressures. He has included creep in his analysis. This vessel has also been considered for elasto-plastic and cracking effects by Bangash [749]. The deformation of the vessel at the cracking stage under the influence of creep and with the steel liner anchored to concrete is shown in Fig. VIII.117. The analyses for these cases are given in Chapters VI and VII.

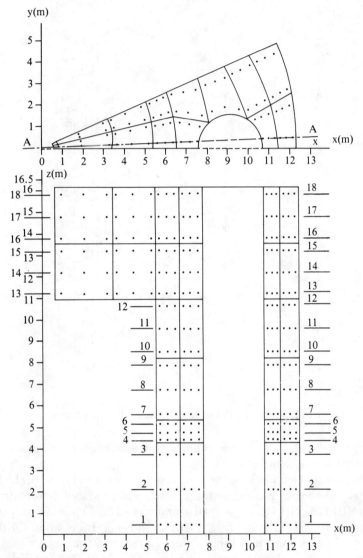

Fig. VIII.113. Gauss point locations in the reactor vessel [262] (all dimensions in mm).

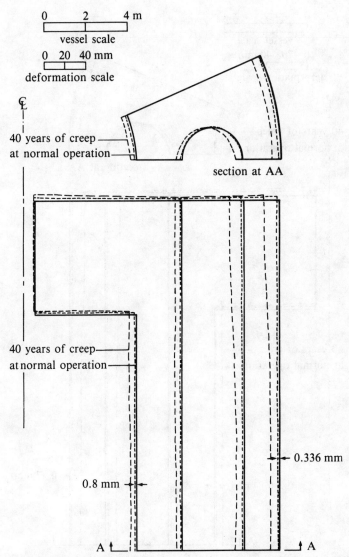

Fig. VIII.114. Deformations under normal operation and after 40 years of creep [262]—vessel type, bonded (with Ahmlink element).

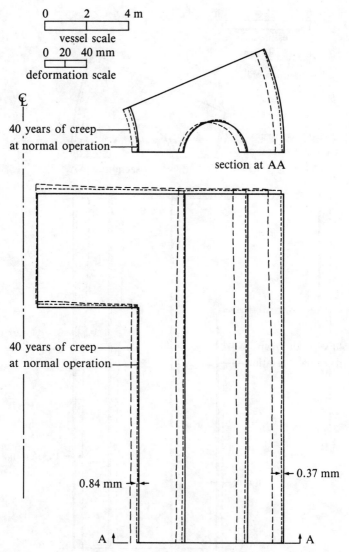

Fig. VIII.115. Deformations under normal operation and after 40 years of creep [262]—vessel type, perfectly bonded.

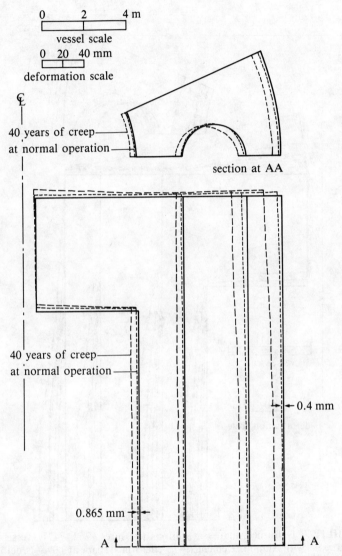

Fig. VIII.116. Deformations under normal operation and after 40 years of creep [262]—vessel type, unbonded.

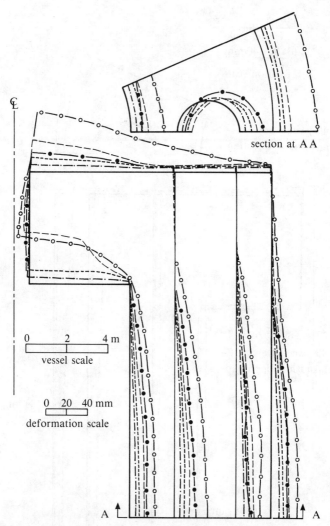

Fig. VIII.117. Deformed shapes at various pressures [262, 749]—vessel type, bonded vessel with Ahmlink element. (—·—, deformed shape at 3·0; ---, deformed shape at 3·3; ———, deformed shape at 3·5; —○—, FE (four-parameter, Bangash); —●—, FE (five-parameter, Bangash).)

The stress trajectories at normal operation and at 2·5 times design pressures can be obtained in the same manner as for vessels discussed previously. The vessel design pressure P_{GD} here is 5·68 MN/m². At a pressure of 7·48 MN/m² (1·315P_{GD}), this vessel's behaviour (Fig. VIII.118) is identical to that of the Hartlepool vessel [418–428]. The only change is the plastic zone developed around the gas inlet duct under mark 6.

At a pressure of 9·28 MN/m² (1·635P_{GD}) the behaviour of this vessel (Fig. VIII.119) in most zones is identical to that of the Hartlepool vessel [423, papers H3/5, H5/3] and others [423, 424]. When the pressure was raised to 6 MN/m² the only difference is in the position of cracks (marks 11, 12 and 13) at the top haunch and the extent of the plastic zones (marks 7, 14, 20, 21 and 22). The crack sizes have not been predicted in the top cap. The bottom cap has been affected only at the bottom haunch. Both FE and M are in agreement in most areas, as shown in Fig. VIII.119. Raising the pressure to 11·08 MN/m² (1·95P_{GD}) there is a significant difference in the behaviour of this vessel (Fig. VIII.120).

The HTGCR vessel failure mode is moving more towards flexural criteria than shear criteria. The initiation and propagation of cracks and their positions is very different. The reason is purely the dominant role that the barrel wall/cap slenderness ratio has played in producing flexural cracks. In plan around the boiler and circulator penetrations and in the stand pipe area, most of the plastic zones are identical.

At a pressure of 12·88 MN/m² (2·24P_{GD}) the top cap mode (marks 46 to 53) is given in Fig. VIII.121.

There is a similarity in the development of plastic zones between the two vessels in the bottom caps. In plan, the plastic patch zones are very similar between these two vessels. The only difference is the extent of cracks developed at the equator (mark 49) by FE. At this pressure in the top cap the plug failure (mark 60) is imminent. It is interesting to note that the crack sizes are identical in these zones. Another difference is in the sudden appearance of a plastic zone just above the gas inlet (mark 55) and a flexural crack in the circumferential direction in the top corner (mark 54) at the outside of the boiler penetration. There is also a slight shift in the position of the crack (mark 57) around the circulator. The crack widths have reached about 0·2 mm in most places. In many places strains in the tendons and in the bands do not exceed 0·009.

Raising the pressure to 14·68 MN/m² (2·58P_{GD}), sufficient flexural cracking in the barrel wall (Fig. VIII.122) has occurred.

The nearest case that can be compared is the behaviour of the Hartlepool vessel [418–431] under 2·5P_{GD}. In the top cap, plastic zones have identical

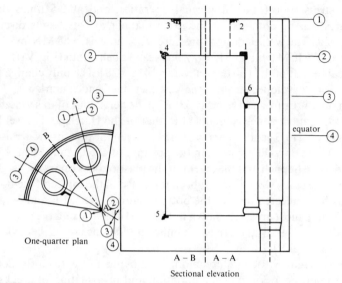

Fig. VIII.118. High-temperature gas-cooled reactor pressure vessel ($P_G = 7.48 \, \text{MN/m}^2$). (See Fig. VIII.88a for key.)

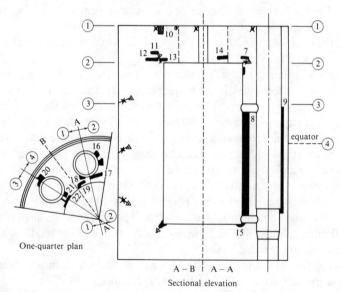

Fig. VIII.119. High-temperature gas-cooled reactor pressure vessel ($P_G = 9.28 \, \text{MN/m}^2$). (See Fig. VIII.88a for key.)

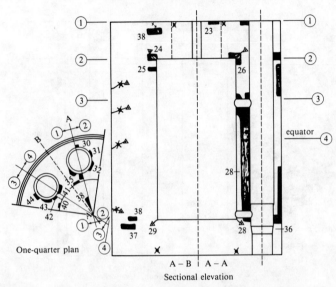

Fig. VIII.120. High-temperature gas-cooled reactor pressure vessel ($P_G = 11.08 \text{ MN/m}^2$). (See Fig. VIII.88a for key.)

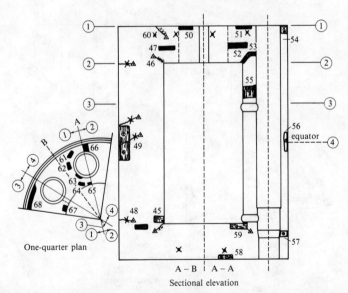

Fig. VIII.121. High-temperature gas-cooled reactor pressure vessel ($P_G = 12.88 \text{ MN/m}^2$). (See Fig. VIII.88a for key.)

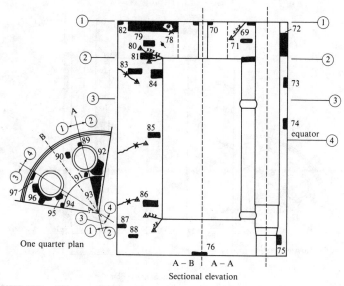

Fig. VIII.122. High-temperature gas-cooled reactor pressure vessel ($P_G = 14{\cdot}68\,\text{MN/m}^2$). (See Fig. VIII.88a for key.)

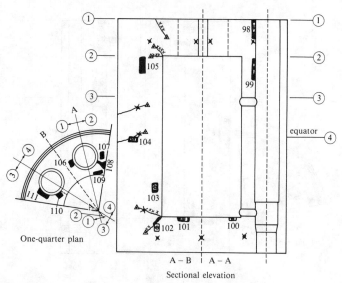

Fig. VIII.123. High-temperature gas-cooled reactor pressure vessel ($P_G = 16{\cdot}48\,\text{MN/m}^2$). (See Fig. VIII.88a for key.)

locations. However, in the HTGCR vessel no internal cracking has been predicted by LS around the boiler and the circulator penetrations.

In plan, the plastic zones are quite similar. There is a change in the plastic zone around the boiler penetration between sections (3)-(3) and (4)-(4). In the top cap a plug failure under mark 69 has been initiated. The strain and crack widths in most places do not exceed 0·01% and 30 mm.

At a pressure of 16·48 MN/m^2 (2·81P_{GD}) the behaviour of this vessel (Fig. VIII.123) is identical in many respects to the behaviour of the Hartlepool vessel [418–431] for 2·82P_{GD}. In plan, the cracks and plastic zones are very similar in both cases. In both the top and bottom caps flexural cracks have been extended in depth. Not enough change is visible of the plug failure line in the top cap. In addition, the barrel (Figs VIII.122 and VIII.123) wall cracks and the haunch cracks have advanced considerably. These flexural cracks are more pronounced than in the case of the Hartlepool vessel. The circumferential prestressing bands have reached the ultimate strain of 1·5% and the cracks in most places are around 50 mm. When the pressure is raised to 18·28 MN/m^2 (3·21P_{GD}), Fig. VIII.124a shows the positions of plastic zones between sections (1)-(1) and (2)-(2).

Additional flexural cracks have appeared in the top cap. It seems that the influence of cracks due to shear compression and plug failure have been minimised. At the equator (marks 115 and 116) a larger plastic zone has been created. The bottom haunch crack (mark 117) position is another difference in the behaviour of these two vessels. The plastic zones also differ from those in the Hartlepool vessel. Their shapes and positions and crack sizes do not match. This again is due to the difference in the dominant role being played by the span/depth ratios. Longitudinal prestressing has failed and the crack sizes are well over 90 mm. Bonded reinforcement has yielded. Although there is sufficient yielding of the liner, it still has not failed.

At a pressure of 20·08 MN/m^2 (3·55P_{GD}) both the top cap and the barrel wall have failed in flexure one after another. All cracks around the boilers and circulators have widened.

Circumferential cracks around the top haunch have joined the longitudinal cracks. The bottom cap has cracked but the various pieces are held together.

It is concluded that the vessel has failed due to the failure of a strand band at a pressure of 3·21P_{GD} and enough cracks have been produced in the top cap to define a premature flexural failure intervening in the shear failure in the top cap. Flexural failure has also occurred at several places in the barrel wall. Figures VIII.124a and b give additional cases for failure modes but these are not governing cases. Figure VIII.125 gives the deflection and

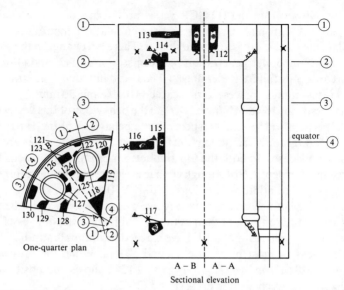

Fig. VIII.124a. High-temperature gas-cooled reactor pressure vessel ($P_G = 18{\cdot}28\,\text{MN/m}^2$). (See Fig. VIII.88a for key.)

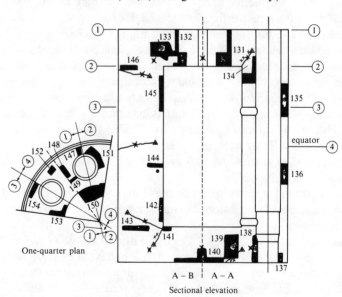

Fig. VIII.124b. High-temperature gas-cooled reactor pressure vessel ($P_G = 20{\cdot}08\,\text{MN/m}^2$). (See Fig. VIII.88a for key.)

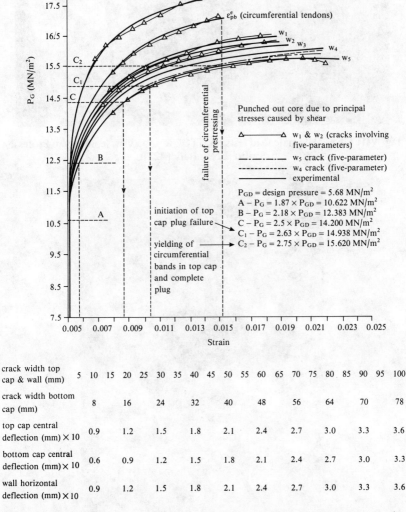

Fig. VIII.125. High-temperature gas-cooled reactor vessel pressure versus cracks and deflections.

Fig. VIII.126a. Ultimate load tests on HTGCR—top view (courtesy of Professor Scotto, ENL, Italy [423]).

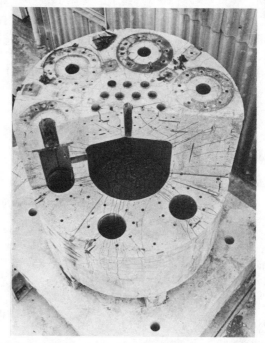

Fig. VIII.126b. Ultimate load tests on HTGCR—cutaway section (courtesy of Professor Scotto, ENL, Italy [423]).

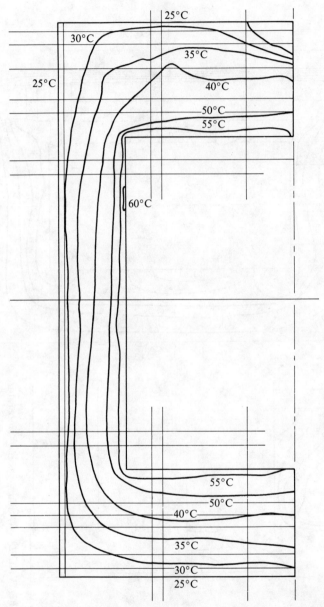

Fig. VIII.127. Dungeness B type vessel—temperature distribution.

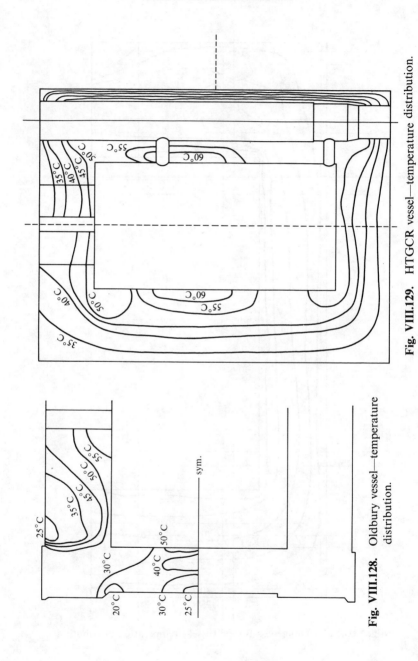

Fig. VIII.128. Oldbury vessel—temperature distribution.

Fig. VIII.129. HTGCR vessel—temperature distribution.

cracking history of this vessel. Figures VIII.126a and VIII.126b show cutaway sections of the HTGCR vessel of ENL (Italy), the cracking zones of which correspond with the final collapse mode of this vessel.

(d) Thermal analysis of vessels. Thermal or temperature analysis has been carried out for all vessels on the lines suggested in Chapters II, VI and VII. The analytical results include the effects of temperature. Figures VIII.127–VIII.130 give the results for the temperature distributions for the

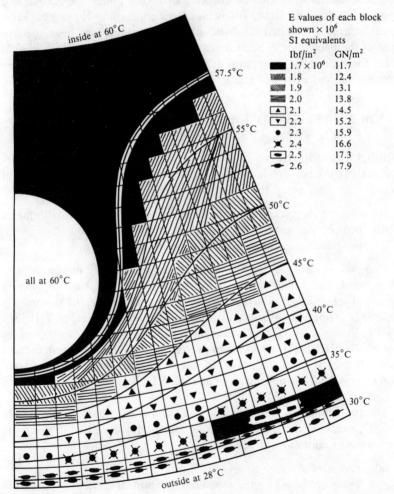

Fig. VIII.130. Temperature and Young's moduli at equator—HTGCR vessel.

vessels. Thermal loads are computed and are put at the nodes of the elements as concentrated loads or patch loads. Alternatively strains are evaluated from these temperatures and they are included in the finite element formulations as initial strains. Stress trajectories can be drawn for the temperature-only case on the lines given for vessels in normal operation—cases which in fact include temperature effects.

(e) General conclusions. Much fuller details are given here at various stages of the deformation and cracking of vessels under operational and overload conditions. In the following pages concerning circular concrete structures, such detailed analysis, although avoided, has been used to predict the behaviour of such structures. This analysis, therefore, should form the basis of investigation of all circular concrete structures under operational and extreme loads.

VIII.8 CASE STUDY No. 7—CONCRETE CONTAINMENT VESSELS

VIII.8.1 Introduction

Bangash [843] has produced a comprehensive analysis of containment vessels. The purpose of the containment is to:

(a) prevent the escape of radioactive materials into the environment;
(b) protect the reactor vessel from external hazards; and
(c) provide biological shielding against nuclear radiation.

Concrete vessels are divided into two classes, namely reinforced and prestressed vessels. The walls are generally of cylindrical shape supporting spherical, elliptical and torospherical domes. Figure VIII.131 shows a typical containment with spherical dome (Sizewell B). Figures VIII.132–VIII.135 show a typical reinforcement system. In many cases a secondary dome (Fig. VIII.131) is provided to protect the primary containment from external hazards and the public from the internal hazards. Some well-known vessel parameters are given in Table VIII.7. Various loading criteria are given in Table VIII.8 [843].

The type of such a vessel depends on the nuclear steam supply system (NSSS), site, environmental and economic conditions. The following are two types of containment vessels:

(a) *Pressurised Water Reactor (PWR)*
These are based on the pressure containment concept. Apart from overpressure conditions, these vessels are checked for high energy

Application to Engineering Problems

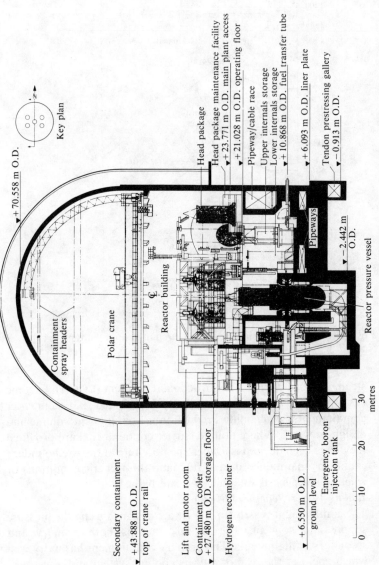

Fig. VIII.131. Sizewell B power station reactor building—sectional elevation on north/south centre line. (With compliments of CEGB, UK)

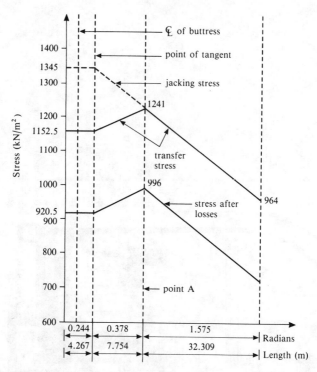

Fig. VIII.132. Post-tensioning loads—circumferential tendons [843].

steam released during a loss-of-coolant accident (LOCA). In some cases pressure suppression capability is provided (ICE-condenser containments) which allows reduction in containment volume and design pressure. Where double barrier containments are provided they have been devised for when the dose rate at the site boundary cannot be maintained during stagnant atmospheric conditions or where danger exists internally or externally.

(b) *Boiling Water Reactors (BWR)*

The containment vessels for BWRs are designed using the pressure suppression concept. The layout contains both primary and secondary containments. The primary containment has a dry well which encloses the reactor and its cooling system and a wet well which channels the steam released during a LOCA into the pool of water where a quenching process is carried out.

Application to Engineering Problems

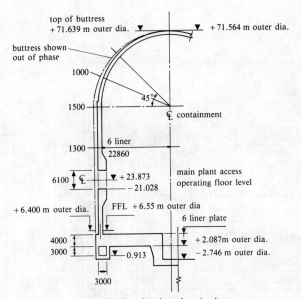

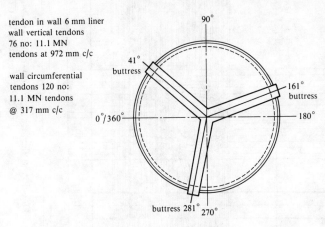

Fig. VIII.133. Sizewell B dimensional layout [599].

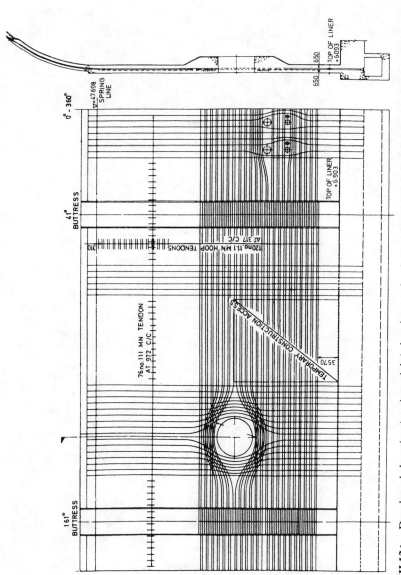

Fig. VIII.134. Developed elevation (section–right) showing tendon layout 0°–180° (all dimensions in mm except reduced level) [599] (courtesy of CEGB and Sizewell Task Force).

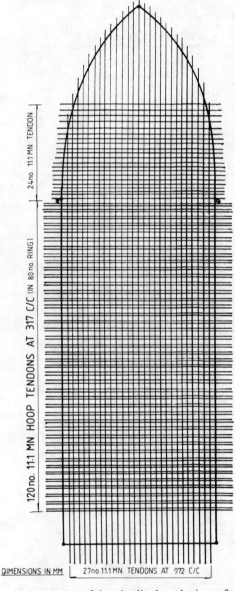

Fig. VIII.135. Arrangement of longitudinal and circumferential tendons (all dimensions in mm).

Table VIII.7
Available data for major vessels [843]

Reactor	Containment dimensions	Available data
CVTR	Concrete containment 34·565 m high 17·632 m diameter	Liner 6·4–12·7 mm Wall thickness 1·19 m Factored load of 1·5 × design and accident pressure
Enricu Fermi	Steel containment 30·4 m high 21·9 m diameter	ANS and ASME Codes
San Onofre	Steel containment 42·56 m diameter sphere	ANS and ASME Codes
EGCR	Concrete containment 60·8 m high 30·4 m diameter	Liner 6·5 mm Wall thickness 1·19 m ACI/ASME Codes
Stone and Webster	Concrete containment 56·39 m high 38·40 m inside diameter	Liner 6·4–12·7 mm
Westinghouse 900 MWe	Concrete containment 53·10 m high 37·0 mm inside diameter	0·90 m thick wall 0·75 m dome thickness Prestress 1000 tonne/m of cylinder
1 350 MWe	69·0 m high 42·0 m inside diameter	Factored load of 1·5 × design and accident pressure Vertical prestress 50 bar
Rancho Seco	56·43 m high inside 39·65 m inside diameter	1·19 m wall thickness; pressure 414 kN/m². Four groups of dome tendons 81 tendons total divided into three groups, 126 vertical tendons, 117 hoop tendons, 6 mm liner thickness
Palisades	58·825 m high 35·264 m inside diameter	1·18 m wall thickness Six buttresses, 165 dome tendons divided into three groups, 180 vertical tendons, 6 mm liner thickness
Trojan	63·232 m high 37·696 m inside diameter	1·18 m wall thickness 3 buttresses, 150 hoop tendons, 70 vertical tendons, 6 mm liner thickness
Doel (Belgium)	Concrete containment 63·3 m high 42·5 inside diameter	1·20 m wall thickness 2·9 m annulus for outer containment 2 buttresses
Ringhals II	Concrete containment 63·4 m high 35·40 inside diameter	4 buttresses 1·10 m wall thickness
Gentilly	Concrete containment 56·0 m high 36·48 m inside diameter	1·216 m thick wall 0·508 m dome thickness 4 buttresses
Ginna	Concrete containment 46·056 m high 31·92 m inside diameter	1·18 m wall thickness 0·976 m dome thickness 3 mm liner thickness 19 mm wire tendons vertical 3 No. 18 S bars hoop steel
Mol (Belgium)	Steel containment 32·3 m high 16·5 m diameter	30 mm thickness
Seebrook	57·456 m high 21·28 m inside diameter	1·368 m wall thickness No. 18 at 300 mm hoop and cylinder vertical bars No. 18 at 550 mm at 45° angles
Bellefonte	67·285 m high 41·14 m inside diameter	1·067 m wall thickness
Fessenheim and Bugey	Concrete containment 56·6 m high 37·0 m inside diameter	0·9 m wall thickness 6·0 mm liner
Creys-Malville	Concrete containment 58·5 m high 42·5 m inside diameter	0·85 m wall thickness 6·0 mm liner
Three Mile Island Sizewell B	54·465 m high 45·0 m inside diameter	2·5 m thick wall cylinder 1·5 m thick dome 6·0 mm liner

Table VIII.8
Partial load factors

Category	D'	L	F	P_t	P_o	T_t	T_o	T_a	E_o	SSE	W	W_t	R_o	R_a	R_r	P_r
Service																
Test	1·0	1·0	1·0	1·0	—	1·0	—	—	—	—	—	—	—	—	—	—
Construction	1·0	1·0	1·0	—	—	—	1·0	—	—	—	—	—	—	—	—	—
Normal	1·0	1·0	1·0	—	—	—	1·0	—	—	—	—	—	1·0	—	—	1·0
Severe environmental	1·0	1·0	1·0	—	—	—	1·0	—	—	—	1·0	—	1·0	—	—	1·0
Factored																
Severe environmental	1·0	1·3	1·0	—	—	—	1·0	—	1·5	—	—	—	1·0	—	—	1·0
Severe environmental	1·0	1·3	1·0	—	—	—	1·0	—	—	—	1·5	—	1·0	—	—	1·0
Extreme environmental	1·0	1·0	1·0	—	—	—	1·0	—	—	1·0	—	1·0	1·0	—	—	1·0
Abnormal	1·0	1·0	1·0	—	1·5	—	—	1·0	—	—	—	—	—	1·0	—	—
Abnormal	1·0	1·0	1·0	—	1·0	—	—	1·0	—	—	—	—	—	1·25	—	—
Abnormal/severe environmental	1·0	1·0	1·0	—	1·25	—	—	1·0	1·25	—	1·25	—	—	1·0	—	—
Abnormal/severe environmental	1·0	1·0	1·0	—	1·25	—	1·0	1·0	1·0	—	1·0	—	—	1·0	—	—
Abnormal/extreme environmental	1·0	1·0	1·0	—	1·0	—	1·0	1·0	—	1·0	—	—	—	—	1·0	—

D' Dead loads, including hydrostatic and permanent equipment loads
E_o Loads generated by the operating basis earthquake
SSE Loads generated by the safe shutdown earthquake (weights considered will be the same as for E_o)
F Loads due to prestress
L Live loads
P_o Design pressure load within the containment generated by the design basis accident (DBA)
P_t Pressure during the structural integrity and leak rate tests
P_r External pressure loads (resulting from pressure variation either inside or outside the containment)
R_o Pipe reaction from thermal conditions due to the DBA, including R_o
R_r The local effects on the containment due to DBA
R_o Pipe reactions during normal operating or shutdown conditions
T_a Thermal loads generated by the DBA, including T_o
T_o Thermal loads during normal operating or shutdown conditions
T_t Thermal effects and loads during the test
W_t Tornado loading, including the effects of missile impact
W Design wind load

VIII.8.2 Loads and Stresses
VIII.8.2.1 Dead and Service Loads
(a) Normal loads.

(i) Dead load (D')—weight of all permanent structural and non-structural components.

(ii) Operating live load—live load is the load superimposed (L, LL) on the vessel and its components.

(iii) Construction loads—construction loads are large heavy loads due to erection of equipment such as those from cranes, fork lift forms, and transport and temporary storage of equipment.

(iv) Concentrated loads (L_C)—rail and truck supports (C_F, H_{2O}), craneways (25% of wheel load), and ordinary machinery impact loads.

(b) Environmental loads.

(i) Wind loads (W) (excluding loads induced by tornadoes)—wind loads are the same as those used in conventional structural design; these are based on statistical data for the vicinity of the structure, ground topography and general climate.

The pressure (P) on any part of the vessel surface exposed to wind varies from point to point over the surface, depending on the direction of the wind and flow pattern. The pressure (P) can be expressed by $P = C_p q$, where C_p is the pressure coefficient and q the dynamic pressure. In the calculation of the wind it is essential to take into account the pressure difference between opposite faces. It is convenient to use distinguishing pressure coefficients:

$$F = (C_{pe} - C_{pi})qA$$

where F = resultant force, C_{pe} = external pressure coefficient, C_{pi} = internal pressure coefficient, and A = surface area. A positive value for F would mean the pressure is inwards, and a negative value that pressure is outwards. In addition to the existing formula, the total wind load on any section of the vessel may be found using

$$F = C_f q A_e$$

where C_f = force coefficient and A_e = effective frontal area of the section.

(ii) Snow loads (S)—complete data on snow load requirements are given by Bangash [843, 105] and by many other researchers [853, 1014–1020].

(c) Soil and hydrostatic pressure (E_p) and buoyancy (B).

(i) Lateral pressures of adjacent soil including active, passive and at-rest pressures.
(ii) Dynamic pressures due to earthquakes.
(iii) At the free water surface the weight of the soil diminished by buoyancy plus hydrostatic pressure.

(d) Piping equipment reaction load (R_0). Piping systems attached through hangers, struts, lugs, restraints and anchors.

(i) Thermal expansion and dynamic effects; together with
(ii) imposed restraints such as changes in momentum, water and steam hammer in the equipment.

(e) Operating pressure and temperature loads (P_0, T_0).

(i) Differential pressure ranges applied to the structures.
(ii) Loads from differential thermal expansion or stresses from thermal gradients within the structure.

(f) Operating basis earthquake (E_0). The minimum acceptable acceleration for E_0 is at least one half of the safe shut-down earthquake.

(g) Extreme loads. Nuclear structures are designed against a much broader range of loads such as major earthquakes, tornadoes, environmental hazards and military missiles. Nuclear structures are also designed for a third level of load known as the extreme load. Details of extreme loads are:

(i) The maximum earthquake potential for a site, taking into consideration regional and local geology, seismology, local foundation conditions, tornadoes, winds and associated airborne missile loads.
(ii) Postulated design basis accidental loads due to high-energy system ruptures resulting in pipe break action, impingement loads, pipe whip and associated accidentally generated missiles, flooding and high thermal transients.

(h) Safe shut-down earthquake (SSE). The items listed briefly below should be considered to establish the ground motion input associated with the SSE:

(i) Geological conditions at the site under consideration.
(ii) Tectonic structure evaluation.
(iii) Identification of effects of prior earthquakes.
(iv) Determination of static and dynamic characteristics of the underlying materials of containment vessels.
(v) Determination of fault locations:

(1) The vibratory ground motion associated with the SSE shall be considered to be acting at the ground surface in that portion of the site not affected by adjacent topographical structures.
(2) For a vessel founded on competent rock or other equally competent material, the soil–structure interaction shall preferably be considered.
(3) The minimum acceleration for the vibratory ground motion associated with the SSE shall be $0.10g$ within a 200-mile radius of the site.
(4) The values determined for the vertical motion associated with the SSE should not be less than two-thirds of the maximum horizontal ground acceleration for the SSE.
(5) The strength of each horizontal component shall be equal to the SSE intensity defined for the site.

(i) Tornado loads (W_t). Structures should be designed to resist maximum tornado conditions for a given plant site.

The effects of a tornado are generated by three separate phenomena:

(i) Wind.
(ii) Differential atmospheric pressure.
(iii) Missiles interacting with the vessel structure causing damage through three principal failure modes:

(1) pressure forces created by drag and lift as air flows around and over the vessel;
(2) pressure forces due to rapid changes in atmospheric pressure resulting in differential pressure between the interior and exterior of the building; and
(3) penetration, spalling and impact forces created by environmental and other missiles.

The criteria suggest that, at the point in a tornado where the wind velocity is a maximum, the atmospheric pressure change is taken as one-half of its maximum value, if a Rankine vortex is assumed. Combining the effects, the velocity pressures shall be multiplied by the appropriate pressure coefficients, and the design loads are determined for:

(i) maximum atmospheric pressure change acting alone; and
(ii) maximum wind velocity plus one-half total atmospheric pressure change combined with tornado missile loads.

(j) Hurricane loads. A hurricane is a cyclonic storm having rotational wind velocities in excess of 119 km/h. The basic characteristics of hurricanes are:

(i) Wind distributions—maximum wind velocities:

(1) The vectorial addition of the translational and rotational components of the wind.
(2) The direction of the wind inclined 20–30° towards the centre of the hurricane.
(3) Wind gusts within the hurricane exceeding sustained winds by 30–50%.

(ii) Atmospheric pressure change.
(iii) Size—hurricanes vary in diameter from 24 km to more than 160 km. Gale force winds (64 km/h) exist within a 560–640 km diameter.

(k) Tsunami. Tsunami are long ocean water-waves generated by earthquakes or underwater explosions, which impinge on coastal areas and structures located therein. Factors for consideration are:

(i) Location and records of tsunami and their mechanism.
(ii) Size of an offshore earthquake or underwater explosion.
(iii) Tsunami wave height.
(iv) Influence of hydrography, harbours and breakwaters.

(l) Missile loads (Y_m). In nuclear plant design, components and equipment should be protected against loss of function due to plant-generated and extreme environmental missiles. Where a containment building is close to a major airport, the damage caused by a potential missile such as an aircraft crash should be considered.

The effect of a missile impact on a target depends on the material and the geometrical properties of the missile and its target. For a rapid impact, local effects must also be considered. These include local structural failures, such as penetration, perforation, scabbing and spalling. In other cases, for long-duration impacts in which momentum exchange takes place, the global response of the containment should be considered. The energy absorption capacity of the containment should be included in the load effects.

The generated missiles may vary widely in size, weight and impact velocity. Extreme environmental missiles include tornado-generated missiles and aircraft.

Factors to consider for tornado-generated missiles are:

(i) Designed tornado wind field.
(ii) Identification of typical objects located at or near a plant site.
(iii) Analytical approaches for the motion of an object should derive the motion of the object until the end of its flight (the point where such missiles are ejected from the tornado wind field).

In the study of aircraft impacts the distance of an airport from a particular power plant site must be considered. This is normally done for the analysis and design on the basis of studies of probability of occurrence.

VIII.8.2.2 Stresses

The allowable stresses are adopted according to the relevant codes of practice, including those given in Section VIII.7 for pressure vessels.

VIII.8.3 Containment Vessel Analysis

A finite element analysis has been carried out using solid, panel and line elements as explained in Chapter VI. Figure VIII.136 shows a typical finite element mesh for a seismic case. Where seismic analysis and aircraft impact analysis are needed, the soil strata are also idealised as layered elements together with a spring and dash-pot system, given in Details 1 and 2 of Fig. VIII.136. Table VIII.9 shows safety margins for containment vessels subjected to internal pressures up to $344 \cdot 75 \, kN/m^2$ along with dead, live and prestressing loads. The idealised strata in case of the Bellefonte vessel are given in Fig. VIII.137a for the vessel parameters shown in Figs VIII.137b and VIII.137c.

VIII.8.3.1 Aircraft Impact

The design and analysis of a containment vessel to withstand the effect of an aircraft crash or impact involves many complex parameters, including

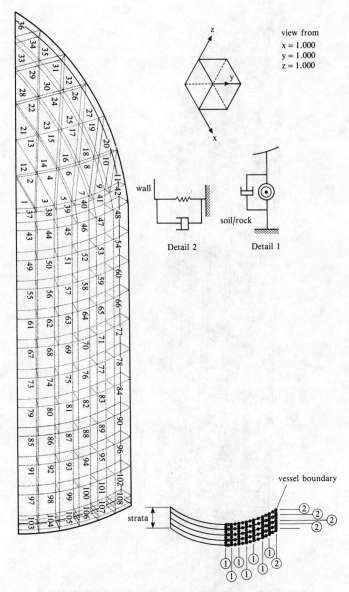

Fig. VIII.136. Nonlinear model—Sizewell B vessel.

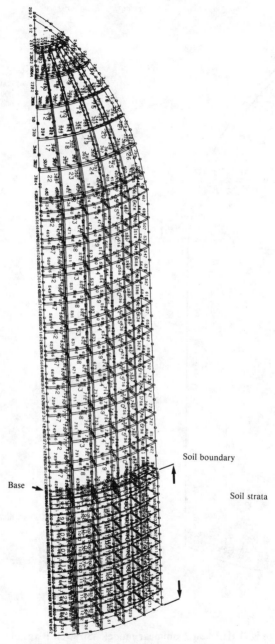

Fig. VIII.137a. Finite elements for the Bellefonte vessel.

Application to Engineering Problems

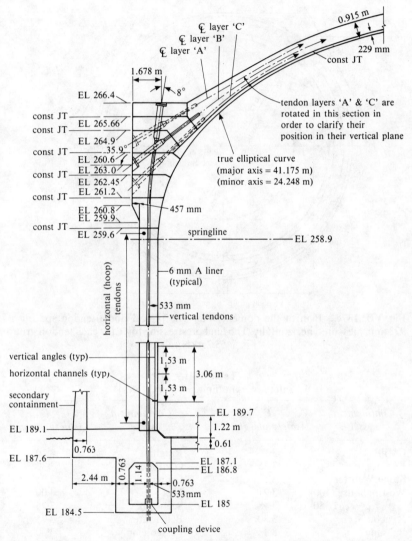

Fig. VIII.137b. Typical section—wall and dome (courtesy of TVA and Bellefonte NPS).

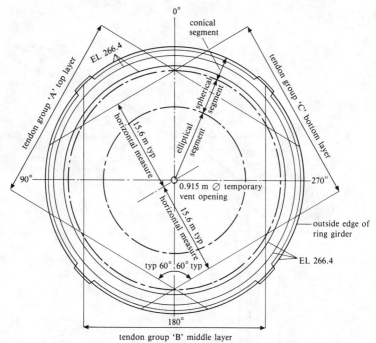

Fig. VIII.137c. Plan of the dome. (Note: The typical dome tendon spacing is 978 mm, measured horizontally. The final prestressing force for each tendon group is 5453·25 kN/m.)

Table VIII.9
Safety margin (finite element) [843]

Containment vessel	F_s (safety margin)	Containment vessel	F_s (safety margin)
CVTR	3·1	Ringhals	3·10
EGCR	3·25	Gentilly	3·30
Stone/Webster	3·13	Ginna	2·95
Westinghouse (400)	3·21	Mol	4·00
(1 350)	3·18	Seebrook	3·75
Rancho Seco	3·30	Bellefonte	3·75
Palisade	3·4	Fessenheim	3·00
Trojan	3·23	Creys-Malville	2·90
Doel	3·00	Sizewell B	2·5

$$F_s = \frac{\text{final pressure at complete cracking}}{\text{operating pressure}}$$

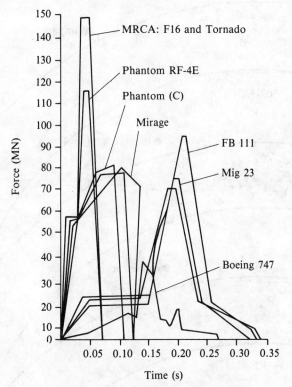

Fig. VIII.138. Time–impact function.

loadings and durations. Bangash [843] has produced modified force–time impact functions for various aircraft, and these are shown in Fig. VIII.138. First, the Bellefonte vessel parameters were chosen for the analysis (Figs VIII.137b and VIII.137c). The impact point is chosen to be the apex of the vessel and at the junction of the dome and the cylinder. Results are produced in the form of acceleration–time histories, displacements and stress states for selected times and acceleration response spectra. Newmark and Wilson θ direct integration methods (discussed in Chapter VI) are used for the analysis. Crack propagation, and linear and nonlinear displacements for four aircraft are shown in Figs VIII.139 to VIII.141a. The impact zones are shown in Fig. VIII.141b. Table VIII.10 shows comparative results for penetrations and perforations based on both empirical (Section II.2.3) and finite element analysis for a number of aircraft impact loads. The endochronic model is used throughout the analysis. A similar analysis was

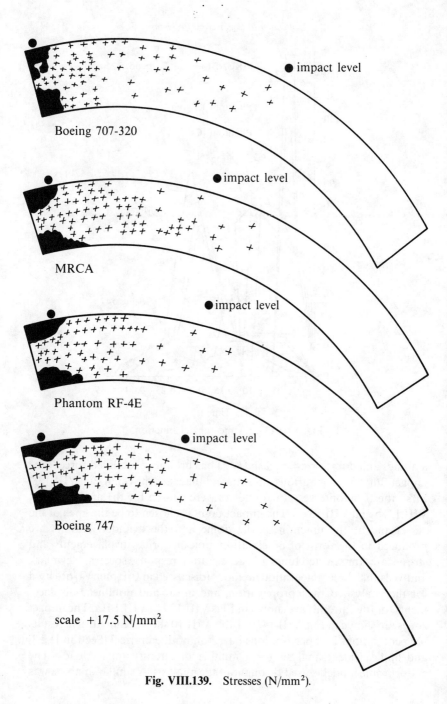

Fig. VIII.139. Stresses (N/mm²).

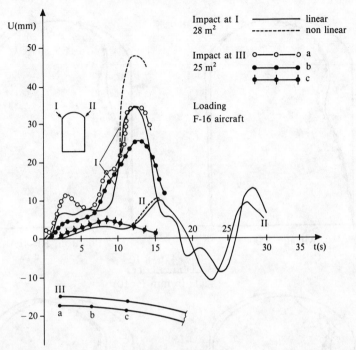

Fig. VIII.140. Linear and nonlinear displacement as a function of time.

carried out for the Sizewell B vessel using load–time functions for Phantom and Tornado aircraft. The 'post-mortem' of the vessel is shown in Fig. VIII.142. The load–displacement as a function of time is given in Fig. VIII.143.

VIII.8.3.2 Seismic Analysis

A nonlinear dynamic analysis was carried out for the containment vessel. A direct approach was adopted in which the whole system (containment–soil/bedrock) was modelled and analysed in the time domain using finite elements. Inertia forces were reproduced by distributed masses by involving consistent mass matrices. In general the equation of motion contains a mass matrix (M), a damping matrix (C) and a stiffness matrix (K). The excitation is taken in the form of a base motion. As shown in Figs VIII.136 and VIII.137, a 20-noded isoparametric element together with line and panel elements representing concrete, prestressing tendons or reinforcement and steel liner respectively are used to model the structure

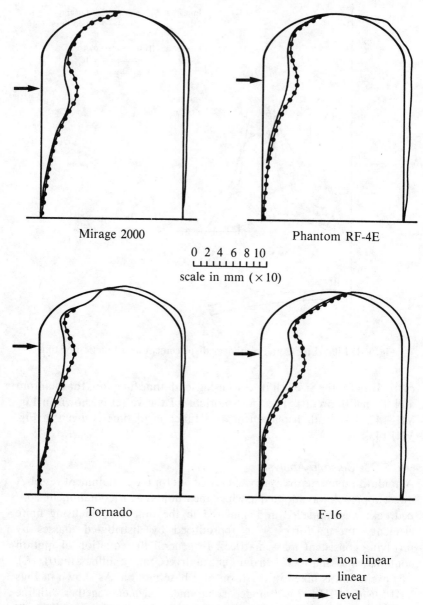

Fig. VIII.141a. Linear and nonlinear displacements for impacts due to four aircraft.

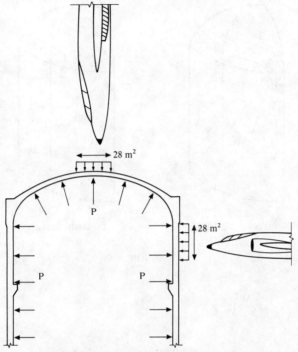

Fig. VIII.141b. Extreme loading condition.

and the supporting medium. This method allows treatment of nonlinear behaviour and material properties which are adjusted at each step. Special care is exercised in using the numerical integration procedure together with time steps in order to achieve stability in the solution. In the stiffness matrix evaluation of the vessel, the four-parameter model is considered. The cracking cases are evaluated using the endochronic cracking model given in Chapters VI and VII and associated appendices. Table VIII.10b shows the foundation parameters for impact and earthquake problems.

The program ISOPAR has been written to perform seismic analysis incorporating the effects of circumferential cracking. The program uses numerical integration rather than the method of normal modes. The cracks were obtained when the overall stiffness of the vessel was no longer constant but was a function of the shearing stresses in the vessel

The equations of motion were solved using the Wilson and Newmark θ methods. The convergence criterion for ending the iterations was 0·0001.

The size of the time step (t) is critical for obtaining accuracy and rapid

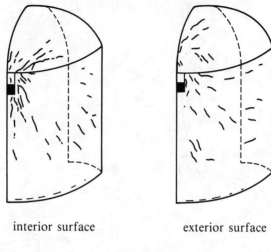
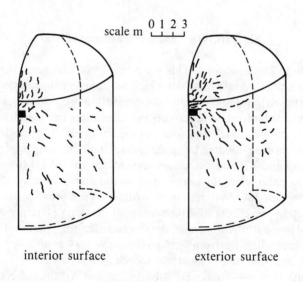

Fig. VIII.142. Phantom and Tornado aircraft impact 'post-mortem' [1007].

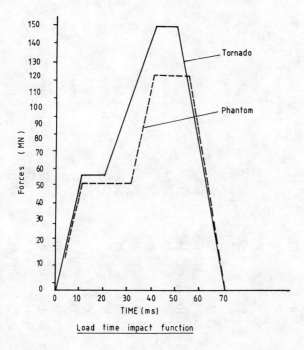
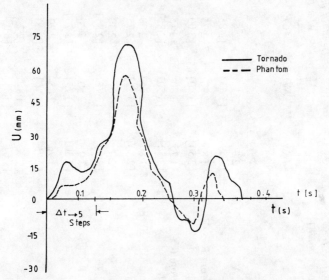

Fig. VIII.143. Load–displacement as a function of time [1007].

Table VIII.10a
Aircraft impact load versus comparative vessel thicknesses

Aircraft [Speed (mph)]	Depth of penetration (m) [Thickness for no perforation (m)]						Finite element thickness[a]/Displacement (mm)	
	IRS	HN–NDRC	ACE	DF–BRL	BRL	DF–ACE	Depth of penetration (m)	Thickness[b]
MRCA	2·0 [2·59]	3·1 [3·75]	2·5 [2·85]	1·57 [2·30]	1·59 [2·75]	2·5 [2·95]	1·80 (55 max)	3·0
Phantom RF-4E [482]	1·85 [2·50]	2·90 [3·70]	2·25 [3·00]	1·25 [2·00]	1·35 [1·75]	2·10 [2·75]	1·55 (55 max)	2·35
Phantom (C) [482]	1·75 [2·50]	2·80 [3·50]	2·15 [2·85]	1·10 [1·75]	1·35 [1·75]	2·0 [2·75]	1·55 (60 max)	2·35
Boeing 707-320 (E) [230]	1·80 [2·50]	2·75 [3·25]	2·00 [2·75]	1·20 [1·75]	1·50 [2·0]	2·0 [2·75]	1·10 (55 max)	2·00
Boeing 747 [250]	2·0 [2·50]	3·0 [3·75]	2·30 [2·85]	1·55 [2·30]	1·85 [2·50]	2·30 [2·85]	1·90 (65 max)	2·75
FB-111 (CH) [200]	2·0 [2·50]	2·50 [3·75]	2·00 [2·75]	1·05 [2·00]	1·35 [2·00]	1·75 [2·25]	1·25 (45 max)	2·30
Mirage [470]	1·75 [2·75]	2·85 [3·75]	2·20 [3·00]	1·25 [2·00]	1·50 [2·20]	2·50 [3·00]	1·45 (40 max)	2·35
Mig-23 [480]	1·2 [1·6]	1·8 [5·0]	2·2 [6·0]	1·0 [3·0]	1·0 [1·15]	2·1 [3·10]	1·80 (45 max)	2·85

[a] All thicknesses mentioned under finite element are computed thicknesses.
[b] For wide perforation.

Table VIII.10b
Foundation parameters (major data)

Foundation radius $R = 24.86$ m

Foundation thickness $t_f = 3.76$ m

Seismic shear force $= 1822.52$ kN/m

Uplift from vertical earthquake component $= 470$ kN/m

Tangential shear from earthquake $= 1272.96$ kN/m

$K_u =$ translational spring constant $= \dfrac{32(1-v)GR_0}{(7-8v)}$

$K_z =$ vertical constant $= \dfrac{4GR}{(1-v)}$

$K_\phi =$ rotational spring constant $= \dfrac{8GR_0^3}{3(1-v)}$

$K_t =$ torsional spring constant $= \dfrac{16GR^3}{3}$

$v = 0.14$

$\beta' = 0.5$

Material damping: 8% for the soil and 2% for the vessel

---

iteration convergence (optimum $t = 0.0025$ s). A method given by Elwi and Murray [5] was considered along with that presented by Bangash [843].

Stiffness changes will occur during some time steps. A typical subroutine that keeps track of where each crack is on the hysteresis loop is shown in Figs VIII.144 and VIII.145, and makes changes in the crack stiffness when necessary. A time step is repeated only when stiffness change from line to line occurs. This is because the high velocities which occur during the unloading from certain lines can cause the crack tip to go far below the specified slips at certain points, which causes the loop to grow much wider than originally specified. Shear–crack relations for a typical vessel are given in Fig. VIII.146.

At time t, one of the cracks is at the position marked on a line by the hysteresis loop. At time $(t + \Delta t)$, a certain point can be missed by a significant amount. The program at this point goes back to time t and refines the time step and computes the shear stress and crack displacement with the new refined time step. If the shear stress is not within a specified limit then the time step is refined further and the process is repeated until

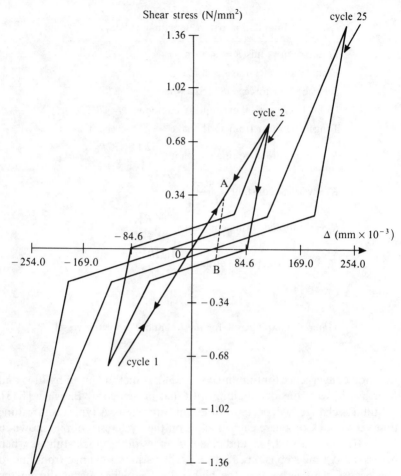

Fig. VIII.144. Shear stress versus slip.

the specified limits are met. The same process is performed for the stiffness change from line to line.

The changing of hysteresis loops due to cycling is shown in Figs VIII.144 and VIII.145. All cracks start on the cycle 1 line. Once a shear stress is exceeded, unloading starts and proceeds along a line parallel to the dotted line A–B. The second cycle is reached when A–B intercepts a line. This process continues.

The circumferential (horizontal) and longitudinal (vertical) cracks in the vessel have a significant effect on the dynamic response of the vessel due to

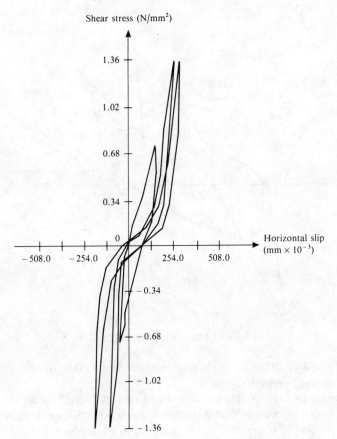

Fig. VIII.145. Shear stress versus horizontal slip.

the SSE. The important design parameters which are affected are the liner distortion and the maximum shear stress in the concrete vessel.

Current design criteria specify that the nuclear containment vessel must be able to withstand the simultaneous occurrence of a LOCA with strong (SSE) earthquake motions. The internal pressure creates tensile stresses in both the longitudinal and circumferential directions, while the earthquake causes inertia forces which in turn cause shearing stresses and bearing stresses in the vessels. These stresses cause forces which must be transferred across horizontal and vertical cracks. The crack patterns, crack widths and spacing are computed. Crack widths vary between 0·25 and 0·38 mm.

The shearing stiffness of the cracks due to aggregate interlock affect the

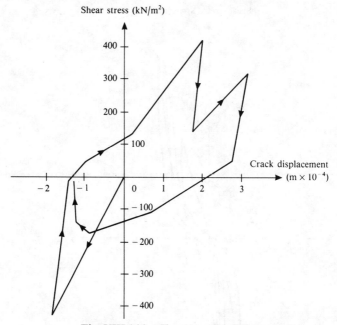

Fig. VIII.146. Shear–crack relation.

shear stress distribution in the containment vessel after immediate short-term internal pressurisation.

During an earthquake, the stresses caused by inertia forces cause the crack width to change. Where the crack was too narrow over the whole of some portion of its circumference, it results in alteration in the shear stress distributions. Computer results show that for a total unbonded length at the crack of 64 mm the change in crack width is small compared to the initial crack width. Therefore the shear stress distribution is not significantly altered from the sinusoidal distribution. The vertical cracks present may effectively decrease the flexural and shear stiffness of the containment vessel.

Figure VIII.147 shows a mass displacement–time relation for the vessel. Figure VIII.148 shows response spectra under various conditions and at the top and bottom of the vessel. The control motion is specified at the free surface of the soil deposit. The curves show the motion specified at bedrock and at the foundation level, respectively. Figure VIII.149 shows a 'post-mortem' of the Sizewell B vessel at 0·3g. Throughout the vessel excitation the damping for the vessel was 2% and that of the soil was 8%.

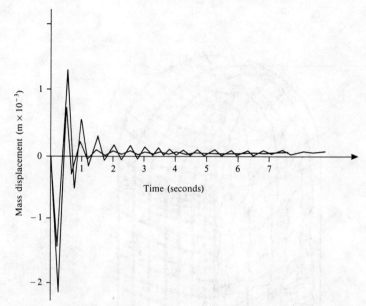

Fig. VIII.147. Mass displacement–time relation for a typical containment vessel.

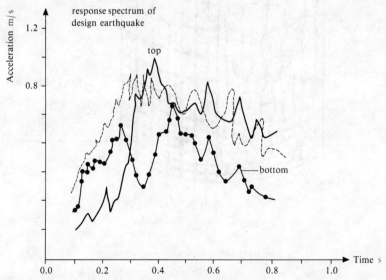

Fig. VIII.148. Containment vessel response spectrum. (---, Measured; ———, nonlinear; —●—, linear.)

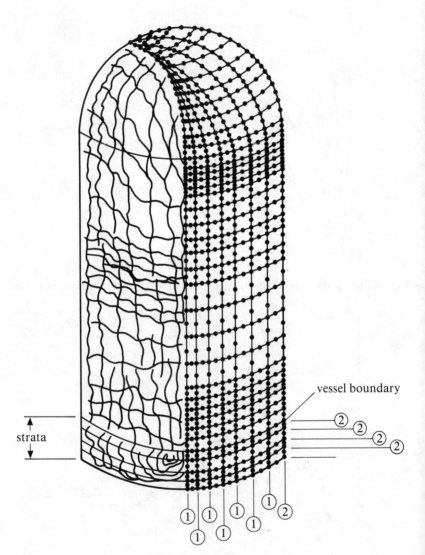

Fig. VIII.149. Cracked containment.

Fig. VIII.150a. One-sixth scale RC model containment (courtesy of Lawrence Livermore Laboratory [1006]).

VIII.8.3.3 Containment Overpressurisation Analysis

A 1/6th scale reinforced concrete model [1006] (as shown in Figs VIII.150a and VIII.150b) has been designed for a design pressure of 317 kPa. The reinforcement layouts are shown in Figs VIII.151 and VIII.152. A pre-test analysis has been carried out using the program ISOPAR. The endochronic model along with cracking criteria proposed in this text have been adopted.

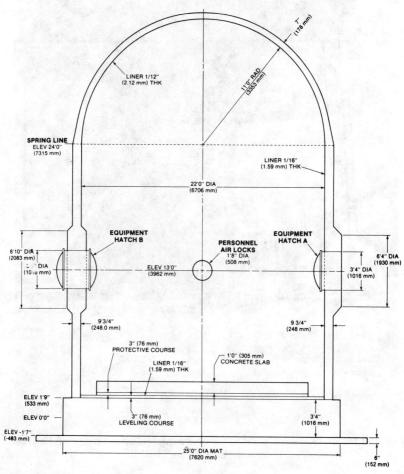

Fig. VIII.150b. Schematic of the 1/6 scale reinforced concrete containment model—elevation view (courtesy of Lawrence Livermore, USA).

REINFORCING BAR DETAILS

LAYER	TYPE	SIZE
1	HOOP	#4
2	MERIDIONAL	#4
3	HOOP	#4
4	HOOP	#4
5	MERIDIONAL	#4
6	HOOP	#4
7	SEISMIC (45° DIAG)	#4
8	SEISMIC (45° DIAG)	#4

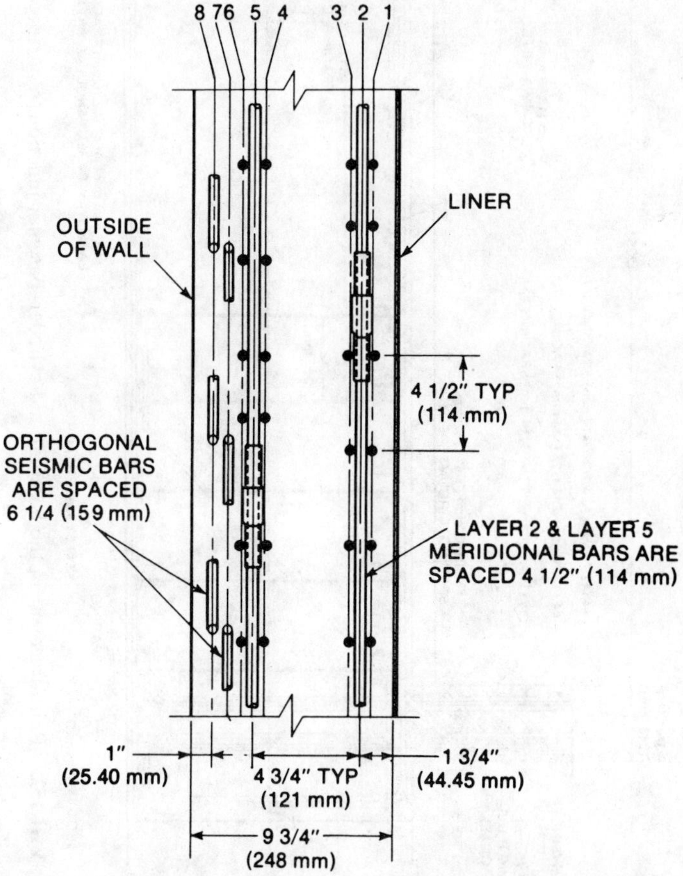

Fig. VIII.151. Reinforcement in the cylinder (free field) (courtesy of Lawrence Livermore, USA).

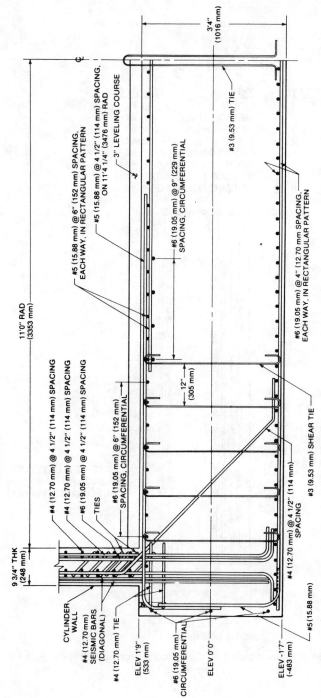

Fig. VIII.152. Reinforcement in the basemat and cylinder–basemat junction. (courtesy of Lawrence Livermore, USA).

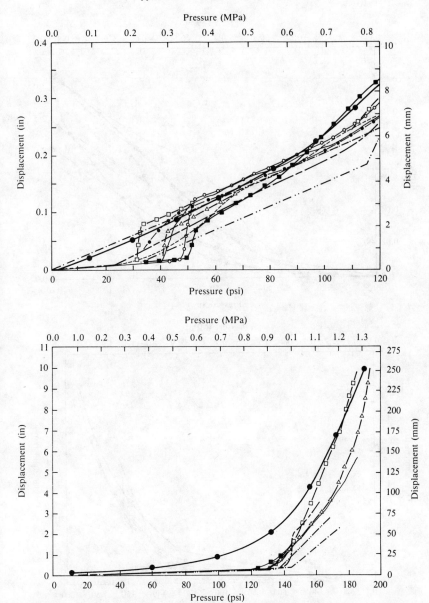

Fig. VIII.153. Comparative Study—1. (———, SNL; ———, NIL; ———, SERD; —■—, CEGB; —○—, EPRI; ---, BNL; —□—, ENEA; ———, ANL; •—•, CEA; —△—, GRS; —●—, Bangash.) (Courtesy of Lawrence Livermore, USA).

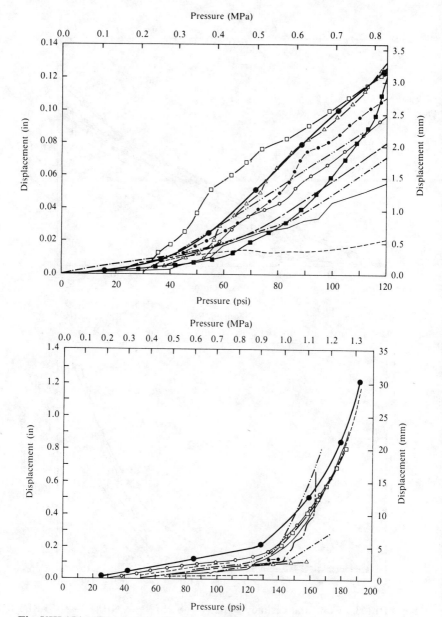

Fig. VIII.154. Comparative Study—2. (See Fig. VIII.153 for key.) (Courtesy of Lawrence Livermore, USA).

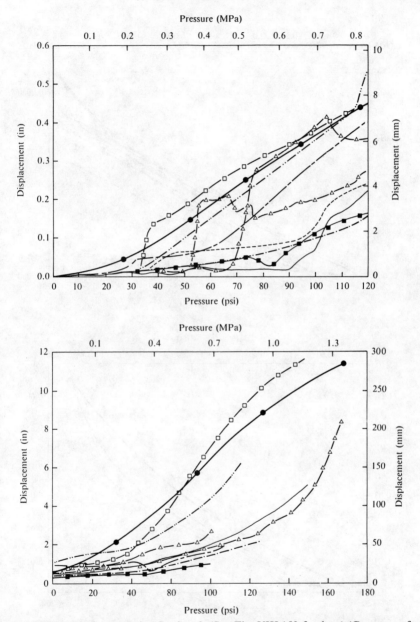

Fig. VIII.155. Comparative Study—3. (See Fig. VIII.153 for key.) (Courtesy of Lawrence Livermore, USA).

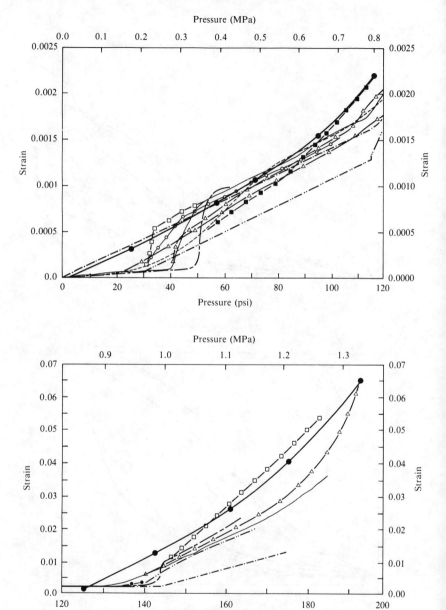

Fig. VIII.156. Comparative Study—4. (See Fig. VIII.153 for key.)

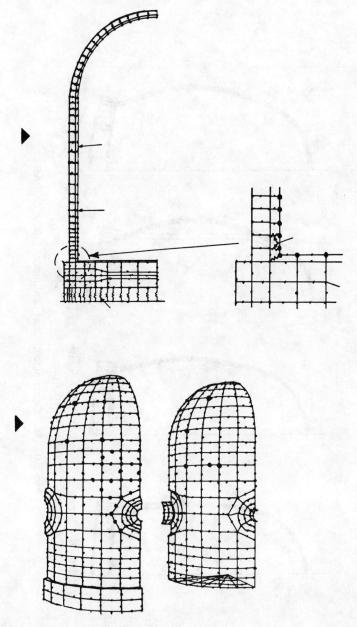

Fig. VIII.157. One-sixth scale RC containment—finite element mesh.

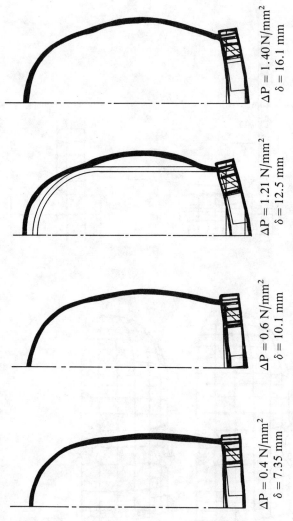

Fig. VIII.158. Containment vessel final modes [1006; ISOPAR].

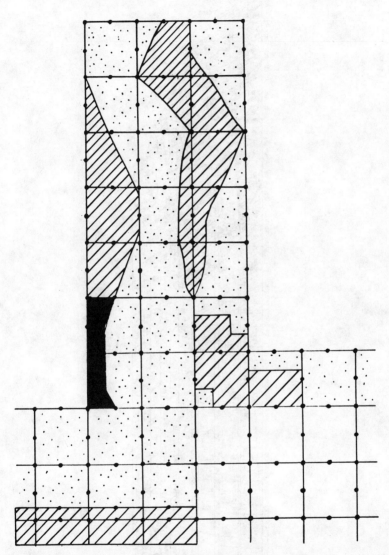

Fig. VIII.159. Major crack zones [1006]. (▨, denotes zero tensile stress; ■, denotes crushing; ▨, denotes radial vertical cracking.)

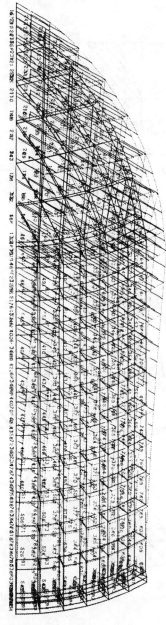

Fig. VIII.160a. Sizewell B overpressurisation analysis—final mode.

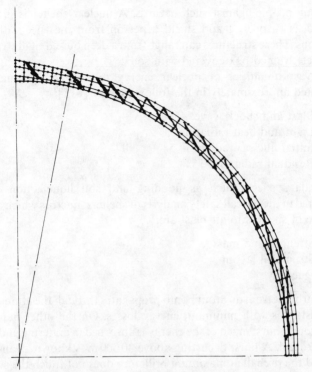

Fig. VIII.160b. Cracking of Sizewell B dome.

The results are compared with others in Figs VIII.153–VIII.156 using the finite element mesh of Fig. VIII.157. The final mode is shown in Fig. VIII.158 and some crack zones are indicated in a portion of the wall in Fig. VIII.159.

The Sizewell B model was overpressurised. Figure VIII.160a shows the displacement pattern. The vessel failed at 3·15 by rupturing of tendons and excessive cracking. The mode is of the type indicated in Figs VIII.159 and VIII.160b.

VIII.9 CASE STUDY No. 8—CONCRETE NUCLEAR SHELTERS

VIII.9.1 Introduction

There is increasing current concern about safety from nuclear hazards, including nuclear blasts and radiation. There will be greater involvement in protecting people against such hazards. A nuclear shelter is just one of many ideas to protect and shield a person from the effects of nuclear explosions. These structures can range from a deep buried rigid structure to a concrete framed box covered with soil.

The vast quantities of nuclear energy released by detonation are distributed approximately in the following way [617]:

> Blast and shock wave: 45%
> Light and heat radiation: 35%
> Initial nuclear radiation: 5%
> Residual radiation: 15%

Secondary effects such as flooding and soil liquefaction are also considered in the overall safety analysis of shelters. Factors which affect the response of shelters to air blast are:

(a) Strength and mass
(b) Structural layout
(c) Ductility

The airburst from an atom bomb propagates through the atmosphere to great distances with minimum energy losses. On the other hand, in an underground burst much of the energy is absorbed in cratering and melting of the ground. A burst occurring above 30 000 m is known as an airburst provided the fireball at maximum brilliance does not touch the surface of the earth. If it touches, it becomes a 'surface burst'. These cases together with their loading conditions are described in this section. A typical concrete shelter design for a six-person family is chosen for finite element analysis using the five-parameter constitutive law. Analytical results show agreement with those provided by the various codes [639, 640, 667].

VIII.9.2 Characteristics of the Blast Wave in Air

Most of the material damage from an airburst (nuclear weapon) is caused mainly by the shock (or blast) wave which accompanies the explosion. The majority of structures will suffer some damage from air blast when the overpressure in the blast wave, i.e. the excess over atmospheric pressure ($101 \cdot 3 \, kN/m^2$ at sea level), is about $3 \cdot 5 \, kN/m^2$ or more. The distance to which this overpressure level will extend depends on the yield of the explosion and on the height of the burst.

A difference in air pressure acting on separate surfaces of a structure produces a force on that structure; the size of these forces is dependent on the difference between ambient air pressure and the overpressure. The maximum value of overpressure, known as the peak overpressure, occurs at the front of the blast wave.

As the blast wave travels in the air away from its source, the overpressure at the front steadily decreases, and the pressure behind the front falls off in a regular manner. After a short time, when the shock front has travelled a certain distance from the fireball, the pressure behind the front drops below that of the surrounding atmosphere and a so-called 'negative phase' of the blast wave forms, as given in Fig. VIII.161.

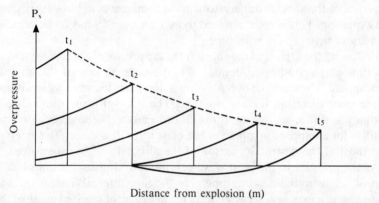

Fig. VIII.161. Variation of overpressure with distance at successive times.

For the curves marked t_1 to t_4 the pressure in the blast wave has not fallen below atmospheric pressure, but on the curve t_5 it is seen that at some distance behind the shock front the overpressure is below that of the original atmosphere so that an 'underpressure' exists.

During the negative (rarefaction or suction) phase a partial vacuum is produced and the air is sucked in, instead of being pushed away, as it is when the overpressure is positive.

The peak values of the underpressure are generally small compared with the peak overpressures, the former having a maximum value of 27·5 kN/m^2 below the ambient pressure. With increasing distance from the explosion, both peak values decrease, the positive more rapidly than the negative, but they do not approach equality until the peak pressures have decayed to a very low level (Fig. VIII.162).

For a short interval after the detonation, point 1, there will be no increase in pressure since it takes the blast wave some time to travel the distance from the point of explosion to the given location. Point 2 indicates the time

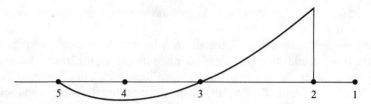

Fig. VIII.162. Variation of pressure with time at a fixed location.

of arrival of the shock front; a strong wind commences to blow away from the explosion. This is often referred to as a 'transient' wind, as its velocity decreases fairly rapidly with time.

Following the arrival of the shock front the pressure falls rapidly, and at the time corresponding to point 3 it is the same as that of the original atmosphere. Although the overpressure is now zero, the wind will continue in the same direction for a short time. The interval from points 2 to 3 (roughly 2–4 s for a 1 megaton explosion) represents the passage of the positive (or compression) phase of the blast wave. It is during this interval that most of the destructive action of the airburst will be experienced.

As the pressure in the blast wave continues to decrease, it sinks below that of the surrounding atmosphere. In the time interval from point 3 to point 5, which may be several seconds, the negative (or suction) phase of the blast wave passes the given location. For most of this period the transient wind blows towards the explosion. As the negative phase passes, the pressure at first decreases below ambient and then increases towards that of the ambient atmosphere which is reached at the time represented by point 5. The blast wind has then effectively ceased and the direct destructive action of the airblast is over.

VIII.9.2.1 Dynamic Pressure

Although the destructive effects of the blast wave are usually related to values of the peak overpressure, there is another quantity of equivalent importance called the 'dynamic pressure'.

The dynamic pressure is proportional to the square of the wind velocity and to the density of the air behind the shock front. For very strong shocks the dynamic pressure is larger than the overpressure but below $480 \, kN/m^2$ (4·7 atmospheres) overpressure at sea level; the dynamic pressure is smaller. The peak overpressure dynamic pressure decreases with increasing distance from the explosion centre, although at a different rate.

Some indication of the corresponding values of peak overpressure, peak

Table VIII.11
Overpressure, dynamic pressure and wind velocities in air at sea level for an ideal shock front

Peak overpressure, kN/m^2 (atmos.)	Peak dynamic pressure, kN/m^2 (atmos.)	Maximum wind velocity, km/h
1 378 (13·78)	2 274 (22·74)	3 347
1 034 (10·34)	1 537 (15·37)	2 860
689 (6·89)	848 (8·48)	2 275
496 (4·96)	551 (5·51)	1 883
345 (3·45)	275 (2·75)	1 512
207 (2·07)	110 (1·10)	1 078
138 (1·38)	55 (0·55)	756
69 (0·69)	15 (0·14)	467
34 (0·34)	5 (0·05)	257
14 (0·14)	0·7 (0·007)	97

dynamic pressure and maximum blast wind velocity for an ideal shock front in air at sea level are given in Table VIII.11.

When the shock front reaches a given point, both the overpressure and the dynamic pressure increase almost immediately from zero to their maximum values and then decrease. The dynamic pressure (and wind velocity) will fall to zero somewhat later than the overpressure because of the momentum of the air in motion behind the shock front, but for purposes of estimating damage the difference is not significant. During the negative (suction) phase of the blast wave the dynamic pressure is very small and acts in the opposite direction.

VIII.9.2.2 Arrival Time and Duration

There is a finite time interval required for the blast wave to move out from the explosion centre to any particular location. This time interval is dependent upon the energy yield of the explosion and the distance involved. At 1·6 km from a 1 megaton burst, the arrival time would be about 4 s. Initially, the velocity of the shock front is quite high, many times the speed of sound, but as the blast wave progresses outward, so it slows down as the pressure at the front weakens. Finally, at long ranges, the blast wave becomes essentially a sound wave and its velocity approaches ambient sound velocity.

The positive phase duration is shortest at close ranges and increases as the blast wave moves outward. At 1·6 km from a 1 megaton explosion, the duration of the positive phase of the blast is about 2 s. The period of time over which the positive dynamic pressure is effective may be taken as essentially the positive phase duration of the overpressure.

VIII.9.2.3 Reflection of the Blast Wave at a Surface

When the incident blast wave from an explosion in air (Fig. VIII.163) strikes a more dense medium, such as the Earth's surface, either land or water, it is reflected.

When such reflection occurs, an object precisely at the surface will experience a single pressure increase, since the reflected wave is formed instantaneously. The value of the overpressure thus experienced at the surface is generally considered to be entirely a reflected pressure. In the region near ground zero, this total reflected overpressure will be more than twice the value of the peak overpressure of the incident blast wave.

There are two important destructive aspects of the blast wave involved with reflection:

(i) Only a single pressure increase is experienced in the Mach region below the triple point as compared to the separate incident and reflected waves in the region of regular reflection.

(ii) (a) Since the Mach stem is nearly vertical, the accompanying blast wave is travelling in a horizontal direction at the surface.

(b) The transient winds are approximately parallel to the ground, thus, in the Mach region, the blast forces on aboveground structures and other objects are directed nearly horizontally, so that vertical surfaces are loaded more intensely than horizontal surfaces.

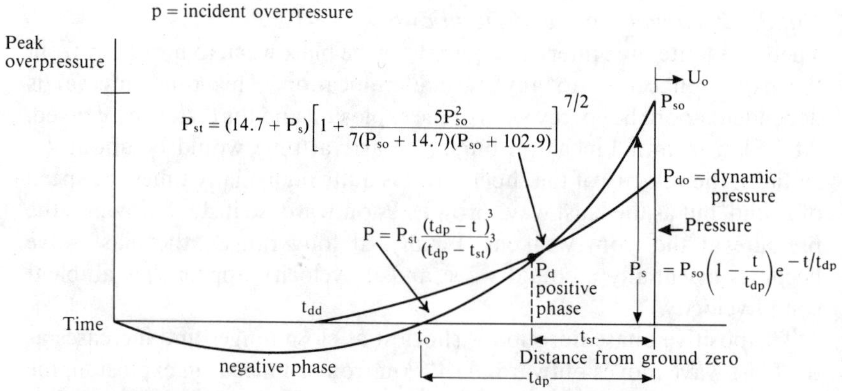

$$P_{st} = (14.7 + P_s)\left[1 + \frac{5P_{so}^2}{7(P_{so} + 14.7)(P_{so} + 102.9)}\right]^{7/2}$$

$$P = P_{st}\frac{(t_{dp} - t)^3}{(t_{dp} - t_{st})^3}$$

$$P_s = P_{so}\left(1 - \frac{t}{t_{dp}}\right)e^{-t/t_{dp}}$$

P_{do} = dynamic pressure

Fig. VIII.163. Variation of overpressure with time at a fixed location.

VIII.9.2.4 Blast from a Surface Burst

In a surface explosion, the front of the blast wave in air is hemispherical in form (Fig. VIII.164), there is no region of regular reflection, and all objects and structures on the surface, even close to ground zero, are subjected to air blast similar to that in the Mach region below the triple point of an airburst.

The diameter of the rupture zone D_r is $1.5D_a$; the overall diameter, including the lip D_t, is $2D_a$; and the height of the lip H_t is $0.25H_a$.

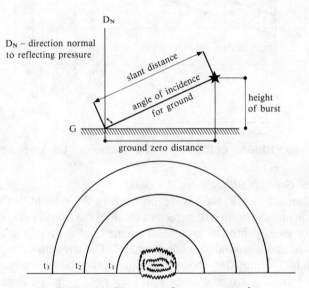

Fig. VIII.164. Blast wave from a contact burst.

The dependence of crater radius and crater depth upon the depth of burst for a 1 kiloton explosion in dry soil is shown in Fig. VIII.165. Also shown are the range of crater dimensions possible from a surface burst to the approximate maximum for an underground burst, for any explosion energy yield from 1 kiloton to 10 megatons. Therefore a 1 megaton blast wave front may be assumed to be vertical for most structures, with both overpressure and dynamic pressure decaying at different rates behind the blast wave as previously stated.

All the above descriptions are based on an ideal blast wave in ideal atmospheric conditions; unfortunately, nothing is perfect, especially meteorological conditions, which can greatly alter the expected performances of explosions, either increasing them or decreasing them.

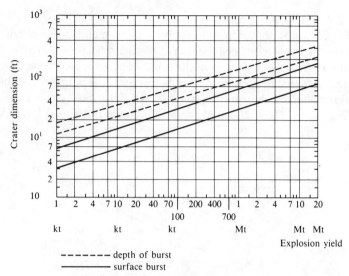

Fig. VIII.165. Crater versus yield (courtesy of US Airforce).

VIII.9.2.5 Ground Shock from Air Blast

Another aspect of the blast wave problem is the possible effect of an airburst on underground structures as a result of the transfer of some of the blast wave energy into the ground. A minor oscillation of the surface is experienced and a ground shock is produced. The strength of the shock at any point is determined by the overpressure in the blast wave immediately above it. For large overpressures with long positive phase duration, the shock will penetrate some distance into the ground; but blast waves which are weaker and of shorter duration are attenuated more rapidly. The major principal stress in the soil will be nearly vertical and about equal in magnitude to the air blast overpressure.

For high airbursts, where relatively large blast pressures are not expected at ground level, the effects of ground shock induced by the air blast will be negligible, but if the overpressure at the surface is large there may be damage to buried structures. However, even if the structure is strong enough to withstand the effect of the ground shock, the sharp jolt resulting from the impact of the shock front can cause injury to occupants and damage to loose equipment.

VIII.9.2.6 Technical Aspects of a Blast Wave

The basic relationships among the properties of a blast wave having a sharp

Table VIII.12
Blast loadings[a]

p_{d0} = peak dynamic pressure	U_0 = velocity of the shock front
p_s = overpressure at time	t = time after detonation of yield 'y'
t_{dd} = delayed time (dynamic)	t_{dp} = duration of dynamic overpressure
$U_0 = [U_s = 1117 \times \sqrt{1 + 6p_{s0}/7p_0}]$	p_0 = atmospheric pressure = 14·7

$$p_r = \text{reflected pressure} = 2p_{s0}\left(\frac{102\cdot 9 + 4p_{s0}}{102\cdot 9 + p_{s0}}\right)$$

t_c = clearing time of the reflection effect
  = 3 × height of the reflecting surface above ground/U_0

Scaling laws (for two bombs)

$R_2 = R_1(y_2/y_1)^{1/3}$ R, R_1, R_2 = radii or distances from explosions

$$t_{dd} = t_{dp}(y_2/y_1)^{1/3}$$

Reference should be made to Fig. VIII.163 for other details

Additional loadings

$$\frac{p_{s0}}{p_0} = 3\cdot 2 \times 10^6 R^{-3} \sqrt{1 - \left(\frac{R}{87}\right)^2} \times \left(\frac{800 + R}{800}\right)$$

$$\text{time in seconds} = t_{dp} = y^{1/3}\left\{\frac{180[1 + (0\cdot 01R)^3]^{1/2}}{1 + (R/40)\sqrt[6]{1 + (R/285)^5}\sqrt[6]{(1 + 0\cdot 0002R)}}\right\}$$

[a] Imperial units.

front at which there is a sudden pressure discontinuity, i.e. a true shock front, are given in Tables VIII.12–VIII.15 and Fig. VIII.166.

VIII.9.3 Air Blast Loading and Target Response

The behaviour of an object or structure exposed to the blast wave from a nuclear explosion may be considered under two main headings.

(1) Loading, the forces which result from the action of that pressure.
(2) Response, or distortion of the structure due to the particular loading.

As a general rule, response may be taken to be synonymous with damage, since permanent distortion of a sufficient amount will impair the usefulness of a structure.

Table VIII.13

Characteristics of blast waves from 1- and 10-megaton nuclear burst optimum height air blasts
(UK Home Office data)

Distance from ground zero, km (miles)	Peak overpressure, kPa (psi)		Windspeed, m/s (mph)		Arrival time, s		Duration, s	
	1 Mt	10 Mt	1 Mt	10 Mt	1 Mt	10 Mt	1 Mt	10 Mt
1 (0·6)	980 (140)	4 200 (600)	980 (2 200)	—	0·9	—	0·9	1
2 (1·25)	200 (29)	1 050 (150)	400 (900)	980 (2 200)	4·0	1·8	1·8	2
3 (1·9)	130 (18)	380 (54)	180 (400)	805 (1 800)	6·0	4·7	2·3	2·4
4 (2·5)	63 (12)	220 (31)	128 (285)	490 (1 100)	9·0	8	2·6	3·7
5 (3·1)	56 (8)	170 (24)	106 (238)	277 (620)	12	11	2·8	4·2
6 (3·8)	42 (6)	135 (19)	81 (182)	200 (450)	14	12·5	3·0	4·7
7 (4·4)	33 (4·7)	110 (16)	69 (155)	170 (380)	18	16	3·2	5·0
8 (5)	27 (3·8)	90 (13)	56 (125)	143 (320)	22	18	3·4	5·3
9 (5·6)	24 (3·4)	77 (11)	48 (108)	125 (280)	26	20	3·5	5·5
10 (6·2)	21 (3)	63 (9)	40 (90)	114 (255)	28	22	3·6	5·9
12 (7·5)	15 (2·2)	46 (6·5)	31 (70)	96 (215)	31	30	3·7	6·3
15 (9·4)	10 (1·5)	34 (4·8)	24 (53)	72 (160)	45	38	3·8	6·7
20 (12·5)	7 (1)	22 (3·3)	17 (37)	46 (103)	59	53	4·0	7·5
25 (16)	5 (0·75)	15 (2·2)	9 (20)	32 (72)	70	67	4·2	7·8

Table VIII.14
Blast loads on shallow buried and surface shelters

Member	Shallow buried		Surface
	In dry ground	High water level	
Roof and floors	p_{s0}	p_{s0}	p_{s0}
Walls	$0.5p_{s0}$	p_{s0}	$2.3p_{s0}$

p_{s0} = overpressure.
Dead loads, soil and water loads should be added to the blast loads given above.

Direct damage to structures attributable to air blast can take various forms; the blast may deflect structural steel frames, collapse roofs, dish-in walls, shatter panels and break windows.

VIII.9.3.1 Loading

(a) Diffraction loading. When the front of an air blast wave strikes the face of a building, reflection occurs. As a result the overpressure builds up rapidly to at least twice that in the incident wave front; the actual pressure

Table VIII.15
Generalised data in relation to loads

1-megaton burst at a distance of 1·6 km (1 mile) from ground zero
Velocity of shock front 500 m
Ductility ratio $\mu = 5$
Maximum allowable time = 0·15 s
Where applicable, drag coefficients: roof = −0·4, wall = +0·9
Yield strength of steel reinforcement = 410 or 425 N/mm^2
f_{cu} (dynamic) = $1.25 f_{cu}$ (static)
f_y (dynamic) = $1.10 f_y$ (static)
Main reinforcement $\not< 0.25\% bd$ (b = width of section, d = effective depth of section)
Secondary reinforcement $\not< 0.15 bd$
The ultimate shear stress $\not> 0.04 f_{cu}$
The dynamic shear stress for mild steel $\not> 172$ N/mm^2
Bolts should be black bolts to BS 4190
Allowable dynamic stresses:
Bolts: tension   275 N/mm^2 Welds: tension/compression 275 N/mm^2
       shear   170 N/mm^2 shear   170 N/mm^2
       bearing   410 N/mm^2

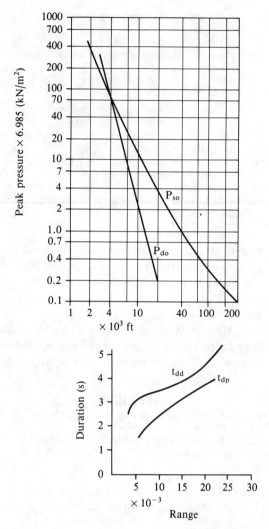

Fig. VIII.166. Megaton weapon curves (US Department of Defense and Atomic Energy Commission). For all other weapon yields Table VIII.11 should be used. For example:

$$1\,\text{Mt}: P_{s0} = 20\,\text{lbf/in}^2 \to 7100\,\text{ft}$$

$$\tfrac{1}{2}\,\text{Mt}: \frac{R_1}{7100} = \left(\frac{0\cdot 5}{1}\right)^{1/2}$$

therefore $\quad R_1 = 5650\,\text{ft}$

(Note: 1 ft = 0·3048 m.)

attained is determined by various factors, such as the peak overpressure of the incident wave and the angle between the direction of motion of the wave and the face of the building. As the wave front moves forward, the reflected overpressure on the face drops rapidly to that produced by the blast wave without reflection, plus an added drag force due to the wind (dynamic) pressure. At the same time, the air pressure wave bends, or 'diffracts' around the structure so that the structure is eventually engulfed by the blast (Fig. VIII.167), and approximately the same pressure is exerted on the side walls and the roof.

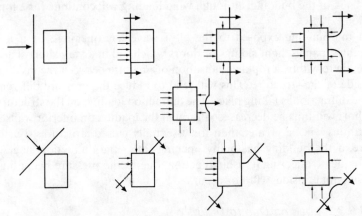

Fig. VIII.167. Stages in diffraction (plan view) without openings (courtesy of US Department of Defense and Atomic Energy Commission).

The front wall, however, is still subject to wind pressure, although the back wall is shielded from it.

The pressure differential between the front and back faces will have its maximum value when the blast wave has not yet completely surrounded the structure. In this case, such a pressure differential will produce a lateral (or translational) force, tending to cause the structure to deflect and thus move bodily, usually in the same direction as the blast wave. This force is known as the 'diffraction loading' because it operates while the blast wave is being diffracted around the structure. The extent and nature of the actual motion will depend upon the size, shape and weight of the structure, and how firmly it is attached to the ground.

When the blast wave has engulfed the structure, the pressure differential is small, and the loading is almost entirely due to the drag pressure exerted on the front face. The actual pressures on all faces of the structure are in excess of the ambient atmospheric pressure and will remain so, although decreasing steadily, until the positive phase of the blast wave has ended, thus the diffraction loading on a structure without openings is eventually replaced by an inwardly directed pressure.

The damage caused during the diffraction stage will be determined by the magnitude of the loading and by its duration. The loading is related to the peak overpressure in the blast wave and this consequently is an important factor. If the structure under consideration has no openings, as has been assumed for this discussion, the duration of the diffraction loading will be very roughly the time required by the wave front to move from the front to the back of the building, although wind loading will continue for a longer period.

If the building exposed to the blast wave has openings, or if it has windows, panels, light sidings or doors which fail in a very short space of time, there will be a rapid equalisation of pressure between the inside and outside of the structure. This will tend to reduce the pressure differential while diffraction is taking place. The diffraction loading on the structure as a whole will thus be decreased, although the loading on interior walls and partitions will be greater than for essentially closed structures. Furthermore, if the building has many openings after the diffraction stage, the subsequent squeezing (crushing) action, due to the pressure being higher outside than inside, will not occur.

VIII.9.3.2 Loading Due to Drag
During the whole of the overpressure positive phase, a structure will be subjected to the dynamic pressure loading or drag loading caused by transient winds behind the blast wave front. Like the diffraction loading, the drag loading, especially in the Mach region, is equivalent to a lateral (air translational) force acting upon the structure.

VIII.9.4 Loading Versus Structural Characteristics
There are basically two types of building when considering blast wave loading:

(1) Diffraction type
(2) Drag type

As these names imply, in a nuclear explosion, the former would be affected mainly by diffraction loading and the latter by drag loading. While it is true that some structures will respond mainly to diffraction forces and others mainly to drag forces, all such buildings will respond to both types of loading.

Application to Engineering Problems

A diffraction-type building is one that is primarily sensitive to the peak overpressure in the shock wave, e.g. reinforced concrete buildings with small window area and large wall-bearing structures.

When the pressures on different areas of a structure are quickly equalised, because of its small size, the characteristics of the structure or the rapid formation of numerous openings by the action of blast, the diffraction forces operate for a very short time. The response of the structure is then mainly due to the dynamic pressures (or drag forces) of the blast wind, e.g. for telephone poles, radio and television transmitter towers, and tall chimneys.

VIII.9.5 Damage Classification

Damage to structures or objects above ground can be divided into three categories, as follows.

Severe damage: A degree of damage that precludes further use of the structure or object for its intended purpose, without essentially complete reconstruction. For a structure or building, collapse is generally implied.

Moderate damage: A degree of damage to principal members that precludes effective use of the structure or object for its intended purpose, unless major repairs are made.

Slight damage: A degree of damage to buildings resulting in broken windows, slight damage to roofing and cladding, blowing down of light interior partitions, and slight cracking of certain walls in buildings.

VIII.9.6 Blast Loads and Stresses

Table VIII.12 and Fig. VIII.166 give a summary of the blast loads generated from a bomb burst. Table VIII.15 gives generalised data, including allowable stresses used later on in the analysis.

Based on HMSO reports [639, 640], the American Defence Agency [642–646] carried out limit state analysis for a six-person domestic shelter, as given in Figs VIII.168a and VIII.168b. This was designed for a 1 megaton weapon yield, for which the data are given in Table VIII.13.

VIII.9.7 Finite Element Analysis of a Domestic Nuclear Shelter

Figure VIII.169 shows a typical finite element mesh for the dynamic analysis of this shelter. Again 20-noded elements representing concrete in the walls and slabs with smeared reinforcements are used to analyse the

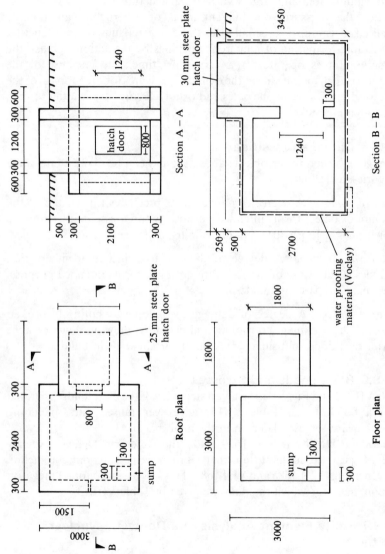

Fig. VIII.168a. Domestic nuclear shelter—general arrangement.

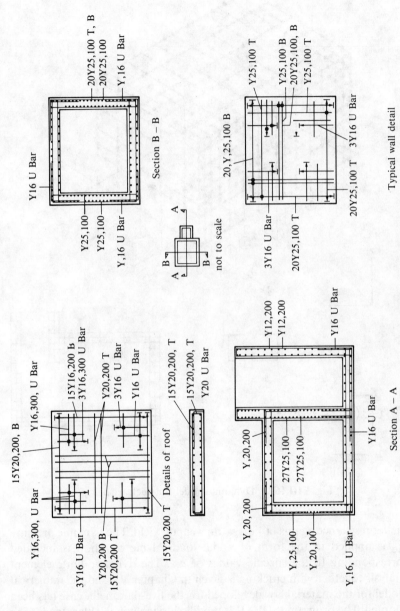

Fig. VIII.168b. Domestic nuclear shelter (reinforced concrete)—detail.

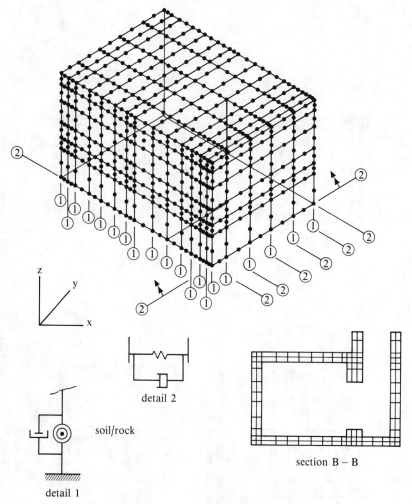

Fig. VIII.169. Dynamic model for nuclear shelter.

shelter resistance against a 1 megaton weapon yield. The dynamic pressure p_{d0} is applied in the form of nodal loads. Time duration is included corresponding to the specific values of p_{d0}. The dynamic finite element analysis together with cracking is given in Chapter VI, and the numerical model for the material law developed for the five-parameter case has been adopted. The program ISOPAR is modified to include the blast loads. The

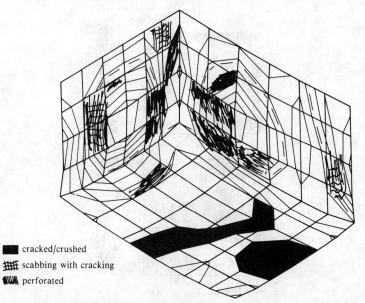

Fig. VIII.170a. Stress trajectories/cracking.

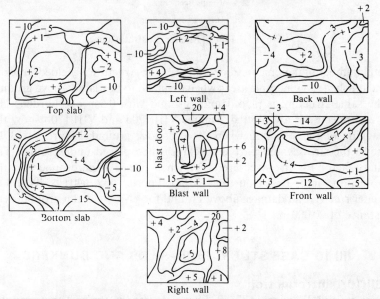

Fig. VIII.170b. Principal stresses for a nuclear shelter (all stresses are maximum principal stresses).

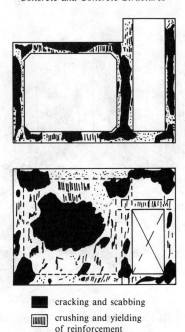

- ■ cracking and scabbing
- ▥ crushing and yielding of reinforcement
- ▦ crushing-cum-spalling

Fig. VIII.170c. Shelter 'post-mortem'.

total number of increments adopted is 12. The structure is yielded, cracked and crushed in vulnerable areas when assumed to be placed above ground. The same structure is placed underground with 0·5 m overburden. The stress trajectories are plotted. Figures VIII.170a and VIII.170b show the cracking of the structure when placed above ground. Figure VIII.170c shows the stresses and partial cracking when placed underground. The safety factor for the structure is 3·1 when placed underground and the protection factor based on the HMSO reports [639, 640] is 4000. The shelter received no damage above ground for $\frac{1}{2}$ Mt yield at a ground–zero distance of 20 000 m.

VIII.10 CASE STUDY No. 9—SILOS AND BUNKERS

VIII.10.1 Introduction

Analysts and designers [677–702] have assumed in the past that granular material behaves quasi-statically, and that the pressures exerted by such a material could be calculated similarly to hydrostatic pressure. Such assumptions are only useful for static horizontal loads and do not consider the material friction in the vertical direction along the wall. Janssen [cited in Ref. 692] derived formulae for granular material pressures on silo walls and bottoms. Many others came forward [677–702] using modified Janssen's formulae. These formulae are summarised in Table VIII.16. Janssen's and Airy's formulae have been derived on the lines of Rankine and Coulomb's theories of earth pressure. Figure VIII.171 gives a summary of silo wall pressures based on various codes of practice. These loads are converted to nodal and patch loads and are applied at appropriate positions on the finite element mesh chosen for a silo. The load is applied incrementally using the material constitutive law based on bulk and shear moduli given in Chapters II and VII. A time dependent analysis is also carried out in which the interaction between the materials and the silo's wall and floor is included. The details of the time dependent dynamic analysis are fully described in Chapter VI. A typical flow pattern is given in Fig. VIII.172a.

VIII.10.2 Loads on Silos

The design loads include silo wall loadings caused by stored material, thermal loads, wind loads, prestressing loads, wall-base-restraint loads and seismic loads. The dominating load for the silo wall design is the lateral load from the stored material. Figure VIII.171 shows several curves. Janssen's formula is modified to take various factors into consideration. The most important of these factors is the pressure increase, in relation to the static load, experienced during overloading of a silo and caused by dynamic effects associated with the flow of materials. These modifications are carried out in numerous ways including overpressure factors prescribed by ACI 313-77 for the formula. Adjustments to the basic parameters, such as density and angle of friction, in the formula have caused the difficult curves given in Fig. VIII.171. These details are given in Refs 678 and 679.

Dead, live, prestressing, thermal and wind loads have already been discussed in greater length in earlier chapters. Tables VIII.16 and VIII.17 give a summary of load computations for silos and bunkers with reference to Fig. VIII.172b.

VIII.10.3 Material Properties

The properties of concrete, conventional and prestressing steels are given full treatment in Chapters I, II and VII, together with the relevant appendices.

Table VIII.16
Formulae for silos and bunkers (reference is made to Fig. VIII.75)

(A) Janssen's theory

$$\delta P_V = \left[D_c - \left(\frac{\mu P_H}{A/p} \right) \right] \delta z \qquad P_H = c P_V$$

at $z = 0$
$$P_V = \frac{D_c A/p}{c\mu} [1 - e^{-c\mu z p/A}]$$

Total pressure

$$P = \sum_0^z P_H \delta z = \frac{D_c A/p}{\mu} \left[z - \left(\frac{A/p}{c\mu} \right) \{ 1 - e^{-c\mu z p/A} \} \right]$$

(B) Airy's theory

$$\mu R = \tan \bar{\theta}_1 W \qquad P_F = \tan \bar{\theta}_2 W = \mu p$$

$\bar{\theta}_1, \bar{\theta}_2$ = angles of friction

A = sectional area

p = perimeter

P_F = frictional force

P_V = vertical pressure/unit area on top

c = constant = 0·5 during filling
= 1·0 during emptying

(i) For shallow bins/silos

$$\text{Depth} = D \tan \theta'$$

$$P = \tfrac{1}{2} W H_T^2 \left\{ 1 \Big/ \left[\mu R \left(\mu R + \frac{P_F}{P_V} \right) + \sqrt{(1 + (\mu R)^2)} \right]^2 \right\}$$

where maximum depth $\not> D \left[\mu R + \left(\mu R + \sqrt{\left(\mu R \frac{1 + (\mu R)^2}{\mu R + P_F/P_V} \right)} \right) \right]$

$$P_H = W H_T \Big/ \left[\mu R \left(\mu R + \frac{P_F}{P_V} \right) + \sqrt{1 + (\mu R)^2} \right]^2$$

at depth H_T below

Table VIII.16—contd

(ii) For deep silos

$$P = \frac{\frac{1}{2}WD^2}{(\mu R + P_F)} \left[\sqrt{\left\{\frac{2H_T}{D}(\mu R + P_F) + 1 - \mu R \frac{P_F}{P_V}\right\}} - \sqrt{\{1 + (\mu R)^2\}} \right]^2$$

$$P_H = \frac{WD}{(\mu R + P_F)} \left[1 - \frac{\sqrt{(1 + (\mu R)^2)}}{\left\{\frac{2H_T}{D}(\mu R + P_F) + 1 - \mu R \frac{P_F}{P_V}\right\}} \right]$$

at depth H_T below

(C) British code (material handling board)

$$P_{zf} = P_{zfV} = P_{V(at\ z)} = 0.6 g D_c \left(z + \frac{H_s}{2 + m}\right) \quad \text{(in Pa)}$$

$$P_{zf} = \frac{D_c g R}{\mu_1} \times \left[1 - \exp\left\{-\frac{\mu_1 K_u}{R}\left(z + \frac{H_s}{2+m}\right)\right\}\right] \quad \text{(in Pa)}$$

(parallel section wall pressure)

P_{zf} = the pressure normal to the wall surface at a depth z (Pa) or
P_H = horizontal pressure/unit area on a vertical surface
P_{zfV} = the vertical pressure at height z (Pa)
D_c = the bulk density (kg/m^3)
g = acceleration due to gravity (m/s^2)
z = the distance from the lowest point of the top surface of the material in the parallel section (m)
μ_1 = lower bound wall friction coefficient
H_s = the height of the surcharge (m)
K_u = the upper bound pressure ratio
$m = 0$ for rectangular silo bins ($L = 3D$)
   $= 1$ for axisymmetric silo bins
R = hydraulic radius of the parallel section
   $= \dfrac{\text{internal cross-sectional area}}{\text{internal perimeter}}$ (m)
D = width or diameter

Material	Unit weight (kg)	$\bar{\theta}$ (degrees)	μR	$P_F/P_V = \mu p$
Wheat	800	25	0·466	0·444
Cement	1 442	17·5	0·316	0·554
Coal	800	35	0·716	0·700

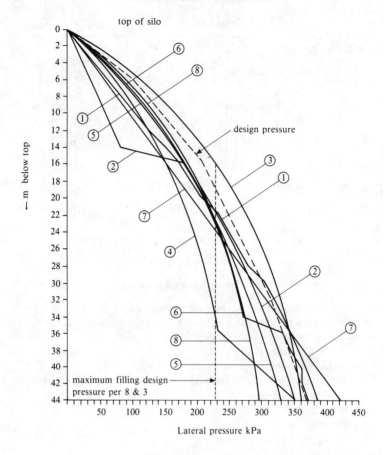

NA	• 1	ACI 313-77 (US Code)
NA	•• 2	CH-302-65 (Soviet Code)
NA	• 3	DIN 1055-Blatt 6, 1977 (German Code)
A	• 4	Peter Martens, Braunschweig
A	5	R. T. Jenkyn
A	• 6	Regles de Conception et de Calcul des Silos en Beton (French Code)
A	7	$.6 \, \gamma \, Z \, (\gamma = 1.6)$
NA	•• 8	DIN 1055-Blatt 6, 1964

• incl. effect of eccentric discharge
•• excl. effect of eccentric discharge
NA = no aeration
A = aeration

Fig. VIII.171. Silo wall pressures [699].

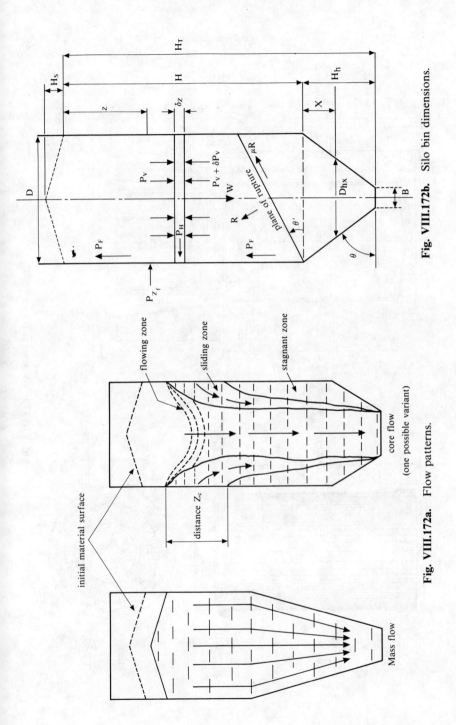

Fig. VIII.172b. Silo bin dimensions.

Fig. VIII.172a. Flow patterns.

424   Concrete and Concrete Structures

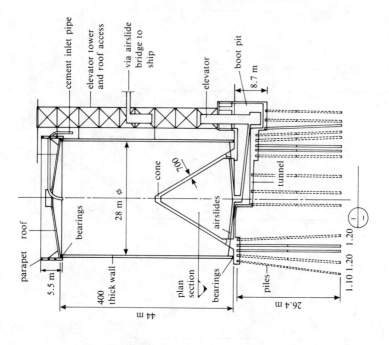

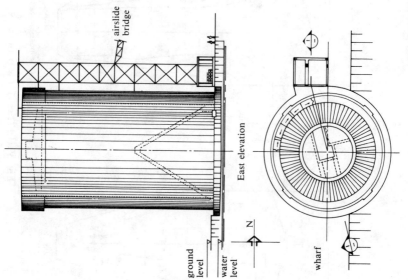

Fig. VIII.173. A general layout [699].

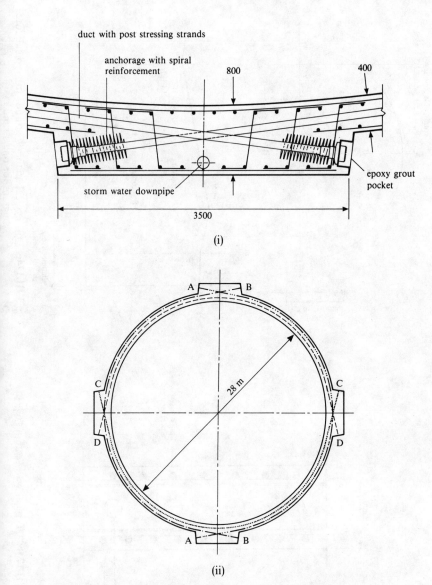

Fig. VIII.174a. (i) Plan of typical buttress; (ii) diagram of typical duct layout [699].

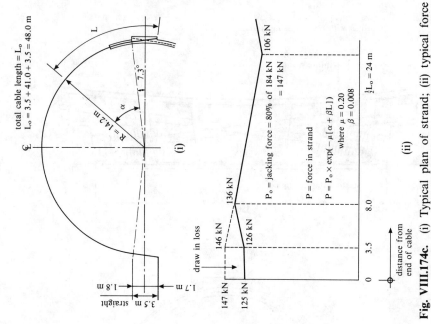

Fig. VIII.174c. (i) Typical plan of strand; (ii) typical force diagram for one strand.

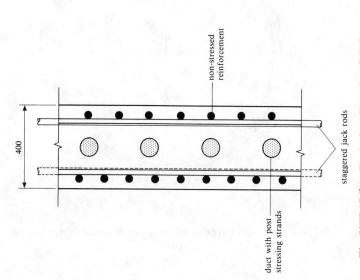

Fig. VIII.174b. Typical wall section [699].

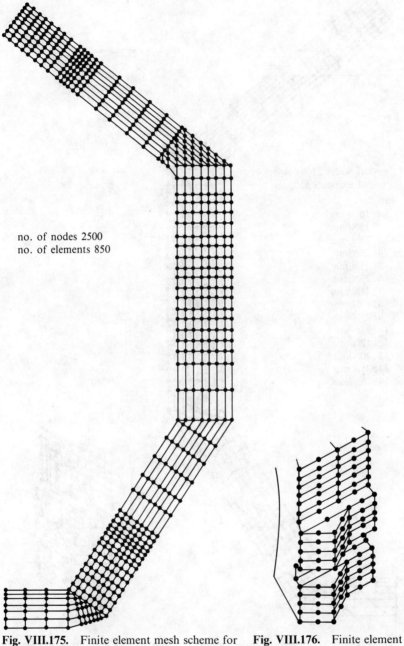

no. of nodes 2500
no. of elements 850

Fig. VIII.175. Finite element mesh scheme for silos (exaggerated).

Fig. VIII.176. Finite element mesh for buttresses.

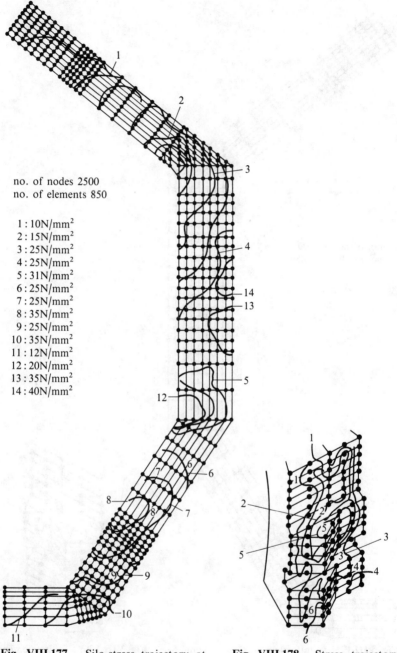

Fig. VIII.177. Silo-stress trajectory at $3.5 \times$ design pressure (exaggerated section).

Fig. VIII.178. Stress trajectory buttresses at $3.5 \times$ design pressure.

no. of nodes 2500
no. of elements 850

1 : 10 N/mm^2
2 : 15 N/mm^2
3 : 25 N/mm^2
4 : 25 N/mm^2
5 : 31 N/mm^2
6 : 25 N/mm^2
7 : 25 N/mm^2
8 : 35 N/mm^2
9 : 25 N/mm^2
10 : 35 N/mm^2
11 : 12 N/mm^2
12 : 20 N/mm^2
13 : 35 N/mm^2
14 : 40 N/mm^2

Application to Engineering Problems 429

no. of nodes 2500
no. of elements 850

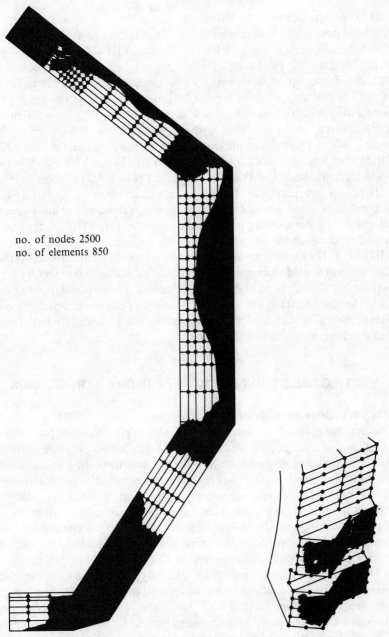

Fig. VIII.179. Silo 'post-mortem' (exaggerated sections).

VIII.10.4 Application to Silos

A typical prestressed concrete silo [699], 28 m internal diameter and 44 m wall height, is shown in Fig. VIII.173. Figures VIII.174a, VIII.174b and VIII.174c show a typical reinforcement plan and elevation. The strand is 12·5 mm of low relaxation wire with a guaranteed minimum breaking force of 184 kN that was adopted for the VSL multi-strand system having 19 prestressing anchorages. The total losses covering elastic shortening friction, creep, shrinkage, wobble effects and relation of steel are of the order of 40%. This cement silo capacity is currently located on a wharf in Birkinhead, South Australia [699]. Figures VIII.175 and VIII.176 show the finite element mesh for this silo. The ACI 313-77 curve given in Fig. VIII.171 is adopted for the finite element analysis. A 20-noded isoparametric element is adopted for concrete. A three-noded line element represents the prestressing tendons. As shown in Fig. VIII.176, the mesh layout is identical to that adopted for the Dungeness B vessel. Figures VIII.177–VIII.179 show the stresses and corresponding failure mode of this silo using 14 load increments. Assuming a 3 mm crack width as a permissible value, the safety margin computed prior to collapse for the design load is 3·65. The detailed cracking analyses given in Chapters VI and VII together with the relevant appendices have been adopted for both static and dynamic investigations for this silo.

VIII.11 CASE STUDY No. 10—OFFSHORE STRUCTURES

VIII.11.1 Concrete Gravity Platforms

The continuing search for petroleum has led to exploration and production of oil and gas in deeper waters and in areas subject to severe storm conditions, such as in the North Sea and the northwestern Atlantic. Since the periods of good weather that permit heavy construction at sea in these hostile offshore environments are relatively short, there is considerable interest in production of structures that can be constructed onshore or in sheltered waters, towed semi-submerged to location on completion and then installed on location in a short time by ballasting or flooding. Concrete gravity platforms meet this requirement [703–735, 1008–1029].

These structures are supported by relatively large foundation elements bearing on the near-seafloor soils, and depend principally on their weight and shape to resist vertical and horizontal forces and overturning moments. Concrete gravity platforms require less time and have reduced demands for costly marine equipment during installation than those for conventional pile-supported steel jacket platforms. In addition, oil storage facilities can easily be incorporated in the cellular configuration of many of the proposed gravity platforms. Unlike steel structures, concrete structures present no severe maintenance and corrosion problems, and are virtually free from problems of fatigue.

Traditionally, offshore platforms are designed on a basis of a static application of a design wave. Hostile sea environments require that the dynamic behaviour of structures in deep water be studied. In addition to the dynamic nature of the environmental loads on the platform, these loads are random in nature. Therefore methods of analysis and design which reflect the dynamic and random nature of the environmental loads on deep water platforms are needed. However, such methods are usually very complex and a large amount of computer time is involved. Nevertheless, the dynamic analysis given in this book can easily be applied for random loads using the ISOPAR routines.

Many concrete gravity platforms have been developed during the past few years. Some examples of these platforms are given in Appendix IX. The first concrete gravity structure installed in the North Sea was the Ekofisk Doris tank for Phillips Petroleum, placed in the summer of 1973. The next two were the Condeep Production Platform Beryl A for Mobil and Brent B for Shell/Esso, both installed in the summer of 1975. A typical case study of the Condeep Platform is examined in this section. A similar procedure is adopted for the other concrete platforms.

VIII.11.2 Design of the Condeep Platform

The concrete part of the Condeep structure (Figs VIII.180a and VIII.180b) consists of 19 cylindrical cells, of which three have been extended upwards to form towers to support the deck. Two of the towers are used for drilling production wells, while the third is filled with pipework and equipment. When the platform is in production the cells will always be filled either with oil or with seawater.

VIII.11.2.1 Environmental Loads
The design environmental loads used for offshore structures are normally based on environmental conditions, with a recurrence period of 50–100 years while the design lifetime is usually 20–30 years. For a concrete gravity platform, the wave load accounts for the major part of the environmental loads. The design wave loads are usually determined on the basis of regular waves by means of analyses and model tests. Loads due to wind and current are usually a relatively small fraction of the total forces. They are, however,

432 Concrete and Concrete Structures

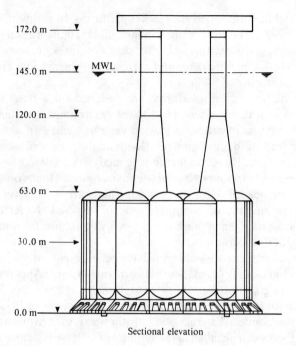

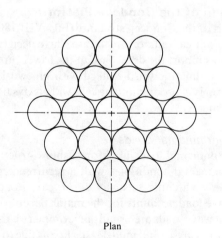

Fig. VIII.180a. Condeep platform.

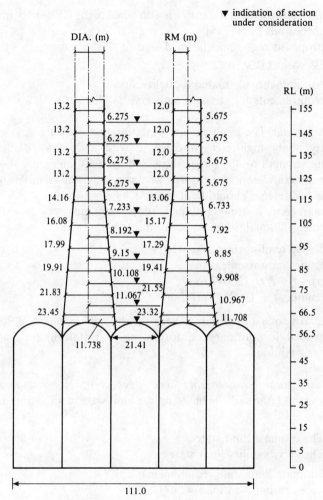

Fig. VIII.180b. Details of Condeep input data.

important and must be accounted for. For further details refer to Appendix IX.

VIII.11.2.2 Requirements of Concrete Structures

The methods of design of offshore concrete structures will in principle be the same as those used for structures on land. Therefore recognised standards for design and construction of reinforced concrete structures such as BS 8110 are used. Reference is also made to the CEB–FIB and other recommendations [703–735].

The proposed requirements for design of offshore concrete structures may be placed in two categories:

(a) requirements for material quality, and
(b) requirements for strength and serviceability.

(a) Materials. The main requirements for materials are strength and durability. Only high-strength concrete with a 28-day characteristic minimum strength of 45 N/mm² measured on a 100 mm cube should be used for structural concrete exposed to marine or crude oil environments. For details reference should be made to Chapter I.

The characteristic strength of the reinforcement may vary depending on its purpose. The qualities most used are:

—ordinary reinforcement $f_y = 410$ N/mm²
—prestress reinforcement
   bars $f_y = 800$–1250 N/mm²
   tendons $f_y = 1200$–1800 N/mm²

In the splash zone a minimum cover over the reinforcement of 70 mm is recommended for ordinary reinforcement, and 100 mm for ducts for prestressing tendons.

(b) Structural requirements. The structural design of reinforced concrete should normally be based on the limit state method and the limit states are:

(i) The ultimate limit states.
(ii) The serviceability limit states.

(i) Ultimate limit state requirements:
   design imposed load $= 1\cdot 6 Q_k$
   design dead load $= 1\cdot 4 G_k$
   design environmental load $= 1\cdot 0 V_k$
   the factor of safety for material $\gamma_m = 1\cdot 5$ for concrete and $1\cdot 15$ for steel

(ii) Serviceability limit state requirements:
   design imposed load $= 1\cdot 0 Q_k$
   design dead load $= 1\cdot 0 G_k$
   design environmental load $= 1\cdot 0 V_k$
   the factor of safety for material $\gamma_m = 1\cdot 0$ for both concrete and steel

Table VIII.17
Design loads

The design requirements for offshore gravity structures are covered by several codes. In the light of their recommendations, the following load combinations are recommended for the design of gravity platforms:

Case I	$G + Q + H + C_1 + BO$	100% unit stress*
Case II	$G + Q + BO + WF + W$	125% unit stress
Case III	Case I + 50%W + WL	130% unit stress
Case IV	Case I + SM + 50%W	125% unit stress
Case V	Case II + SM + 50%W	130% unit stress
Case VI	Case III + SM + 50%W	140% unit stress
Case VII	$G + BO + H + WF + EQ$	150% unit stress
Case VIII	Case II + C_1 + H	140% unit stress
Case IX	Case I + C_1 + 50%W	130% unit stress
Case X	Case I + W + SM	150% unit stress

(*stresses for service load)

where

G = dead load
Q = live load
H = hydrodynamic load due to wave
C_1 = load due to ice and other debris
BO = buoyancy
SM = seabed movement
WF = wave force steady state or slamming forces
W = wind forces
EQ = earthquake effects

These factors vary from code to code. For other conditions reference should be made to load combinations based on data collected for other cases [703–735]. Table VIII.17 gives the best possible load combination for the design of offshore facilities.

(c) Cracking. Cracking of concrete should not adversely affect the durability of the structure. The engineer should satisfy himself that any cracking will not be excessive. For reinforced concrete, the surface crack width should not exceed 0·2 mm, or 0·004 times the nominal coverage of the main reinforcement.

The concrete strength and cracking theories and numerical models are fully described in Chapters I, II, VI and VII. Any one of these can be examined in relation to the behaviour of such structures.

VIII.11.2.3 Implosion

Implosion phenomena in concrete gravity platforms are extremely important. They can occur very rapidly and result in catastrophic failure. The cells of the caissons can be vulnerable to implosion. A particular check is required against implosion.

Several analysis methods will be presented here, because different design approaches were found appropriate for cylindrical structures of different geometries. The following four cases are treated separately: (a) short, (b) moderately long and (c) long thin-walled cylinders, and (d) thick-walled cylinders of all lengths.

The separation point between thin-walled and thick-walled cylinders is selected at $t/D_0 = 0.063$, where t is the thickness and D_0 is the outer diameter.

Initial implosion pressure formulae are given in Appendix IX.

VIII.11.3 Data for Condeep

Steel stress $f_y = 410 \, \text{N/mm}^2$
Concrete stress $f_{cu} = 45 \, \text{N/mm}^2$ or $f_c = 0.87 f_{cu}$
Modulus of elasticity of steel reinforcement $E_s = 200 \, \text{kN/mm}^2$
Minimum cover to reinforcement $= 75 \, \text{mm}$
Main reinforcement bar diameter $= 20 \, \text{mm}$
Steel ratio $p = 0.01$
Thickness of tower $= 600 \, \text{mm}$
Height of tower $= 100 \, \text{m}$
Diameter of dome on caisson cell $= 28 \, \text{m}$
Diameter of caisson cylindrical wall $= 24 \, \text{m}$
Thickness of dome $= 600 \, \text{mm}$
Height of caisson cell $= 60 \, \text{m}$
Area of foundation base $= 6400 \, \text{m}^2$
Equivalent side of base $= 80 \, \text{m}$
Weight of superstructure $= 175\,000 \, \text{t}$

A section and the diameters of the towers and caisson cell are given in Figs VIII.180a and VIII.180b. The depths of the sections below mean seawater level are also given in the figures.

Wave forces and wave moments are obtained using various theories [703–720, 1008–1029] including the Hogber and Standing diffraction wave theories. Appendix IX gives some basic wave theories based on these, for maximum forces and wave moments at phase angle 90°.

Maximum wave force $= 0.865 \times 10^6 \, \text{kN}$
Maximum wave moment $= 0.396 \times 10^8 \, \text{kN m}$

VIII.11.3.1 Soil Information
The soil conditions in the North Sea have the following approximate properties:

Stiff clay:
Modulus of elasticity of soil = 500 t/m^2 (5 × 10^5 kN/m^2)
Coefficient of consolidation $(M_v) = 0$

Soft clay:
Modulus of elasticity of clay = 1 × 10^5 kN/m^2
Coefficient of consolidation $(M_v) = 1\cdot7 \times 10^{-5}$ kN/m^2

The minimum factors of safety against failure of the foundations are 1·5 and 2·0 for extreme and operating conditions, respectively.

VIII.11.3.2 Waves
(a) Extreme conditions. Extreme sea conditions are the highest sea state having a duration of 6 h during a once-in-a-hundred-years storm. This sea state is represented by a sea spectrum with a significant wave height (H_s) of 15·00 m and a significant period (T_s) of 15 s. The spectrum is defined by the spectral density function:

$$S(w) = H_s^2/4w \times A(w) \times e^{-A(w)}$$

where

H_s = significant wave height from crest to trough

w = circular wave frequency in radians per second

$A(w) = 690/(wT)$ ($T = 0\cdot88 \times$ significant wave period (T_s))

(b) Hydrodynamic coefficients. For the calculation of wave forces using Morison's theory (Appendix XII), a drag coefficient of 0·6 and an inertia coefficient of 2·0 is used.

VIII.11.3.3 Wind Velocities
(a) Storm wind. The maximum one minute sustained wind velocity of 125 mph is used in conjunction with the storm wave.

(b) Operating wind. A wind velocity of 60 mph is used in conjunction with the operating wave.

(c) Instantaneous gust. A maximum instantaneous gust velocity of 160 mph is used throughout.

VIII.11.3.4 Tides

(a) Storm tide. The maximum tide of 2·15 m is assumed and finally combined with a storm surge of 0·6 m to produce a total storm tide rise of 2·75 m.

(b) Operating tide. The maximum tide of 2·15 m is combined with a storm surge of 0·12 m to produce a total operating tide rise of 2·27 m.

VIII.11.3.5 Currents

(a) Storm current. The maximum storm current velocity shall vary with water depth according to the following profile:

(a) At the water surface = 1·38 m/s
(b) Above bottom = 0·54 m/s at 15 m
(c) At bottom = 0·24 m/s

(b) Operating current. The maximum current velocity under operating conditions shall vary with water depth according to the following profile:

(a) At water surface = 1·17 m/s
(b) Above bottom = 0·45 m/s
(c) At bottom = 0·24 m/s

VIII.11.3.6 Temperatures

Minimum water temperature is assumed as $+7·0°C$. Crude oil temperature in storage is taken as 40°C and in production wells it is taken as 75°C.

VIII.11.4 Ship Impact on Concrete Gravity Platforms

VIII.11.4.1 Introduction

There is growing public interest in the effects of collisions involving ships [1030–1038] and tankers carrying particularly hazardous material (including the impact of nuclear vessels on vital installations such as the gravity platforms in the sea). Table VIII.18 gives useful formulae and data for the ship impact phenomenon. A theoretical model based on finite elements is fully described in earlier chapters. The level at which the kinetic energy of impact is lost depends on the ship/tanker/vessel angle of encounter, mass and speed, location of the platform and of the impacter-surrounding water, shock effects, etc. The analysis carried out assumes throughout that the collision force, owing to its quasi-static nature, is constant with the depth of penetration.

Table VIII.18
Ship impact data

$$F_I = \text{impact force} = \frac{E_D}{e} = \tfrac{1}{2}m_s v_s^2 \left[\frac{m_s + m_t}{m_t} \bigg/ 1 + \frac{m_s + m_t}{m_t}\right]\bigg/ e$$

m_s, v_s = mass and velocity of the striking ship
m_t = equivalent added mass of the target
e_t = penetration
E_D = energy of deformation = $\tfrac{1}{2}m_s v_s^2 \phi_1 \phi_2 \phi_4 + \tfrac{1}{2}m_t v_t^2 \phi_5 - v_s v_t m_t \phi_6$

ϕ represents the non-dimensional functions representing the colliding structure/vessel, the impact location, the angle of encounter, the speed of the target, and the product of both speeds.

(a) Duration for constant collision force:

$$T = \phi_1 \phi_2 \frac{m_t}{F_I} v_s$$

(b) Linear collision force—duration:

$$T = \frac{\pi}{2}\sqrt{\frac{\delta(2-\delta)}{\gamma\eta^2 \tan\theta}}\,\phi_1\phi_2 m_t$$

Acceleration at the time of collision

$$\ddot{\delta}'_{\max_{(x)}} = \frac{F_{I(x)}}{m_t} \qquad \ddot{\delta}_{(y)} = \frac{F_{I(y)}}{m_t + m_s} + \frac{\delta x(F_I)_y e}{I'}$$

where

x = the distance between the CG of the target and the point at which acceleration is computed.

$F_{I(x)}, F_{I(y)}$ = the maximum components of the collision force in the transverse and longitudinal directions

I' = the mass moment of inertia of the target

VIII.11.4.2 Present Methods of Assessment

Frequent impacts against offshore platforms may come from vessels operating close to the platform. There is growing concern that an impact may occur due to navigational errors by heavy vessels. The DnV rules for mobile offshore units specify a ship of 5000 tons displacement with impact speed of 2 m/s. This is treated as the design limit state. Another method

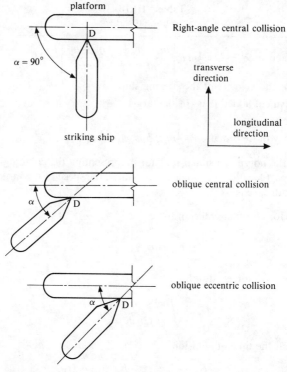

Fig. VIII.181. Ship impact data—typical collision cases.

adopted recently is based on the energy absorption capability rather than the design limit state, and this DnV impact energy is given below.

(a) 14 megajoules (MJ) for a side collision equal to 40% of added mass.
(b) 11 megajoules (MJ) for a bow or stern collision equal to 10% of added mass.

VIII.11.4.3 Impact Mechanics

The ship/platform impact mechanics are based on two criteria, namely conservation of momentum and conservation of energy. For a short impact duration (Fig. VIII.181)

$$E_s + E_p = \tfrac{1}{2}(m_s + \Delta m_s)v_t^2 \frac{(1 - v_t/v_s)^2}{1 + \dfrac{(m_s + \Delta m_s)}{(m_t + \Delta m_t)}}$$

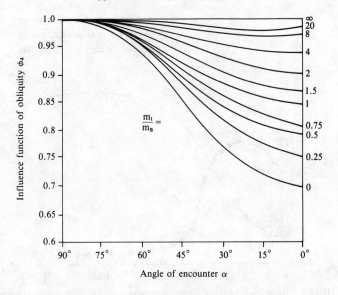
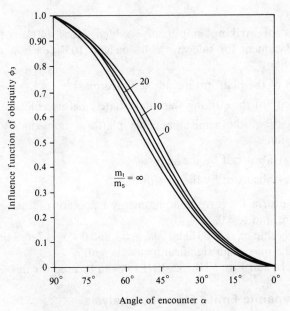

Fig. VIII.182. Ship impact data—ϕ_3, ϕ_4 versus α.

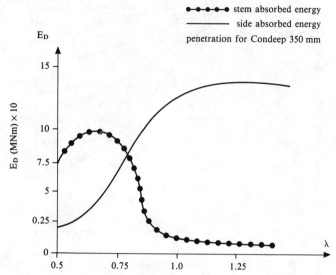

Fig. VIII.183. Ship impact data—relations between strength ratio (λ) of stern to side and absorbed energy (E_D).

where

m_s = mass of striking ship (Δm_s = added mass (40% of vessel displacement for sideways collision and 10% for bow or stern collision))

m_t = mass of the platform including added mass

v_s = velocity of the striking ship immediately before collision

v_t = velocity of the semisubmersible platform immediately before collision

E_s = energy absorbed by the ship

E_p = energy absorbed by the platform

For a fixed platform, the corresponding energy expression is obtained using $(m_t + \Delta m_t) = \infty$ and $v_t = 0$.

For a supply ship m_t is 5000 t and $\Delta m_t = 0.1$ and 0.4; v_s is 2 m/S or $0.5 H_s$, where H_s is the maximum significant wave height.

Table VIII.18 and Figs VIII.182–VIII.184 can be used for this analysis.

VIII.11.5 Dynamic Finite Element Analysis

An offshore platform oscillating in the ocean has been discretised by

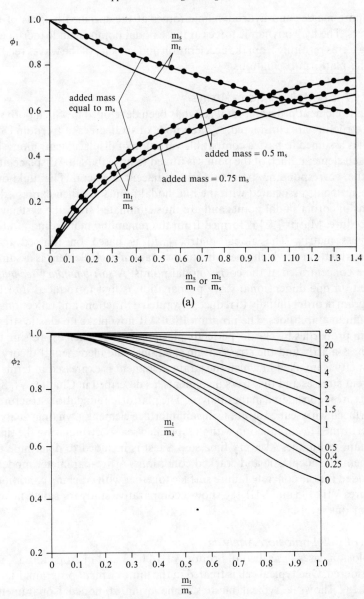

Fig. VIII.184. (a) Ship impact data—influence of mass ratio ϕ_1; (b) influence function of eccentricity ϕ_2.

lumping masses and hydrodynamic forces at the specified heights of the towers. The hydrodynamic forces at these nodal points include inertia and drag terms resulting from the acceleration and velocity of the waves relative to the platform oscillations.

VIII.11.5.1 Main Platform Analysis

A finite element program, ISOPAR, has been developed in which both soil and structure are considered. A mesh node of six degrees of freedom [375, 749] is assumed to have a compatible variation of displacements alongside of each element. Each of the points is forced to have the same displacement as the corresponding points of the adjacent element. The unknown displacements associated with the mid nodal points are then expressed in terms of corner nodal points and are thus eliminated from the systematic procedure. Matrix $[K]$ is formed from the remaining matrix and a direct stiffness matrix. The global matrix $[M]$ is based on lumped mass approximation in which one-third of the mass of each element is assumed to be concentrated at the corner nodal points. A *soft boundary technique* based on one-dimensional wave propagation is first introduced into the program in order initially to reduce unwanted defractions and reflections to minimum amplitudes. The program ISOPAR now picks up elastic stress–strain properties by elastic–perfectly plastic relations when expressing the stiffness matrix and the stress–strain relations. The incremental theory of plasticity is invoked in which the strain increment is expressed in terms of current stresses and the stress increments, as described in Chapter VI. Soil structure is considered in the analysis. The platform foundation structure is interfaced with seabed strata using an interface element. Dynamic analysis is performed in conjunction with the specified sea spectrum using the time-domain technique. At every time step a test is included to determine the element elastic, plastic and cracked conditions. A five-parameter model is included for the concrete failure surface together with cracking conditions. Figures VIII.185 and VIII.186 show a comparative study for safety factors at various levels.

VIII.11.5.2 Implosion Analysis

Implosion cases described in Section VIII.11.2.3 and in Appendix IX are considered. One typical cell is treated in the finite element program. Figure VIII.187 shows a typical mesh scheme using 20-noded isoparametric elements for the concrete and three-dimensional line elements for the reinforcement in the vertical, radial and horizontal directions. Based on implosion pressures, the damaged part is shown in Figs VIII.188 and

Application to Engineering Problems

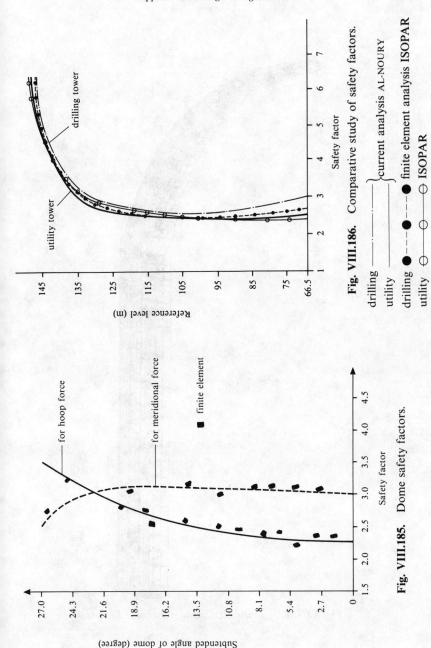

Fig. VIII.185. Dome safety factors.

Fig. VIII.186. Comparative study of safety factors.

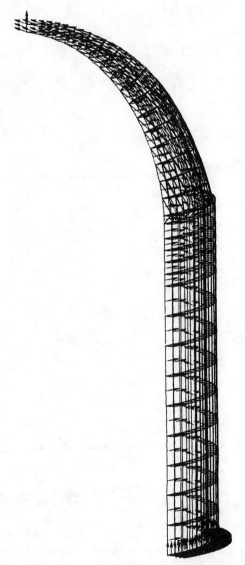

Fig. VIII.187. Finite element mesh scheme for a cell.

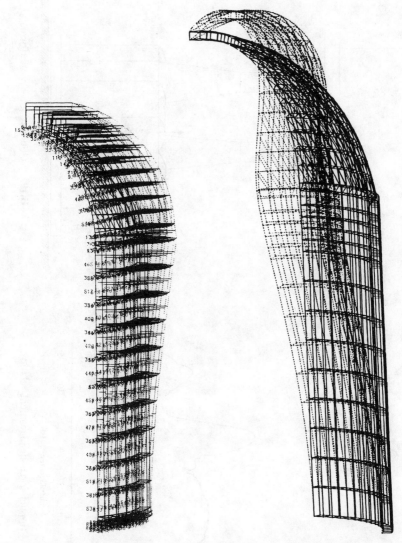

Fig. VIII.188. Damage caused by implosion at a level 2·75 times extreme loads (minimum safety factor 2·5).

Fig. VIII.189. Damage caused by implosion at a level 3·5 times extreme loads (concrete cracked/scabbed; steel yielded/plastic).

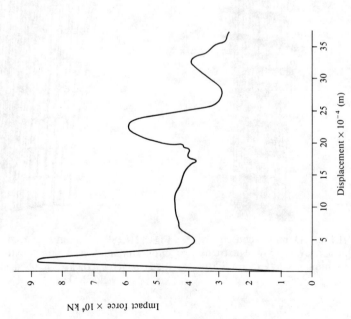

Fig. VIII.190. Impact force–displacement relation.

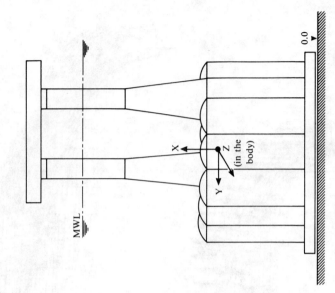

Fig. VIII.191. Concrete gravity platform.

VIII.189 when the final incremental failure pressure is 3·5 times the wave pressure based on extreme load cases. For Condeep the safety factor is thus 3·5 against these types of pressure. For further details see Appendix IX.

VIII.11.5.3 Ship Impact Analysis

The ship impact case given in Section VIII.11.4.3 is examined using the loading cases and material criteria for the Condeep platform. The platform is analysed using a given supply ship. The load–displacement curve is given in Fig. VIII.190. Three failure modes are given in Figs VIII.191 and VIII.192, and the cracking pattern of the shafts is given in Figs VIII.193 and

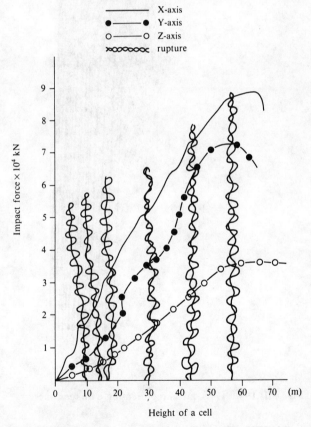

Fig. VIII.192. Failure modes of a cell (see Fig. VIII.191).

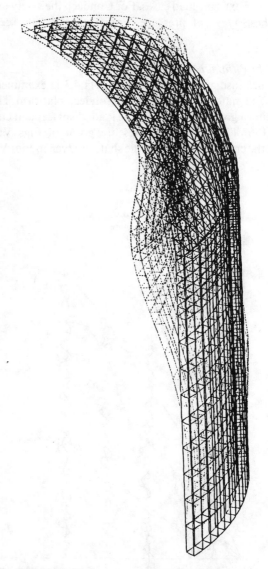

Fig. VIII.193. Dark areas representing damage zones.

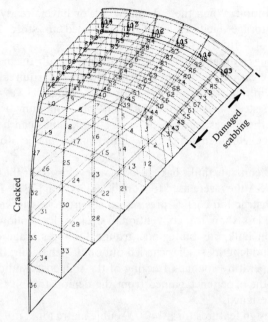

Fig. VIII.194. Junction between the dome and the wall.

VIII.194 for the cell. The major results associated with final mode of failure are:

	Type	Maximum displacement (m)	Time (s)
(a)	Base of the shaft	0·02	3·50
(b)	Top of the cell	0·018	3·10
(c)	Major part of the cell	0·0159	3·03

VIII.12 CASE STUDY No. 11—LIQUEFIED NATURAL GAS (LNG) TANKS

VIII.12.1 Introduction

Due to the growth and expansion of industrial natural gas facilities the storage of large volumes of gas is needed to meet winter demands. In many areas there exist underground reservoirs which provide a satisfactory and economic solution. When these do not exist, or have already been used, alternative storage facilities must be provided to store LNG at a temperature of $-165°C$.

In natural gas storage systems the container is usually the main expense. Storage of the gas in liquid form is preferable since it requires less storage space. Conventional tanks are flat-bottomed cylindrical vessels with an external mild steel shell and internal tank of aluminium or steel, with insulation in the spaces between the walls. These tanks can be expensive, but recent developments in alloy tanks have made storage more economical.

Prestressed concrete tanks have also been developed for the industry; in 1953 in the USA they were used to store liquid oxygen [737]. The running costs and construction cost of prestressed concrete vessels were, in 1961, substantially less than for conventional steel tanks. Following this, a demonstration tank was built underground for the storage of liquefied natural gas and lengthy studies carried out on its behaviour; these studies included temperature changes, freezing of the soil, costs and evaporation rates. Using the experience gained from the demonstration tanks, larger tanks were designed.

General design features for a 140 000 cubic metre tank, prestressed and situated below ground, include the following features:

(a) The base of the tank can be a reinforced concrete slab covered with 300 mm of foamglass insulation within a steel liner.
(b) The concrete walls on the outside may have a steel liner wrapped with prestressed wire and protected with a special mortar.
(c) The roof is a dome supported by a prestressed concrete ring beam covered with a thin steel liner and foamglass insulation.
(d) Heating coils are installed 600 mm below the tank base and halfway up the side of the underground walls in order to counter freezing and frost heaving in the winter months.

Properties of the structural components of such tanks were studied and found to be adequate for use at cryogenic temperatures, both the compressive strength of moist concrete and the tensile strength of high carbon steel prestressing wire increased considerably at temperatures below room temperature.

In a cryogenic tank there are two tension-producing effects: thermal and hydrostatic. Since concrete has good compressive strength, it is vital that the concrete is kept in compression so that no tension-producing force can ever quite overcome the initial compression.

The tank can be designed either as a rigid structure, that is with the base and the walls fixed, or as a free-sliding [736–748] wall-to-base structure, the latter design being more feasible for the pressures exerted on the walls. This design using independent movement for the side walls relative to the base is made possible by the introduction of an expansion joint in the steel liner at the bottom of the tank wall.

The data obtained were adequate to be scaled up from the demonstration model to the prototype. The data also suggested a change in the insulation of the base. In order to improve the performance of tanks subject to such temperatures it is necessary to design the base ring for varying temperatures.

Extensive tests on the expansion joint using liquid nitrogen proved its adequacy and showed a large safety factor. Checks were made for leaks. Tests showed that there were 220 leaks in the structure of the tank.

Experimental tests showed no extensive vertical movement of the tank during cool-down contraction of the circumference. The maximum movement that occurred was around 25 mm. When the tank was filled again no noticeable movement was observed.

Tests were also conducted on the effectiveness of the foamglass insulation used. The calculated values compared with measured values during use of the tank showed about 83% efficiency. Tests were also carried out on various soils and their freezing zones. It was observed that this zone moved further away when the inside temperature was brought up to room temperature.

Conclusions from all tests showed that prestressed concrete tanks were a viable proposition for the storage of liquid gases. A typical concrete tank prestressed by Dykman (BBR Tank) is shown in Fig. VIII.195.

VIII.12.2 Methods of Analysis and Design [736–748, 1039]

VIII.12.2.1 Methods of Analysis and Design Considerations

There are many considerations which have to be taken into account even before design can be carried out. The four basic methods of design are:

(a) *Empirical*—Promoters of this concept do not go into deep theoretical considerations for their design. They rather try to make corrections in future designs, responding to problems from earlier structures.

(b) *Crack control*—This theory is largely based on the idea of eliminating all straight tensions through prestressing and reinforcing the concrete with mild steel against bending stresses. It is then assumed that the crack developments in concrete will not be

Fig. VIII.195. A typical natural gas reservoir (San Clemente, California). Tank diameter 35·4 m (116 ft), wall height 7·93 m (26 ft). (Courtesy of BBRV Prestress Tanks Inc., California, USA.)

necessary provided that the crack size can be controlled using smaller diameter reinforcing bars with particular bar spacings and concrete cover. Under this concept, stresses are permitted close to the yield point of the steel.

(c) *Crack prevention*—Most 50-year-rated designs are now prepared with this concept in mind. A truly non-cracked tank condition can most easily be accomplished by prestressing the concrete in two directions. To minimise leakage the crack prevention concept seems more reliable for the design of prestressed concrete tanks. In order to prevent leakage, the strength of linings may be introduced into the design. For lighter fluids such as kerosene, gasoline and jet fuels, it is necessary to provide a steel, plastic or rubber lining.

(d) *Three-dimensional numerical modelling with well-disposed cracks*—Analyses include the prediction of cracks under increasing pressures and cryogenic temperatures for both short- and long-term conditions.

VIII.12.2.2 Design Consideration

(a) *Life span*—The demand for fully prestressed concrete structures has increased rapidly in the last decade. Around 5000 circular prestressed concrete structures are now in service on a global basis, some of which have already given over 30 years service. On the basis of what is known today, it is possible to build concrete storage structures with substantial trouble-free performance for periods of at least 50 years. It is expected that fully prestressed systems will outlast ordinary reinforced structures by a considerable margin.

(b) *Corrosion*—Special precautions are needed for the storage of acids, brine and salts. It is generally not sufficient to rely on the concrete to protect the steel. Thus it is necessary to resort to steel, plastic or rubber linings, or use suitable stainless steel reinforcement throughout. Particular attention must be given to preventing sulphates and chlorides from getting to carbon steel. Provision for cathodic protection is generally ignored when proper design and construction techniques are followed. The use of galvanised wire for prestressed reinforcement is particularly advantageous for a long life of the tank.

(c) *Leakage*—Concrete storage structures have a negligible loss factor when such liquids as water and oils are stored. For lighter fluids such as kerosene, gasoline and jet fuels, it is necessary to provide a steel, plastic or rubber lining.

(d) *Safety against failure*—Concrete storage structures have a good operational record and a relatively small number of failures are on record, all of which are the result of poor detailing and inadequate design and specification. The minimum factor of safety for design shall not be less than 2·5.

(e) *Formulae for preliminary sizes*—For dome roofs having diameters up to 60 m and live loads to 195 kg/m² the following empirical formulae are suggested:

Shell thickness d	$0·00417 R_d$ (m)
Fillet thickness t	$d + 0·00078 W$ (m)
Fillet length F	$0·2023 R_d$ (m)
Dome radius R_d	$2·125 R_d$ (m)
Dome shell weight	$8025·657 R_d \times d$ (kg)
Dome fillet weight	$1·216 R_d^2 w$ (kg)
Dead weight (metric) of circumference	$0·1592 W/R_d$ (kg)
Dead weight of projected dome area	$2553·48 d + 0·377 W$ (kg m)
Shell reinforcing A_s	$625 d$ (mm²/m)
Fillet reinforcing A_s	$0·221 R_d$ (m)
Top and bottom reinforcement At	$635 t$ (mm²/m)

The wall has a uniform thickness:

$M/bt^2 = 0·35$ N/mm for grade 30 where $d/t = 0·75$
$M/bt^2 = 0·32$ N/mm for grade 20 where $d/t = 0·75$

The assumptions give the moment at the bottom of the wall.

VIII.12.2.3 LNG Tank Data

Chapter I gives the material properties of concrete under cryogenic conditions. In addition, some data are given here for both concrete and tendons. Figures VIII.196a and VIII.196b give the thermal coefficient relationships for steel.

Design calculations for a prestressed concrete cylindrical tank (Figs VIII.197 and VIII.198) with a spherical dome are carried out using the numerical models described in this book. The main dimensions are shown in the figures.

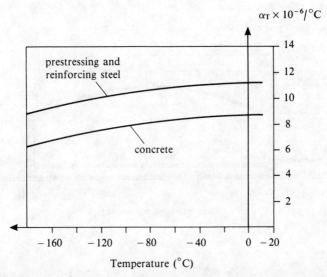

Fig. VIII.196a. Coefficient of thermal expansion (α_T) for dry to moderately moist concrete and steel (data collected from [743–746]).

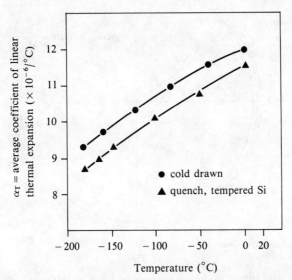

Fig. VIII.196b. Relation between temperature and coefficient of thermal expansion for several steels for prestressing (data collected from [743–746]).

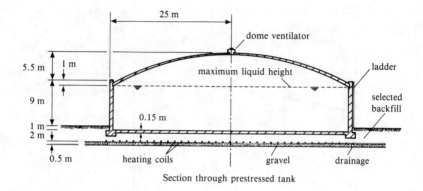

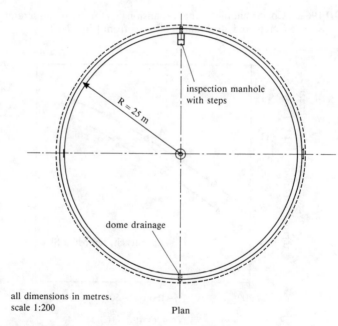

Fig. VIII.197. LNG tank parameters (all dimensions in metres)

(a) Design parameters for prestressing steel.

$P = 413 \cdot 7 \, \text{kN/m}^2$
$f'_c = 41 \cdot 37 \, \text{N/m}^2$
$f_y = 413 \cdot 7 \, \text{N/m}^2$
$f_{pu} = 16 \cdot 548 \times 10^5 \, \text{kN/m}^2$
$E_c = 28 \cdot 27 \, \text{N/m}^2$
$E_s = 200 \, \text{GN/m}^2$
$\mu = 0 \cdot 14$
$k = 0 \cdot 0003$

and the assumed prestress level $= 1 \cdot 2PR$.
In addition, long-term losses are taken as:

Creep and shrinkage $= 500 \times 10^{-6}$
Tendon relaxation $= 8\%$ at $0 \cdot 7 f_{pu}$
Tendon sheathing diameter $= 150 \, \text{mm}$
Wall (hoop) prestressing $FHW = 1 \cdot 2PR$

(a) Tendon stress after friction loss:

$$f_{pu} = f_{px} e^{(kl + \mu\theta)} \qquad f_{ps} = 0 \cdot 8 f_{pu} \qquad f_{px} = f_{ps} e^{-(kl + \mu\theta)}$$

(b) Tendon stresses at stress transfer after friction loss:

$$f_{pA} = 0 \cdot 75 f_{pu} \qquad f_{ps} = f_{pA} e^{(kl + \mu\theta)}$$

(c) Average tendon stress and stress transfer due to friction loss: Since the hoop tendons are anchored at the buttresses alternately, it may be assumed that the stresses at the wall are uniformly distributed. The average stress, f_{avg}, is written as

$$f_{avg} = \frac{\text{area under the stress curve at transfer}}{\text{length of tendon}}$$

(d) Loss of tendon stress due to elastic shortening of concrete:

$$f_{ces} = \tfrac{1}{2} f_c \frac{E_s}{E_c}$$

assuming $f_c = 0 \cdot 3 f'_c$

$$f_{ces} = \tfrac{3}{20} f'_c \frac{E_s}{E_c}$$

(e) Loss of tendon stress due to creep and shrinkage of concrete:

$$f_{ces} = \varepsilon_{cs} E_s$$

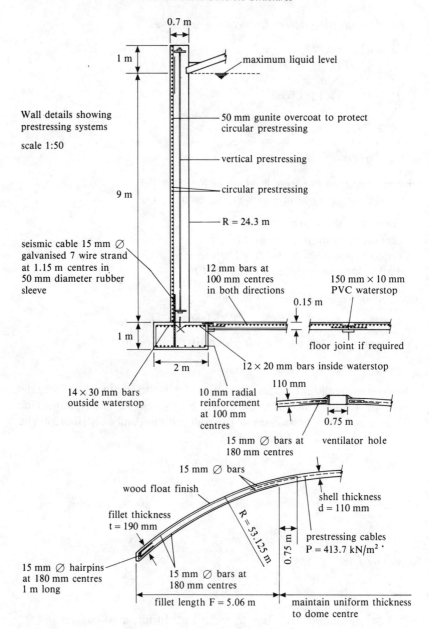

Section through dome Scale 1:50

(f) Loss of tendon stress due to steel relaxation:
$$f_{sr} = 0.08 \times 0.7 f_{pu}$$

(g) Average effective tendon stress after all losses:
$$f_{se} = f_{avg} - f_{sr}$$

(h) Tendon requirements:
$$N = \frac{Fl}{\frac{2}{3} A_{ps} f_{se}}$$

(b) Tank dome (meridional) and wall (vertical) prestressing. The losses due to elastic shortening and concrete creep are assumed to be the same in the vertical and dome tendons as in the wall hoop tendons. At the apex of the dome $FM = 1.2(PR/2)$.

(a) Tendon stresses after a friction loss:
$$\text{apex of the dome } f_{ps} = f_{pxe}(kl + \mu\theta)$$

(b) Tendon stress at stress transfer after friction loss:
$$f_{pA} = 0.7 f_{pu}$$

Fig. VIII.198. LNG tank—data and structural details. General design parameters:

1. Capacity			15 000 m³
2. Dimensions	(a)	Inside diameter	50 m
	(b)	Liquid height	8 m
	(c)	Free height	1 m
3. Loadings	(a)	Live load on roof	100 kg/m²
	(b)	Specific gravity	1·5
	(c)	Liquid temperature:	
		Summer average	60°F
		Winter average	50°F
	(d)	Air temperature:	
		High summer average	70°F
		Low winter average	40°F
		Temperature peaks	±15°F
	(e)	Seismic acceleration	10% g
	(f)	Backfill height	1 m
	(g)	Backfill density	1700 kg/m³
	(h)	Allowable soil pressure	2·5 × 10⁴ kN/m²
4. Roof type			Dome

(c) Loss of tendon stress due to elastic shortening of concrete:

$$f_{ces} = \tfrac{1}{2} f_c \frac{E_s}{E_c}$$

assuming

$$f_c = 0.3 f'_c$$

(d) Loss of tendon stress due to creep and shrinkage of concrete:

$$f_{cses} = \varepsilon_{cs} E_s$$

(e) Loss of tendon stress due to steel relaxation:

$$f_{sr} = (0.08)(0.7 f_{pu})$$

(f) Effective tendon stress after all losses:

$$\text{losses other than friction} = (f_{ces} + f_{cses} + f_{sr})$$

(g) Tendon requirements:

$$N = \frac{Fl}{A_{ps}(F_{se})}$$

(h) Tank dome (hoop) prestressing:

$$FHD = \frac{1 \cdot 2 PR}{2}$$

Assuming the hoop tendon stresses in the dome are the same as the hoop tendon stresses in the cylindrical wall.

Tendon requirement—hoop tendons are provided for the hemispherical dome to a line which forms the 90° solid angle of the dome

$$N = \frac{Fl}{\tfrac{2}{3} A_{ps} f_{se}}$$

Tendon spacing $= 1/N$.

(c) Concrete material characteristics. Reference is made to material properties of concrete in Chapter I. Recent tests in different countries, to measure the compressive strength of concrete at temperatures down to −196°C, indicate that compressive strength increases with reducing

temperature. The following are concrete material properties used in the design of a number of tanks.

(a) Coefficient of contraction:
Below $-70°C = 53 \times 10^{-6}/°C$.

(b) Compressive strength:
Strength increment at temperatures below $-120°C = T°C$

$$= \frac{-T \times m_m}{12} \text{ N/mm}^2$$

(m_m = moisture content, 5%). The compressive strength is independent of the strength grade in this case.

(c) Tensile strength:
Tensile strength increases as the temperature is reduced down to about $-70°C$. At temperatures lower than $-70°C$, there is a small reduction. At $-165°C$ there is an increase in residual strength up to 4 N/mm^2 for moist concrete and 2.5 N/mm^2 for dry concrete.

(d) Elastic moduli:
Reference is made to the relevant values in Chapter I.

(e) Creep:
At a temperature of $-100°C$, 60% reduction of creep is adopted.

(f) Poisson ratio:
At temperatures of $-165°C$ and below a value of 0·22 is adopted. For details refer to Chapter I.

(g) Thermal contraction:
The coefficient of contraction is $5-6 \times 10^{-6}/°C$ at $-165°C$. A value of 0·13% is chosen for the range 20 to $-165°C$.

(h) Thermal conductivity:
An increased value of thermal conductivity: moist concrete, 60% at $-165°C$; partially dried concrete, 20%.

(i) Impact resistance:
Reference should be made to earlier chapters for impact on concrete.

VIII.12.3 Finite Element Analysis of LNG Tanks

Finite element analysis is carried out on this tank using three-dimensional eight-noded isoparametric elements for concrete with two-noded line elements, representing steel, placed within the body and on top of the solid concrete elements. The choice of eight-noded elements was made on economic grounds. Shape functions and stiffness matrices for these elements are fully discussed in Chapter VI and associated appendices. Figure VIII.199 shows a typical $22\frac{1}{2}°$ mesh for the tank. The total numbers of solid elements and line elements were 230 and 375, respectively. A 2×2 Gaussian quadrature is adopted. The prestressing and the LNG pressure (Fig. VIII.200) loads are computed. The prestressing loads were transferred to the nodal points while LNG pressures are assumed to act on the solid element surfaces. The total number of load increments chosen was 15. Figures VIII.201–VIII.203 show principal stress trajectories and crack positions in the walls and dome of the tank. The procedure for analysing

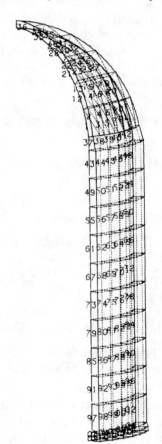

Fig. VIII.199. Finite element mesh scheme for an LNG tank.

Application to Engineering Problems

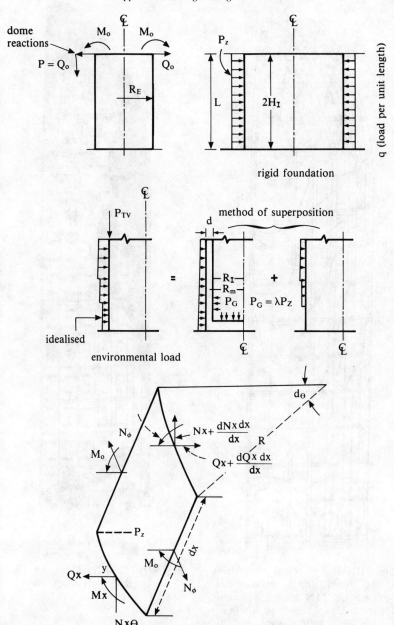

Fig. VIII.200. Forces on barrel wall.

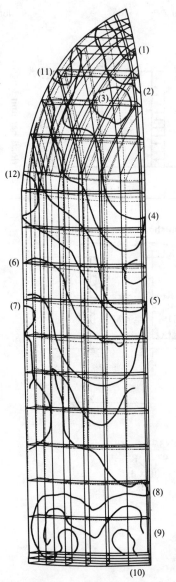

Fig. VIII.201. Principal tensile stress at 3·5 times design pressure. (1, 30 N/mm^2; 2, 24 N/mm^2; 3, 5 N/mm^2; 4, 10 N/mm^2; 5, 5 N/mm^2; 6, 8 N/mm^2; 7, 11 N/mm^2; 8, 10 N/mm^2; 9, 11 N/mm^2; 10, 12·0 N/mm^2; 11, 16 N/mm^2; 12, 23 N/mm^2.)

■ heavy compressive and tensile stresses

Fig. VIII.202. Cracking (3·5 times design pressure).

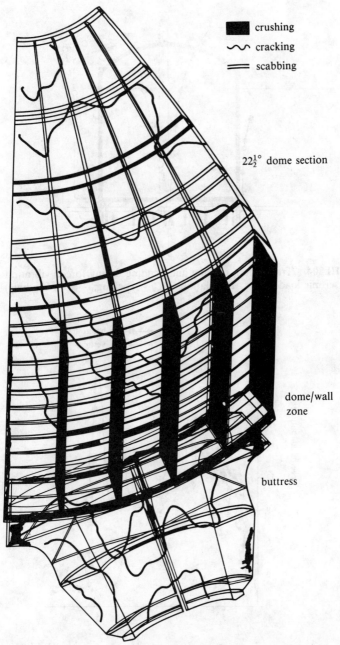

Fig. VIII.203. Top view of a dome with a buttress—cracks (3.5 times design pressure).

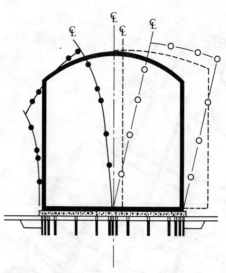

Fig. VIII.204. Translation, rocking and bending during forced vibration under seismic loads. (---, Translation; —○—, rocking; —●—, bending.)

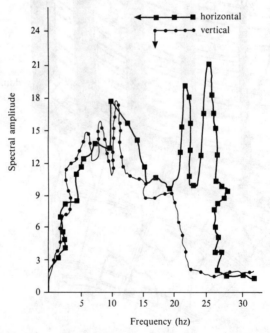

Fig. VIII.205. Forced vibration spectral data for an LNG tank.

crack sizes under incremental loads is the same as described for pressure vessels in this chapter and in Chapter VI. The concrete of the tank is assumed to behave as a three-parameter model as described in Chapter VII. The safety margin computed over the design loads is 2·91.

Seismic analysis was carried out on this tank. For soils a layered finite element scheme with springs (Table VIII.10b) was chosen. A seismic acceleration of 0·25 g was considered for modes such as translation, rocking and bending given in Fig. VIII.204; recorded acceleration was plotted in real time. The forced vibration spectral data is plotted in Fig. VIII.205. The computed frequencies are given below:

Vertical	8·8 Hz
Horizontal bending	2·5 Hz
Rocking	3·5 Hz
Translation	5·5 Hz

The data are useful for an earthquake safety review of the tank. The program ISOPAR can use two levels of exciting force in order to make it possible to distinguish between steady-state motions from nearby mechanical equipment and the transient response together with associated mode shapes.

Bibliography

1. Hognestad, E., A study of combined bending and axial load in reinforced concrete members. Bulletin No. 399, Engineering Experiment Station, University of Illinois, Urbana, Illinois, Vol. 49, No. 22, November 1951.
2. Desayi, P. & Krishnan, S., Equation for stress strain curve of concrete. *J.ACI Proc.*, **3** (1964).
3. British Standard CP110, Code of practice for the structural use of concrete, Part 1, design materials and workmanship. BSI London, November, 1972.
4. Saenz, L. P., Equation for stress strain curve of concrete in uniaxial and biaxial compression of concrete. *J.ACI*, **61** (1965) 1229–35.
5. Elwi, A. & Murray, D. W., A three-dimensional hypo-elastic concrete constitutive relationship. *J. Eng. Mech. Div.*, *ASCE*, **105** (1976) 623–41 (No. EM4, Proc. paper 14746).
6. Bangash, Y., Discussion of Ref. 20. Paper 7865, Institute Civil Engineers, 1971.
7. Darwin, D. & Pecknold, D. A. (a) Analysis of cyclic loading of plane reinforced concrete structures. *J. Computers and Structures*, **7** (1977) 137–47. (b) Non-linear biaxial stress–strain law for concrete. *J. Eng. Mech. Div., ASCE*, **103** (1977) 229–49 (No. EM2, Proc. paper 12839).
8. Karson, I. D. & Jirsa, J. O., Behaviour of concrete under compressive loadings. *J. Struct. Div., ASCE*, **95** (1969) 2543–63.
9. Welch, P. F., Fracture of plain concrete. *Indian Concrete Journal*, **46** (1972) 468–79.
10. Evans, R. H., Effect of rate of loading on the mechanical properties of some materials. *J.ICE*, **18** (1962) 296.
11. Hughes, B. P. & Chapman, G. P. The deformation of concrete and microconcrete and tension with particular reference to aggregate size. *Magazine of Concrete Research*, **18**(54) (1966) 19–24.
12. Popovics, S., A numerical approach to the complete stress–strain curve of concrete. *Cem. Concr. Res.*, **3** (1973) 583–99.
13. Johnson, R. D., *Structural Concrete*. McGraw Hill, London, 1969.
14. Richart, F. E., Brandzaeg, A. & Brown, R. L., A study of the failure of concrete under combined compressive stresses. Bulletin No. 185, Engineering Experiment Station, University of Illinois, Vol. 26, 1928.

15. Glucklich, J., On the compression failure of plain concrete T & A M. Report No. 125, University of Illinois, March 1962.
16. Balmer, G. G., Shearing strength of concrete under high triaxial stress-computation of Mohr's envelope as a curve. Struct. Research Lab., Report No. SP-23, United States Bureau of Reclamation (USBR), 1949.
17. Westergaard, H. M., On the resistance of ductile materials to combined stresses in two or three directions perpendicular to one another. *J. Frankline Inst.*, **189** (1920) 627–40.
18. Nadai, A., *Theory of Flow and Fracture of Solids*, Vol. 1. McGraw Hill, New York, 1950.
19. Merkle, J. G., An engineering approach to multiaxial plasticity. ORNL-4138, 1968.
20. Sandbye, P., A plastic theory for plain concrete. In: *Bygnings-Statiske Meddelelser*, Vol. 36. Teknisk Forlag, Copenhagen, 1965, pp. 41–62.
21. Newman, K. & Newman, J. B., Failure Theories and Design Criteria for Plain Concrete Structure. In: *Solids, Solid Mechanics*, Wiley Interscience, London, 1971, pp. 963–95.
22. Wastlung, G., Nye ron angaende betongens grundlaggende hallfasthet-segenskabar. Betong Haft 3, Stockholm, 1937.
23. Bresler, B. & Pister, K. S., Failure of plain concrete under combined stresses. *Trans. ASCE*, **122** (1957) 1049–68.
24. Ottoson, N. S., A failure criteria for concrete. *J. Eng. Mech. Div., ASCE*, **103** (1977) 527–35 (No. EM4, Proc. paper 13111).
25. Ottoson, N. S., Constitutive model for short-time loading of concrete. *J. Eng. Mech. Div., ASCE*, **105** (1979) 127–41 (No. EM1, Proc. paper 14375).
26. Ottoson, N. S. Structural failure of thick walled concrete elements. 4th International Conference on Structural Mechanics in Reactor Technology (SMiRT), Paper H4/3, San Francisco, August 1977.
27. Suidan, M. & Schnobrich, W., Finite element analysis of reinforced concrete. *J. Struct. Div., ASCE*, **99** (1973) 2109–22.
28. Cervenka, V., Inelastic finite element analysis of reinforced concrete panels under in-plane loads. PhD Thesis, University of Colorado, Boulder, Colorado, 1970.
29. Kupfer, H. B., Hilsdorf, H. K. & Rüch, H., Behaviour of concrete under biaxial stresses. *ACI Journal*, **66** (1969) 556–66.
30. Liu, T. C., Nilson, A. H. & Slate, F. O., Stress strain response and fractures of concrete in uniaxial and biaxial compression. *J.ACI Proc.*, **69** (1972) 291–5.
31. Kupfer, H. B. & Gerstle, K. H., Behaviour of concrete under biaxial stresses. *J. Eng. Mech. Div., ASCE*, **99** (1973) 552–866 (EM4, Proc. paper 9917).
32. 1st International Conference on Structural Mechanics in Reactor Technology (SMiRT), Berlin, September 1971.
33. 2nd International Conference on Structural Mechanics in Reactor Technology (SMiRT), Berlin, 1973.
34. 3rd International Conference on Structural Mechanics in Reactor Technology (SMiRT), London, September 1975.
35. Von Mises, R., Mechanik der festen Koerper im plastisch deformablen Zustant, *Getlinger Nachr., Math. Phys.*, **K1** (1913) 582–92.

36. Prandtl, L. & Reuss, A., (a) *The Mathematical Theory of Plasticity*, ed. R. Hill. Oxford University Press, London, 1950. (b) *Plasticity, Theory and Application*, ed. A. Mendelson. Macmillan, London, 1968.
37. Hencky, H. Z. & Nadai, A., Zur Theorie Plastischer Deformationen und der hierdurch im Material hervorgerufenen Nachspannungen, Report, 1924, pp. 323–34. (Available National Lending Library.)
38. Sturman, G. M., Shah, S. P. & Winter, G., Effect of flexural strain gradients on micro-cracking and stress–strain behaviour of concrete. *J.ACI*, **62** (1965) 805–22.
39. Schleicher, F., Der Spannungszustand an der Fliessgrenze (Plasticitatsbedonging). *Zietschrift fur Angewandte Mathematik und Mechanik*, Band 6 Heft 3. Berlin, June 1926.
40. Vile, G. W. D., The strength of concrete under short term static biaxial stress. In: *The Structure of Concrete and its Behaviour under Load*, Proc. Int. Conf., London, 1968, pp. 275–88.
41. Smith, G. M., Failure of concrete under combined tensile and Compressive stresses. *J.ACI*, **50** (1953) 137–40.
42. Bresler, B. & Pister, K. S., Strength of concrete under combined stresses. *J.ACI*, **55** (1958) 321–45.
43. Bellamy, C. J., Strength of concrete under combined stresses. *J.ACI*, **58** (1961) 367–80.
44. Bazant, Z. P., Comment on orthotropic models for concrete and geomaterials, *J. Eng. Mech. Div.*, *ASCE*, **109** (1983) 849–65.
45. Bazant, Z. P. & Cedolin, L., Fracture mechanics of reinforced concrete. *J. Eng. Mech. Div.*, *ASCE*, **106** (1980) 1287–307.
46. Bazant, Z. P. & Kim, S. S., Plastic-fracturing theory for concrete. *J. Eng. Mech. Div.*, *ASCE*, **105** (1979) 407–28.
47. Bazant, Z. P. & Shieh, C., Hysteretic fracturing endochronic theory for concrete. *J. Eng. Mech. Div.*, *ASCE*, **106** (1980) 929–50.
48. Akroyd, T. N. W. Concrete under triaxial stress. *Magazine of Concrete Research*, **13**(39) (1961) 111–18.
49. Nashizawa, N., Strength of concrete under combined tensile and compressive loads. *Japan Cement Engineering Association Review*, May 1961, 126–31.
50. Campbell Allen, D., Strength of concrete under combined tensile and compressive loads. *Constructional Review*, **35** (1962) 29–37.
51. Sigvaldason, O. T., Failure characteristics of concrete. PhD Thesis, University of London, June 1965.
52. Baker, A. L. L. (a) An analysis of deformation and failure characteristics of concrete. *Magazine of Concrete Research*, **11**(33) (1959) 119–28. (b) A criterion of concrete failure. *Proc. Inst. Civ. Eng.* London, (1970) 269–78.
53. Anson, M. An investigation into a hypothetical deformation and failure mechanism for concrete. PhD Thesis, University of London, November 1962.
54. Robinson, G. S., The failure mechanism of concrete under particular reference to the biaxial compressive strength. PhD Thesis, University of London, November 1964.
55. Rüsch, H., Der Zuzammenhang Zwischen Risbilding und Haftfestigkeit

unter besonderer Berucksichtingung der Anwendunghober Stahlspannungen, 5. Konferenz der Int. Vereiningung fur Brucken und Hochbau, Lissabon, 1956.
56. Fumagalli, E., *Strength Characteristics of Concrete under Conditions of Multi-Axial Compression.* Cement and Concrete Association, London, (Translation No. CU 128), 1975.
57. Weigher, H. & Becker, G., Untersuchungen uber das Bruch und verformungsverhalten von Beton bei Zweiachsiger Beausprachung. *Deutscher Ausschuss fur Stahlbeton*, Heft 157. Berlin, 1963.
58. Hannant, D. J. & Frederick, C. D., Failure for concrete in compression. *Magazine of Concrete Research*, **20** (1968) 137–44.
59. Link, J., Numerical analysis oriented biaxial stress–strain relation and failure criterion of plain concrete. 3rd International Conference on Structural Mechanics in Reactor Technology (SMiRT), Paper H1/2, London, September, 1975.
60. Baker, A. L. L., Safety of pressure vessels, group B. International Conference on PCPV, Paper 8, Inst. Civ. Eng., London, 1967.
61. Chawalla, E., *Ein Fuhrung in die Baustatik*, Stahlbauverlag, Koln, 1954.
62. Chinn, J. & Zimmerman, R. M., Behaviour of plain concrete under various high triaxial compression loading conditions. Technical Report No. WL TR 64-163 (AD468460), Air Force Weapon Laboratory, New Mexico, August 1965.
63. Reimann, H., Kritische spannungszustande der Betons bei mehrachsiger, ruhender Kurzzeitbelastung. *Deutscher Ausschuss fur Stahlbeton*, Heft 175, Berlin, 1965.
64. Argyris, J. H., Recent developments in the finite element analysis of prestressed concrete reactor vessels. IABSE Seminar, ISMES, Bergamo, Italy, May 1974.
65. Willam, K. J., Constitutive model for the triaxial behaviour of concrete. IABSE Seminar, ISMES, Bergamo, Italy, May 1974.
66. Launay, P. & Gachon, H. (a) Strain and ultimate strength of concrete under triaxial stress. 1st International Conference on Structural Mechanics in Reactor Technology (SMiRT), Paper H1/3, Berlin, September 1971. (b) Strain and ultimate strength of concrete under triaxial stress. ACI SP-34, Paper 13, 1970. (c) Deformation et resistance ultime du beton sous etreinte triaxiale. Annales de L'Institut Technique dur Batiment et du Travaux Publics, No. 269, 1970.
67. York, G. P., Kennedy, T. W. & Perry, E. S., An experimental approach to the study of the creep behaviour of plain concrete subjected to triaxial stresses and elevated temperatures. Union Carbide, Report No. 28642, The University of Texas at Austin, August 1971.
68. Garas, F. K., Contributions to paper 6, pp. 100–1. International Conference on Model Techniques for Prestressed Concrete Pressure Vessels, BNES, London, 1969.
69. Liu, T. C. Y., Stress–strain response and fracture of concrete in biaxial compression. Research Report No. 339, Department of Structural Engineering, Cornell University, February 1971.

70. Liu, T. C. Y., Nilson, A. H. & Slate, F. O., Biaxial stress–strain relation of concrete. *J. Struct. Div., ASCE*, **98** (1972) 1025–34.
71. Buyukozturk, O., Nonlinear analysis of reinforced concrete structures. *J. Computers and Structures*, **7** (1977) 149–56.
72. Popovics, S., A numerical approach to the complete stress–strain curve of concrete. *Cement and Concrete Research*, **3** (1973) 538–99.
73. Tasuji, M. E., Nilson, A. H. & Slate, F. O., Biaxial stress–strain relationships for concrete. *Magazine of Concrete Research*, **31**(109) (1979) 217–24.
74. Tasuji, M. E., The behaviour of plain concrete subject to biaxial stress. Research Report No. 360, Department of Structural Engineering, Cornell University, March 1976.
75. Tassoulas, J. L., Inelastic behaviour of concrete in compression. MS Thesis, Department of Civil Engineering, MIT, Cambridge, Mass., 1979.
76. Hussain, M. & Saugy, S., Elevation of the influence of some concrete characteristics on non-linear behaviour of a prestressed concrete reactor vessel. Paper SP-34-8, Concrete for Nuclear Reactors, American Concrete Institute, Detroit, Michigan, October 1970.
77. Linse, D., Aschl, H. & Stockl, S., Concrete for PCRV's: strength of concrete under triaxial loading and creep at elevated temperatures. 3rd International Conference on Structural Mechanics in Reactor Technology (SMiRT), Paper H1/3, London, September 1975.
78. Gardner, J. J., Triaxial behaviour of concrete. *Proc. Am. Concr. Inst.* No. 2, (1969) 136–47.
79. Ross, A. D., Shrinkage in concrete pressure vessels. *Nucl. Eng. Des.*, **5** (1967) 150–60.
80. Kasami, H., Properties of concrete exposed to sustained elevated temperature. 3rd International Conference on Structural Mechanics in Reactor Technology (SMiRT), Vol. 3, Paper H1/5, London, September 1975.
81. Nasser, K. W. & Neville, A. M., Creep of concrete at elevated temperatures. *Proc. Am. Concr. Inst.*, **12** (1965) 1567–79.
82. Irving, J. & Carmichael, G. D. T., The influence of creep on the behaviour of concrete structures subjected to cyclic heating. Conference Solid Mechanics and Engineering Design in Civil Engineering Materials, Paper No. 107, Southampton University, April 1969.
83. Karni, J. & McHenry, D., Strength of concrete under combined tensile and compressive stresses, *J.ACI* (April 1958).
84. Browne, R. D., Properties of concrete in reactor vessels. ICE Conference Prestressed Concrete Pressure Vessels, Institution of Civil Engineers, London, 1958, pp. 131–51.
85. Taylor Woodrow Construction, Tests on Wylfa models. Private communications, 1968.
86. England, G. L. & Phok, M., Time-dependent stresses in a long thick cylindrical prestressed concrete vessel subjected to sustained temperature crossfall. *Nucl. Eng. Des.*, **9** (1969) 488–95.
87. Zienkiewicz, O. & Watson, M., Some creep effects in stress analysis with particular reference to concrete pressure vessels. *Nucl. Eng. Des.*, No. 4 (1966).

88. Hannant, D. J., Creep and creep recovery of concrete subjected to a multiaxial compressive stress. *Proc. Am. Concr. Inst.*, **66** (1969) 391–4.
89. British Standards CP110, Parts 1 and 2, 1972, and CP8110, 1985.
90. Truesdell, C., Hypo-elasticity. *J. Ration., Mech. Analysis*, **13** (1955) 1019–20.
91. Rivlin, A. & Erickson, J. L., Stress–deformation relations for isotropic materials. Report, Graduate Institute for Maths and Mechanics, Indiana University, December 1954.
92. Ross, A. D., Creep of concrete under variable stress. *ACI Journal*, **54** (1958) 739–58.
93. England, G. L. & Jardaan, I. J., Time-dependent and steady state stresses in concrete structures with steel reinforcement at normal and raised temperatures. *Magazine of Concrete Research*, **27** (1975) 131–42.
94. Bangash, Y. & England, G. L., The influence of thermal creep on the operational behaviour of complex structures. International Conference on Fundamental Creep and Shrinkage, Lausanne, Switzerland, 1980.
95. 4th International Conference on Structural Mechanics in Reactor Technology (SMiRT), Berlin, 1977.
96. 5th International Conference on Structural Mechanics in Reactor Technology (SMiRT), Berlin, 1979.
97. 6th International Conference on Structural Mechanics in Reactor Technology (SMiRT), Paris, 1981.
98. Nayak, G. C. & Zienkiewicz, O. C., Note on the 'alpha' constant stiffness method for analysis of non-linear problems. *Int. J. Num. Meth. Eng.*, (1972) 579–82.
99. Vallapan, S. & Doolan, T. F., Non-linear stress analysis of reinforced concrete. *J. Struct. Div., ASCE*, **98** (1972) 885–98.
100. Schrobrich, W. C. *et al.*, Discussion on Ref. 134. *J. Struct. Div., ASCE*, **98** (1972) 2419–32.
101. Timoshenko, S., *Theory of Elasticity*. McGraw Hill, New York, 1951.
102. Marin, J., *Mechanical Behaviour of Engineering Materials*. Prentice Hall, New York, (1962) p. 117.
103. Lame, A., In: *Theory of Flow and Fracture of Solids*, Vol. 1, ed. A. Nadai, McGraw Hill, New York, 1950.
104. Clapeyron, Z., *Theory of Flow and Fracture of Solids*. (2nd edn) ed. A. Nadai, McGraw Hill, New York, 1950.
105. Rankine, T., Soil mechanics and plastic analysis or limit design. *Q. Appl. Math.*, **10** (1952).
106. Poncelot, F., *Foundations of the Non-Linear Theory of Elasticity*, ed. V. V. Novozhilov, Greylock Press, USA, 1953.
107. St Venant, *Mem. Savants Etrangers*, **14** (1855). (A copy of this reference is held at the British Museum Library, London).
108. Tresca, H., Sur l'ecoulement des corps solids soumis a de fort pression. *Comp. Rend.*, **59** (1864) 754.
109. Coulomb, C. A., Essai sur une application des regles des maximis et minimis a quelques problems de statique. Royale Academie des Sciences, Paris, Vol. 7, 1776.
110. Navier, C. F., In: *A Theory of Deformation and Failure of Concrete*, ed. Reinius, Stockholm, Vol. 40, No. 1, 1955.

111. Mohr, O., *Abhandlungen aus dem Gebiet der technischen Mechanik* (2nd edn), Wilhelm Ernst und Sohn, Berlin, 1914, p. 192.
112. Sell, R., Investigation into the strength of concrete under sustained load. RILEM Bulletin No. 5, December 1959, pp. 5–13.
113. Kupfer, H., *Das Verhalten des Betons Unter Mehrachsigen Kurzzeitbelastung det Zweiachsigen Beanspruchung Deutscher Ausshuss fur Stahlbeton*, Heft 229, Berlin, 1973.
114. Evans, R. J. & Pister, K. S., Constitutive equations for a class of on-linear elastic solids. *Int. J. Solids Structures*, **2** (1966) 427–45.
115. Chiorino, M. A. & Levi, R., Influence de l'Elasticite Differee sur le Regime des Contraintes des Constructions en Beton. Cahiers de la Recherche No. 24, Institut Technique du Batiment et des Travaux Publics, Eyrolles, Paris, France, 1967 (see also *Giornale del Genio Civile*, 1967; and Academia Nazionale dei Lincei, Fasc. 5, Series 8, Vol. 38, May 1965).
116. Chinn, J. & Zimmerman, R. M., Behaviour of plain concrete under various high triaxial compression loading conditions. Technical Report No. WL TR 64-163 (AD4684460), Air Force Weapon Laboratory, New Mexico, 1965.
117. Mills, L. L. & Zimmerman, R. M., Compressive strength of plain concrete under multiaxial loading conditions. *ACI Journal*, **67** (1970) 802–7.
118. Sozen, M. A. & Corley, W. G., Time-dependent deflection of reinforced concrete beams. *ACI Journal*, March 1966, 374–85.
119. Conference on Prestressed Concrete Pressure Vessels, Inst. Civ. Eng., London, 1967.
120. Bangash, Y., The automated three-dimensional cracking analysis of prestressed concrete vessels. Sixth International Conference on Structural Mechanics in Reactor Technology (SMiRT), Paper H3/2, Paris, August 1981.
121. Chen, E. S. & Buyukozturk, O., Damage model for concrete in multiaxial cyclic stress. Department of Civil Engineering, MIT, Cambridge, Massachusetts, 1983.
122. Bakker, W. T., Properties of refractory concretes. Publication SP-57, American Concrete Institute, Detroit, Michigan, 1978, pp. 11–52.
123. Bakker, W. T., Recent advances in refractory concrete technology, Publication SP-74. American Concrete Institute, Detroit, Michigan, 1982, pp. 1–16.
124. Calvo, J. J., Design of reinforced concrete containment walls under combined axial and shear forces. MS Thesis, Department of Civil Engineering, MIT, Cambridge, Massachusetts, 1982.
125. Langan, D. & Garas, F. K., Behaviour of end slabs in cylindrical prestressed concrete pressure vessels. 2nd International Conference on Structural Mechanics in Reactor Technology (SMiRT), Paper H3/4, Berlin, 1973.
126. Bangash, Y., Reactor pressure vessel design and practice. *Progress in Nuclear Energy*, **10** (1982) 69–124.
127. Bangash, Y., PWR steel pressure vessel design and practice. *Progress in Nuclear Energy*, **16** (1985) 1–40.
128. Saugy, B., Three-dimensional rupture analysis of a prestressed concrete pressure vessel including creep effects. 2nd International Conference on Structural Mechanics in Reactor Technology (SMiRT), Berlin, 1973.
129. Saouma, V., Automated nonlinear finite element analysis of reinforced

concrete: a fracture mechanics approach. Thesis presented to Cornell University, Ithaca, NY, in partial fulfilment of the requirements for the degree of Doctor of Philosophy, 1981.
130. Willam, K. J. & Wanke, E. P., Constitutive nodal for triaxial behaviour of concrete. IABSE Seminar on Concrete Structures Subjected to Triaxial Stresses, Part III-I, Bergamo, Italy, May 1974.
131. Hobbs, D. W., Design stresses for concrete structures subjected to multiaxial stresses. Struct. Eng., **55** (1977) 157–64.
132. Kotsovos, M., Failure criteria for concrete under generalised stress states. PhD Thesis, University of London, 1974, p. 284.
133. Langan, D. & Garas, F. K., Capacity of prestressed vessels to resist shear. *ACI Journal* (*Seminar on Concrete for Nuclear Reactors*), (October 1970) SP34-12.
134. Langan, D. & Garas, F. K., The design of pod boiler pressure vessels with particular reference to nuclear power stations. 2nd International Conference on Structural Mechanics in Reactor Technology (SMiRT), Paper H4/4, Berlin, 1973.
135. Elwi, A. A. & Murray, D. W., A three-dimensional hypo-elastic concrete constitutive relationship. Private communication, 1980.
136. Turner, F. H., *Concrete and Cryogenics*. View Point, New York, 1978.
137. Indian Standard IS 456, Delhi, 1979.
138. British Standard BS8110. The structural use of concrete, 1985.
139. ACI 318. Building code requirements for reinforced concrete, 1971.
140. DIN 4227. Standard specification for reinforced concrete, Germany, 1978.
141. Ross, A. D., Private communications, Kings College, London, 1968.
142. Buyukozturk, O., A constitutive model for concrete in compression. Proc. Third ASCE Engineering Division Specialty Conference, Austin, September 1979.
143. Kasami, H. *et al.*, Private communications.
144. USBR (United Bureau of Reclamation), Private communications.
145. Green, S. J. & Swanson, S. R., Static constitutive relations for concrete. Tech. Report AFWC-TR-72-2, 1973.
146. Zienkiewicz, O. C., Watson, M. & King, I. P., A direct method of visco-elastic creep. *Int. J. Mech. Sci.*, **10** (1968) 807–27.
147. Dibsi, J., PhD Thesis, Imperial College, London, 1979.
148. Chen, A. C. T. & Chen, W. F., Constitutive relations for concrete. *J. Eng. Mech. Div. ASCE*, **101** (1975) 465–81.
149. Kojic, M. & Cheathem, J. B., Theory of plasticity of porous media with fluid flow. *Soc. Pet. Eng. J.*, **257** (1974) 263.
150. Hsu, T. T. C., Slate, F. O., Sturman, G. M. & Winter, G., Microcracking of plain concrete and the shape of the stress–strain curve. *J. Am. Concr. Inst.*, **60** (1963) 209–24.
151. Chen, W. F. *et al.*, *Constitutive Equations for Engineering Materials*, Wiley, New York, 1983.
152. Suh, N. P., A yield criterion for plastic frictional, work hardening granular materials, *Int. J. Powder Materials* (1969) 69.
153. Zienkiewicz, O. C. & Pande, G. C., Private communication, 1983.
154. Drucker, D. C. & Prager, W., A more fundamental approach to plastic

stress–strain relation. Proc. Natl. Cong. US Inst. Appl. Mech., Chicago, 1951, pp. 487–91.
155. England, G. L., The direct calculation of stresses in creeping concrete and concrete structures. *Proc. Conference on Structure, Solid Mechanics and Design, Southampton*, Part 2, ed. M. Te'eni, 1969.
156. England, G. L. & Illston, J. M., Method of computing stress in concrete from history to measured strain. *Civil Engineering and Public Review*, London, **60** (1965).
157. Parrot, L. J., Some observations on the components of creep in concrete. *Magazine of Concrete Research*, **22** (1970).
158. Bazant, Z. P., Endochronic inelasticity and incremental plasticity. *Int. J. Solids and Structures*, **14** (1978) 691–714.
159. Coons, M. D. & Evans, R. J., On the recoverable deformation of granular media. Private report, University of Washington, Seattle, 1972.
160. Love, A. E. H., *Mathematical Theory of Elasticity*, McGraw Hill, New York, 1939.
161. Lekhnitskii, S. G., *Theory of Elasticity of an Anisotropic Elastic Body*, Holden-Day, San Francisco, California, 1963.
162. Nye, J. F., *Physical Properties of Crystals*, Clarendon Press, Oxford, 1957.
163. Green, A. E., Hypo-elasticity and plasticity. *Proc. Roy. Soc. London*, **A234** (1956) 46–57.
164. Hughes & Chapman, J. C., Private communications, University of Birmingham, 1970.
165. Charlton, S. et al., Prestressing and proof pressure testing of the Wylfa prestressed concrete pressure vessels and comparison of measured data with predicted performance. 1st International Conference on Structural Mechanics in Reactor Technology (SMiRT), Berlin, September 1971.
166. Noll, W. J., On the continuity of solids and fluid states rational mechanics. *Anal.*, **4** (1955) 3.
167. Phillips, D. V. et al., Finite element methods in the analysis of reactor vessels. *Nucl. Eng. and Design*, **20** (1972).
168. Phillips, D. V. & Zienkiewicz, O. C., Finite element non-linear analysis of concrete structures. *Proc. Inst. Civ. Eng. Research and Theory*, **61** (1976).
169. Cedolin, L., Crutzen, Y. R. J. & Dei Foli, S., Triaxial stress–strain relationships for concrete. *J. Eng. Mech. Div., ASCE*, **103** (1977) 423–39. No. EM3, Proc. paper 12969.
170. Saugy, B. et al., Private communication, 1979.
171. Valanis, K. C., A theory of visco-plasticity without a yield surface. *Archiwum Mechaniki Stosowanej* (Warsaw), **3** (1971) 517–51.
172. Bazant, Z. P. & Bhat, P. D., Endochronic theory of inelasticity and failure of concrete, *J. Eng. Mech. Div., ASCE*, (1976) 701–21.
173. Bazant, Z. P. & Shieh, H., Endoelectronic model for non-linear triaxial behaviour of concrete. *Nucl. Eng. and Des.*, **47** (1978) 305–15.
174. Bangash, Y., A report on the progress of G/H Division, SMiRT 7. BNES, 1984.
175. Pfiffer, P. A. et al., Blunt crack propagation in finite element analysis for concrete structures. 7th International Conference on Structural Mechanics in Reactor Technology (SMiRT), North Holland, Amsterdam, Paper H/2, 1983.

176. Gopalakrishnan, K. S., Neville, A. M. & Ghali, A., Creep Poisson's ratio of concrete under multiaxial compression. *J. Am. Conc. Inst.*, **66** (1969) 1008–19.
177. Gambarora, P. G. et al., Rough cracks in reinforced concrete. *J. Struct. Div., ASCE*, **106** (1980) 819–42.
178. Rebora, B., Zimmermann, T. H. & Wolf, J. P., Dynamic rupture analysis of reinforced concrete shells. *Nucl. Eng. Des.*, **37** (1976) 269–97.
179. Ree, H. V. & Hoek, M. J. V. D., Analysis of the behaviour of concrete structure due to an airplane. Private communication, 1979.
180. Zerna, W. et al., Optimized reinforcement of nuclear power plant structures for aircraft impact forces. *Nucl. Eng. Des.*, **37** (1976) 313–20.
181. Davies, I. L., Damaging effects from the impact of missiles against reinforced concrete structures. *Nuclear Energy*, **19** (1980) 199–205.
182. Zukas, J. A., *Impact Dynamics*. Wiley, New York, 1981.
183. Wolf, J. P., Bucher, K. M. & Skirikerud, P. E., Response of equipment to aircraft impact. *Nucl. Eng. Des.*, **47** (1978) 169–93.
184. Rice, J. S. & Bahar, L. Y., Reaction-time relationship and structural design of reinforced concrete slabs and shells for aircraft impact. 3rd International Conference on Structural Mechanics in Reactor Technology (SMiRT). Paper J5/3, London, 1975.
185. Riera, J. D., On the stress analysis of structures subjected to aircraft impact forces. *Nucl. Eng. Des.*, **8** (1968) 415–26.
186. Rock, T. A., Dynamic analysis of civil engineering structures. PhD Thesis, University of Wales, Swansea, 1974.
187. Ronstad, K. M., Taylor, M. A. & Harmann, L. R., Numerical biaxial characterization for concrete. *J. Eng. Mech. Div., ASCE*, **100** (1974) 935–48.
188. Rynen, J., Villafane, E. & Crutzen, Y., Impulsive loading on concrete structures, 6th International Conference on Structural Mechanics in Reactor Technology (SMiRT), Paper J10/1, Paris, 1981.
189. Smith, G. M., Failure of concrete under combined tensile and compressive stresses. *J. Am. Concr. Inst.*, **50** (1953) 137–47.
190. Stephenson, A. E., Full-scale tornado-missile impacts tests. EMPRI NP-440, Sandia Laboratories, 1977.
191. Suidan, M. & Scnobrich, W. C., Finite element analysis of reinforced concrete. *J. Struct. Div., ASCE*, **99** (1973) 2109–22.
192. Varpasuo, P. & Kenttala, J., The analysis of the containment building for global effects of an aircraft crash. International Conference on Structural Mechanics in Reactor Technology (SMiRT), 1977. Paper J9/10.
193. Kar, A. E., Impactive effects of tornado missiles and aircrafts, *Trans. ASCE*, **105** (1979) 2243–60.
194. Kar, A. E., Impact load for tornado generated missiles. *Nucl. Eng. Des.*, **47** (1978) 107–14.
195. Kar, A. E., Loading time history for tornado generated missles. *Nucl. Eng. Des.*, **51** (1979) 487–93.
196. Zapp, F. C., Testing of containment systems used in light-water-cooled power reactors. Report ORNL-NSIC-26, United States Atomic Energy Commission, 1968.
197. Kennedy, R. P., Effects of an aircraft crash into a concrete reactor

containment building. Holmes and Narver Inc., Oakland, California, July 1966.
198. Yeh, G. C. K., Probability and containment of turbine missiles. *Nucl. Eng. Des.*, **37** (1976) 307–12.
199. Zienkiewicz, O. C., *The Finite Element Method*. McGraw Hill, New York, 1980.
200. General design criteria for nuclear power plant construction permit. *Federal Register*, 33 FR 19213, 11 July 1967.
201. Gwaltney, R. C., Missile generation and protection in light-water-cooled power reactor plants. Report ORNL-NSIC-22 United States Atomic Energy Commission, 1968.
202. Burns, W. J., Kennedy, W. J. L. & Weinzimmer, F., Shoreham nuclear power station with concrete containment. *Proc. Am. Power Conf.*, **30** (1968) 365–70.
203. Owen, D. R., Implicit finite element methods for the dynamic transient analysis of solids with particular reference to non-linear situations. In: *Advanced Structural Dynamics*, ed. J. Donea. Applied Science Publishers, London, 1978, pp. 123–52.
204. Nasser, K. W. & Neville, A. M., Creep of concrete at elevated temperatures. *ACI Journal*, **62** (1965) 1567–79; and Private communication, 1965.
205. Hannant, D. J., The strain behaviour of concrete under compressive stresses at elevated temperatures. CERI, Note. No. RD/LIN 67/66, 1967.
206. Arthanari, S., Influence of temperature and biaxial stresses on creep in concrete. PhD Thesis, University of London, 1966.
207. Ross, A. D., Concrete creep data. *Proc. Inst. Struct. Eng.*, **15** (1957) 314–25.
208. Ross, A. D., Creep and shrinkage in plain, reinforced and prestressed concrete. *Proc. Inst. Civ. Eng.*, **21** (1943).
209. Zienkiewicz, O. C., Watson, M. & King, I. P., A numerical method of viscoelastic stress analysis. *Int. J. Mech. Sci.*, **10** (1968) 807–27.
210. Zienkiewicz, O. C., Gago, J. P. S. E. & Kelly, D. W., The hierarchical concept of finite element analysis. *Computers and Structures*, **16** (1983).
211. Thorpe, D. E., Elastic analysis of the Dungeness B vessel. Conference on PCPV, Inst. Civ. Eng., London, Paper 36, March 1967.
212. Indian Standard Code of Practice for Prestressed Concrete, IS:1345, October 1975.
213. Hilsdorf, H. K. & Muller, H. S., Comparison of prediction methods for creep coefficient of structural concrete with experimental data. In: *Fundamental Research on Creep and Shrinkage*, Sidhoff and Nishoff, The Hague, 1982.
214. ACI prediction of creep shrinkage and temperature effects. Draft, ACI209, October 1978.
215. CEB–FIP Model Code, 1978.
216. Hansen, T. C., Creep of concrete. A discussion of some fundamental problems. Bulletin 33, Swedish Cement and Concrete Institute, 1975.
217. Zienkiewicz, O. C. & Cormeau, I. C., Viscoplasticity–plasticity and creep in elastic solids. *Int. J. Num. Meth.*, (1974) 821–45.
218. Hilsdorf, H. K., *Drying and Shrinkage of Concrete and Reinforced Concrete Structures*. Ernst and Sohn, Verlag, 1969.
219. ACI prediction of creep, shrinkage and temperature effects. ACI 209, 1970.
220. CEB–FIP 70, method for shrinkage. FIP, 1978.

221. Bangash, Y., Movements in prestressed concrete reactor vessels. *Nuc.. Eng. Des.*, **50** (1978) 462–73.
222. Ross, A. D., Illston, J. M. & England, G. L., Short and long-term deformations of concrete as influenced by its physical structure and state. International Conference on the Structure of Concrete, London, 1965.
223. Sharp, T. J., The influence of elevated temperature on the physical behaviour of water in concrete. PhD Thesis, University of London, August 1971.
224. Schimmelwitz, P. & Hundt, J., A note on moisture condition of concrete. SP-34, Vol. 2, American Concrete Institute, Detroit, Michigan, 1972.
225. Davis, R. E., A summary of the results of investigations having to do with volumetric changes in cements, mortars and concretes due to causes other than stress. *ACI Journal, Proc.*, **26** (1930) 407–43.
226. Mitchell, L. J., Thermal expansion tests on aggregates, neat cement, and concretes. *Proc. ASTM*, **53** (1953) 963–77.
227. Meyers, S. L., Thermal expansion characteristics of hardened cement paste and of concrete. *Proc. Highway Research Board*, **30** (1950) 193–203.
228. Bonnel, D. G. R. & Harper, F. C., Thermal expansion of concrete. *J. Inst. Civ. Eng.*, **33** (1950) 320–30.
229. Hatt, W. K., The effect of moisture on concrete. *Trans. ASCE*, **89** (1926) 270–370.
230. Walker, K., Stanton, L., Bloem, D. L. & Mullen, W. G., Effects of temperature changes on concrete as influenced by aggregates. *ACI, Journal, Proc.*, **48** (1952) 661–80.
231. Berwanger, C., The modulus of concrete and the coefficient of expansion of concrete and reinforced concrete at below normal temperatures. *Temperature and Concrete*, SP-25, American Concrete Institute, Detroit, Michigan, 1971, pp. 191–233.
232. Emanuel, H. J. & Hulsey, L. J., Prediction of the thermal expansion of concrete. *ACI Journal*, (1977) 149–55.
233. Philleo, R., Some physical properties of concretes at high temperatures. *ACI Journal, Proc.*, **54** (1958) 857–64.
234. Malhotra, H. L., The effect of temperature on the compressive strength of concrete. *Magazine of Concrete Research*, **8** (1956) 63–8.
235. Sullivan, P. J. & Poucher, M. P., The influence of temperature on the physical properties of concrete and mortar in the range 20°C to 400°C. *Temperature and Concrete*, SP-25, American Concrete Institute, Detroit, Michigan, 1971, pp. 103–35.
236. Ali, I. & Kesler, Å., Rheology of concrete. A review of research. The Engineering Experiment Station, University of Illinois Bulletin, Vol. 62, No. 68, 1965.
237. Newman, B. J., Deformational behaviour, failure mechanisms and design criteria for concrete under combinations of stress. PhD Thesis, University of London, 1973.
238. Buettner, D. R. & Hollrah, R. L., Creep recovery of plain concrete. *J. Am. Concr. Inst.*, **64** (1968) 452–61.
239. Brice, L. P., Etude des conditions de formation des fissures de glissement et de decohesion dans les solides, *Travaux*, June (1964).
240. Saliger, I., *Der Stahlbetonbau*, Vol. 7, Auglage, 1956.

241. Dressel, J., Square mesh reinforcement of plane-supporting membrane structures. Results of Wastlund Saliger Technical Report No. 71, Technical University, Dresden, 1964.
242. Somayaji, S. & Shah, S. P., Bond stress versus slip relationship and cracking response of tension members. *J.ACI* (May/June, 1981) 217–25.
243. Somayaji, S., Composite response, bond stress–slip relationships and cracking in ferrocement and reinforced concrete. PhD Thesis, University of Illinois, 1979.
244. Marshall, W. T. & Krishnamurthy, P., Design of end zone reinforcement for pretensioned concrete beam. Proc. 6th FIP, Prague, June 1970.
245. Hoyer, E. *et al.*, Beitrag zur frage der Haftspannung in Eisten beton bauteilen. *Beton und Eisen, Berlin*, **38** (1939) 107–10.
246. Marshall, G. G., End anchorage and bond stresses in prestressed concrete. *Magazine of Concrete Research*, **1**, (1949) 123–7.
247. Marshall, W. T., A theory for end zone stresses in pretensioned concrete beams. *J.ACI*, **11** (1966) 45–51.
248. Bangash, Y., The riddle of bonded and unbonded tendons. TPI Report, 1978; also *Concrete*, **8**(6) (1974).
249. Edward, A. D. & Picard, D., Bonding properties of $\frac{1}{2}$ inch diameter strand. *ACI Journal* (November 1972) 684–9.
250. Yannopoulos, P. J., Fatigue, bond and cracking characteristics of reinforced concrete tension members. PhD Thesis, University of London, 1976.
251. Nilson, A. H., Bond stress–slip relations in reinforced concrete. Report No. 345, Department of Structural Engineering, Cornell University, December 1970.
252. Rehm, G., The basic principles of bond between steel and concrete. Translation No. 134, Cement and Concrete Association, London, 1968, p. 66.
253. Evans, R. H. & Robinson, G. W., Bond stresses in prestressed concrete from X-ray photographs. *Proc. ICE, Part 1*, **4** (1955).
254. Hanson, N. W. & Kaar, P. H., Flexural bond tests of pretensioned prestressed beams, *ACI Journal*, **55** (1959).
255. Dably, L. J., Static tests on prestressed concrete beams using 7/16 inch strands. Progress Report No. 13 on Prestressed Concrete Bridge Members, Leigh University, Fritz Engineering Laboratory, November 1956.
256. Base, G. D., An investigation of transmission length in pretensioned concrete. Congress of Federation Internationale de la Precontrainte, Session 3, Paper 9, Berlin, 1958, pp. 603–23.
257. Ratz, E. H. *et al.*, The transmission of prestressed concrete by bond. Congress of Federation Internationale de la Precontrainte, Session 3. Paper 10, Berlin, 1958, pp. 624–40.
258. Dinshore, G. A. *et al.*, Anchorage and bond in pretensioned prestressed concrete members. Report No. 223–19, Leigh University, Fritz Engineering Laboratory, December 1953.
259. Ruch, H. & Rehm, G., Tests for determination of transfer lengths of prestressing steel. *Deutscher Ausschuss fur Stahlbeton*, Heft 147, Berlin, 1963, pp. 1–38.
260. Kaar, P. H. *et al.*, Influence of concrete strength and strand transfer length. *J. Prestressed Concrete Institute (PCI)*, (October 1963).

261. Preston, H. K., Characteristics of 15 per cent stronger seven wire strand. *PCI J.*, **8** (1963) 39–45.
262. Ahmed, M., Bond strength history in prestressed concrete reactor vessels. R-84, Thames Polytechnic, 1984; PhD Thesis, (CNAA), 1983.
263. Hanson, J. M. & Hulsbos, C. L., Overload behaviour of pretensioned prestresses concrete I-beams with web reinforcement. *Highway Record*, No. 76, (1965) 1–33.
264. Over, R. S. & Au, T., Prestressed transfer bond of pretensioned strands in concrete. *ACI Journal*, **62** (1965).
265. Stocker, M. F. & Sozen, M. A., Investigation of prestressed reinforced concrete for highway bridges, Part 5: bond characteristics of prestressing strand. University of Illinois, College of Engineering, Bulletin No. 503, 1970.
266. Lutz, L. A. & Gergely, P., Mechanics of bond and slip for deformed bars in concrete. *ACI Journal*, **64** (1967) 711–21.
267. Wahl, A. et al., Direct measurement of bond–slip in reinforced concrete. *Proc. Am. Soc. Agr. Eng.*, (1969).
268. Tanner, J. A., An experimental investigation of bond–slip in reinforced concrete. MS Thesis, Cornell University, November 1971.
269. Morris, S. & David, W. J., Bond development length tests of grouted 54 strand post-tensioned tendons. *ACI Journal* (October, 1974) 522–5.
270. Morris, S., Large post-tensioned tendons, *ACI Journal* (May/June, 1972).
271. Morris, S., Grouting tests on large post-tensioned tendons for secondary nuclear containment structures. *ACI Journal* (March/April, 1971).
272. Ngo, D. & Scorcelis, A. S., Finite element analysis of reinforced concrete beams. *ACI Journal*, **64** (1967).
273. Mirza, M. S. & Houde, J., Study of bond stress–slip relationships in reinforced concrete. *ACI Journal, Proc.*, **76** (1979) 19–146.
274. Naus, W., A new method for the measurement of local slip between steel and concrete. PhD Dissertation, Technical University, Darmstadt, 1979.
275. Shah, S. P., Workshop on high strength concrete, Department of Materials Engineering, University of Illinois at Chicago Circle, December 1979.
276. Holman, J. O., Fatigue of concrete by constant and variable amplitude loading. Recent research in fatigue of concrete structures, ACJ SP-75, 1982, pp. 72–110.
277. Kakuta, Y. et al., New concepts for concrete fatigue design procedures in Japan. *Proc. IABSE, Lausanne*, **37** (1982) 51–8.
278. Aas-Jacobson, K., Fatigue of concrete beams and columns. Norwegian Institute of Technology, Bulletin No. 70-1, Trondheim, 1979, p. 148.
279. Miner, M. A., Cumulative damage in fatigue. *Trans. Am. Soc. Mech. Eng.*, **67** (1945) 159–64.
280. Waagaard, K., Design recommendations for offshore concrete structures. *Proc. IABSE, Lausanne*, **37** (1982) 59–67.
281. Japan Society of Civil Engineers, Trans. 122, October 1965.
282. Proc. CEB–FIP, Recommendation on fatigue, 1982.
283. American Association of State Highway and Transportation Officials, Report No. 4, Publication No. 953, 1962, p. 217.
284. NCHR (The National Cooperative Highway Research). Private communication, Transportation Research Board, Washington, 1984.

285. Wilson, A., Durability of concrete and concrete structures by slow cycle fatigue. SP-47, No. 12, American Concrete Institute, Detroit, Michigan, 1975, pp. 259–88.
286. Cook, R. D., *Concepts and Applications of Finite Element Analysis* (2nd edn), John Wiley and Sons, New York, 1981.
287. Zienkiewicz, O. C., *The Finite Element in Engineering Science*. McGraw Hill, London, 1977.
288. Martin, H. C. & Carrey, G. F., *Introduction to Finite Element Analysis*. McGraw Hill, New York, 1973.
289. Bathe, K. J. & Wilson, E. L., *Numerical Methods in Finite Element Analysis*. Prentice-Hall, Englewood Cliffs, New Jersey, 1972.
290. Desai, C. S. & Abel, J. F., *Introduction to the Finite Element Method*. Van Nostrand Reinhold, New York, 1972.
291. Ahmad, S., Curved finite element in the analysis of solid shell and plate structures. PhD Thesis, University of Wales, Swansea, 1969.
292. Clough, R. W., Comparison of three-dimensional finite elements. Proc. Symposium on Application of Finite Element Methods in Civil Engineering, Vanderbilt University, 1969.
293. Ngo, D., Scordelis, A. C. & Franklin, H. A., Finite element study of reinforced concrete beams with diagonal tension cracks. University of California, Berkeley, UC-SESM Report No. 70-19, December 1970.
294. Marcal, P. V., Finite element analysis with material nonlinearities—theory and practice. Conference on Finite Element Methods in Civil Engineering, Montreal, 1972, pp. 71–113.
295. ASCE Committee on Concrete and Masonry Structures. State-of-the-Art Report on Finite Element Analysis of Reinforced Concrete, Task Committee on Finite Element Analysis of Reinforced Concrete Structures, ASCE Special Publication, 1982.
296. Argyris, J. H. *et al.*, Recent development in the finite element analysis of prestressed concrete reactor vessels. *Nucl. Eng. Des.*, **28** (1974).
297. Bicanic, N. & Zienkiewicz, O. C., Constitutive models for concrete under dynamic loading. *Earthquake Engineering and Structural Dynamics*, **11** (1983) 689–710.
298. Argyris, J., Faust, Willam, Limit load analysis of thick-walled concrete structures—a finite element approach to fracture. *Computer Methods in Applied Mechanics and Engineering*, **8** (1976) 215–43.
299. Mohraz, B. *et al.*, Crack development in a prestressed concrete reactor as determined by lumped parameter method. *Nucl. Eng. Des.* (1970).
300. Sozen, M. A. & Paul, S. L., Structural behaviour of small scale prestressed concrete reactor vessels. *Nucl. Eng. and Des.* (1969) 403–14.
301. Campbell-Allen, D. & Low, E. W. E., Pressure tests on end slabs for prestressed concrete pressure vessels. *Nucl. Eng. and Des*, **6** (1967) 345–59.
302. Rashid, Y. R., Ultimate strength analysis of prestressed concrete pressure vessels. *Nucl. Eng. and Des.*, **7** (1968).
303. Rashid, Y. R., Analysis of concrete composite structures by the finite element method. *Nucl. Eng. and Des.*, **3** (1966).
304. Rashid, Y. R. & Rockenhauser, W., Pressure vessel analysis by finite element

techniques. *Conference on Prestressed Concrete Pressure Vessels*, Inst. Civ. Eng., London, 1968.
305. Connor, J. J. & Sarne, Y., Non-linear analysis of prestressed concrete reactor pressure vessels. 3rd International Conference on Structural Mechanics in Reactor Technology (SMiRT), Paper H2/2, London, 1975.
306. Takeda, T., Yamaguchi, T. & Imoto, K., Inelastic analysis of multi-cavity PCRV under internal pressure. 3rd International Conference on Structural Mechanics in Reactor Technology (SMiRT) Paper H2/2, London, 1975.
307. Akyuz, F. A. & Merwin, J. E., Solution of non-linear problems of elastoplasticity by finite element method. *AIAA J.*, **6** (1968).
308. Alexander, J. M. & Gunasekera, J. S., On the geometrically similar expansion of a thin infinite plate. *Proc. Roy. Soc. London*, **A326** (1972) 361–72.
309. Anand, S. C., Lee, S. L. & Rossow, E. C., Finite element analysis of elastic–plastic plane stress problems based on Tresca's yield criterion. *Ingenieur Archiv*, **39** (1970).
310. Argyris, J. H., *Energy Theorems and Structural Analysis*. Butterworths, London, 1960.
311. Argyris, J. H., Three-dimensional anisotropic and inhomogenous elastic media, matrix analysis for small and large displacements. *Ingenieur Archiv*, **34** (1965).
312. Argyris, J. H., Elasto-plastic matrix displacement analysis of three-dimensional continua. *J. RAeS*, **69** (1965).
313. Argyris, J. H., Continua and discontinua. Proc. Conference Matrix Methods in Structural Mechanics, Ohio, 1965.
314. Argyris, J. H., Application of the matrix displacement method to the analysis of pressure vessels. *ASME, J. Eng. Industry*, (1970).
315. Argyris, J. H. *et al.*, *Matrix Methods of Structural Analysis*. Pergamon Press, Oxford, 1964.
316. Argyris, J. H. *et al.*, Some general considerations of the natural mode technique, Part 1. *J. RAeS*, **73** (1969).
317. Argyris, J. H. *et al.*, Some general considerations of the natural mode technique, Part 2. *J. RAeS*, **73** (1969).
318. Argyris, J. H. *et al.*, Method of elasto-plastic analysis. Symposium of Finite Element Techniques at the Institute fur Statik und Dynamik cer Luft und Raumfahrtstrucktionen, University of Stuttgart, 1969.
319. Armen, H. *et al.*, Discrete element methods for the plastic analysis of structures subjected to cyclic loading. *Int. J. Num. Meth. Eng.*, **2** (1970).
320. Bogner, F. K. *et al.*, A cylindrical shell discrete element. *AIAA J.*, **5** (1967).
321. Brebbia, C. *et al.*, Geometrically non-linear finite element analysis. *J. Eng. Mech. Div., ASCE*, **95** (1969).
322. Bushnell, D., Non-linear axisymmetric behaviour of shells of revolution. *AIAA J.*, **5** (1967).
323. Chan, A. S. L. *et al.*, The analysis of cooling towers by the matrix finite element method. *J.RAeS*, **74** (1970).
324. Dupuis, G. A., Incremental finite element analysis of large elastic deformation problems. Brown University, May 1971.
325. Ford, H. & Alexander, J. M., *Advanced Mechanics of Materials*. Longman, London, 1963.

326. Gallagher, R. H. et al., Discrete element procedure for thin shell instability analysis. *AIAA J.*, **5** (1967).
327. Gallagher, R. H. et al., Discrete element approach to structural instability analysis. *AIAA J.*, **1** (1963).
328. Gunasekera, J. S., A preliminary study of closed die forging. MSc Thesis, University of London, 1969.
329. Gunasekera, J. S. et al., Matrix analysis of the large deformation of an elastic–plastic axially symmetric continuum. Proc. Symposium on Foundations of Plasticity, Warsaw, September 1972.
330. Green, A. E. et al., A general theory of an elastic plastic continuum. *Arch. Ration. Mech. Anal.*, **18** (1965).
331. Green, A. E. & Zerna, W., *Theoretical Elasticity*, Clarendon Press, Oxford, 1954.
332. Hartley, H. O., The modified Gauss–Newton method for the fitting of non-linear regression functions by least squares. *Technometrics*, **3** (1961).
333. Hartz, B. J. et al., Finite element formulation of geometrically non-linear problems of elasticity. Proc. Japan–US Seminar on Matrix Methods of Structural Analysis and Design, Tokyo, 1969.
334. Hibbit, H. D. et al., Hybrid finite element analysis with particular reference to axisymmetric structures. *AIAA J.* (8th Aerospace Sciences Meeting, New York, January 1970).
335. Hibbit, H. D. et al., A finite element formulation for problems of large strain and large displacement. *Int. J. Solids Struct.*, **6** (1970).
336. Hill, R., *The Mathematical Theory of Plasticity*. Clarendon Press, Oxford, 1950.
337. Mallet, R. H. et al., Finite element analysis of non-linear structures. *J. Struct. Div., ASCE*, **94** (1968).
338. Marcel, P. V., A comparative study of numerical method of elastic–plastic analysis. *AIAA J.*, **6** (1968).
339. Marcel, P. V., Finite element analysis of combined problems of non-linear material and geometric behaviour. Brown University, March 1969.
340. Marcel, P. V. et al., Elastic–plastic analysis of two-dimensional stress systems by the finite element method. *Int. J. Mech. Sci.*, **9** (1967).
341. Martin, H. C., Derivation of stiffness matrices for the analysis of large deflection and stability problems. Proc. 1st Conf. on Matrix Methods in Structural Mechanics, Ohio, October 1965.
342. Washizu, K., *Variational Methods in Elasticity and Plasticity*. Pergamon Press, Oxford, 1968.
343. Webster, G. A., Iterative procedures for elastic, plastic and creep deformation in beams. *J. Mech. Eng. Sci.*, **9** (1967).
344. Wilson, E. L., Structural analysis of axisymmetric solids. *AIAA J.*, **3** (1965).
345. Wissman, J. W., Non-linear structural analysis; tensor formulation. Proc. 1st Conf. on Matrix Methods in Structural Mechanics, Ohio, October, 1965.
346. Yamada, Y. et al., Plastic stress–strain matrix and its application for the solution of elastic–plastic problems by the finite element method. *Int. J. Mech. Sci.*, **10** (1968).
347. Zienkewicz, O. C., Valliappan, S. & King, I. P., Stress analysis of rock as a no tension material. *Geotechnique* (1968).

348. Murray, D. W. *et al.*, An approximate non-linear analysis of thin-plates. *Proc. 2nd Conf. on Matrix Methods in Structural Mechanics,* December 1969.
349. Argyris, J. H. & Sharpe, D. W., Methods of elasto–plastic analysis. *Proc. ISA–ISSC Symposium on Finite Element Technology,* Stuttgart, 1969, pp. 381–416.
350. Bangash, Y., The structural integrity of concrete containment vessels under external impacts. *6th International Conference on Structural Mechanics in Reactor Technology (SMiRT),* J7/6, Paris, August 1981.
351. Bangash, Y., The automated three-dimensional cracking analysis of prestressed concrete vessels. *6th International Conference on Structural Mechanics in Reactor Technology (SMiRT),* H Paper 3/2, Paris, August 1981.
352. Bathe, K. J. & Wilson, E. L., Stability and accuracy analysis of direct integration method. *Int. J. of Earthquake Engineering and Structural Dynamics,* **1** (1977) 283–91.
353. Armen, H. *et al.*, Discrete element methods for the plastic analysis of structures subjected to cyclic loading. *Int. J. Num. Meth. Eng.*, **2** (1970).
354. Bogner, F. K. *et al.*, A cylindrical shell discrete element. *AIAA J.*, **5** (1967).
355. Brebbia, C. *et al.*, Geometrically non-linear finite element analysis. *J. Eng. Mech. Div., ASCE,* **95** (1969).
356. Bushnell, D., Non-linear axisymmetric behaviour of shells of revolution. *AIAA J.*, **5** (1967).
357. Chan, A. S. L. *et al.*, The analysis of cooling towers by the matrix finite element method. *J.RAeS,* **74** (1970).
358. Dupuis, G. A., Incremental finite element analysis of large elastic deformation problems. Brown University, May 1971.
359. Newmark, N. M., A method for computation of structural dynamics. *Proc. AM Soc. Civ. Eng.*, **85** (1959) 67–94.
360. Akyuz, F. A. & Merwin, J. E., Solution of non-linear problems of elastoplasticity by finite element method. *AIAA J.*, **6** (1968).
361. Alexander, J. M. & Gunasekera, J. S., Private communication.
362. Anand, S. C. *et al.*, Finite element analysis of elastic–plastic plane stress problems based upon Tresca's yield criterion. *Ingenieur Archiv,* **39** (1970).
363. Bathe, K. J. *et al.*, Finite element formulation and solution of non-linear heat transfer. *Nucl. Eng. Des.*, **51** (1979) 389–405.
364. Chan, R. C. Y., Concrete filled tubular columns. Project Report Civil Engineering, Middlesex Polytechnic, May 1984.
365. Knowles, R. B. & Park, R., Strength of concrete filled steel tubular columns. *J. Struct. Div., ASCE,* **95** (1969).
366. Knowles, R. B. & Park, R., Axial load design for concrete filled steel tubes, *J. Struct. Div., ASCE,* **96** (1970).
367. Neogi, P. K., Sen, H. K. & Chapman, I. C., Concrete filled tubular columns under eccentric loading. *ACI Journal,* **47** (1969) 187–95.
368. Russel, W. A., Structural properties of light-gage tubular columns. Housing Research Paper No. 21, Housing and Home Finance Agency, October 1953.
369. Salani, H. J. & Sims, J. R., Behaviour of mortar filled steel tubes in compression. *ACI Journal Proc.*, **61** (1964) 1271–83.

370. Sen, H. K. & Chapman, J. C., Ultimate load tables for concrete filled steel tubular. CIRIA Technical Note 13, August 1970.
371. Furlong, R. W., Strength of steel-encased concrete beam columns. *J. Struct. Div., ASCE*, **93** (1967) 113–24.
372. Furlong, R. W., Design of steel-encased concrete beam columns. *J. Struct. Div., ASCE*, **94** (1968) 267–81.
373. Furlong, R. W., Concrete-encased steel columns—design tables. *J. Struct. Div., ASCE*, **100** (1964).
374. Gardner, N. J. & Jacobson, R., Structural behaviour of concrete filled steel tubes. *ACI Journal* (1967) 404–13.
375. Bangash, Y., The finite element analysis of concrete filled tubes under complex load—Program CONFT. Middlesex Polytechnic, 1984.
376. Sen, H. K. & Chapman, J. C., Ultimate load tables for concrete filled tubular steel columns. CIRIA Tech. Note 13, August 1970.
377. BS 5950. The structural use of steel, 1985.
378. Basu, A. & Sommerville, J., Derivation of formulae for the design of rectangular composite columns. ICI Paper No. 7206S, May 1969.
379. BS 4360., Weldable structural steels, 1968.
380. BS 4: Part 2, Structural steel sections—hot rolled hollow sections, 1969.
381. Pink, A., *Winter Concreting*. Cement and Concrete Association, London, 1967.
382. BS 938. General requirements for metal—arc welding of structural steel tubes to BS 1775, 1962.
383. *ACI Journal*, January 1971 to February 1983.
384. BS 449: Part 1. The use of structural steel in building (Imperial Units), 1970.
385. Portland Cement Association. Private communication, Concrete Filled Tubes, Skokie, USA, 1981.
386. British Steel Corporation. Private communication, Literature on concrete filled tubes, 1981.
387. Timoshenko, S. P. & Gere, J. M., *Theory of Elastic Stability*. McGraw Hill, New York, 1958.
388. Bangash, Y., The ultimate load analysis of containment vessels under accidental pressures. International Conference on Structural Mechanics in Reactor Technology (SMiRT), August 1987, pp. 275–80.
389. Terzaghi, K., *Theoretical Soil Mechanics*. John Wiley, New York, 1943.
390. Ghana Department of Irrigation. Private communication, Akra, Ghana, 1973.
391. Brink, A., Recent developments in the design of submerged tunnels. *Struct. Engr.*, **44** (1966).
392. Bickel, J. O., Trench type subaqueous tunnels: design and construction. *Struct. Engr*, **44, 45** (1966, 1967).
393. Pequignot, C. A., Selective bibliography on immersed tubes. *Tunnels and Tunnelling*, **1** (1969).
394. Palmer, W. F. *et al.*, Developments in trench-type tunnel construction. *J. Const. Div., ASCE*, **101** (1975).
395. Glerum, A. *et al.*, Motorway tunnels built by the immersed tube method. *Rijkswaterstaat Comm.*, **25** (1976).
396. Brakel, J., Submerged tunnelling. Technische Hogeschool, Delft, 1978.

397. United States Bureau of Reclamation. Hoover Dam Tunnels & Spillways, Washington DC, 1955.
398. WAPDA. Reports on tunnels for Mangla and World Bank. Tarbela dams. West Pakistan Water and Power Development Authority, Reports 2, 3, 1975.
399. Waters, T. C. & Barrett, N. T., Prestressed concrete pressure vessels for nuclear reactors. *J. BNES* (July 1963).
400. Marsh, R. O. & Melese, G. B., Prestressed concrete pressure vessels. *Nucleonics*, **23** (1975) 63–74.
401. Frankline Institute Laboratories. State of Art of Prestressed Concrete Pressure Vessels for Nuclear Power Reactors—A Critical Review of Literature. USEAC Report ORNL-TM-812, Oak Ridge National Laboratory, June 1964.
402. Bender, M., A status report on prestressed concrete reactor vessels, pressure vessel technology, Parts 1 and 2. *Nuclear Structural Engineering* (1965) 83–90; 206–23.
403. Jaeger, T. A., Note on stress analysis of prestressed concrete reactor vessels. *Nuclear Structural Engineering*, **1** (1965) 133–6.
404. Rashid, Y. R., Analysis of axisymmetric composite structures by the finite element. *Nucl. Eng. Des.*, **3** (1966) 163–82.
405. Tottenham, H. & Kanchi, M. B., Structural characteristics of cylindrical pressure vessels of medium thickness. *Nucl. Eng. Des.* (1966) 177–92.
406. Kanchi, M. B., Elastic theories of thick plates and shells. CE Dept. Report No. CE11-65, University of Southampton, 1965.
407. Reisnner, E., Stress–strain relations in the theory of thin elastic shells. *J. Math. Physics*, **31** (1952) 109.
408. Orr, R. S. & Holland, D. A., Theoretical analysis of model spherical vessels. Conf. on PCPV, London, Paper 33, March 1967.
409. UKAEA, Private communication, 1967.
410. Common, D. K. & Hannah, I. W., Specifications of concrete vessels for gas cooled reactors. Conference on PCPV, London, Paper 7, March 1967.
411. Finigen, A., Ultimate analysis of Dungeness B vessels. Conference on PCPV, London, Paper 31, Group F, March 1967.
412. Zbirohowski-Kuscia, K. & Carlton, D., Analysis of vessel structures with particular reference to Wylfa. Conference on PCPV, London, Paper 30, March 1967.
413. Morice, P. B., Discussion on group F papers on design and analysis of vessel structures. Conference on PCPV, London, March 1967, pp. 397–434.
414. Zudnas, Z. & Tans, C. P., Feasibility study of prestressed concrete pressure vessel design. USAEC Report CRNL-TM-813, Oak Ridge National Laboratory, June 1967.
415. (a) Brown, A. H., The Oldbury vessels. Conference on PCPV, London, Paper 1, Group A, March 1967. (b) Harris, A. J. & Hay, J. D., Rupture design of the Oldbury vessels. Conference on PCPV, London, Paper 29, Group F, March 1967.
416. Rashid, Y. R., Ultimate strength analysis of prestressed concrete pressure vessels. *Nucl. Eng. Des.*, **7** (1968) 334–44.
417. Gomez, A. E. *et al.*, Lumped parameter analysis of cylindrical prestressed concrete reactor vessels. University of Illinois, Vols. 1 & 2, 1968.

418. Papers on Ultimate Load Tests, Conference on PCPV Model Techniques, British Nuclear Energy Society, The Institution of Civil Engineers, London 1968. (a) Paper 3—Deflection measurement technique, Brading, K. F. & McKillen, R. R. (b) Paper 4—Techniques for rupture testing of prestressed concrete vessel models, Scotto, F. (c) Discussion on papers 2 and 4, Bangash, Y. (d) Paper 6—The use of simplified models for the design of end slabs, Campbell-Allen, D. & Low, E. W. E. (e) Paper 7—small scale model prestressed concrete pressure vessel tests at Foulness, Davidson, I. (f) Paper 8—An experimental investigation into the load behaviour of perforated end slabs for concrete pressure vessels under temperature and external load, Stefanou, G. D. *et al.* (g) Paper 9—Model philosophy in relation to prestressed concrete pressure vessel design problems, Langan, D. (h) Discussion on Papers 7 and 9, Bangash, Y. (i) Paper 11—The behaviour of the Oldbury model vessel with time under thermal and pressure loadings, Hornby, T. W. (j) Paper 12—The two Bugey 1/5 models, Launay, P. (k) Paper 13—Elastic and ultimate pressure tests on a one-tenth scale model of the Dungeness B concrete pressure vessel, Brading, K. F. & Finigin, A. (l) Paper 14—Testing the one-tenth scale model of the Hinkley B and Hunterston B power station prestressed concrete pressure vessels, Eadie, D. McD. & Bell, D. J. (m) Discussion Session C including P on Paper 13, Bangash, Y.
419. Sozen, M. A. *et al.*, Strength and behaviour of prestressed concrete vessels for nuclear reactors. Vols 1, 2 & 3, University of Illinois, 1969.
420. Koerner, R. J., Over pressure analysis of prestressed concrete pressure vessels. *Nucl. Eng. Des.* (December 1970).
421. ACI-BAM Concrete for Nuclear Reactors. Proc. International Seminar at Bundesantalt fur Materialsprufungin, Berlin, October 1970, ACI Special Publication SP-34, Vols 1, 2 & 3. (a) Paper SP34-12, Capacity of prestressed concrete pressure vessels to resist shear, Garas, F. K. & Langan, D. (b) Paper SP34-67, Concrete behaviour under continued stresses up to failure: test results on small dimension prestressed concrete pressure vessel models, Scotto, F. L. (c) Paper SP34–68 Testing of a perforated prestressed concrete pressure vessel top slab model, Meerwald, K. & Schwiers, G. (d) Paper SP34-69, Apparatus, instrumentation and concrete models of Bugey 1 prestressed concrete pressure vessel.
422. ACI-BAM Concrete for Nuclear Reactors. Proc. International Seminar at Bundesantalt fur Materialsprufungin, Berlin, October 1970. ACI Special Publication SP-34, Vols 1, 2 & 3. (a) Paper SP34-7, PCRV problems of calculation, material properties and evaluation of results, Hofmann, B. H. H. (b) Paper SP34-9, Discussions on finite element analysis of prestressed concrete reactor vessels, Argyris, J. H. *et al.* (c) Paper SP34-10, Analysis of prestressed concrete pressure vessels for nuclear reactors in Czechoslovakia, David, M. (d) Paper SP34-70, Strain behaviour of the Oldbury concrete pressure vessel under operating pressure and temperature, Carmichael, G. D. T. *et al.* (e) Paper SP34-39, An investigation of the time dependent behaviour of prestressed concrete pressure vessels, Whitman, G. D. *et al.*
423. CEC & BAM Proc. 1st International Conference on Structural Mechanics in Reactor Technology, 20–24 September 1971, Berlin, Vol. 4, Reactor Pressure

Vessels, Part H, Prestressed Concrete Pressure Vessels. (a) Paper H3/4, Behaviour of eng. slabs in cylindrical prestressed concrete pressure vessels, Langan, D. & Garas, F. K. (b) Paper H3/5, Analyse tridimensionelle du comportement non lineaire d'un caisson de reacteur nucleaire en beton precontraint, Hussain, M. K., et al. (c) Paper H3/7, Allegeineine Berechnung Von Spannbeton-Reaktordruck behaltern unter Berucksichtigung von Nichtlinearen Spannungs-Dehnungs- gesetzen Nach Der Methode Der Dynamischen Relaxation, Schnellenbach, G. (d) Paper H4/3, Der Spannberunbehalter des THTR-300-MWe PrototypKernkraftwerkes, Lotz, H. *et al*. (e) Paper H4/4, The design of pod boiler pressure vessels with particular reference to Hartlepool nuclear power station, Langan, D. *et al*. (f) Paper H5/3, The Scandinavian PCRV model project stress calculation and experimental verification, Andersen, S. I. *et al*. (g) Paper H5/4, die Dreidimensionalen Elastischen Spannungszustande im Bereich der Offnungen des THTR-Spannbeton behalters Ein Vergleich Zwischen Numerischer Berechnung und Messergebnissen Eines Elastischen Modells, Hansson, V. (h) Paper H5/5, Modellversuche zur Vertiefung der Kenntnisse Uber das Tragverhalten des Spannberonbehalters fur das THTR-300-MWe-Prototyp-Kernkraftwerk, Lotz, H. & Meerwald, K. (i) Paper H5/6, Thin-walled 1:20 prestressed concrete pressure vessel model for THTR reactor type, Scotto, F. L. (j) Paper H5/7, An investigation on a prestressed concrete reactor vessel under internal gas pressure mode of rupture, Lenschow, R. (k) Paper H5/8, Theoretische und Experimentale Untersuchungen des Spannungs-Dehnungs Verhaltens der Druckbehalter aus Vorgespanntem Beton, Servit, R.

424. CEC & BAM Proc. 2nd International Conference on Structural Mechanics in Reactor Technology, September 1973. (a) Paper H2, Simple approximate method for the analysis of the ultimate carrying capacity of prestressed concrete reactor pressure vessels, Huber, A. E. & Hoffman, H. H. (b) Paper H2/2, On the behaviour of PCPV under ultimate load conditions and its influence on calculations and design especially applied to the THTR vessel, Binseil, P. (c) Paper H2/3, Investigation on ultimate load design of prestressed concrete reactor pressure vessels, Schimmelpfennig, K. & Hansen, V. (d) Paper 2/4, Inelastic behaviour and failure modes of prestressed concrete vessels, Fluge, F., Gausel, E. & Lenschow, R. (e) Paper 2/5, Three-dimensional rupture analysis of a prestressed concrete pressure vessel including creep effects, Sang, B. *et al*.

425. CEC & BNES Trans. 3rd International Conference on Structural Mechanics in Reactor Technology, 1–5 September 1975, London, Vol. 3, Reactor Vessels, Part H, Structural Analysis of Prestressed Concrete Reactor Pressure Vessels. (a) Paper H2/2, Non-linear analysis of prestressed concrete reactor pressure vessels, Connor, J. J. & Sarne, Y. (b) Paper H2/3, Experience in the application of a finite element system to the analysis of complex prestressed concrete pressure vessels, Wade, M. J. & Henrywood, R. K. (c) Paper H2/4, Analysis of prestressed concrete reactor vessel under high thermal gradient, Gulkan, P. & Akay, H. U. (d) Paper H2/5, Inelastic analysis of a multicavity PCRV under internal pressure, Takeda, T. *et al*. (e) Paper

H2/6, The method of slice substructures in the analysis of boiler-podded PCPV, Kawamata, S. *et al.* (f) Paper H2/9, Probabilistic determination of partial safety factor for the design of prestressed concrete high pressure vessels, Rackwitz, R. (g) Paper H3/6, Theoretical and experimental studies for optimization of PCRV top closures, Ottasen, N. S. & Andersen, S. I. (h) Paper H4/1, The analysis of crack structures, Davidson, I. (i) Paper H4/2, Crack analysis of multicavity prestressed concrete reactor vessels, Gallix, R. *et al.* (j) Paper H4/3, Failure mode analysis of PCRV influence of some hypothesis, Zimmermann, Th. *et al.* (k) Paper H4/4, Comparison of two methods of analysing three-dimensional cracking in the upper section of the THTR prestressed concrete pressure vessel, Bindseil, P. & Hansson, V. (i) Paper H4/5, Three-dimensional analysis of cracked concrete in prestressed concrete pressure vessels and comparison with experimental results, Hansson, V. & Stover, R. (m) Paper H4/6, Ultimate load analysis of prestressed concrete reactor pressure vessels considering a general material law, Schimmelpfenning, K. (n) Paper H4/7, Inelastic behaviour, failure modes and ultimate load design of prestressed concrete pressure vessels, Berquam, T. *et al.* (o) Paper H4/8, Ultimate strength design of prestressed concrete reactor vessels, Meyer, C. & Goldman, B. I.

426. Irving, J. *et al.*, A full-scale model test of hot spots in the prestressed concrete pressure vessels of Oldbury nuclear power station. *Proc. Inst. Civ. Eng. Part 2*, **57** (1974) 331–51 & 795–8 (Discussion).
427. Brunton, J. D. *et al.*, Wire-winding of Hartlepool and Heysham reactor pressure vessels. International Conference in Design, Construction and Operation PCPV's and Containments for Nuclear Reactors, Paper 147/75, University of York, September 1975.
428. Cahill, T. & Branch, G. D., Long term relaxation behaviour of stabilized prestressing wires and strands. Inst. Civ. Eng., March 1967.
429. Bangash, Y., A circumferential prestressing load analysis due to wire/strand winding systems on prestressed concrete reactor vessels. *Proc. Inst. Civ. Eng.*, **157** (1974) 437–50 (Discussion).
430. Hruska, F. H., Radial forces in wire ropes. *Wire and Wire Products* (May 1952).
431. Leissa, A. W., Contact stresses in wire ropes. *Wire and Wire Products*, **34** (1959) 307–14 & 372–73.
432. Luskin, J. L., *Contact problems*. In: *Handbook of Engineering Mechanics*, W. Flugge, ed. McGraw Hill, New York, 1962.
433. Starkey, W. L. & Cress, H. A., Analysis of critical stress and mode of failure of a wire rope. *J. Eng. for Industry*, *ASME*, **81**, (1959) 307–16.
434. Durelli, A. J. & Machida, S., Response to oversized epoxy strand models to axial, torsional and bending loads. Catholic University of America, Washington DC. Technical Report 72-2, 1972, p. 17.
435. Machida, S. & Durelli, A. J., Response of a strand in twisted wire cables. Catholic University of America, Washington DC, Technical Report 72-1, 1972, p. 22.
436. Hertz, H., *Uber die Behruhrung Fester Elastische Korper*, Leipzig, 1965, pp. 174–96 & 155–73.

437. Prandtl, L., Spannungs Verteilung in Plastischen Koerpern. Proc. 1st International Congress Applied Mechanics, Delft, Technische Boekhandel en Druckerij, J. Waltman Jr., 1925, pp. 43–54.
438. Reusse, E., Beruecksichtigung der Elastischen Formaenderungen in der Plastizitaetstheorie. *Z. Angew. Math. Mech.*, **10** (1930) 226–74.
439. Bland, D. R., The associated flow rule of plasticity, *J. Mech. Phys. Solids*, **5** (1957) 71–8.
440. Budiansky, B., A reassessment of deformation theories of plasticity, *J. Appl. Mech.*, **26** (1959) 259–64.
441. Johnson, W. & Mellor, P. B., *Plasticity of Mechanical Engineers*. Van Nostrand, Princeton, NJ, 1962.
442. Mendelson, A. & Manson, S. S., Practical solution of plastic deformation problems in elastic–plastic range, NASA Tech. Report R-28, 1959.
443. Berg, O. Y., The problem of strength and plasticity of concrete. *Doklady Akademii Nauk*, **70** (1950) 617–20.
444. Hill, R., *The Mathematical Theory of Plasticity*. Oxford University Press, Oxford, 1956, pp. 34–45.
445. Green, A. E. & Adkins, J. E., *Large Elastic Deformations*. Oxford University Press, Oxford, 1960.
446. Marcal, P. V. & King, I. P., Elastic–plastic analysis of two-dimensional stress systems by the finite element method. *Int. J. Mech. Sci.*, **9** (1967) 143–55.
447. Eringen, A. C., Linear theory of micropolar elasticity, *J. Math. and Mech.*, **15** (1966).
448. Yamada, Y. *et al.*, Plastic stress–strain matrix and its application for the evolution of elastic–plastic problems by the finite element method. *Int. J. Mech. Sci.*, **10** (1969) 343–54.
449. Zienkiewicz, O. C. *et al.*, Finite element methods in the reactor vessels. *Nucl. Eng. Des.*, **20** (1972) 507–41; Private communication, 1972.
450. Schalkwijk, R. & Konter, A., Dynamic analysis with MARC. Zoetermeer (1986).
451. Simon Carves and Atomic Power Construction. BBRV tests on axial and circumferential tendons for short and long term conditions. Private communication, 1973.
452. Warner, P. C., The Dungeness B vessels. Conference on PCPV. Inst. Civ. Eng., Paper 3, Group A, London, March 1967.
453. Thorpe, D. E., Elastic analysis of the Dungeness B vessel. Conference on PCPV, Inst. Civ. Eng., Paper 36, London, March 1967.
454. Cable Covers & Taylor Woodrow. Cable covers multiforce tendons, Private communication, 1973.
455. Dawson, P., Development of the prestressing system for the Wylfa vessel, Conference on PCPV, Paper 23, March 1967.
456. Burrow, R. E. D., Prestressing tendon systems. Conference on PCPV, Inst. Civ. Eng., Group E, Paper 22, London, 1967.
457. Sokolikoff, I. S. & Redheffer, R. M., *Mathematics of Physics and Modern Engineering*. McGraw Hill, New York, 1958.
458. Taylor, S. J. *et al.*, Development, testing and installation of prestressing of the PCPV's at Hinkley B and Hunterston B. International Conference on Experience in Design, Construction and Operation of PCPV's and

Containments for Nuclear Reactors, Inst. Mech. Eng. & University of York, Paper C152/75, September 1975.
459. Thorpe, W., BBRV post-tensioning systems as applied to reactor containments and prestressed concrete pressure vessels. International Conference on Experience in Design, Construction and Operation of PCPV's and Containments for Nuclear Reactors, Inst. Mech. Eng. & University of York, Paper C145/75, September 1975.
460. Ross, A. D., The elasticity, creep and shrinkage of concrete. Proc. Conference Mechanical Properties of Non-Metallic Brittle Materials, London, April 1958.
461. Hardingham, R. R., Parker, J. V. & Spruce, T. W., Liner design and development for the Oldbury vessel. Conference on PCPV, Inst. Civ. Eng., Paper 56, London, March 1967.
462. Young, A. G. & Tate, L. A., Design of liners for reactor vessels. Conference on PCPV, Inst. Civ. Eng., London, Paper 57, March 1967.
463. Chapman, J. C. & Carter, A., Interaction between a pressure vessel and its liner. Conference on PCPV, Inst. Civ. Eng., Paper 58, London, March 1967.
464. Bishop, R. F. *et al.*, Liner design and construction. Conference on PCPV, Inst. Civ. Eng., Paper 59, London, March 1967.
465. Parker, J. V., Stress analysis of liners for prestressed concrete pressure vessels. Proc. 1st International Conference on Structural Mechanics in Reactor Technology, Paper H6/1, Berlin, September 1971.
466. White, C. M. *et al.*, Basis of design of liners for prestressed concrete pressure vessels and practical examples of the application of the design basis. Proc. 1st International Conference on Structural Mechanics in Reactor Technology, Paper 6/8, Berlin, September 1971.
467. Chan, H. C. & McMinn, S. J., The stabilisation of the steel liner for a prestressed concrete pressure vessel. *Nucl. Eng. Des.*, 3 (1966) 66–73.
468. Peinado, C. O., Private communications, 1977.
469. Kitcher, T. P., Buckling of a stud supported thin cylindrical shell encased in concrete. Report prepared for Nelson Stud Welding Co., Lorain, Ohio, August 1969.
470. Tate, L. A. *et al.*, The design of penetration—liner junctions for prestressed concrete pressure vessels. 2nd International Conference on Structural Mechanics in Reactor Technology, Paper H5/4, Berlin, September 1973.
471. NSW Design Data, Nelson concrete anchor studs. Nelson Stud Welding, Division of Gregory Industries, Ohio, Manual No. 21, 1961.
472. Nemet, J., Reaktordruck Behalter aus Spannbeton mit Heiber Dichthaut. 1st International Conference on Structural Mechanics in Reactor Technology, Paper H4/6, Berlin, 1971.
473. Nemet, J., The Austrian PCRV project with a hot liner—A status Report. 2nd International Conference on Structural Mechanics in Reactor Technology, Paper 3/8, Berlin, September 1973.
474. Howarth, H., Inelastic buckling of the lining units of a prestressed concrete pressure vessel. MSc Thesis, The Victoria University of Manchester, May 1969.
475. Oberpichler, R. *et al.*, Stress analysis of liners for prestressed concrete reactor pressure vessels with regard to non-linear behaviour of liner material and of

anchor characteristics. 3rd International Conference on Structural Mechanics in Reactor Technology, Paper H3/8, London, September 1975.
476. Yang, T. Y., A matrix displacement matrix on pre and post buckling analysis of liners for reactor vessels. 1st International Conference on Structural Mechanics in Reactor Technology, Paper H6/2, Berlin, 1971.
477. Schnelenbach, G., Beitrag zur Numerischen Berechung des Raumlichen Spannungs-Zustandes in Hohlzylindern mit Ausschnitten. PhD University of Bochum, 1969.
478. BS 4975. Specification for prestressed concrete pressure vessels for nuclear reactors, 1973.
479. Doyle, J. M. & Chu, S. L., Liner plate buckling and behaviour of stud and rip type anchors. 1st International Conference on Structural Mechanics in Reactor Technology, Paper H6/3, Berlin, September 1971.
480. Proposed standard code for concrete reactor vessels and containments. ACI-HSME Technical Committee on Concrete Pressure Components for Nuclear Service, August 1972.
481. Structural analysis of critical areas of cavity liner. Fort St. Vrain, Gulf General Atomic Report GADR-17, May 1972.
482. Containment Building Liner Plate Design Report. Bechtel Corporation, San Francisco, December 1972.
483. Tan, C. P., A study of the design and construction of particles of prestressed concrete and reinforced concrete containment vessels. Clearing House for Federal Scientific and Technical Information, Report No. TID-25176, 1969.
484. Burdett, E. G. & Rogers, L. W., Liner anchorage tests. *J. Struct. Div., ASCE*, (July 1975).
485. Mutzl, J. *et al.*, Buckling analysis of the hot liner of the Austrian PCRV Concept. 3rd International Conference on Structural Mechanics in Reactor Technology, Paper H3/9, London, 1975.
486. Levy, S., Bending of rectangular plates with large deflections. NACA TR-737, 1942.
487. Preliminary safety analysis report. Bellefonte Nuclear Plant, Docket Nos. 50-438 and 50-439, Tennessee Valley Authority, Knoxville, Tennessee, 1974.
488. Onat, E. T. & Drucker, D. C., Inelastic stability and incremental theories of plasticity. *J. Aeronautical Sci.*, **20** (1953) 181–6.
489. Murphy, L. M. & Lee, L. H. N., Inelastic buckling process of axially compressed cylindrical shells subject to edge constraints. *Int. J. Solids and Structures*, **7**, (1971) 1153–70.
490. Shanley, F. R., Inelastic column theory. *J. Aeronautical Sci.*, **14** (1947) 261–7.
491. Hill, R., The essential structure of constitutive laws for metal composites and polycrystals. *J. Mech. Phys. Solids*, **15** (1967) 79–95.
492. Haghdi, P. M. Stress–strain relations in plasticity and thermoplasticity. Proc. 2nd Symposium Naval Structural Mechanics, Brown University, 1960, pp. 121–67.
493. Sewell, M. J., A yield surface corner lowers the buckling stress of an elastic/plastic compressed plate. MRCTR 1226, University of Wisconsin, 1973.
494. Gerard, G., Plastic stability theory of stiffened cylinders under hydrostatic pressure. *J. Ship Research*, **6** (1962) 1–7.

495. Koiter, W. T., The effect of axisymmetric imperfections on the buckling of cylindrical shells under axial compression. *Proc. Koninkl. Nederl. Akademic van Wenschappen*, **66** (Series B) (1963).
496. Odquist, F. K. G., *Mathematical Theory of Creep and Creep Rupture*. Oxford University Press, Oxford, 1966.
497. Alnajafi, A. M. J., Non-linear behaviour of cylindrical shells under lateral pressure loading. Report TT7003, Loughborough University of Technology, October 1970.
498. Crose, J. G. & Ang, A. H. S., A large deflection analysis method for elastic–perfectly plastic circular plates. Civ. Eng. Studies, Struct. Research Series No. 323, University of Illinois, 1967.
499. Yaghmai, S., Incremental analysis of large deformations in mechanics of solids with applications to axisymmetric shells of revolution. PhD Thesis University of California, Berkeley, 1969.
500. Irons, B. M., Quadrature rule of brick based finite-elements. *Int. J. Num. Methods Engineering*, **3** (1971) 293–4.
501. Hellen, T. K., Effective quadrature rules for quadratic solid isoparametric finite elements. *Int. J. Num. Methods. Engineering*, **4** (1972) 597–600.
502. Yamada, Y., Yoshimura, N. & Sakurai, T., Plastic stress–strain matrix and its application for the solution of elastic–plastic problems by the finite-element method. *Int. J. Mech. Sci.*, **10** (1968).
503. Owen, D. & Hinton, E., *Finite Elements in Plasticity Theory and Practice*. Pineridge Press, 1980.
504. Nayak, G. C. & Zienkiewicz, O. C., A generalization of various constitutive relations including strain softening. *Int. J. Num. Methods Engineering*, **5** (1972) 113–35.
505. Yagawa, G. & Ando, Y., Three-dimensional finite-element method of thermoelastoplasticity with creep effect. 2nd International Conference on Structural Mechanics in Reactor Technology (SMiRT), Berlin, September 1973.
506. Drucker, D. C., A more fundamental approach to plastic stress–strain relations. 1st US Congress of Applied Mechanics, ASME, New York, 1952, pp. 487–91.
507. Reckling, K. A., *Plastizitaetstheorie und ihre Anwendung auf Festigkeitsprobleme*. Springer Verlag, Berlin/Heidelberg/New York, 1967.
508. Foye, R. L., Theoretical post-yielding behaviour of composite laminates Part 1—inelastic micromechanics. *J. Comp. Mats*, **7** (1973) 178–93.
509. Lin, T. H., Salinas, D. & Ito, Y. M., Elastic–plastic analysis of unidirectional composites. *J. Comp. Mats*, **6** (1972) 48–60.
510. Heide & Zumsteg, Mentat A System for Finite Element Pre- and Post-processing. MARC, Zentech, London, 1968.
511. Davidson, I. & Barrett, N. T., Design philosophy and safety. International Conference on PCPV, Inst. Civ. Eng., London, Paper 6, 1967.
512. Freudenthal, A. M., Safety and probability of structural failure. *Trans. ASCE*, **121** (1956) 1337.
513. Freudenthal, A. M. et al., The analysis of structural safety. *J. Struct. Div., ASCE*, **92** (1966) 267–325.

514. Pugsley, A., Report on structural safety. *The Structural Engineer*, **34** (1955) 141.
515. Baker, M. J., The evaluation of safety factors in structures. CIRIA, Project 72, Final Report, Imperial College, London, 1970.
516. Baker, A. L. L., Safety pressure vessels. International Conference on PCPV, Inst. Civ. Eng., London, Group B, Paper 8, 1967.
517. Bangash, Y., Riddle of bonded and unbonded tendons in prestressed concrete reactor vessels. *Concrete*, **8** (1974) 46–8.
518. Paul, S. L., Structural behaviour of prestressed concrete reactor vessels. *J. Struct. Div., ASCE* (July 1971) 1897–6.
519. Ravindra, M. K. *et al.*, Illustration of reliability-based design. *J. Struct. Div., ASCE* (September 1974) 1789–811.
520. Lind, N. C., Deterministic formats for the probabilistic design of structures. University of Waterloo, Report 1, 1969.
521. Rosenblueth, E. *et al.*, Reliability basis for some Mexico codes. Probabilistic design of reinforced concrete buildings, Publication SP31, ACI, 1972.
522. GKN, Tests on GKN reinforcements. Private communication, 1970.
523. Baker, A. L. L., *The Ultimate Load Theory Applied to the Design of Reinforced and Prestressed Concrete Frames*. Concrete Publications Ltd, London, 1956.
524. Benjamin, J. R. & Cornell, A., *Probability, Statistics and Decisions for Civil Engineers*. McGraw Hill, New York, 1970.
525. European Code Bulletin d'Information CEB. (a) No. 50, 201–226 (July 1965); (b) No. 50, 174–200 (July 1965); (c) No. 56 (August 1966).
526. Corso, J. M., Discussions on earthquake codes. *Proc. ASCE*, **80** (1954).
527. Cave, M., Atomic Power Construction, Sutton, Surrey. Private communication on reactor safety, 1969.
528. Mischke, C. R., A method of relating factors of safety and reliability. *J. Engineering for Industry* (August 1970) 537–42.
529. Mischke, C. R., Implementing mechanical design to a reliability specification. Des. Eng. Tech. Conf. New York, CASME Publications, October 1974.
530. Mischke, M. R., Rationale for mechanical design to a reliability specification. Iowa State University, Ames, October 1974.
531. Drucker, D. C., Limit analysis of cylindrical shells under axially symmetric loadings. Proc. First Mid-Western Conf. on Solid Mechanics, USA 1953.
532. Onat, E. T., Plastic collapse of cylindrical shells under axially symmetrical loading. *Quart. App. Math.*, **13** (1959) 63–72.
533. Hodge, P. G., The rigid–plastic analysis of symmetrically loaded cylindrical shell. *J. Appl. Mech.*, **21** (1954) 336–42.
534. Hopkins, J. & Prager, W., The load carrying capacities of circular plates. *J. Mech. Phys. Solids*, **2** (1954).
535. Bangash, Y., Prestressed concrete reactor vessel. Time-saving ultimate load analysis. *J. Inst. Nucl. Eng.*, **13** (1972).
536. Bangash, Y., A basis for the design of bonded reinforcement in the prestressed concrete reactor vessels. Paper 7478S, Supplement (8), Inst. Civ. Eng., 1972.
537. Smee, D. J., The effect of aggregate size and concrete strength on the failure of concrete under multiaxial compression. Civil Engineering Transaction, Inst. of Engineers, Australia, Vol. CE9, No. 2, October 1967, pp. 339–44.

538. The Nuclear Power Group Company, Risley, Lancashire. Private communication, 1968.
539. Corum, J. M. & Smith, J. E., Use of small models in design and analysis of prestressed concrete reactor vessels. ORNL-4346, USA, May 1970.
540. Smith, J. R., Problems of assessing the correlation between the observed and predicted behaviour of models. Model Techniques for Prestressed Concrete Pressure Vessels, BNES, London, Paper 10, 1969.
541. Treharne & Davies, Tests on samples and calculation of standard deviations for bonded reinforcement. Private communication, 1969.
542. Paul, S. L. *et al.*, Mortar models of prestressed concrete reactor vessels. *J. Struct. Div.*, ASCE (1969).
543. Brown, G., Advances in the gas-cooled reactor. Lecture Joint Session of IEE, I MechE and BNES, University of Strathclyde, Glasgow, 16 October 1968 (available from British Nuclear Energy Society).
544. Cluckman, A. L., Reactor containment structures abroad. *J. Power Div.*, ASCE (1969) 61–76.
545. Lorenz, H., Design of the concrete containment vessel for the R.E. GINNA nuclear power plant. *Nucl. Eng. Des.*, **6** (1967) 360–6.
546. Wahl, W. W. *et al.*, Design and construction aspects of large prestressed concrete (PWR) containment structures. *ACI Journal* (May 1969) 400–12.
547. Halligan, D. W., Structural design criteria for secondary containment structures. *Nucl. Eng. Des.*, **8** (1968) 427–34.
548. Kulka, F., Prestressing systems for secondary and primary containment structures. *Nucl. Eng. Des.*, **8** (1968) 435–9.
549. Hofman, H., Spannbeton Reaclordruckbehalter mit Stutzbeton. *Nucl. Eng. Des.*, **8** (1968) 467–70.
550. Aymot, P., Construction of the containment building for the Gentilly nuclear power reactor. *Nucl. Eng. Des.*, **9** (1969) 479–87.
551. Harstead, G. A. *et al.*, Grouted tendons for nuclear containments. *J. Power Div.*, ASCE (October 1969) 277–92.
552. Chauvin, G. A., Post-Tensioned Concrete Nuclear Structures. Private Report No. 615, Fressynet, 1971.
553. Attala, I. & Nowatony, B., Missile impact on a reinforced concrete structure. *Nucl. Eng. Des.*, **37** (1976) 321–32.
554. Bannerjee, A. K. *et al.*, Design of reinforced concrete containments. 3rd International Conference on Structural Mechanics in Reactor Technology (SMiRT), Paper J3/1, London, 1975, pp. 1–12.
555. Green, D. E. *et al.*, Design of concrete containments for tangential shear loads. 3rd International Conference on Structural Mechanics in Reactor Technology (SMiRT), Paper J3/4, London, 1975, pp. 1–12.
556. Buchart, K. P. *et al.*, Analysis of reinforced concrete containment vessels considering concrete cracking, 3rd International Conference on Structural Mechanics in Reactor Technology (SMiRT), Paper J4/1, London, 1975, pp. 101–10.
557. Singh, M. P. *et al.*, Statistical aspects of containments tendon surveillance. Am. Soc. Civ. Eng., PO1, 1975, pp. 23–33.
558. Chu, K. Y., Prestress system for nuclear concrete containment domes. Am. Soc. Civ. Eng., PO1, 1974, pp. 111–23.

559. Bahar, L. Y. & Rice, J. S., Simplified derivation of the reaction time history in aircraft impact on a nuclear power plant. *Nucl. Eng. Des.*, **49** (1978) 263–8.
560. Bangash, Y., The structural integrity analysis of concrete containment vessels under external impacts. 6th International Conference on Structural Mechanics in Reactor Technology (SMiRT), Paper J7/6, Paris, 1981.
561. Bangash, Y., The automated three-dimensional cracking analysis of prestressed concrete containment vessels. 6th International Conference on Structural Mechanics in Reactor Technology (SMiRT), Paper H3/2, Paris, 1981.
562. Barr, P., Brown, M. L., Carter, P. G., Howe, W. P., Jowet, J., Neilson, A. J. & Young, R. L. D., Studies of missiles impact with reinforced concrete structures. BNES Symposium, 1980.
563. Bartlett, W. R. & Davies, I., Aircraft impact design for SGHWR containment. International Conference on Experience in the Design, Construction and Operation of PCPV, York, England, Paper 128/75, 1975.
564. Bathe, K. J. & Wilson, E. L., *Numerical Methods in Finite Element Analysis*. Prentice Hall, New York, 1976.
565. Brais, A. & Godbout, P., Modelling and simulation of local failures for reinforced concrete structures under concentrated impulsive loads. Private communication, 1978.
566. Bresler, B. & Pister, K. S., Strength of concrete under combined stresses. *J. Am. Concr. Inst.*, **55** (1957) 321–45.
567. Bresler, B. & Pister, K. S., Failure of plain concrete under combined stresses. *Trans. ASCE*, **122** (1957).
568. Brown, M. K., Curtsen, N. & Jowett, J., Local failure of reinforced concrete under missile impact loading. 5th International Conference on Structural Mechanics in Reactor Technology (SMiRT), Berlin, 1979.
569. Carlton, D. & Bedi, A., Theoretical study of aircraft impact on reactor containment structures. *Nucl. Eng. Des.*, **45** (1978) 197–206.
570. Cedolin, L., Crutzen, R. J. & Poli, J. D., Triaxial stress–strain relationship for concrete. *J. Eng. Mech. Div., ASCE*, **103** (1977) 423–39.
571. Kameswara, S. & Prasad, S., Impact loads on beams on elastic foundations. 3rd International Conference on Structural Mechanics in Reactor Technology (SMiRT), Paper J5/8, London, 1975.
572. Davis, I., Studies of the response of reinforced concrete structures to short duration dynamic loads. In: *Design for Dynamic Loading*, Construction Press, UK, 1982.
573. Chinn, J. & Zimmermann, R. M., Behaviour of plain concrete under various high triaxial compression loading conditions. Technical Report No. WLTR 64-163 (AD468460), Air Force Weapon Laboratory, New Mexico, 1965.
574. Constantopoulos, I. V. & Vardanega, C. & Attalla, I., Dynamic response to aircraft impact of reactor building with protective shell on independent foundation. 7th International Conference on Structural Mechanics in Reactor Technology (SMiRT), Paper J9/6, 1981.
575. Degen, P., Furrer, H. & Jemielewski, J., Structural analysis and design of a nuclear power plant building for aircraft crash effects. *Nucl. Eng. Des.*, **37** (1976) 241–68.
576. Dritler, K. & Guner, P., Calculation of the total force acting upon a rigid wall by projectiles. *Nucl. Eng. Des.*, **47** (1976) 231–44.

577. AEC takes long hard look at its regulation policies, *Nuclear News* (August 1969) 21–2.
578. Amirikian, A., Design of protective structures. Bureau of Yards and Docks, Publication No. NavDocks P-51, Department of the Navy, Washington DC, 1950.
579. A Seismic Design and Testing of Nuclear Facilities. Technical Reports, Series No. 88, International Atomic Energy Agency, June 1968.
580. Berg, G. V. & Stratta, J. L., Anchorage and the Alaska earthquake of March 27 1968. American Iron and Steel Institute, 1968.
581. Brown, T., Large capacity unbonded tendons. Presented at ACSE National Meetings on Water Resources Engineering, New Orleans, Louisiana, Preprint No. 841, 3–8 February 1969.
582. Burns, W. J., Kennedy, W. J. L. & Weinzimmer, F., Shoreham nuclear power station with concrete containment. *Proc. Am. Power Conf.*, **30** (1968) 365–70.
583. Cornell, C. A., Design seismic inputs. Seminar on Seismic Design for Nuclear Power Plants, MIT, 31 March–3 April 1969.
584. Fundamentals of protective design. Report AT 1297821, Army Corps of Engineers, Office of the Chief of Engineers, 1966.
585. General design criteria for nuclear power plant construction permit. Published in the *Federal Register*, 33 F.R. 19213, July 11 1967.
586. Gwaltney, R. C., Missile generation and protection in light-water-cooled power reactor plants. United States Atomic Energy Commission Report, ORNL-NSIC-22, 1968.
587. Kennedy, R. P., Effects of an aircraft crash into a concrete reactor containment building. Holmes and Narver Inc., Oakland, California, June 1966.
588. National Defence Research Committee. Effects of impact and explosion. Summary Technical Report of Division 2, Vol. 1, Washington DC, 1946.
589. Newmark, N. M., Design criteria for nuclear reactors subjected to earthquake hazards. Unpublished paper, 25 May 1967.
590. Report of pressure testing of reactor containment for Connecticut Yankee atomic power plant. Connecticut Yankee Atomic Power Company, Stone and Webster Engineering Corporation, October 1967.
591. Review of methods of mitigating spread of radioactivity from a failed containment system. United States Atomic Energy Commission, Report ORNL-NSIC-27, September 1968.
592. Riera, J. D., On the stress analysis of structures subjected to aircraft impact forces. *Nucl. Eng. Des.*, **8** (1968) 415–26.
593. Sharpe, R. L., Earthquake engineering for nuclear power plants. *Civil Engineering, ASCE* (March 1969) 39–43.
594. Spohn, H. R. & Waite, P. J., Iowa tornadoes. *Monthly Weather Review* (September 1962) 398–406.
595. Tan, C. P., A study of the design and construction practices of prestressed concrete and reinforced concrete containment vessels. United States Atomic Energy Commission, Report No. TID-24176, 1969.
596. Transcript of nuclear containment. Seminar co-sponsored by Power Engineering Magazine and Graver Tank and Manufacturing Co., Washington DC, 21 March 1969, Unpublished report.

597. Marti, J., Kalsi, G. & Attalla, I., Three-dimensional aircraft impact analysis. 7th International Conference on Structural Mechanics in Reactor Technology (SMiRT), Paper J9/7, 1983.
598. Labra, J., Protective structures response to vehicle impact. *J. Struct. Div., ASCE* (1979).
599. CEGB, A case for Sizewell B, Vol. 1. Sizewell Inquiry, 1981.
600. Zapp, F. C., Testing of containment systems used in light-water-cooled power reactors. United States Atomic Energy Commission, Report ORNL-NSIC-26, 1968.
601. Zudans, Z., Dynamic response to shell type structures subjected to impulsive mechanical and thermal loadings. *Nucl. Eng. Des.*, **3** (1966) 117–37.
602. Halligan, D. W., Structural design criteria for secondary containment structures. *Nucl. Eng. Des.*, **8** (1968) 427–34.
603. Richardson, K. et al., Dynamic analysis of footing on layered media. *J. Eng. Mech. Div., ASCE*, **101** (1975) 167–88.
604. Newmark, N. M. & Rosenblueth, E., *Fundamentals of Earthquake Engineering*, Prentice Hall, New York, 1971.
605. Hudson, D. E., Response spectrum techniques in engineering seismology. Proceedings of the World Conference on earthquakes engineering, Earthquake Engineering Research Institute, June 1956.
606. Housner, G. W., Behaviour of structures during earthquakes. *Proc. ASCE*, **85** (1959).
607. Newmark, N. M., Hall, W. J. & Noraz, B., Seismic design spectra for nuclear power plants. *J. Power Div., ASCE*, **99** (1973).
608. Carmichael, G. D. T. & Irving, J., Calculation of PCPV prestressing forces. *Nucl. Eng. Des.* (1968).
609. Elwi, A. A. & Murray, D. W., A three-dimensional hypo-elastic concrete constitutive relationship. *J. Eng. Mech. Div., ASCE*, **105** (1979) 623–41.
610. Filho, F. V., Coombi, R. F. & Beerreto, L. C., Design of the reinforced concrete containment shell of a nuclear reactor for aircraft impact. 7th International Conference on Structural Mechanics in Reactor Technology (SMiRT), Paper J9/11, 1979.
611. Gardner, N. J., Triaxial behaviour of concrete. *Proc. Am. Concr. Inst.*, **2** (1969) 136–47.
612. Gupta, Y. M. & Seaman, L., Local response of reinforced concrete to missile impacts. *Nucl. Eng. Des.*, **45** (1978) 507–14.
613. Halder, A., Turbine missile—A critical review. *Nucl. Eng. Des.*, **55** (1979) 293–304.
614. Hammel, J., Aircraft impact on a spherical shell. *Nucl. Eng. Des.*, **37** (1976) 206–23.
615. Hammel, J., Impact loading on a spherical shell. 5th International Conference on Structural Mechanics in Reactor Technology (SMiRT), 1979, Berlin.
616. Hannant, D. J. & Frederick, C. O., Failure criteria for concrete in compression. *Magazine of Concrete Research*, **20** (1969) 134–44.
617. Barber, E. M. et al., A study of air distribution in survival shelters using a small-scale modelling technique. USA National Bureau of Standards, Report 10689, 1972.

618. Barneby, H. L. et al., Toxic gas protection. Paper to ASHRAE Symposium on Survival Shelters, 1962; ASHRAE, 1963.
619. Barthel, R., Research on the climate in an underground shelter *Ingenieur*, **77** (1965) 143–7 (in Dutch).
620. Barthel, R., Theoretical and experimental research regarding the indoor climate in an underground shelter. *Ingenieur*, **77** (1965) 6143–7 (in Dutch).
621. Baschiere, R. J. et al., Analysis of above-ground fallout shelter ventilation requirements. *ASHRAE Journal*, **7** (1965) 88–90.
622. Baschiere, R. J. et al., Control of fallout shelter environments. ASHRAE Symposium on Survival Shelter Problems, 1970; ASHRAE, 1971.
623. Baschiere, R. J. et al., Development of manually powered pedal ventilator for shelter use. ASHRAE Symposium on Survival Shelter Problems, 1970; ASHRAE, 1971.
624. Breistein, M., Climate conditions in a survival shelter—ventilation and filtration. *Norsk VVS*, **10** (1967) 353–64 (in Norwegian).
625. Broido, A., Environmental hazards of fire. ASHRAE Symposium on Survival Shelters, 1962; ASHRAE, 1963.
626. Coal mine rescue survival system. *Heating and Ventilating Engineer*, **45** (1972) 353–4.
627. Cooper, J., After the bomb. *Journal of the Chartered Institution of Building Services*, **2** (1980) 48–9.
628. Dasler, A. R. & Minrad, D., Environmental physiology of shelter habitation. *ASHRAE Transactions*, **71** (1965) 115–24.
629. Drucker, E. E. & Cheng, H. S., Analogue study of heating in survival shelters. ASHRAE Symposium on Survival Shelters, 1962; ASHRAE, 1963.
630. Bigg, J. M., *Introduction to Structural Dynamics*. McGraw Hill, New York, 1964.
631. Ducar, G. J. & Engholm, G., Natural ventilation of underground fallout shelters. *ASHRAE Journal*, **7** (1965) 91–2.
632. Engle, P. M., Air filtration—biological and radiological. ASHRAE Symposium on Survival Shelters, 1962; ASHRAE, 1963.
633. Everetts, J. et al., Conditional air for improving shelter habitability. ASHRAE Symposium on Survival Shelter Problems, 1970.
634. Flannigan, F. M. & Morrison, C. A., Moisture storage in survival shelter structures. ASHRAE Symposium on Survival Shelter Problems, 1971.
635. Gates, A. S., Air revitalization in sealed shelters. ASHRAE Symposium on Survival Shelters, 1962; ASHRAE, 1963.
636. Gessner, H., The ventilation of air-raid shelters. *Schweizerische Blatter fur Heizung und Luftung*, **28** (1961) 1–12 (in German).
637. Hanna, G. M., Ventilation design for fallout and blast shelters. *Air Engineering*, **4** (1962) 19–21.
638. Heating and ventilation of nuclear fallout shelters. *Heating and Ventilating Engineer*, **36** (1962) 221–9.
639. Home Office, Domestic nuclear shelters (DNS)—Technical Guidance. HMSO, 1981.
640. Home Office, Domestic Nuclear Shelters. HMSO, 1981.
641. Hummell, J. D. & Beck, W. D., The study of a shelter cooling system using

methanol as the heat sink and as the fuel. Battelle Memorial Institute, AD 637953, 1966.
642. Hummell, J. D. et al., Survival shelter cooling: conventional and novel systems. *ASHRAE Journal*, **7** (1965) 35–42, 87.
643. Humphreyes, C. M. et al., Sensible and latent heat losses from occupants of survival shelters. *ASHRAE Transactions*, **72** (1966) 255–63.
644. Improving the conditions in fallout shelters. *Engineering* (26 January 1962) 158–9.
645. Kusuda, T. & Achenbach, P. R., Outdoor air psychrometric criteria for summer ventilation of protective shelters. *ASHRAE Transactions*, **71** (1965) 76–87.
646. Laine, P., Air conditioning system for a new type bomb shelter. *Rakennustekniikka*, **9128** (1972) 205–7 (in Finnish).
647. Moore, M., Building for the bomb. *Building*, **239** (1980) 39–42.
648. Morrison, C. A. & Flanigan, F. M., A graphical method for determining ventilation requirements for underground shelters. ASHRAE Symposium on Survival Shelter Problems, 1971.
649. Murphy, H. C., Ventilation on survival areas for civil defence. *Heating, Piping and Air Conditioning*, **23** (1951) 97–9; **23** (1951) 85–6, 90.
650. Nevins, R. G., Physiological aspects of survival. ASHRAE Symposium on Survival Shelters, 1962; ASHRAE, 1963.
651. Nykanen, J., Air and water services in air-raid shelters. *LVI*, **23** (1971) 8–12 (in Finnish).
652. Parry, T. H., Mechanical requirements of bomb shelters. *Air Conditioning, Heating and Ventilating*, **65** (1980) 108–10.
653. Rathman, C. E., Air distribution in shelters due to Kearney pumps and pedal ventilators. General American Research Division, Report No. 1476-1, 1970.
654. Rehm, F. R., Radiological instruments for shelters. *Air Engineering*, **4** (1962) 32–5.
655. Rickenbach, H., Heating, ventilation, cooling and sanitary engineering for large air-raid shelters. *Schweizerische Blatter fur Heizung und Luftung*, **28** (1961) 13–22 (in German).
656. Rosell, A., How underground workshops are made livable. *Heating, Piping and Air Conditioning*, **30** (1958) 118–20.
657. Rosell, A., Swedish air-raid shelter ventilation designed for atomic war emergency. *Heating, Piping and Air Conditioning*, **30** (1958) 136–9.
658. Rossetti, G., Technical installations in civil defence shelters. *Schweizerische Blatter fur Heizung und Luftung*, **37** (1970) 119–29 (in French).
659. Roubinet, M., Air conditioned anti-atomic shelters. *Industries Thermiques et Aerauliques*, **12** (1966) 301–31 (in French).
660. Scripture, D. G., How to engineer air for a municipal emergency center. *Air Engineering*, **4** (1962) 25–31.
661. Sisson, G. N., Underground for nuclear protection. *Underground Space*, **4** (1980) 341–8.
662. Small, S., Tests disclose serious defects in air-raid shelters. *Norsk VVS*, **15** (1972) 539–47 (in Norwegian).
663. Taylor, D. W. & Gonzales, J. O., Air distribution in multi-room shelter using

a package ventilation kit. US Army Florida Engineering Industry Experimental Station, AD 633916, 1965.
664. Trayser, D. A. & Flanigan, L. J., Shelter power systems: their effect on environmental control. *ASHRAE Transactions*, **72** (1966) 246–54.
665. Urdahl, T. H., Success of underground structures depends on air conditioning. *Heating, Piping and Air Conditioning*, **30** (1958) 108–11.
666. Van Biesen, J. A. H., Calculation of the climate in an underground shelter when the period of occupancy is limited. *Ingenieur*, **77** (1965) 6133–41 (in Dutch).
667. Ventilation for the piping in private air-raid shelters. *Sanitar Heizungs Technik*, **36** (1971) 732–3 (in German).
668. Viessman, W., Environment protection in subterranean shelters. *Refrigerating Engineering*, **59** (1951) 1175–8.
669. Viessman, W., How to plan air conditioning for protective shelters. *Heating Piping and Air Conditioning*, **26** (1954) 122–7; **26** (1954) 84–8.
670. Viessman, W., Protective shelters in modern warfare. *Heating, Piping and Air Conditioning*, **23** (1951) 77–82; **23** (1951) 111–16.
671. Weise, E., The time of residence inside ventilated shelters. *Sanitar Heizungs Technik*, **35** (1970) 143–4 (in German).
672. Wood, R., Blast protection valves shield building's Achilles heel. *Process Engineering* (October 1977) 88–9.
673. Wright, M. D. & Hill, E. L., Ventilation kit application studies. ASHRAE Symposium on Survival Shelter Problems, 1970; ASHRAE, 1971.
674. Yaglou, C. P., Limits for cold, heat and humidity in underground shelters. *Archives of Environmental Health*, **2** (1961) 110–15.
675. Moazami, M., Course work on nuclear shelters. Thames Polytechnic, London, 1982/83.
676. American Concrete Institute, ACI 313: recommended practice for design and construction of concrete bins, bunkers and silos for storing granular material, 1984.
677. Arnold, P. C., McLean, A. & Roberts, A., Bulk solids: storage, flow and handling. TUNRA Bulk Solids Handling Associates, 1982.
678. ACI Standard 313-77. Recommended practice for design and construction of concrete bins bunkers for storing granular materials. Second Printing 1982.
679. Brown, R. L. & Hawksley, P. G. W., The flow of granular material. *Fuel*, No. 26-159 (1947).
680. Butters, G., *Plastics Pneumatic Conveying and Bulk Storage*. Applied Science Publishers, 1981.
681. Durand, R. & Condolios, E., The flow of granular material through a circular orifice. *Revue Travaux (m)* (1980).
682. Enstad, G., On the theory of arching in mass flow hoppers. *Chem. Eng. Science*, **30** (1975) 1273–83.
683. Faber, J. & Alsop, D. J. A., Economics of reinforced concrete multibin grain silo configurations. *The Structural Engineer*, **60A** (1982) 121.
684. ISO 8456, Storage equipment for loose bulk materials (safety code), 1980.
685. Jenike, A. W., Johanson, J. R. & Carson, J. W., Bin loads—Part 2; concepts. *ASME J. Eng. for Ind.*, **95** (1973) 1–5.

686. Jenike, A. W., Johanson, J. R. & Carson, J. W., Bin loads—Part 3; mass flow bins. *ASME J. Eng. for Ind.*, **95** (1973) 6–12.
687. Jenike, A. W., Effect of solid flow properties and hopper configurations on silo loads. In: *Unit and Bulk Material Handling*. J. F. Loefler and C. R. Proctor, eds. ASME, New York, 1980.
688. Leonhardt, F., Boll, K. & Spiedel, E., The safe design of cement silos. Translation by Amerongen, C & CA, Translation No. 94, London, 1961.
689. McLean, A. G., Arnold, A. G. & Martin, A. G., Draft code for the evaluation of bin wall design. Association of Consulting Structural Engineers, Australia, September 1983.
690. Nguyen, T. V., Brenner, C. E. & Sabersky, R. H., Funnel flow in hoppers. *ASME J. Appl. Mech.*, **47** (1980) 729–35.
691. Nedderman, R. M. & Thorpe, R. B., The rate of discharge of powders. Seminar on Bunker/Feeder Interface Problems, Institution of Mechanical Engineers, September 1983.
692. Reimbert, M. & Reimbert, A., *Silos—Theory and Practice*. Trans. Tech. Publications, 1976.
693. Jenike, A. W., Johnson, J. R. & Carson, J. W., Bin loads, Part 1, concepts. ASME Paper No. 72-MH-1, 1975.
694. Jenike, A. W., Johnson, J. R. & Carson, J. W., Bin loads, Part 3, Mass Flow Bins. ASME Paper No. 72-HM-2, 1975.
695. Emery, R. B., Aeration apparatus converts bin from funnel flow to mass flow characteristics. ASME Paper No. 72-MH-5, ACI Standard 318 and 313, 1975.
696. API-620, Recommended rules for design and construction of large, welded, low pressure storage tanks, 1980.
697. IS-4996, Part 1 and Part 2, Criteria for design of reinforced concrete bins for the storage of granular and powdery materials. Indian Code, 1980.
698. Proposed ACI Standard (P 529), Recommended practice for design and construction of concrete bins and bunkers for storing granular materials and its commentary, October 1975.
699. Janson, Birkenhead Project, Private communication, 1983.
700. Janson, Analysis. Private communication, 1985.
701. DIN 1055, German codes on loads on bins and storage, 1964.
702. Soviet Code 313-77, Soviet code on loads on bins and storage, 1980.
703. Myers, J. J., Holm, C. H. & McAllister, R. F., *Design Considerations*, Handbook of Ocean and Underwater Engineering, McGraw Hill, New York, 1970.
704. Browne, R. D. & Domone, P. L. J., The long term performance of concrete in the marine environment. Off-shore Structures, Inst. Civ. Eng., Paper 5, 1949.
705. Bury, M. R. C. & Domone, P. L. J., The role of research in the design of concrete off-shore structures. Off-shore Technology Conference, Houston, Texas, Paper No. OTC 1949, 1974.
706. Netherlands Committee for Concrete Research, Cases of damage to corrosion of prestressing steel. CUR, July 1971.
707. Hove, K. & Foss, I., Quality assurance for off-shore concrete gravity structures. Off-shore Technology Conference, Houston, Texas, Paper No. OTC 2113, 1974.

708. API, Planning, designing and constructing fixed off-shore platforms, 1976.
709. Bishop, R. E. D. & Hassan, Y., The lift and drag forces on a circular cylinder in a flowing fluid. *Proc. Roy. Soc.*, **A277** (1963) 32–50.
710. Det Norske Veritas, Rules for the design, construction and inspection of fixed offshore structures. Oslo, Norway, 1974.
711. La Fond, E. C., Deep current measurements with the bathyscaphe 'Trieste'. *Deep Sea Res.*, **9** (1962) 115–16.
712. Nowroozi, A. *et al.*, Deep ocean current and its correlation with the ocean tide off the coast of Northern California. *J. Geophys. Res.*, **73** (1968) 1921–32.
713. Condeep safely on the bottom. *NCE* (July 1975) 11–13.
714. Cracks damage Condeep pride but not design integrity. *NCE* (July 1974) 10.
715. Special feature: North Sea oil. *NCE* (1973).
716. Special review: North Sea oil. *NCE* (May 1974).
717. Hirayama, K. I., Schwarz, J. & Wu, H. C., Effects of the ice thickness on ice forces. Off-Shore Technology Conference, Houston, Texas, Paper No. OTC 1974, 1974.
718. McDowell, D. N. & Holmes, P., General research problems in the design of off-shore structures. Off-Shore Structures, ICE, Paper No. 13, 1975.
719. Department of Energy, Guidance on the design and construction of off-shore installations. HMSO, 1974.
720. Shell Exploration, Design of floating concrete platform for deep water oil production. Private communication, London, 1974.
721. Skjelbreia, L. & Hendrickson, J., Fifth order gravity wave theory. Proc. 7th Conf. on Coastal Engineering, Vol. 1, Chapter 10, 1961.
722. Garrison, C. J. & Rao, V. S., Interaction of waves with submerged objects. *Journal of the Waterways, Harbours and Coastal Engineering Division, Proc. ASCE* (May 1961).
723. Boreel, L. J., Wave action on large off-shore structures. Off-Shore Structures, ICE, Paper No. 2, 1975.
724. Zienkiewicz, O. C. *et al.*, *Numerical Methods in Offshore Engineering*. Wiley, New York, 1982.
725. BBR Tanks Inc., BBR tanks in operation. In: *Prestressed Concrete Tanks*. Vols. 1 and 2—BBR Tanks Inc., USA.
726. Morrison, J. R., O'Brien, M. P., Johnson, J. W. & Scnaaf, S. A., The force exerted by surface wave on piles. *Petroleum Trans., AIME*, **189** (1950).
727. Hogben, N. & Standing, R. G., Wave loads on large bodies. Int. Symp. on the Dynamics of Marine Vehicles and Structures in Waves, London, April 1974.
728. Garrison, C. J., Torum, A., Iversen, C. & Leivseth, S., Wave forces on large volume structures—a comparison between theory and model tests. Off-shore Technology Conference, Houston, Texas, Paper No. OTC 2137, 1974.
729. Valenta, O., Kinetics of water penetration into concrete as an important factor of its deterioration and of reinforcement corrosion. Proc. International Symposium Durability of Concrete, Prague, A177-A193, 1969.
730. Taylor, R. E., Paper No. 5: Structural dynamics of off-shore platforms. Off-Shore Structures, Inst. Civ. Eng., Paper No. 5, 1975.
731. Horie, F. & Odaka, T., Consideration on effect of hysteresis damping for composite systems. Proc. 5th World Conference on Earthquake Engineering, Vol. II, Rome, 25–29 June 1973, pp. 1734–43.

732. Clausen, C. J. F., Stability problems related to off-shore gravity structures. Norwegian Geotechnical Institute, 1983.
733. Mansour, A. E. & Millman, D. N., Dynamic random analysis of fixed offshore platforms. Off-Shore Technology Conference, Houston, Texas, Paper No. OTC 2049, 1974.
734. Biggs, M. J., *Introduction to Structural Dynamics*, McGraw Hill, New York, 1964.
735. Fertis, D. G., *Dynamics and Vibration of Structures*, John Wiley & Sons, New York, 1973.
736. Preload Co. London, Design of LNG tanks. Private communications, 1971.
737. First, Second and Third Conferences on LNG Available, American Petroleum Institute, 1975, 1976, 1979.
738. J. Zarafa, The design of concrete LNG tanks. Partial fullfilment for the degree of BSc, CNAA, 1973.
739. Adesola, O. T., A feasibility study of non-conventional storage of liquefied natural gas. Final Year BSc (Hons) Project, Thames Polytechnic, London, 1983.
740. Bruggeling, A. S. G., *Prestressed Concrete for the Storage of Liquified Gases*. Viewpoint Publication No. 12.083, Eyre & Spottiswoode, London, 1981.
741. Trotter, H. G., Turner, F. H., Sullivan, P. J. E. & Brooks, W. T., Behaviour of prestressed concrete materials at very low temperatures. *Build International*, **8** (1975) 85–91.
742. Anchor, R. O., *Handbook on BS 5337; The Structural Use of Concrete for Retaining Aqueous Liquids*. Viewpoint Publications, 1976.
743. Bridon Wire, Cryogenic temperatures; the behaviour of carbon steel prestressing wire and strand. Technical Note No. 13, Bridon Wire, Doncaster, May 1970.
744. FIP, *Cryogenic Behaviour of Material for Prestressed Concrete*. Viewpoint Publications, 1983.
745. FIP, *Recommendations for the Design of Prestressed Concrete Oil Storage Tanks*. Viewpoint Publications, 1983.
746. *Prestressed Concrete for Storage of Liquified Gases*. Viewpoint Publications, 1982.
747. Turner, F. H., *Concrete and Cryogenics*. Viewpoint Publications, 1979.
748. Marchel, J. C., Variations in modulus of elasticity and Poissons ratio with temperature. ACI Publication, Sp. 35–27, 1972.
749. Bangash, Y., A three-dimensional rupture analysis of steel liners anchored to concrete pressure and containment vessels. *Eng. Fract. Mech.*, **28** (1987) 157–85.
750. Bangash, Y., The simulation of endochronic model in the cracking analysis of PCPV. 9th International Conference on Structural Mechanics in Reactor Technology (SMiRT), Lausanne, 1987, Vol. 4, pp. 333–40.
751. Al-Noury, S. I. & Bangash, Y., Prestressed concrete containment structures—circumferential hoop tendon calculation. 9th International Conference on Structural Mechanics in Reactor Technology (SMiRT), Lausanne, 1987, Vol. 4, pp. 395–401.
752. Petersson, J., Crack growth and development of fracture zones in plain concrete and similar materials. PhD Thesis, University of Lund, 1981.

753. Eibl, J., Behaviour of critical regions under soft missile impact and impulsive loading. IBAM, June 1982.
754. Hughes, G. & Speirs, D., An investigation of the beam impact problem. Technical Report 546, C and CA, 1982.
755. Bate, S., The effect of impact loading on prestressed and ordinary reinforced concrete beams. National Building Studies Research Paper 35, 1961.
756. Billing, I., *Structure Concrete*. MacMillan, London, 1960.
757. Watson, A. & Ang, T., Impact resistance of reinforced concrete structures. In: *Design of Dynamic Loading*, Construction Press, 1982.
758. Watson, A. & Ang, T., Impact response and post-impact residual strength of reinforced concrete structures. International Conference and Explosion of Structural Impact and Crashworthiness, July 1984.
759. Perry, S. & Brown, I., Model prestressed slabs subjected to hard missile loading. In: *Design for Dynamic Loading*, Construction Press, 1982.
760. Perry, S., Brown, I. & Dinic, G., Factors influencing the response of concrete slabs to impact. International Conference and Exposition of Structural Impact and Crashworthiness, July 1984.
761. Kufuor, K. & Perry, S., Hard impact of shallow reinforced concrete domes. International Conference and Exposition of Structural Impact and Crashworthiness, July 1984.
762. Burgess, W. & Campbell-Allen, D., Impact resistance of reinforced concrete as a problem of containment. School of Civil Engineering, The University of Sydney, Research Report No. R251, 1974.
763. Stephenson, A., Tornado-generated missile full-scale testing. Proc. Symposium on Tornadoes, Assessment of Knowledge and Implications for Man, Texas University, June 1976.
764. Jankov, Z., Turnahan, J. & White, M., Missile tests of quarter-scale reinforced concrete barriers. Proc. Symposium on Tornadoes, Assessment of Knowledge and Implications for Man, June 1976.
765. Stephen, A. & Silter, G., Full-scale tornado-missile impact tests. 4th International Conference on Structural Mechanics in Reactor Technology (SMiRT). Paper J10/1, Berlin, 1977.
766. Jonas, W. & Rudiger, E., Experimental and analytical research on the behaviour of reinforced concrete slabs subjected to impact loads. 4th International Conference on Structural Mechanics in Reactor Technology (SMiRT), Paper J7/6, Berlin, 1977.
767. Beriaud, C. *et al.*, Local behaviour of reinforced concrete walls under hard missile impact. 4th International Conference on Structural Mechanics in Reactor Technology (SMiRT), Paper J7/9, Berlin, 1977.
768. Gupta, Y. & Seaman, L., Local response of reinforced concrete to missile impacts. 4th International Conference on Structural Mechanics in Reactor Technology (SMiRT), Paper J10/4, Berlin, 1977.
769. Barr, P. *et al.*, An experimental investigation of scaling of reinforced concrete structures under impact loading. In: *Design for Dynamic Loading*, Construction Press, 1982.
770. Barr, P. *et al.*, Experimental studies of the impact resistance of steel faced concrete composition. 7th International Conference on Structural Mechanics in Reactor Technology (SMiRT), Paper J8/4, 1983.

771. Barr, P. et al., Studies of missile impact with reinforced concrete structures. Nuclear Energy, **19** (1980).
772. Neilson, A. Missile impact on metal structures. Nuclear Energy, **19** (1980).
773. Anderson, W., Watson, A. & Armstrong, P., Fibre reinforced concrete for the protection of structures against high velocity impact. International Conference and Exposition of Structural Impact and Crashworthiness, July 1984.
774. Berriaud, C. et al., Test and calculation of the local behaviour of concrete structures under missile impact. 5th International Conference on Structural Mechanics in Reactor Technology (SMiRT), Paper J7/1, Berlin, 1979.
775. Brandes, K., Limberger, E. & Hertes, J., Experimental investigation of reinforced concrete behaviour due to impact load. 5th International Conference on Structural Mechanics in Reactor Technology (SMiRT), Paper J7/3, Berlin, 1979.
776. Sage, E. & Pfeiffer, A., Response of reinforced concrete targets to impacting soft missile. 5th International Conference on Structural Mechanics in Reactor Technology (SMiRT), Paper J8/4, Berlin, 1979.
777. Jonas, W. et al., Experimental investigations to determine the kinetic ultimate capacity of reinforced concrete slabs subject to deformable missiles. 5th International Conference on Structural Mechanics in Reactor Technology (SMiRT), Paper J8/4, Berlin, 1979.
778. Dulac, J. & Giraud, J., Impact testing of reinforced concrete slabs. 6th International Conference on Structural Mechanics in Reactor Technology (SMiRT), Paper J7/1, Paris, 1981.
779. Chiba, N. et al., Nonlinear dynamic analysis of steel plates subjected to missile impact. 6th International Conference on Structural Mechanics in Reactor Technology (SMiRT), Paper J7/4, Paris, 1981.
780. Ohte, S. et al., The strength of steel plates subjected to missile impact. 6th International Conference on Structural Mechanics in Reactor Technology (SMiRT), Paper J7/10, Paris, 1981.
781. Yamamoto, S. et al., Nonlinear analysis of missile impact on steel plates. 7th International Conference on Structural Mechanics in Reactor Technology (SMiRT), Paper J7/3, 1983.
782. Woodfin, R. & Sliter, G., Modelling and contact of turbine missile concrete impact experiments. 6th International Conference on Structural Mechanics in Reactor Technology (SMiRT), Paper J8/1, Paris, 1981.
783. Woodfin, R. & Sliter, G., Results of full-scale turbine missile concrete impact experiments. 6th International Conference on Structural Mechanics in Reactor Technology (SMiRT), Paper J8/2, Paris, 1981.
784. Rudiger, E. & Riech, H., Experimental and theoretical investigations on the impact of deformable missiles onto reinforced concrete slabs. 7th International Conference on Structural Mechanics in Reactor Technology (SMiRT), Paper J8/3, 1983.
785. Brandes, K., Limberger, E. & Herter, J., Strain rate dependent energy absorption capacity of reinforced concrete members under aircraft impact. 7th International Conference on Structural Mechanics in Reactor Technology (SMiRT), Paper J9/5, 1983.
786. Gueraud, R. et al., Study of the perforation of reinforced concrete slabs by

rigid missiles—general introduction and experimental study, Part 1. *Nucl. Eng. Des.*, **41** (1977).
787. Figuet, G. & Dacquet, S., Study of the perforation of reinforced concrete slabs by rigid missiles—experimental study, Part II. *Nucl. Eng. Des.*, **41** (1977).
788. Goldstein, S., Berriaud, C. & Labrot, R., Study of the perforation of reinforced concrete slabs by rigid missiles—experimental study, Part III. *Nucl. Eng. Des.*, **41** (1977).
789. Davidson, I. & Bradbury, J., The analysis of dynamically loaded non-linear structures. *Nucl. Eng. Des.*, **45** (1978).
790. Gupta, Y. & Seamae, L., Local response of reinforced concrete to missile impact. *Nucl. Eng. Des.*, **45** (1978).
791. Limberger, E., A simple mode for predicting energy dissipation of thin plates being perforated by hard missiles. *Nucl. Eng. Des.*, **51** (1979).
792. Stefanou, G., An investigation into the behaviour of perforated slabs for concrete reactor vessels under temperature and external load. *Nucl. Eng. Des.*, **52** (1979).
793. Bignon, P. & Riera, J., Verification of methods of analysis of soft missile impact problems. *Nucl. Eng. Des.*, **60** (1980).
794. Albertini, C. & Montagnani, M., Constitutive laws of materials in dynamics—outline of a program of testing on small and large specimens for containment of extreme dynamic loading conditions. *Nucl. Eng. Des.*, **68** (1981).
795. Nachtsheim, W. & Stangenberg, F., Investigation of results of Meppen slab tests—comparison with parametric investigation. *Nucl. Eng. Des.*, **75** (1982).
796. Kussmaul, K., The investigation of the tensile and notch impact bend test into an experimentally validated fracture mechanics concept. *Nuc. Eng. Des.*, **72** (1982).
797. Riera, J., Basic concept and load characteristic in impact problem. IBAM, June 1982.
798. Degen, P., Perforation of reinforced concrete slabs by rigid missiles. *J. Struct. Div., ASCE*, **106** (1980).
799. Ghaboussi, J., Millavec, W. & Jsenberg, J., Reinforced concrete structures under impulsive loading. *J. Struct. Div., ASCE* (March 1984).
800. Zerna, W. & Stangenberg, F., On the shock behaviour of reinforced concrete structural systems. IBAM, June 1982.
801. Douglas, R. & Bingham, W., Strains and wave velocities in high velocity impact. Proc. Second Annual Engineering Mechanics Division Speciality Conference, May 1977.
802. Broman, R. *et al*, Report on the ASCE committee on impactive and impulsive loads. *Civil Engineering and Nuclear Power*, **V** (1980).
803. Broman, R. *et al.*, Analysis and design for impact loads. *Civil Engineering and Nuclear Power*, **V** (1980).
804. Rotz, J., Evaluation of tornado missile impact effects on structures. Proc. Symposium on Tornadoes, Assessment of Knowledge and Implications for Man, June 1976.
805. Baker, A., Partition of impact energy between local deformation and target response. *Nuclear Energy*, **19** (1980).

806. Jonker, J., Dynamic response of a clamped/free hollow circular cylinder under travelling torsional impact loads. *Nucl. Eng. Des.*, **67** (1981).
807. Broman, R. *et al.*, Analysis and design for impulsive loads. *Civil Engineering and Nuclear Power*, **V** (1980).
808. Chang, W., Burdette, E. & Barnett, R., Missile penetration. *J. Struct. Div.*, *ASCE* (1976).
809. Chang, W., Impact of solid missiles on concrete barriers. *J. Struct. Div.*, *ASCE* (February 1981).
810. Kar, A., Barrier design for tornado generated missiles. 4th International Conference on Structural Mechanics in Reactor Technology (SMiRT), Paper J10/3, Berlin, 1977.
811. Kar, A., Local effects of tornado generated missiles. *J. Struct. Div. ASCE* (May 1978).
812. Kar, A., Projectile penetration into steel. *J. Struct. Div.*, *ASCE* (October 1979).
813. Kar, A., Impactive effects of tornado missiles and aircrafts. *J. Struct. Div.*, *ASCE* (November 1979).
814. Kar, A., Projectile penetration into buried structures. *J. Struct. Div.*, *ASCE* (January 1978).
815. Kar, A., Impact load for tornado-generated missiles. *Nucl. Eng. Des.*, **47** (1978).
816. Kar, A., Loading time history for tornado-generated missiles. *Nuc. Eng. Des.*, **51** (1979).
817. Sliter, G., Assessment of empirical concrete impact formulae. *J. Struct. Div.*, *ASCE* (May 1980).
818. Halder, A. & Miller, F., Penetration depth in concrete for non-deformable missiles. *Nucl. Eng. Des.*, **71** (1982).
819. Haldar, A., Hatami, M. & Miller, F., Concrete structures: penetration depth estimation for concrete structures. *J. Struct. Div.*, *ASCE* (January 1983).
820. Haldar, A., Hatami, M. & Miller, F., Penetration and spalling depth estimation for concrete structures. 7th International Conference on Structural Mechanics in Reactor Technology (SMiRT), Paper J7/2, 1983.
821. Berriaud, C. *et al.*, Local behaviour of reinforced concrete walls under missile impact. *Nucl. Eng. Des.*, **45** (1978).
822. Dubois, J., Chedmail, J. & Bianchini, J., Numerical analysis of impact-penetration problems for nuclear reactor safety. 4th International Conference on Structural Mechanics in Reactor Technology (SMiRT), Paper J7/4, Berlin, 1977.
823. Drittler, K., Gruner, P. & Krivy, J., Calculation of forces arising from impacting projectiles upon yielding structures. 4th International Conference on Structural Mechanics in Reactor Technology (SMiRT), Paper J7/4, Berlin, 1977.
824. Alderson, M. *et al.*, Reinforced concrete behaviour due to missile impact. 4th International Conference on Structural Mechanics in Reactor Technology (SMiRT), Paper J7/7, Berlin, 1977.
825. McMahon, P., Sen, S. & Meyers, B., Behaviour of reinforced concrete barriers subject to the impact of turbine missiles. 5th International

Conference on Structural Mechanics in Reactor Technology (SMiRT), Paper J7/6, Berlin, 1979.
826. Stangenberg, F. & Buttmann, P., Impact testing of steel fibre reinforced concrete slabs with liner. 5th International Conference on Structural Mechanics in Reactor Technology (SMiRT), Paper J7/5, Berlin, 1979.
827. Haldar, A., Turbine missile—a critical review. *Nucl. Eng. Des.*, **55** (1979).
828. Haldar, A., Impact loading—damage predicting equations. 6th International Conference on Structural Mechanics in Reactor Technology (SMiRT), Paper J8/4, Paris, 1981.
829. Davis, I., Design and analysis of concrete structures under impact and impulsive loading. IBAM, June 1982.
830. Healing, J., Local effects of missile impact on protective barriers. Proc. Symposium on Tornadoes, Assessment of Knowledge and Implications for Man, June 1976.
831. Yakeda, J., Crashworthiness of concrete structures subject to impact or explosion. International Conference and exposition of Structural Impact and Crashworthiness, July 1984.
832. Zielinski, A., Fracture of concrete under impact loading. International Conference and Exposition of Structural Impact and Crashworthiness, July 1984.
833. Zorn, N. & Reinhardt, H., Concrete structures under high intensity tensile waves. International Conference and Exposition of Structural Impact and Crashworthiness, July 1984.
834. Tulacz, J. & Smirth, R., Assessment of missile generated by pressure component failure and its application to recent gas-cooled nuclear plant design. *Nuclear Energy*, **19** (1980).
835. Nashtsheim, W. & Stangenberg, F., Impact of deformable missiles on reinforced concrete plates—comparisonal calculation of Meppens tests. 6th International Conference on Structural Mechanics in Reactor Technology (SMiRT), Paper J7/3, Paris, 1981.
836. Nashtshein, W. & Stangenberg, F., Selected result of Meppen slab tests—state of interpretation, comparison with computation investigation. 7th International Conference on Structural Mechanics in Reactor Technology (SMiRT), Paper J8/1, 1983.
837. Davis, I., Damaging effects from the impact of missiles against reinforced concrete structures. *Nuclear Energy*, **19** (1980).
838. Kennedy, R., A review of procedures for the analysis and design of concrete structures to resist missile impact effects. *Nucl. Eng. Des.*, **37** (1976).
839. Attalla, I., Missile impact on reinforced concrete structures. *Nucl. Eng. Des.*, **37** (1976).
840. Hughes, G., Hard missile impact on reinforced concrete. *Nucl. Eng. Des.*, **77** (1984).
841. Reyen, J., Villafune, E. & Crutzen, Y., Impulsive loading on concrete structures. 6th International Conference on Structural Mechanics in Reactor Technology (SMiRT), Paper J10/1, Paris, 1981.
842. Bangash, Y., The limit state analysis of a prestressed concrete containment vessel for PWR. 7th International Conference on Structural Mechanics in Reactor Technology (SMiRT), Paper J3/8, 1983.

843. Bangash, Y., Containment vessel design and practice. *Nuclear Energy*, **11** (1982).
844. Krulzik, N., Analysis of aircraft impact problem. In: *Advanced Structural Dynamics*, Applied Science Publishers, London, 1980.
845. Glasstone, S. & Dolan, P. J., The effect of nuclear weapons (3rd edn). US Govt Printing Office, Washington DC 1977.
846. Rice, J. & Bahar, L., Reaction-time relationship and structural design of reinforced concrete slabs and shells for aircraft impact. 3rd International Conference on Structural Mechanics in Reactor Technology (SMiRT), Paper J5/3, London, 1975.
847. Bahar, Y. & Rice, S., Simplified derivation of the reaction–time history in aircraft impact on a nuclear power plant. *Nucl. Eng. Des.*, **49** (1978).
848. Sharpe, R., Ramil, H. & Scunlan, R., Analysis of aircraft impact on reactor building. 3rd International Conference on Structural Mechanics in Reactor Technology (SMiRT), Paper J5/4, London, 1975.
849. Zerna, W., Schnellenbach, G. & Stangenberg, F., Optimized reinforcement of nuclear power plant structures for aircraft impact forces. *Nucl. Eng. Des.*, **37** (1976).
850. Meder, G., Dynamic response of a soft elastic–plastic systems subjected to aircraft impact pulses. *Nucl. Eng. Des.*, **47** (1982).
851. Schnellenbach, G. & Stangenberg, F., Design of concrete containments for aircraft impact. 4th International Conference on Structural Mechanics in Reactor Technology (SMiRT), Paper J8/4, Berlin, 1977.
852. Riera, J., A critical reappraisal of nuclear power plant safety against accidental aircraft impact. *Nucl. Eng. Des.*, **57** (1980).
853. Steveson, J., Current summary of international extreme load design requirements for nuclear power plant facilities. *Nucl. Eng. Des.*, **60** (1980).
854. Hornyk, K., Analytical modelling of the impact of soft missiles on protective walls. 4th International Conference on Structural Mechanics in Reactor Technology (SMiRT), Paper J7/3, Berlin, 1977.
855. Stoykovich, M., Impact load time histories for viscoelastic missiles. 4th International Conference on Structural Mechanics in Reactor Technology (SMiRT), Paper J7/5, Berlin, 1977.
856. Chiapetta, R. & Costello, J., Automobile impact forces on concrete walls. 6th International Conference on Structural Mechanics in Reactor Technology (SMiRT), Paper J8/8, Paris, 1981.
857. Porter, W., Generation of missiles and destructive shock fronts and their consequences. *Nuclear Energy*, **19** (1980).
858. McMahon, P., Meyers, B. & Buchert, K., The behaviour of reinforced concrete barriers to the impact of tornado generated deformable missiles. 4th International Conference on Structural Mechanics in Reactor Technology (SMiRT), Paper J10/2, Berlin, 1977.
859. Jones, W. *et al.*, Approximate calculation of the impact of missiles onto reinforced concrete structures and comparison of test results. 5th International Conference on Structural Mechanics in Reactor Technology (SMiRT), Paper J8/6, Berlin, 1979.
860. Hanagud, S. & Bernard, R., A penetration theory for axisymmetric

projectiles. 4th International Conference on Structural Mechanics in Reactor Technology (SMiRT), Paper J7/10, Berlin, 1977.
861. Brown, M., Curtress, N. & Jewett, J., Local failure of reinforced concrete under missile impact loading. 5th International Conference on Structural Mechanics in Reactor Technology (SMiRT), Paper J8/7, Berlin, 1979.
862. Jamet, P. et al., Perforation of a concrete slab by a missile: finite element approach. 7th International Conference on Structural Mechanics in Reactor Technology (SMiRT), Paper J7/6, 1983.
863. Wittmann, F. & Boulahdour, T., Variability of resistance of concrete slabs under impact load. 6th International Conference on Structural Mechanics in Reactor Technology (SMiRT), Paper J7/2, Paris, 1981.
864. Stangenberg, F. & Zerna, W., Extreme load resistance design of nuclear power plant structures. 4th International Conference on Structural Mechanics in Reactor Technology (SMiRT), Paper J8/5, Berlin, 1977.
865. Munday, G., Initial velocities attained by plant generated missile. *Nuclear Energy*, **19** (1980).
866. Yen, G., Probability of a containment of turbine missile. *Nucl. Eng. Des.*, **37** (1976).
867. Twisdale, L., Dunn, W. & Davies, T., Tornado missile transport analysis. *Nucl. Eng. Des.*, **51** (1979).
868. Goodman, J. & Koch, J., The probability of a tornado missile hitting a target. *Nucl. Eng. Des.*, **75** (1983).
869. Haldar, A. & Hamieh, H., Local effects of solid missiles on concrete structures. *J. Struct. Div., ASCE* (May 1984).
870. Costello, J., Simiu, E. & Cordes, M., Tornado borne missile speed. *J. Struct. Div., ASCE* (June 1977).
871. Watwood, V., Jr., The finite element method for prediction of crack behaviour. *Nucl. Eng. Des.*, **11** (1969).
872. Hopkirk, R., Lympany, S. & Marti, J., Three-dimensional numerical predictions of impact effects. *Nuclear Energy*, **19** (1980).
873. Ahmad, M. & Bangash, Y., A three-dimensional bond analysis using finite element. *Computers and Structures*, **25** (1987) 281–96.
874. Hodge, P. G., The theory of piece-wise linear isotropic plasticity. IUTAM Colloquium, Madrid, 1955.
875. Martin, H. C. & Carrey, G. F., *Introduction to Finite Element Analysis*. McGraw Hill, New York, 1973.
876. Bathe, K. J. & Wilson, E. L., *Numerical Methods in Finite Element Analysis*. Prentice-Hall, Englewood Cliffs, New Jersey, 1972.
877. Desai, C. S. & Abel, J. F., *Introduction to the Finite Element Method*. Van Nostrand Reinhold, New York, 1972.
878. Ahmad, S., Curved finite element in the analysis of solid shell and plate structures. PhD Thesis, University of Wales, Swansea, 1969.
879. Clough, R. W., Comparison of three-dimensional finite elements. Proc. Symposium on Application of Finite Element Methods in Civil Engineering, Vanderbilt University, 1969.
880. Scotto, F. L., Thin walled 1:20 prestressed concrete pressure vessel model for THTR reactor type. 2nd International Conference on Structural Mechanics in Reactor Technology (SMiRT), Paper H5/6, Berlin, 1973.

881. Ottosen, N. S. & Andersen, S.I., Theoretical and experimental studies for optimisation of PCRV top closures. 3rd International Conference on Structural Mechanics in Reactor Technology (SMiRT), Paper H3/6, London, 1975.
882. Langan, D. & Smith, J. R., 1/10th scale model of Hartlepool pressure vessel—introduction statement. Structural Research Laboratory, Taylor Woodrow Construction Ltd, November 1968.
883. Houde, J., Study of force–displacement relationships for the finite element analysis of reinforced concrete. Report No. 73-2, Department of Civil Engineering and Applied Mechanics, McGill University, Montreal, December 1973.
884. Houde, J. & Mirza, M. S., A finite element analysis of shear strength of reinforced concrete beams. In: *Shear in Reinforced Concrete*, Vol. 1, Special Publication SP-42, American Concrete Institute, Detroit, Michigan, 1974.
885. Houde, J. & Mirza, M. S., Investigation of shear transfer across cracks by aggregate interlock. Research Report No. 72-06, Ecole Polytechnique de Montreal, Department of Genie Civil, Division de Structures, 1972.
886. Fort St. Vrain Nuclear Generating Station, Final Safety Analysis Report, Appendix E. PCRV Data, Public Service Co. of Colorado Report, 1979.
887. Freudenthal, A. M. & Roll, F., Creep and creep recovery of concrete under high compressive stress. *J. Am. Concr. Inst.*, **54** (1958) 1111–42.
888. Hungspreug, S., The local interaction between reinforcing bars and concrete under high-level reversed cyclic loading. Thesis presented to Cornell University in partial fulfillment of the requirements for the degree of Doctor of Philosophy, 1981.
889. Herrmann, L. R., Finite element analysis of contact problems. *J. Eng. Mech. Division, ASCE*, **104** (1978).
890. Hatt, W. K., Notes on the effect of time element in loading reinforced concrete beams. *Proc. ASTM*, **7** (1907) 421–33.
891. Van der Houwen, P. J., *Construction of Integration Formulas for Initial Value Problems*. North Holland, Amsterdam, 1977.
892. Huet, C., Application of Bazant's algorithm to the analysis of viscoelastic composite structures. *Materials and Structures* (RILEM, Paris), **13** (1980) 91–8.
893. Gilbert, R. I. & Warner, R. F., Nonlinear analysis of reinforced concrete slabs with tension stiffening. Univiv Report R-167, University of New South Wales, January 1977.
894. Gilbert, R. I. & Warner, R. F., Time dependent behaviour of reinforced concrete slabs. Univiv Report No. R-173, University of New South Wales, October 1977.
895. Grootenboer, H. J., Finite element analysis of two dimensional reinforced concrete structures, taking account of nonlinear physical behaviour and the development of discrete cracks. Doctoral Dissertation, Technical University of Delft, March 1979.
896. Fanning, D. N. & Dodge, W. G., Methods of calculating inelastic response of prestressed concrete pressure vessels. 5th International Conference on Structural Mechanics in Reactor Technology (SMiRT), Berlin, August 1979.
897. Echeverria, A., Mohraz, B. & Schnobrich, W. C., Crack development in a

prestressed concrete reactor vessel as determined by a lumped parameter method. *Nucl. Eng. Des.*, **11** (1970) 286–94.
898. Bertero, V. V., Bresler, B. & Liao, H., Stiffness degradation of reinforced concrete members subjected to cyclic flexural loads. Earthquake Engineering Research Center, Report No. EERC 69-12, University of California, Berkeley, 1969.
899. Bertero, V. V., Experimental studies concerning reinforced, prestressed and partially prestressed concrete structures and their elements. Symposium on Resistance and Ultimate Deformability of Structures Acted on by Well Defined Repeated loads, IABSE, Lisbon, 1973.
900. Bertero, V. V., Popov, E. P. & Wang, T. Y., Hysteretic behaviour of reinforced concrete flexural members with special web reinforcement. Earthquake Engineering Research Center, Report No. EERC 74–9, University of California, Berkeley, 1974.
901. Bertero, V. V. & Popov, E. P., Hysteretic behaviour of ductile moment-resisting reinforced concrete frame components. Earthquake Engineering Research Center, Report No. EERC 75–16, University of California, Berkeley, 1975.
902. Bertero, V. V. & Popov, E. P., Seismic behaviour of moment-resisting reinforced concrete frames. In: *Reinforced Concrete Structures in Seismic Zones*. SP-53, American Concrete Institute, Detroit, Michigan, 1977.
903. Backlund, J., Finite element analysis of nonlinear structures. Doctoral Thesis, Chalmers Technical University, Goteborg, Sweden, November 1973.
904. Becker, J. M. & Bresler, B., FIRES-RC—A computer program for the fire response of structures—Reinforced concrete frames. Report No. UCB FRG 74-3, Fire Research Group, Structural Engineering and Structural Mechanics, Department of Civil Engineering, University of California, Berkeley, July 1974.
905. Bell, J. C., A complete analysis of reinforced concrete slabs and shells. PhD Dissertation, Department of Civil Engineering, University of Canterbury, Christchurch, New Zealand, 1970.
906. Bell, J. C. & Elms, D. G., A finite element post elastic analysis of reinforced concrete shells. *Bulletin of the International Association for Shell and Spatial Structures*, No. 54 (April 1974).
907. Bell, J. C. & Elms, D., Partially cracked finite elements, *J. Struct. Div., ASCE*, **97** (1971).
908. Browne, R. D. & Bamforth, P. P., The long term creep of the Wylfa P. V. concrete for loading ages up to $12\frac{1}{2}$ years. 3rd International Conference on Structural Mechanics in Reactor Technology, London, Paper H1/8, September 1975.
909. Browne, R. D. & Blundell, R., The influence of loading age and temperature on the long term creep behaviour of concrete in a sealed moisture stable state. *Materials and Structures* (RILEM, Paris), **2** (1969) 133–44.
910. Sharma, N. K., Splitting failures in reinforced concrete members. PhD Thesis, Department of Structural Engineering, Cornell University, 1969.
911. Taylor, H. P. J., The fundamental behaviour of reinforced concrete beams in bending and shear. In: *Shear in Reinforced Concrete*, Vol. 1, SP-42, American Concrete Institute, Detroit, Michigan, 1974.

912. Taylor, H. P. S., Investigation of the forces carried across cracks in reinforced concrete beams in shear by interlock of aggregate. Cement and Concrete Association, London, Report No. 42-447, November 1970.
913. Taylor, R., A note on the mechanism of diagonal cracking in reinforcement. *Magazine of Concrete Research*, **11** (1959) 151-8.
914. Teller, L. W. & Cashell, H. D., Performance of doweled joints under repetitive loading. Bureau of Public Roads, April 1958.
915. Rostasy, F. S., Teichen, K. Th. & Engleke, H., Beitrag zur Klarung der Zusammenhanges von Kriechen und Relaxation bei Normal-beton. *Amtliche Forschungs-und Materialprufungsanstalt fur das Bauwesen*. Otto-Graf-Institut, Universi*at Stuttgart, Strassenbau und Strassenverkehrstechnick, Heft 139, 1972.
916. Ruetz, W., An hypothesis for the creep of hardened cement paste and the influence of simultaneous shrinkage. International Conference on the Structure of Concrete, London, 1965; Cement and Concrete Association, 1968, pp. 365-87. (See also Deutscher Ausschuss fur Stahlbeton Heft 183, 1966.)
917. Oesterle, R. G. & Russel, H. G., Shear transfer in large scale reinforced concrete containment elements. Report for US Nuclear Regulatory Commission, NUREG/CR-1374, Construction Technology Laboratories, Portland Cement Association, April 1980.
918. Pauley, T., Design aspects of shear walls for seismic areas. Department of Civil Engineering, Research Report No. 71-11, University of Canterbury, Christchurch, New Zealand, October 1974.
919. Pauley, T. & Loeber, P. S., Shear transfer by aggregate interlock. In: *Shear in Reinforced Concrete*, Vol. 1, SP-42, American Concrete Institute, Detroit, Michigan, 1974.
920. Pauley, T., Park, R. & Phillips, M. H., Horizontal construction joints in cast-in-place reinforced concrete. In: *Shear in Reinforced Concrete*, Vol. 2, American Concrete Institute, Detroit, Michigan, 1974.
921. Mahaidi, A. L., Nonlinear finite element analysis of reinforced concrete deep members. Thesis presented to Cornell University, in partial fulfillment of the requirements for the degree of Doctor of Philosophy, 1979.
922. Aktan, H. M. & Hanson, R. D., Nonlinear cyclic analysis of reinforced concrete plane stress members. *J. Am. Concr. Inst.*, SP-63, No. 6 (1980).
923. Arutyunian, N. Kh., *Some Problems in the Theory of Creep*. Technteorizdat, Mozcow, 1952 (in Russian); Engl. trans. Pergamon Press, 1966.
924. Agrawal, A. B., Jaeger, L. G. & Mufti, A. A., Crack propagation and plasticity of reinforced concrete shear-wall under monotonic and cyclic loading. Conference on Finite Element Methods in Engineering, Adelaide, Australia, December 1976.
925. Agrawal, A. B., Nonlinear analysis if reinforced concrete planar structures subjected to monotonic, reversed cyclic and dynamic loads. PhD Thesis, University of New Brunswick, Fredericton, Canada, March 1977.
926. Agrawal, A. B., Mufti, A. A. & Jaeger, L. G., A program for the nonlinear analysis of reinforced concrete planar structures subject to monotonic and reversed cyclic loads. School of Computer Science, Acadia University, Wolfsville, N. S., Canada, Report No. 1, April 1977.

927. Agrawal, A. B., Mufti, A. A. & Jaeger, L. G., A program for the nonlinear analysis of reinforced concrete structures underground accelerations. School of Computer Science, Acadia University, Wolfsville, N.S., Canada, Report No. 2, April 1977.
928. Mikkola, M. J. & Schnobrich, W. C., Material behaviour characteristics of reinforced concrete shells stressed beyond the elastic range. Civil Engineering Studies, SRS No. 367, University of Illinois at Urbana-Champaign, Urbana, Illinois, August 1970.
929. Marcus, H., Load carrying capacity of dowels at transverse pavements joints. *J. Am. Conc. Inst., Proc.*, **48** (1951).
930. Mattock, A. H., Effect of moment and tension across the shear plane on single direction shear transfer strength in monolithic concrete. Report No. SM-74-3, Department of Civil Engineering, University of Washington, Seattle, Washington, October 1974.
931. Mattock, A. H., Shear transfer in concrete having reinforcement at an angle to the shear plane. In: *Shear in Reinforced Concrete*, Vol. 1, SP-42, American Concrete Institute, Detroit, Michigan, 1974.
932. Mattock, A. H., Shear transfer under cyclically reversing loading across an interface between concretes cast at different times. Report No. SM-77-1, Department of Civil Engineering, University of Washington, Seattle, Washington, November 1974.
933. Mattock, A. H., Shear transfer under monotonic loadings across an interface between concrete cast at different times. Report No. SM-76-3, Department of Civil Engineering, University of Washington, Seattle, Washington, September 1976.
934. Mattock, A. H., The shear transfer behaviour of cracked monolithic concrete subject to cyclically reversing shear. Report No. SM-74-4, Department of Civil Engineering, University of Washington, Seattle, Washington, November 1974.
935. Mattock, A. H. & Hawkins, N. M., Shear transfer in reinforced concrete—recent research. *J. Prestressed Concrete Institute*, **17** (1972) 55–75.
936. McDonald, J. E., An experimental study of multiaxial creep in concrete. In: *Concrete for Nuclear Reactors*, SP-34, American Concrete Institute, Detroit, Michigan (1972) pp. 732–68.
937. McHenry, D., A new aspect of creep in concrete and its application to design. *Proc. ASTM*, **43** (1943) 1069–86.
938. Macmillan, M., Evolution de fluage et des properietes de beton. *Annales, Inst. Tech. du Batiment et des Travaux Publics*, **21** (1969) 1033; and **13** (1960) 1017–52.
939. Mamillan, M. & Lelan, M., Le fluage de beton. *Annales, Inst. Techn. du Batiment et des Travaux Publics* (Supplement) **23** (1970) 7–13; and **21** (1968) 847–50.
940. Mandel, J., Sur les corps viscoelastiques lineaires dont les proprietes dependent de l'age. *Comptes Rendus des Seances de l'Academie des Sciences*, **247** (1958) 175–8.
941. Marechal, J. C., Contribution a l'etude des properietes thermiques et mecaniques du beton en fonction de la temperature. *Annales, Inst. Tech. du Batiment et des Travaux Publics*, **23** (1970) 123–45.

942. Marechal, J. C., Fluage de beton en fonction de la temperature. *Materials and Structures* (RILEM, Paris), **2** (1969) 111–15; see also *Materials and Structures*, **3** (1970) 395–405.
943. Marechal, J. C., Fluage du beton en fonction de la temperature. *Annales, Inst. Tech. du Batiment et des Travaux Publics*, **23** (1970) 13–24.
944. Maslov, G. N., Thermal stress states in concrete masses, with account of concrete creep. *Izvestia Nauchno-Issledovatelskogo Instituta Gidrotekhniki, Gosenergoizdat*, **28** (1940) 175–88 (in Russian).
945. McNeice, G. M., Elastic–plastic bending of plates and slabs by finite element method. Thesis presented to the University of London in partial fulfillment of the requirements for the degree of Doctor of Philosophy, 1967.
946. Saouma, V. E., Interactive finite element analysis of reinforced concrete: a fracture mechanics approach. PhD Thesis, Cornell University, January 1981.
947. Sandler, I. S., On the uniqueness and stability of endochronic theories of material behaviour. *J. Appl. Mech.*, **45** (1978) 263–6.
948. Schapery, R. A., A thermodynamic constitutive theory and its applications to various nonlinear materials. *Proceedings IUTAM Symposium, Kilbride, June 1968*, B. A. Boley, ed. Springer Verlag, New York, 1968.
949. Sinha, B. P., Gerstle, K. H. & Tulin, L. G., Stress–strain relations for concrete under cyclic loading. *J. Am. Concr. Inst.*, **61** (1964) 195–211.
950. Valanis, K. C. & Fan, J., Endochronic analysis of cyclic elastoplastic strain fields in a notched plate. *ASME, J. Appl. Mech.*, **50** (1983) 789–94.
951. Valanis, K. C. & Fan, J., A numerical algorithm for endochronic plasticity and comparison with experiment. *Computers and Structures*, **19** (1984) 717–24.
952. Van der Houwen, P. J., *Construction of Integration Formulas for Initial Value Problems*. North Holland, Amsterdam, 1977.
953. Van Greunen, J., Nonlinear geometric, material and time dependent analysis of reinforced and prestressed concrete slabs and panels. PhD Thesis, Division of Structural Engineering and Structural Mechanics, University of California, Berkeley, Report No. UC-SESM 79-3, October 1979.
954. Van Zyl, S. F., Analysis of curved segmentally erected prestressed concrete box girder bridges. PhD Dissertation, Division of Structural Engineering and Structural Mechanics, University of California, Berkeley, Report No. UC-SESM 78-2, January 1978.
955. Van Zyl, F. S. & Scordelis, A. C., Analysis of curved prestressed segmental bridges. *J. Struct. Div., ASCE*, **105** (1979).
956. Volterra, V., *Lecons sur les Fonctions de Ligne*. Gauthier-Villars, Paris, 1913; and *Theory of Functionals and of Integral and Integrodifferential Equations*. Dover, New York, 1959.
957. Wegner, R. & Duddeck, H., Der Gerissenen Zustand Zweiseitig Gelagerter Platten unter Einzellasten-Nichtlineare Berechnung mit Finiten Elementen. *Beton und Stahlbetonbau*, **70** (1975).
958. Wegner, R., Harbord, R. & Duddeck, H., Flach-und Pilzgecken im Ungerissenen und gerissenen Zustand. *Bauingenieur*, **50** (1975).
959. Wanchoo, M. K. & May, G. W., Cracking analysis of reinforced concrete plates. *J. Struct. Div., ASCE*, **101** (1975).

960. Wittmann, F. H., Vergleich einiger Kriechfunktionen mit Versuchsergebnissen. *Cement and Concrete Research*, **1** (1971) 679–90.
961. Yuzugullu, O. & Schnobrich, W. C., A numerical procedure for the determination of the behaviour of a shear wall frame system. *J. Am. Conc. Inst.*, **70** (1973) 474–9.
962. Wittmann, F., Einfluss des Feuchtigkeitsgehaltes auf das Kriechen des Zementsteines. *Rheol. Acta*, **9** (1970) 282–7.
963. Wittmann, F., Kriechen bei Gleichzeitigem Schwinden des Zementsteins. *Rheol. Acta*, **5** (1966) 198–204.
964. Wittmann, F., Kreichverformung des Betons Unter Statischer und unter Dynamischer Belastung. *Rheol. Acta*, **10** (1971) 422–8.
965. Wittmann, F., Surface tension, shrinkage and strength of hardened cement paste. *Materials and Structures* (RILEM, Paris), **1** (1968) 547–52.
966. Wittmann, F., Bestimmung Physikalischer Eigenschauften des Zementsteins. *Deutscher Ausschuss fur Stahlbeton*, Heft 232, W. Ernst, Berlin, 1974.
967. Isenberg, J., Wojcik, G. L. & Nikooyeh, H., Dynamic analysis of buried structures including catastrophic failure. Weidlinger Associates, Final Report under Contract DNA001-79-C-0227 to Defense Nuclear Agency, 1 December 1979.
968. Isenberg, J., Vaughan, D. K. & Wong, F. S., Effects of scale in the response of MX vertical shelters. Weidlinger Associates, Final Report under Contract DNA001-79-C-0253 to Defense Nuclear Agency, 15 April 1980.
969. Ingraffea, A. R., On discrete fracture propagation in rock loaded in compression. Proc. First International Conference on Numerical Methods in Fracture Mechanics, Swansea, 9–13 January, 1978, pp. 235–48.
970. Jimenez, R., Gergely, P. & White, R. N., Shear transfer across cracks in reinforced concrete. Report 78-4, Department of Structural Engineering, Cornell University, August 1978, p. 357.
971. Jimenez, R., White, R. N. & Gergely, P., Bond and dowel capacities of reinforced concrete. *J. Am. Concr. Inst.*, **76** (1979) 73–92.
972. Jones, R., The ultimate strength of reinforced concrete beams in shear. *Magazine of Concrete Research*, **8** (1956).
973. Kesler, C. E., Wallo, E. M., Yuan, R. L. & Lorr, J. L., Sixth Progress Report, Prediction of Creep in Structural Concrete From Short Time Tests. Department of Theoretical and Applied Mechanics, University of Illinois, Urbana, Illinois, August 1965.
974. Kimishima, H. & Kitahara, H., Creep and creep recovery of mass concrete. Technical Report C-64001, Central Research Institute of Electric Power Industry, Tokyo, Japan, September 1964.
975. Klug, P. & Wittmann, T., The correlation between creep deformation and stress relaxation in concrete. *Materials and Structures* (RILEM, Paris), **3** (1970) 75–80.
976. Komendant, G. J., Polivka, M. & Pirtz, D., Study of concrete properties for prestressed concrete reactor vessels. Final report Part II, Creep and strength characteristics of concrete at elevated temperatures. Report No. UCSEMS76-3, Structures and Materials Research, Department of Civil Engineering, Report to General Atomic Company, San Diego, California, Berkeley, California, April 1976.

977. Lin, C., Hseih, B. J. & Valentin, R. A., The application of endochronic plasticity theory in modelling and dynamic inelastic response of structural systems. *Nucl. Eng. Des.*, **66** (1981) 213–21.
978. Lauer, R. K. & Slate, F. J., Autogeneous healing of cement paste. *J. Am. Concr. Inst.*, **52** (1956) 1083–98.
979. LeCamus, B., Recherches experimentales sur le deformation du beton et du beton arme, Part II. *Annales Inst. Tech. du Batiment et des Travaux Publics*, (1947).
980. Lambotte, H. & Mommens, A., L'evolution du fluage du beton en fonction de sa composition, du taux de contrainte et de l'age. Groupe de travail GT 22, Centre National de Recherches Scientifiques et Techniques pour l'Industrie Cimentiere, Bruxelles, July 1976.
981. Leombruni, P., Buyukozturk, O. & Connor, J. J., Analysis of shear transfer in reinforced concrete with application to containment wall specimens. Report MIT-CER-79-26, Department of Civil Engineering, Massachusetts Institute of Technology, June 1979.
982. Mufti, A. A., Mirza, M. S., McCutcheon, J. O. & Houde, J., A study of the behaviour of reinforced concrete elements using finite elements. Structural Concrete Series No. 70-5, McGill University, Montreal, September 1970.
983. Mufti, A. A., Mirza, M. S., McCutcheon, J. O. & Spowski, R. W., A finite element study of reinforced concrete structures. Structural Concrete Series No. 71–8, McGill University, Montreal, October 1971.
984. Mikkola, M. J. & Schnobrich, W. C., Material behaviour characteristics for reinforced concrete shells stressed beyond the elastic range. Civil Engineering Studies, SRS No. 367, University of Illinois at Urbana-Champaign, Urbana, Illinois, August 1970.
985. Mroz, Z., Mathematical models of inelastic concrete behaviour. In: *Inelasticity and Non-Linearity in Structural Concrete*, M. Z. Cohn, ed. University of Waterloo Press, Waterloo, Ontario, Study No. 8, 1972, pp. 47–72.
986. Mueller, G., Numerical problems in nonlinear analysis of reinforced concrete. US-SESM Report No. 77-5, University of California, Berkeley, September 1977.
987. Murray, D. W., Octahedral based incremental stress–strain matrices. *J. Eng. Mech. Division, ASCE*, **105** (1979) 501–13.
988. Murray, D. W., Chitnuyanondh, L., Agha, K. Y. & Wong, C., A concrete plasticity theory for biaxial stress analysis. *J. Eng. Mech. Division, ASCE*, **105** (1979) 989–1006.
989. Ma, S.-Y.M., Bertero, V. V. & Popov, E. P., Experimental and analytical studies on the hysteretic behaviour of reinforced concrete rectangular and T-beams. Earthquake Engineering Research Center, Report No. EERC 76-2, University of California, Berkeley, 1976.
990. Mahin, S. A. & Bertero, V. V., Rate of loading effects on uncracked and repaired reinforced concrete members. Earthquake Engineering Research Center, Report No. EERC 72-9, University of California, Berkeley, 1972.
991. Marchertas, A. H., Fistedis, S. H., Bazant, Z. P. & Belytschko, T. B., Analysis and application of prestressed concrete reactor vessels for LMFBR containment. *Nucl. Eng. Des.*, **49** (1978) 155–73.

992. Marchertas, A. H., Belytschko, T. B. & Bazant, Z. P., Transient analysis of LMFBR reinforced/prestressed concrete containment. 5th International Conference on Structural Mechanics in Reactor Technology (SMiRT), Paper H8/1, Berlin, August 1979.
993. dePaiva, H. A. R. & Siess, C. P., Strength and behaviour of deep beams in shear. *J. Struct. Div., ASCE*, **91** (1965) 19–41.
994. Dulacska, H., Dowel action of reinforcement crossing cracks in concrete. *J. Am. Concr. Inst.*, **69** (1972).
995. Bathe, K. J. & Wilson, E. L., *Numerical Methods in Finite Element Analysis*. Prentice-Hall, Englewood Cliffs, New Jersey, 1976.
996. Bathe, K. J. et al., Some computational capabilities for nonlinear finite element analysis. *Nucl. Eng. Des.*, **46** (1978) 429–55.
997. Colville, J. & Abbasi, J., Plane stress reinforced concrete finite elements. *J. Struct. Div., ASCE*, **100** (1974).
998. Crisfield, M. A., A faster modified Newton–Raphson integration. *Computer Methods in Applied Mechanics and Engineering*, **20** (1979) 267–78.
999. Glemberg, R., Dynamic analysis of concrete structures. PhD Thesis Department of Structural Mechanics, Chalmers University of Technology, Stockholm, 1984.
1000. Belytschko, T. & Mullen, R., Stability of explicit–implicit time mesh partitions in time integration. *Int. J. Num. Meth. Eng.*, **12** (1978) 1575–86.
1001. Berwanger, C., The modulus of concrete and the coefficient of thermal expansion below normal temperatures. In: *Temperature and Concrete*, SP-25, American Concrete Institute, Detroit, Michigan, 1971, pp. 181–234.
1002. Ergataudis, J. G., Isoparametric finite elements in two and three dimensional analysis. PhD Thesis, University of Wales, Swansea, 1968.
1003. Fahmi, H. M., Polivka, M. & Bresler, B., Effect of sustained and cyclic elevated temperature on creep and concrete. *Cement and Concrete Research*, **2** (1972) 591–606.
1004. Bangash, Y. & England, G. L., The influence of thermal creep on the operational behaviour of complex structures. *Proc. International Conference on Fundamental Research on Creep and Shrinkage, Lausanne*, Martinus Nijhoff, The Hague, 1982.
1005. Bangash, Y., The structural design of circular diversion, power and pressure tunnels. *Tunnels and Tunnelling*, Part 1, May/June (1971); Part 2, July/August (1971); Part 3, September/October (1971).
1006. Saudia Laboratories, 1:6 scaled model of R.C. containment—a pre-test analysis. Report: N/CR 4913 Sand 87-0891, Saudi Laboratories, 1987.
1007. Bangash, Y., Aircraft crash analysis of the proposed Sizewell B containment vessel. 9th International Conference on Structural Mechanics in Reactor Technology (SMiRT), Vol. J, Lausanne, 1987, pp. 307–14.
1008. Bangash, Y., Loads and moments due to currents and wind on offshore gravity platforms. *Ocean Engineering*, **6** (1979) 285–96.
1009. Nataraja, R. & Kirk, C. L., Dynamic response of a gravity platform under random wave forces. Proc. OTC, Paper OTC-2904, 1977.
1010. J. Mowlem Co., London. Private communications on CONDEEP, 1983.
1011. BS 5337: The structural use of concrete for retaining aqueous liquids, London, 1976.

1012. Dean, R. G., Relative validities of water wave theories. Civil Engineering the Oceans, ASCE Conference, 6–8 September 1967.
1013. FIP, *Recommendations for the design and Construction of Concrete Sea Structures*, Federation Internationale de Constrainte, Paris, France (2nd edn), 1973.
1014. Frimann Clausen, C. J., *The Condeep Story*, John Mowlam & Co., London, 1976.
1015. Hogben, N. & Standing, R. G., Experience in computing wave loads on large bodies. NPL Offshore Technology Conference, Paper No. OTC 2189, Vol. 1, May 1975, & private notes.
1016. Maccamy, R. C. & Fuchs, R. A., Wave forces on piles—a diffraction theory. Beach Erosion Board Technical Memorandum, No. 69, 1954.
1017. Morison, J. R., O'Brien, M. P., Johnson, J. W. & Schaof, S. A., The force exerted by surface waves on piles. *Petroleum Transactions, AIME*, **189** (1950).
1018. Myers, J. J., Holm, C. H. & McCallister, R. F., *Handbook of Ocean and Underwater Engineering*. McGraw Hill, New York, 1969.
1019. Pinfold, G. M., *Reinforced Concrete and Towers*. McGraw Hill, New York, 1969.
1020. Skjelbreia, L. & Hendrickson, J., Fifth order gravity wave theory. Proc. Seventh Conference Coastal Engineering, 1961.
1021. Gibson, J. E., *Thin Shells*. Pergamon Press, Oxford, 1980.
1022. Stephen, P., Timoshenko, S. & Woinowsky-Krieger, S. *Theory of Plates and Shells*. McGraw Hiil, New York, 1981.
1023. Ursell, F., Mass transport in gravity waves. *Proc. Cambridge Phil. Soc.*, 1953.
1024. Werenskiold, S. Private communication 1976.
1025. Al-Noury, S., Three-dimensional analysis of the TLP platform. *Computers and Structures*, **23** (1986) 699–714.
1026. Sachs, P., *Wind Forces in Engineering*. Pergamon Press, Oxford, 1972.
1027. *ABAQUS User's Manual*, University of Manchester, 1986.
1028. Newman, K., The structure and engineering properties of concrete. International Symposium Theory Arch. Dams, Southampton, England, 1965.
1029. Olsen, O., Implosion analysis of concrete cylinders under hydrostatic pressure. *ACI Journal*, March 1978, 83–5.
1030. Det Norske Veritas, Rules for classification of mobile offshore units, 1981.
1031. Minorsky, V. V., An analysis of ship collisions with reference to nuclear power plants. *Journal of Ship Research*, **3** (1959) pp. 1–4.
1032. Søreide, T. H., *Ultimate Load Analysis of Marine Structures*. Tapir Publishing Company, Trondheim, Norway, 1981.
1033. JABSF, Ship collision with an offshore structure. IABSE Colloquium, Copenhagen, 1983.
1034. Woisin, G., Conclusion from collision examinations for nuclear merchant ships in the FRG. Proc. of Symposium on Naval Submarine, Hamburg, 1977.
1035. Reckling, K. A., On the collision protection of ships. PRADS Symposium, Tokyo, 1977.
1036. Nagasawa, H., Arita, K., Tani, M. & Oka, S., A study on the collapse of ship structure in collision with bridge piers. *J. Soc. of Nav. Arch. of Japan*, **142** (1977).

1037. Faulkner, D., A review of effective plating for use in the analysis of stiffened plating in bending and compression. *J. of Ship Research*, **19** (1975).
1038. Ohnishi, T., Kawakami, H., Yasukawa, W. & Nagasawa, H., On the ultimate strength of bow construction. *J. Soc. of Nav. Arch. of Japan*, **151** (1982).
1039. BBRV Tanks Inc., California, Design of prestressed concrete tanks, Vols 1, 2 and 3, 1977.

APPENDIX I

RELEVANT ELEMENT LIBRARY

APPENDIX IA: MATERIAL MATRICES

[D] — Variable Young's Modulus and Constant Poisson's Ratio

$$D_{11} = \frac{E_1(E')^3 - E_{cr}}{v''} \quad D_{12} = \frac{vE_1E_2(E')^2 + E_{cr}}{v''} \quad D_{13} = \frac{vE_1E_3(E')^2 + E_{cr}}{v''} \quad D_{14} = 0 \quad D_{15} = 0 \quad D_{16} = 0$$

$$D_{22} = \frac{E_2E_3(E')^2 + E_{cr}}{v''} \quad D_{23} = \frac{vE_2E_3(E')^2 + E_{cr}}{v''} \quad D_{24} = 0 \quad D_{25} = 0 \quad D_{26} = 0$$

$$D_{33} = \frac{E_3(E')^3 - E_{cr}}{v''} \quad D_{34} = 0 \quad D_{35} = 0 \quad D_{36} = 0$$

$$D_{44} = G_{12} \quad D_{45} = 0 \quad D_{46} = 0$$

$$D_{55} = G_{23} \quad D_{56} = 0$$

$$D_{66} = G_{31}$$

$E_{cr} = v^2 E_1 E_2 E_3 E'$

$E' = (E_1 + E_2 + E_3)/3$

$v'' = (E')^3 - 2E_1E_2E_3v^2 - E'v^2(E_1E_2 + E_1E_3 + E_2E_3)$

$G_{12} = E_{12}/2(1+v)$

$E_{12} = (E_1 + E_2)/2$

$G_{23} = E_{23}/2(1+v)$

$E_{23} = (E_2 + E_3)/2$

$G_{31} = E_{31}/2(1+v)$

$E_{31} = (E_3 + E_1)/2$

[D]—Variable Young's Modulus and Poisson's Ratio

$$
\begin{bmatrix}
D_{11} = \dfrac{(1-\nu_{23}\nu_{32})}{\bar{\nu}}E_1 & D_{12} = \dfrac{(\nu_{12}+\nu_{12}\nu_{32})}{\bar{\nu}}E_2 & D_{13} = \dfrac{(\nu_{13}+\nu_{12}\nu_{23})}{\bar{\nu}}E_3 & D_{14} = 0 & D_{15} = 0 & D_{16} = 0 \\
D_{21} = \dfrac{(\nu_{21}+\nu_{23}\nu_{31})}{\bar{\nu}}E_1 & D_{22} = \dfrac{(1-\nu_{13}\nu_{31})}{\bar{\nu}}E_2 & D_{23} = \dfrac{(\nu_{23}+\nu_{13}\nu_{21})}{\bar{\nu}}E_3 & D_{24} = 0 & D_{25} = 0 & D_{26} = 0 \\
D_{31} = \dfrac{(\nu_{31}+\nu_{21}\nu_{32})}{\bar{\nu}}E_1 & D_{32} = \dfrac{(\nu_{32}+\nu_{12}\nu_{31})}{\bar{\nu}}E_2 & D_{33} = \dfrac{(1-\nu_{12}\nu_{21})}{\bar{\nu}}E_3 & D_{34} = 0 & D_{35} = 0 & D_{36} = 0 \\
D_{41} = 0 & D_{42} = 0 & D_{43} = 0 & D_{44} & D_{45} = 0 & D_{46} = 0 \\
D_{51} = 0 & D_{52} = 0 & D_{53} = 0 & D_{54} = 0 & D_{55} & D_{56} = 0 \\
D_{61} = 0 & D_{62} = 0 & D_{63} = 0 & D_{64} = 0 & D_{65} = 0 & D_{66}
\end{bmatrix}
$$

$$\bar{\nu} = 1 - \nu_{12}\nu_{21} - \nu_{13}\nu_{31} - \nu_{23}\nu_{32} - \nu_{12}\nu_{23}\nu_{31} - \nu_{21}\nu_{13}\nu_{32}$$

Due to symmetry of compliances, the following relations can be written:

$$E_1 v_{21} = E_2 v_{12} \quad D_{55} = G_{23}$$
$$E_2 v_{32} = E_3 v_{23} \quad D_{66} = G_{13}$$
$$E_3 v_{13} = E_1 v_{31}$$

The values of G_{12}, G_{23} and G_{13} are calculated in terms of modulus of elasticity and Poisson's ratio as follows:

$$G_{12} = \frac{1}{2}\left[\frac{E_1}{2(1+v_{12})} + \frac{E_2}{2(1+v_{21})}\right] = \frac{1}{2}\left[\frac{E_1}{2(1+v_{12})} + \frac{E_1}{2\left(\frac{E_1}{E_2}+v_{12}\right)}\right]$$

$$G_{23} = \frac{1}{2}\left[\frac{E_2}{2(1+v_{23})} + \frac{E_3}{2(1+v_{32})}\right] = \frac{1}{2}\left[\frac{E_2}{2(1+v_{23})} + \frac{E_2}{2\left(\frac{E_2}{E_3}+v_{23}\right)}\right]$$

$$G_{13} = \frac{1}{2}\left[\frac{E_3}{2(1+v_{31})} + \frac{E_1}{2(1+v_{13})}\right] = \left[\frac{E_3}{2(1+v_{31})} + \frac{E_3}{2\left(\frac{E_3}{E_1}+v_{31}\right)}\right]$$

For isotropic cases;

$$E_1 = E_2 = E_3 = E$$
$$v_{12} = v_{13} = v_{23} = v_{21} = v_{31} = v_{32} = v$$

[D]—Constant Young's Modulus and Poisson's Ratio

$$[D] = \frac{E}{(1+v)(1-2v)}\begin{bmatrix} 1-v & v & v & 0 & 0 & 0 \\ v & 1-v & v & 0 & 0 & 0 \\ v & v & 1-v & 0 & 0 & 0 \\ 0 & 0 & 0 & \frac{1-2v}{2} & 0 & 0 \\ 0 & 0 & 0 & 0 & \frac{1-2v}{2} & 0 \\ 0 & 0 & 0 & 0 & 0 & \frac{1-2v}{2} \end{bmatrix}$$

Bulk and Shear Moduli

$$[D] = \begin{bmatrix} K+\tfrac{4}{3}G & K-\tfrac{2}{3}G & K-\tfrac{2}{3}G & 0 & 0 & 0 \\ K-\tfrac{2}{3}G & K+\tfrac{4}{3}G & K-\tfrac{2}{3}G & 0 & 0 & 0 \\ K-\tfrac{2}{3}G & K-\tfrac{2}{3}G & K+\tfrac{4}{3}G & 0 & 0 & 0 \\ 0 & 0 & 0 & G & 0 & 0 \\ 0 & 0 & 0 & 0 & G & 0 \\ 0 & 0 & 0 & 0 & 0 & G \end{bmatrix}$$

For Plane Stress

$$[D] = \frac{E}{1-v^2} \begin{bmatrix} 1 & v & 0 \\ v & 1 & 0 \\ 0 & 0 & \dfrac{1-v}{2} \end{bmatrix}$$

For Plane Strain

$$[D] = \frac{E(1-v)}{(1+v)(1-2v)} \begin{bmatrix} 1 & \dfrac{v}{1-v} & 0 \\ \dfrac{v}{1-v} & 1 & 0 \\ 0 & 0 & \dfrac{1-2v}{2(1-v)} \end{bmatrix}$$

For Axisymmetric Cases

$$[D] = \frac{E}{(1+v)(1-2v)} \begin{bmatrix} 1-v & v & v & 0 \\ v & 1-v & v & 0 \\ v & v & 1-v & 0 \\ 0 & 0 & 0 & \dfrac{1-2v}{2} \end{bmatrix}$$

APPENDIX IB: ELEMENT TYPES, SHAPE FUNCTIONS, DERIVATIVES, STIFFNESS MATRICES

[K]—Shear and Torsion Included for Line Element

$$[D] = \begin{bmatrix}
\frac{EA}{L} & & & & & & & & & & & \\
0 & \frac{12EI_\xi}{L^3(1+\bar{\tau}_\eta)} & & & & & & & & & & \\
0 & 0 & \frac{12EI_\eta}{L^3(1+\bar{\tau}_\xi)} & & & & & & & & & \\
0 & 0 & 0 & \frac{GJ}{L} & & & & & & & & \\
0 & 0 & \frac{-6EI_\eta}{L^2(1+\bar{\tau}_\xi)} & 0 & \frac{(4+\bar{\tau}_\xi)EI_\eta}{L(1+\bar{\tau}_\xi)} & & & & & & & \\
0 & \frac{6EI_\xi}{L^2(1+\bar{\tau}_\eta)} & 0 & 0 & 0 & \frac{(4+\bar{\tau}_\eta)EI_\xi}{L(1+\bar{\tau}_\eta)} & & & & & & \\
\frac{-EA}{L} & 0 & 0 & 0 & 0 & 0 & \frac{AE}{L} & & & & & \\
0 & \frac{-12EI_\xi}{L^3(1+\bar{\tau}_\eta)} & 0 & 0 & 0 & \frac{-6EI_\xi}{L^2(1+\bar{\tau}_\eta)} & 0 & \frac{12EI_\xi}{L^3(1+\bar{\tau}_\eta)} & & & & \\
0 & 0 & \frac{-12EI_\eta}{L^3(1+\bar{\tau}_\xi)} & 0 & \frac{6EI_\eta}{L^2(1+\bar{\tau}_\xi)} & 0 & 0 & 0 & \frac{12EI_\eta}{L^3(1+\bar{\tau}_\xi)} & & & \\
0 & 0 & 0 & \frac{-GJ}{L} & 0 & 0 & 0 & 0 & 0 & \frac{GJ}{L} & & \\
0 & 0 & \frac{-6EI_\eta}{L^2(1+\bar{\tau}_\xi)} & 0 & \frac{(2-\bar{\tau}_\xi)EI_\eta}{L(1+\bar{\tau}_\xi)} & 0 & 0 & 0 & \frac{6EI_\eta}{L^2(1+\bar{\tau}_\xi)} & 0 & \frac{(4+\bar{\tau}_\xi)EI_\eta}{L(1+\bar{\tau}_\xi)} & \\
0 & \frac{6EI_\xi}{L^2(1+\bar{\tau}_\eta)} & 0 & 0 & 0 & \frac{(2-\bar{\tau}_\eta)EI_\xi}{L(1+\bar{\tau}_\eta)} & 0 & \frac{-6EI_\xi}{L^2(1+\bar{\tau}_\eta)} & 0 & 0 & 0 & \frac{(4+\bar{\tau}_\eta)EI_\xi}{L(1+\bar{\tau}_\eta)}
\end{bmatrix}$$

where

$\bar{\tau}_\eta = \dfrac{12EI_\xi}{GA_{s\eta}L^2} = 24(1+\nu)\dfrac{A}{A_{s\eta}}\left(\dfrac{T_\xi}{L}\right)^2$ $\bar{\tau}_\xi = \dfrac{12EI_\eta}{GA_{s\xi}L^2} = 24(1+\nu)\dfrac{A}{A_{s\xi}}\left(\dfrac{T_\eta}{L}\right)^2$

A_s = shear area T = torsional moment of inertia ρ = density

ξ, η local axes are parallel to Z and Y axes

Chain Rule

$$\begin{bmatrix} \dfrac{\partial u}{\partial X} \\ \dfrac{\partial u}{\partial Y} \\ \dfrac{\partial u}{\partial Z} \\ \dfrac{\partial v}{\partial X} \\ \dfrac{\partial v}{\partial Y} \\ \dfrac{\partial v}{\partial Z} \\ \dfrac{\partial w}{\partial X} \\ \dfrac{\partial w}{\partial Y} \\ \dfrac{\partial w}{\partial Z} \end{bmatrix} = \dfrac{1}{\det J} \begin{bmatrix} C_{11} & C_{12} & C_{13} & 0 & 0 & 0 & 0 & 0 & 0 \\ C_{21} & C_{22} & C_{23} & 0 & 0 & 0 & 0 & 0 & 0 \\ C_{31} & C_{32} & C_{33} & 0 & 0 & 0 & 0 & 0 & 0 \\ 0 & 0 & 0 & C_{11} & C_{12} & C_{13} & 0 & 0 & 0 \\ 0 & 0 & 0 & C_{21} & C_{22} & C_{23} & 0 & 0 & 0 \\ 0 & 0 & 0 & C_{31} & C_{32} & C_{33} & 0 & 0 & 0 \\ 0 & 0 & 0 & 0 & 0 & 0 & C_{11} & C_{12} & C_{13} \\ 0 & 0 & 0 & 0 & 0 & 0 & C_{21} & C_{22} & C_{23} \\ 0 & 0 & 0 & 0 & 0 & 0 & C_{31} & C_{32} & C_{33} \end{bmatrix} \begin{bmatrix} \dfrac{\partial u}{\partial \xi} \\ \dfrac{\partial u}{\partial \eta} \\ \dfrac{\partial u}{\partial \zeta} \\ \dfrac{\partial v}{\partial \xi} \\ \dfrac{\partial v}{\partial \eta} \\ \dfrac{\partial v}{\partial \zeta} \\ \dfrac{\partial w}{\partial \xi} \\ \dfrac{\partial w}{\partial \eta} \\ \dfrac{\partial w}{\partial \zeta} \end{bmatrix}$$

where

$$C_{11} = \dfrac{\partial Y}{\partial \xi}\dfrac{\partial Z}{\partial \xi} - \dfrac{\partial Z}{\partial \eta}\dfrac{\partial Y}{\partial \xi} \qquad C_{12} = \dfrac{\partial Z}{\partial \xi}\dfrac{\partial Y}{\partial \zeta} - \dfrac{\partial Y}{\partial \xi}\dfrac{\partial Z}{\partial \zeta}$$

$$C_{13} = \dfrac{\partial Y}{\partial \xi}\dfrac{\partial Z}{\partial \zeta} - \dfrac{\partial Z}{\partial \xi}\dfrac{\partial Y}{\partial \zeta} \qquad C_{21} = \dfrac{\partial Z}{\partial \xi}\dfrac{\partial X}{\partial \zeta} - \dfrac{\partial X}{\partial \xi}\dfrac{\partial Z}{\partial \zeta}$$

$$C_{22} = \dfrac{\partial X}{\partial \xi}\dfrac{\partial Z}{\partial \zeta} - \dfrac{\partial Z}{\partial \xi}\dfrac{\partial X}{\partial \zeta} \qquad C_{23} = \dfrac{\partial Z}{\partial \xi}\dfrac{\partial X}{\partial \eta} - \dfrac{\partial X}{\partial \xi}\dfrac{\partial Z}{\partial \eta}$$

$$C_{31} = \dfrac{\partial X}{\partial \eta}\dfrac{\partial Y}{\partial \zeta} - \dfrac{\partial Y}{\partial \eta}\dfrac{\partial X}{\partial \zeta} \qquad C_{32} = \dfrac{\partial Y}{\partial \xi}\dfrac{\partial X}{\partial \zeta} - \dfrac{\partial X}{\partial \xi}\dfrac{\partial Y}{\partial \zeta}$$

$$C_{33} = \dfrac{\partial X}{\partial \xi}\dfrac{\partial Y}{\partial \eta} - \dfrac{\partial Y}{\partial \xi}\dfrac{\partial X}{\partial \eta}$$

det $[J]$ = the determinant of the Jacobian matrix.

Solid Isoparametric Elements

Eight-Noded Solid Element

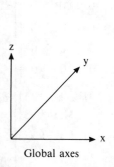
Global axes

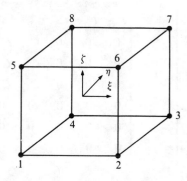

Node i	Shape functions $N_i(\xi,\eta,\zeta)$	Derivatives $\dfrac{\partial N_i}{\partial \xi}$	$\dfrac{\partial N_i}{\partial \eta}$	$\dfrac{\partial N_i}{\partial \zeta}$
1	$\tfrac{1}{8}(1-\xi)(1-\eta)(1-\zeta)$	$-\tfrac{1}{8}(1-\eta)(1-\zeta)$	$-\tfrac{1}{8}(1-\xi)(1-\zeta)$	$-\tfrac{1}{8}(1-\eta)(1-\xi)$
2	$\tfrac{1}{8}(1+\xi)(1-\eta)(1-\zeta)$	$\tfrac{1}{8}(1-\eta)(1-\zeta)$	$-\tfrac{1}{8}(1+\xi)(1-\zeta)$	$-\tfrac{1}{8}(1+\xi)(1-\eta)$
3	$\tfrac{1}{8}(1+\xi)(1+\eta)(1-\zeta)$	$\tfrac{1}{8}(1+\eta)(1-\zeta)$	$\tfrac{1}{8}(1+\xi)(1-\zeta)$	$-\tfrac{1}{8}(1+\xi)(1+\eta)$
4	$\tfrac{1}{8}(1-\xi)(1+\eta)(1-\zeta)$	$-\tfrac{1}{8}(1+\eta)(1-\zeta)$	$\tfrac{1}{8}(1-\xi)(1-\zeta)$	$-\tfrac{1}{8}(1-\xi)(1+\eta)$
5	$\tfrac{1}{8}(1-\xi)(1-\eta)(1+\zeta)$	$-\tfrac{1}{8}(1-\eta)(1+\zeta)$	$-\tfrac{1}{8}(1-\xi)(1+\zeta)$	$\tfrac{1}{8}(1-\xi)(1-\eta)$
6	$\tfrac{1}{8}(1+\xi)(1-\eta)(1+\zeta)$	$\tfrac{1}{8}(1-\eta)(1+\zeta)$	$-\tfrac{1}{8}(1+\xi)(1+\zeta)$	$\tfrac{1}{8}(1+\xi)(1-\eta)$
7	$\tfrac{1}{8}(1+\xi)(1+\eta)(1+\zeta)$	$\tfrac{1}{8}(1+\eta)(1+\zeta)$	$\tfrac{1}{8}(1+\xi)(1+\zeta)$	$\tfrac{1}{8}(1+\xi)(1+\eta)$
8	$\tfrac{1}{8}(1-\xi)(1+\eta)(1+\zeta)$	$-\tfrac{1}{8}(1+\eta)(1+\zeta)$	$\tfrac{1}{8}(1-\xi)(1+\zeta)$	$\tfrac{1}{8}(1-\xi)(1+\eta)$

Twenty-noded Solid Element

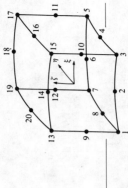

Node i	Shape functions $N_i(\xi,\eta,\zeta)$	Derivatives $\dfrac{\partial N_i}{\partial \xi}$	$\dfrac{\partial N_i}{\partial \eta}$	$\dfrac{\partial N_i}{\partial \zeta}$
1	$\frac{1}{8}(1-\xi)(1-\eta)(1-\zeta)(-\xi-\eta-\zeta-2)$	$\frac{1}{8}(1-\eta)(1-\zeta)(2\xi+\eta+\zeta+1)$	$\frac{1}{8}(1-\xi)(1-\zeta)(2\eta+\xi+\zeta+1)$	$\frac{1}{8}(1-\xi)(1-\eta)(2\zeta+\eta+\xi+1)$
2	$\frac{1}{4}(1-\xi^2)(1-\eta)(1-\zeta)$	$-\frac{1}{2}\xi(1-\eta)(1-\zeta)$	$-\frac{1}{4}(1-\xi^2)(1-\zeta)$	$-\frac{1}{4}(1-\xi^2)(1-\eta)$
3	$\frac{1}{8}(1+\xi)(1-\eta)(1-\zeta)(\xi-\eta-\zeta-2)$	$\frac{1}{8}(1-\eta)(1-\zeta)(2\xi-\eta-\zeta-1)$	$-\frac{1}{8}(1+\xi)(1-\zeta)(2\eta-\xi+\zeta+1)$	$-\frac{1}{8}(1+\xi)(1-\eta)(2\zeta-\xi+\eta+1)$
4	$\frac{1}{4}(1+\xi)(1-\eta^2)(1-\zeta)$	$\frac{1}{4}(1-\eta^2)(1-\zeta)$	$-\frac{1}{2}\eta(1+\xi)(1-\zeta)$	$-\frac{1}{4}(1+\xi)(1-\eta^2)$
5	$\frac{1}{8}(1+\xi)(1+\eta)(1-\zeta)(\xi+\eta-\zeta-2)$	$\frac{1}{8}(1+\eta)(1-\zeta)(2\xi+\eta-\zeta-1)$	$\frac{1}{8}(1+\xi)(1-\zeta)(2\eta+\xi-\zeta-1)$	$-\frac{1}{8}(1+\xi)(1+\eta)(2\zeta-\xi-\eta+1)$
6	$\frac{1}{4}(1-\xi^2)(1+\eta)(1-\zeta)$	$-\frac{1}{2}\xi(1+\eta)(1-\zeta)$	$\frac{1}{4}(1-\xi^2)(1-\zeta)$	$-\frac{1}{4}(1-\xi^2)(1+\eta)$
7	$\frac{1}{8}(1-\xi)(1+\eta)(1-\zeta)(-\xi+\eta-\zeta-2)$	$-\frac{1}{8}(1+\eta)(1-\zeta)(-2\xi+\eta-\zeta-1)$	$\frac{1}{8}(1-\xi)(1-\zeta)(2\eta-\xi-\zeta-1)$	$-\frac{1}{8}(1-\xi)(1+\eta)(2\zeta+\xi-\eta+1)$
8	$\frac{1}{4}(1-\xi)(1-\eta^2)(1-\zeta)$	$-\frac{1}{4}(1-\eta^2)(1-\zeta)$	$-\frac{1}{2}\eta(1-\xi)(1-\zeta)$	$-\frac{1}{4}(1-\xi)(1-\eta^2)$
9	$\frac{1}{4}(1-\xi)(1-\eta)(1-\zeta^2)$	$-\frac{1}{4}(1-\eta)(1-\zeta^2)$	$-\frac{1}{4}(1-\xi)(1-\zeta^2)$	$-\frac{1}{2}\zeta(1-\xi)(1-\eta)$
10	$\frac{1}{4}(1+\xi)(1-\eta)(1-\zeta^2)$	$\frac{1}{4}(1-\eta)(1-\zeta^2)$	$-\frac{1}{4}(1+\xi)(1-\zeta^2)$	$-\frac{1}{2}\zeta(1+\xi)(1-\eta)$
11	$\frac{1}{4}(1+\xi)(1+\eta)(1-\zeta^2)$	$\frac{1}{4}(1+\eta)(1-\zeta^2)$	$\frac{1}{4}(1+\xi)(1-\zeta^2)$	$-\frac{1}{2}\zeta(1+\xi)(1+\eta)$
12	$\frac{1}{4}(1-\xi)(1+\eta)(1-\zeta^2)$	$-\frac{1}{4}(1+\eta)(1-\zeta^2)$	$\frac{1}{4}(1-\xi)(1-\zeta^2)$	$-\frac{1}{2}\zeta(1-\xi)(1+\eta)$
13	$\frac{1}{8}(1-\xi)(1-\eta)(1+\zeta)(-\xi-\eta+\zeta-2)$	$\frac{1}{8}(1-\eta)(1+\zeta)(2\xi+\eta-\zeta+1)$	$\frac{1}{8}(1-\xi)(1+\zeta)(2\eta+\xi-\zeta+1)$	$\frac{1}{8}(1-\xi)(1-\eta)(2\zeta-\eta-\xi-1)$
14	$\frac{1}{4}(1-\xi^2)(1-\eta)(1+\zeta)$	$-\frac{1}{2}\xi(1-\eta)(1+\zeta)$	$-\frac{1}{4}(1-\xi^2)(1+\zeta)$	$\frac{1}{4}(1-\xi^2)(1-\eta)$
15	$\frac{1}{8}(1+\xi)(1-\eta)(1+\zeta)(\xi-\eta+\zeta-2)$	$\frac{1}{8}(1-\eta)(1+\zeta)(2\xi-\eta+\zeta-1)$	$-\frac{1}{8}(1+\xi)(1+\zeta)(2\eta-\xi-\zeta+1)$	$\frac{1}{8}(1+\xi)(1-\eta)(2\zeta+\xi-\eta-1)$
16	$\frac{1}{4}(1+\xi)(1-\eta^2)(1+\zeta)$	$\frac{1}{4}(1-\eta^2)(1+\zeta)$	$-\frac{1}{2}\eta(1+\xi)(1+\zeta)$	$\frac{1}{4}(1+\xi)(1-\eta^2)$
17	$\frac{1}{8}(1+\xi)(1+\eta)(1+\zeta)(\xi+\eta+\zeta-2)$	$\frac{1}{8}\xi(1+\eta)(1+\zeta)(2\xi+\eta+\zeta-1)$	$\frac{1}{8}(1+\xi)(1+\zeta)(2\eta+\xi+\zeta-1)$	$\frac{1}{8}(1+\xi)(1+\eta)(2\zeta+\eta+\xi-1)$
18	$\frac{1}{4}(1-\xi^2)(1+\eta)(1+\zeta)$	$-\frac{1}{2}\xi(1+\eta)(1+\zeta)$	$\frac{1}{4}(1-\xi^2)(1+\zeta)$	$\frac{1}{4}(1-\xi^2)(1+\eta)$
19	$\frac{1}{8}(1-\xi)(1+\eta)(1+\zeta)(-\xi+\eta+\zeta-2)$	$\frac{1}{8}(1+\eta)(1+\zeta)(2\xi-\eta-\zeta+1)$	$\frac{1}{8}(1-\xi)(1+\zeta)(2\eta-\xi+\zeta-1)$	$\frac{1}{8}(1-\xi)(1+\eta)(2\zeta-\xi+\eta-1)$
20	$\frac{1}{4}(1-\xi)(1-\eta^2)(1+\zeta)$	$-\frac{1}{4}(1-\eta^2)(1+\zeta)$	$-\frac{1}{2}\eta(1-\xi)(1+\zeta)$	$\frac{1}{4}(1-\xi)(1-\eta^2)$

Thirty-two-Noded Solid Element

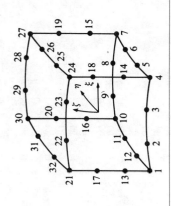

Node i	Shape functions $N_i(\xi, \eta, \zeta)$	$\dfrac{\partial N_i}{\partial \xi}$	$\dfrac{\partial N_i}{\partial \eta}$	$\dfrac{\partial N_i}{\partial \zeta}$
23	$\tfrac{9}{64}(1-\xi^2)(1+3\xi)(1-\eta)(1+\zeta)$	$\tfrac{9}{64}(3-2\xi-9\xi^2)(1-\eta)(1+\zeta)$	$-\tfrac{9}{64}(1-\xi^2)(1+3\xi)(1+\zeta)$	$\tfrac{9}{64}(1-\xi^2)(1+3\xi)(1-\eta)$
24	$\tfrac{9}{64}(1+\xi)(1-\eta)(1+\zeta)[\xi^2+\eta^2-\tfrac{19}{9}+\zeta^2]$	$\tfrac{9}{64}(1+\zeta)(1-\eta)(2\xi+3\xi^2+\zeta^2+\eta^2-\tfrac{19}{9}]$	$\tfrac{9}{64}(1+\xi)(1+\zeta)(2\eta^2-3\eta-\xi^2-\zeta^2+\tfrac{19}{9})$	$\tfrac{9}{64}(1+\xi)(1-\eta)(2\zeta+3\zeta^2+\xi^2+\eta^2-\tfrac{19}{9})$
25	$\tfrac{9}{64}(1-\eta^2)(1-3\eta)(1+\xi)(1+\zeta)$	$\tfrac{9}{64}(1-\eta^2)(1-3\eta)(1+\zeta)$	$\tfrac{9}{64}(9\eta^2-2\eta-3)(1+\xi)(1+\zeta)$	$\tfrac{9}{64}(1-\eta^2)(1-3\eta)(1+\xi)$
26	$\tfrac{9}{64}(1-\eta^2)(1+3\eta)(1+\xi)(1+\zeta)$	$\tfrac{9}{64}(1-\eta^2)(1+3\eta)(1+\zeta)$	$\tfrac{9}{64}(3-2\eta-9\eta^2)(1+\xi)(1+\zeta)$	$\tfrac{9}{64}(1-\eta^2)(1+3\eta)(1+\xi)$
27	$\tfrac{9}{64}(1+\xi)(1+\eta)(1+\zeta)(\xi^2+\eta^2+\zeta^2-\tfrac{19}{9})$	$\tfrac{9}{64}(1+\eta)(1+\zeta)(2\xi+3\xi^2+\eta^2+\zeta^2-\tfrac{19}{9})$	$\tfrac{9}{64}(1+\xi)(1+\zeta)(2\eta+3\eta^2+\xi^2+\zeta^2-\tfrac{19}{9})$	$\tfrac{9}{64}(1+\xi)(1+\eta)(2\zeta+3\zeta^2+\xi^2+\eta^2-\tfrac{19}{9})$
28	$\tfrac{9}{64}(1-\xi^2)(1+3\xi)(1+\eta)(1+\zeta)$	$\tfrac{9}{64}(1+\eta)(1+\zeta)(3-2\xi-9\xi^2)$	$\tfrac{9}{64}(1-\xi^2)(1+3\xi)(1+\zeta)$	$\tfrac{8}{64}(1-\xi^2)(1+3\xi)(1+\eta)$
29	$\tfrac{9}{64}(1-\xi^2)(1-3\xi)(1+\eta)(1+\zeta)$	$\tfrac{9}{64}(1+\eta)(1+\zeta)(9\xi^2-2\xi-3)$	$\tfrac{9}{64}(1-\xi^2)(1-3\xi)(1+\zeta)$	$\tfrac{9}{64}(1-\xi^2)(1-3\xi)(1+\eta)$
30	$\tfrac{9}{64}(1-\xi)(1+\eta)(1+\zeta)(\zeta^2+\eta^2+\xi^2-\tfrac{19}{9})$	$\tfrac{9}{64}(1+\eta)(1+\eta)(2\zeta-3\xi^2-\eta^2-\zeta^2+\tfrac{19}{9})$	$\tfrac{9}{64}(1-\xi)(1+\zeta)(2\eta+3\eta^2+\zeta^2+\xi^2-\tfrac{19}{9})$	$\tfrac{9}{64}(1-\xi)(1+\eta)(2\zeta+3\zeta^2+\xi^2+\eta^2-\tfrac{19}{9})$
31	$\tfrac{9}{64}(1-\eta^2)(1+3\eta)(1-\xi)(1+\zeta)$	$-\tfrac{9}{64}(1-\eta^2)(1+3\eta)(1+\zeta)$	$\tfrac{9}{64}(3-2\eta-9\eta^2)(1-\xi)(1+\zeta)$	$\tfrac{9}{64}(1-\eta^2)(1+3\eta)(1-\xi)$
32	$\tfrac{9}{64}(1-\eta^2)(1-3\eta)(1-\xi)(1+\zeta)$	$-\tfrac{9}{64}(1-\eta^2)(1-3\eta)(1+\zeta)$	$\tfrac{9}{64}(9\eta^2-2\eta-3)(1-\xi)(1+\zeta)$	$\tfrac{9}{64}(1-\eta^2)(1-3\eta)(1-\xi)$

Isoparametric Membrane Elements

Four-Noded Membrane Element

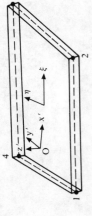

Node i	Shape functions $N_i(\xi, \eta)$	Derivatives $\dfrac{\partial N_i}{\partial \xi}$	$\dfrac{\partial N_i}{\partial \eta}$
1	$\frac{1}{4}(1-\xi)(1-\eta)$	$-\frac{1}{4}(1-\eta)$	$-\frac{1}{4}(1-\xi)$
2	$\frac{1}{4}(1+\xi)(1-\eta)$	$\frac{1}{4}(1-\eta)$	$-\frac{1}{4}(1+\xi)$
3	$\frac{1}{4}(1+\xi)(1+\eta)$	$\frac{1}{4}(1+\eta)$	$\frac{1}{4}(1+\xi)$
4	$\frac{1}{4}(1-\xi)(1+\eta)$	$-\frac{1}{4}(1+\eta)$	$\frac{1}{4}(1-\xi)$

Global cartesian system

Eight-Noded Membrane Element

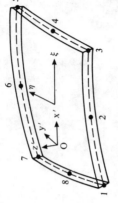

Node i	Shape functions $N_i(\xi, \eta)$	Derivatives $\dfrac{\partial N_i}{\partial \xi}$	$\dfrac{\partial N_i}{\partial \eta}$
1	$\tfrac{1}{4}(1-\xi)(1-\eta)(-\xi-\eta-1)$	$\tfrac{1}{4}(1-\eta)(2\xi+\eta)$	$\tfrac{1}{4}(1-\xi)(2\eta+\xi)$
2	$\tfrac{1}{2}(1-\xi^2)(1-\eta)$	$-\xi(1-\eta)$	$-\tfrac{1}{2}(1-\xi^2)$
3	$\tfrac{1}{4}(1+\xi)(1-\eta)(\xi-\eta-1)$	$\tfrac{1}{4}(1-\eta)(2\xi-\eta)$	$\tfrac{1}{4}(1+\xi)(2\eta-\xi)$
4	$\tfrac{1}{2}(1-\eta^2)(1+\xi)$	$\tfrac{1}{2}(1-\eta^2)$	$-\eta(1+\xi)$
5	$\tfrac{1}{4}(1+\xi)(1+\eta)(\xi+\eta-1)$	$\tfrac{1}{4}(1+\eta)(2\xi+\eta)$	$\tfrac{1}{4}(1+\xi)(2\eta+\xi)$
6	$\tfrac{1}{2}(1-\xi^2)(1+\eta)$	$-\xi(1+\eta)$	$\tfrac{1}{2}(1-\xi^2)$
7	$\tfrac{1}{4}(1-\xi)(1+\eta)(-\xi+\eta-1)$	$\tfrac{1}{4}(1+\eta)(2\xi-\eta)$	$\tfrac{1}{4}(1-\xi)(2\eta-\xi)$
8	$\tfrac{1}{2}(1-\eta^2)(1-\xi)$	$-\tfrac{1}{2}(1-\eta^2)$	$-\eta(1-\xi)$

Twelve-Noded Membrane Element

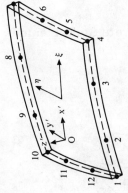

Node i	Shape functions $N_i(\xi, \eta)$	Derivatives $\dfrac{\partial N_i}{\partial \xi}$	$\dfrac{\partial N_i}{\partial \eta}$
1	$\tfrac{9}{32}(1-\xi)(1-\eta)[(\xi^2+\eta^2-\tfrac{10}{9}]$	$\tfrac{9}{32}(1-\eta)[2\xi-3\xi^2-\eta^2+\tfrac{10}{9}]$	$\tfrac{9}{32}(1-\xi)[2\eta-3\eta^2-\xi^2+\tfrac{10}{9}]$
2	$\tfrac{9}{32}(1-\xi)(1-\xi^2)(1-\eta)$	$\tfrac{9}{32}(1-\eta)(3\xi^2-2\xi-1)$	$-\tfrac{9}{32}(1-\xi)(1-\xi^2)$
3	$\tfrac{9}{32}(1-\eta)(1-\xi^2)(1+\xi)$	$\tfrac{9}{32}(1-\eta)(1-2\xi-3\xi^2)$	$-\tfrac{9}{32}(1-\xi^2)(1+\xi)$
4	$\tfrac{9}{32}(1+\xi)(1-\eta)[\xi^2+\eta^2-\tfrac{10}{9}]$	$\tfrac{9}{32}(1-\eta)[2\xi+3\xi^2+\eta^2-\tfrac{10}{9}]$	$\tfrac{9}{32}(1+\xi)[2\eta-3\eta^2-\xi^2-\tfrac{10}{9}]$
5	$\tfrac{9}{32}(1+\xi)(1-\eta^2)(1-\eta)$	$\tfrac{9}{32}(1-\eta^2)(1-\eta)$	$\tfrac{9}{32}(1+\xi)(3\eta^2-2\eta-1)$
6	$\tfrac{9}{32}(1+\xi)(1-\eta^2)(1+\eta)$	$\tfrac{9}{32}(1-\eta^2)(1+\eta)$	$\tfrac{9}{32}(1+\xi)(1-2\eta-3\eta^2)$
7	$\tfrac{9}{32}(1+\xi)(1+\eta)[\xi^2+\eta^2-\tfrac{10}{9}]$	$\tfrac{9}{32}(1+\eta)[2\xi+3\xi^2+\eta^2-\tfrac{10}{9}]$	$\tfrac{9}{32}(1+\xi)[2\eta+3\eta^2+\xi^2-\tfrac{10}{9}]$
8	$\tfrac{9}{32}(1+\eta)(1-\xi^2)(1+\xi)$	$\tfrac{9}{32}(1+\eta)(1-2\xi-3\xi^2)$	$\tfrac{9}{32}(1-\xi^2)(1+\xi)$
9	$\tfrac{9}{32}(1+\eta)(1-\xi^2)(1-\xi)$	$\tfrac{9}{32}(1+\eta)(3\xi^2-2\xi-1)$	$\tfrac{9}{32}(1-\xi^2)(1-\xi)$
10	$\tfrac{9}{32}(1-\xi)(1+\eta)[\xi^2+\eta^2-\tfrac{10}{9}]$	$\tfrac{9}{32}(1+\eta)[2\xi-3\xi^2-\eta^2+\tfrac{10}{9}]$	$\tfrac{9}{32}(1-\xi)[2\eta+3\eta^2+\xi^2-\tfrac{10}{9}]$
11	$\tfrac{9}{32}(1-\xi)(1-\eta^2)(1+\eta)$	$-\tfrac{9}{32}(1+\eta)(1-\eta^2)$	$\tfrac{9}{32}(1-\xi)(1-2\eta-3\eta^2)$
12	$\tfrac{9}{32}(1-\xi)(1-\eta^2)(1-\eta)$	$-\tfrac{9}{32}(1-\eta)(1-\eta^2)$	$\tfrac{9}{32}(1-\xi)(3\eta^2-2\eta-1)$

Three-Dimensional Reinforced Concrete Solid Element

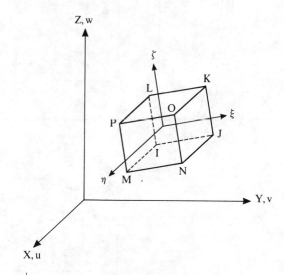

	Shape functions
Stiffness matrix $[K]$	$u = \frac{1}{8}[u_I(1-\xi)(1-\zeta)(1-\eta)$ $+ u_J(1+\xi)(1-\zeta)(1-\eta)$ $+ u_K(1+\xi)(1+\zeta)(1-\eta)$ $+ u_L(1-\xi)(1+\zeta)(1-\eta)$ $+ u_M(1-\xi)(1-\zeta)(1+\eta)$ $+ u_N(1+\xi)(1-\zeta)(1+\eta)$ $+ u_O(1+\xi)(1+\zeta)(1+\eta)$ $+ u_P(1-\xi)(1+\zeta)(1+\eta)]$ $+ u_1(1-\xi^2)$ $+ u_2(1-\zeta^2)$ $+ u_3(1-\eta^2)$ $v = \frac{1}{8}(v_I(1-\xi)\ldots$ (similar to u) $w = \frac{1}{8}(w_I(1-\xi)\ldots$ (similar to u)

Crack Tip Solid Element

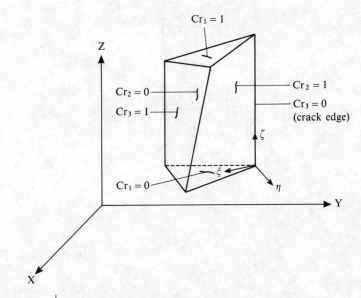

	Shape functions
Stiffness matrix $[K]$	6-node option $\xi = \alpha_1 + \alpha_2\sqrt{Cr_3} + \alpha_3 Cr_2\sqrt{Cr_3}$ $\quad + \alpha_4 Cr_1 + \alpha_5 Cr_1\sqrt{Cr_3} + \alpha_6 Cr_2 Cr_1\sqrt{Cr_3}$ $\eta = \alpha_7 + \alpha_8\sqrt{Cr_3} + \alpha_9 Cr_2\sqrt{Cr_3}$ $\quad + \alpha_{10} Cr_1 + \alpha_{11} Cr_1\sqrt{Cr_3} + \alpha_{12} Cr_2 Cr_1\sqrt{Cr_3}$ $\zeta = \alpha_{13} + \alpha_{14} Cr_3 + \alpha_{15} Cr_2 Cr_3$ $\quad + \alpha_{16} Cr_1 + \alpha_{17} Cr_1 Cr_3 + \alpha_{18} Cr_2 Cr_1 Cr_3$

Shape Function for a Prism Element

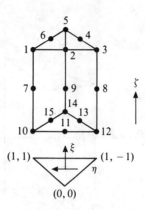

Fifteen Nodes

$N_1(\xi,\eta,\zeta) = -\frac{1}{2}(1+\zeta)(1-\xi^{1/2})(\sqrt{2}\xi^{1/2}-\zeta)$

$N_2(\xi,\eta,\zeta) = +\frac{1}{2}(1+\sqrt{2})(1+\zeta)(1-\xi^{1/2})(\xi^{1/2}+\eta\xi^{-1/2})$

$N_3(\xi,\eta,\zeta) = +\frac{1}{4}(1+\zeta)(\xi^{1/2}+\eta\xi^{-1/2})[(2+\sqrt{2})(\xi^{1/2}-1)-(\xi-\eta-\zeta)]$

$N_4(\xi,\eta,\zeta) = +\frac{1}{2}(1+\zeta)(\xi^{3/2}-\eta^2\xi^{-1/2})$

$N_5(\xi,\eta,\zeta) = +\frac{1}{4}(1+\zeta)(\xi^{1/2}-\eta\xi^{-1/2})[(2+\sqrt{2})(\xi^{1/2}-1)-(\xi+\eta-\zeta)]$

$N_6(\xi,\eta,\zeta) = +\frac{1}{2}(1+\sqrt{2})(1+\xi)(\xi^{1/2}-\eta\xi^{-1/2})(1-\xi^{1/2})$

$N_7(\xi,\eta,\zeta) = (1-\zeta^2)(1-\xi^{1/2})$

$N_8(\xi,\eta,\zeta) = +\frac{1}{2}(1-\zeta^2)(\xi^{1/2}+\eta\xi^{-1/2})$

$N_9(\xi,\eta,\zeta) = +\frac{1}{2}(1-\zeta^2)(\xi^{1/2}-\eta\xi^{-1/2})$

$N_{10}(\xi,\eta,\zeta) = -\frac{1}{2}(1-\zeta)(1-\xi^{1/2})(\sqrt{2}\xi^{1/2}+\zeta)$

$N_{11}(\xi,\eta,\zeta) = +\frac{1}{2}(1+\sqrt{2})(1-\zeta)(\xi^{1/2}+\eta\xi^{-1/2})(1-\xi^{1/2})$

$N_{12}(\xi,\eta,\zeta) = +\frac{1}{4}(1-\zeta)(\xi^{1/2}+\eta\xi^{-1/2})[(2+\sqrt{2})(\xi^{1/2}-1)-(\xi-\eta+\zeta)]$

$N_{13}(\xi,\eta,\zeta) = +\frac{1}{2}(1-\zeta)(\xi^{3/2}-\eta^2\xi^{-1/2})$

$N_{14}(\xi,\eta,\zeta) = +\frac{1}{4}(1-\zeta)(\xi^{1/2}-\eta\xi^{-1/2})[(2+\sqrt{2})(\xi^{1/2}-1)-(\xi+\eta+\zeta)]$

$N_{15}(\xi,\eta,\zeta) = +\frac{1}{2}(1+\sqrt{2})(1-\zeta)(1-\xi^{1/2})(\xi^{1/2}-\eta\xi^{-1/2})$

Right Tetrahedral Element

Four-Noded

Co-ordinates:

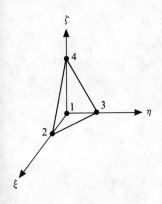

Nodal no	ξ_i	η_i	ζ_i
1	0	0	0
2	1	0	0
3	0	1	0
4	0	0	1

$N_1(\xi,\eta,\zeta) = (1 - \xi - \eta - \zeta)$
$N_2(\xi,\eta,\zeta) = +\xi$
$N_3(\xi,\eta,\zeta) = +\eta$
$N_4(\xi,\eta,\zeta) = +\zeta$

Ten-noded

Co-ordinates:

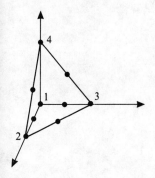

Nodal no,	ξ_i	η_i	ζ_i
1	0	0	0
2	1	0	0
3	0	1	0
4	0	0	1
5	$\tfrac{1}{2}$	0	0
6	$\tfrac{1}{2}$	$\tfrac{1}{2}$	0
7	0	$\tfrac{1}{2}$	0
8	0	0	$\tfrac{1}{2}$
9	$\tfrac{1}{2}$	0	$\tfrac{1}{2}$
10	0	$\tfrac{1}{2}$	$\tfrac{1}{2}$

$N_1(\xi,\eta,\zeta) = 2(1 - \xi - \eta - \zeta)^2 - (1 - \xi - \eta - \zeta)$
$N_3(\xi,\eta,\zeta) = (2\eta - 1)\eta$
$N_5(\xi,\eta,\zeta) = 4\xi(1 - \xi - \eta - \zeta)$
$N_7(\xi,\eta,\zeta) = 4\eta(1 - \xi - \eta - \zeta)$
$N_9(\xi,\eta,\zeta) = 4\xi\zeta$

$N_2(\xi,\eta,\zeta) = (2\xi - 1)\xi$
$N_4(\xi,\eta,\zeta) = (2\zeta - 1)\zeta$
$N_6(\xi,\eta,\zeta) = 4\xi\eta$
$N_8(\xi,\eta,\zeta) = 4\zeta(1 - \xi - \eta - \zeta)$
$N_{10}(\xi,\eta,\zeta) = 4\eta\zeta$

Shell Element

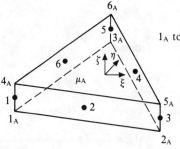

μ_A = area coordinate
1_A to 6_A = modified nodal system

Shape functions

$N_{1_A} = \frac{1}{2}(1 - \mu_A)\xi$ $N_{2_A} = \frac{1}{2}(1 - \mu_A)\eta$
$N_{3_A} = \frac{1}{2}(1 - \mu_A)\zeta$ $N_{4_A} = \frac{1}{2}(1 + \mu_A)\xi$
$N_{5_A} = \frac{1}{2}(1 + \mu_A)\eta$ $N_{6_A} = \frac{1}{2}(1 + \mu_A)\zeta$

Triangular Element

Three-Noded

Co-ordinates:

Nodal no.	ξ_i	η_i
1	0	0
2	1	0
3	0	1

$N_1(\xi, \eta) = 1 - \xi - \eta$ $N_2(\xi, \eta) = \xi$ $N_3(\xi, \eta) = \eta$

$\dfrac{\partial N_1}{\partial \xi} = -1$ $\dfrac{\partial N_1}{\partial \eta} = -1$ $\dfrac{\partial N_2}{\partial \xi} = 1$

$\dfrac{\partial N_2}{\partial \eta} = 0$ $\dfrac{\partial N_3}{\partial \xi} = 0$ $\dfrac{\partial N_3}{\partial \eta} = 1$

Six-Noded

Co-ordinates

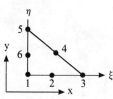

Nodal no.	ξ_i	η_i
1	0	0
2	$\frac{1}{2}$	0
3	1	0
4	$\frac{1}{2}$	$\frac{1}{2}$
5	0	1
6	0	$\frac{1}{2}$

$N_1(\xi, \eta) = 1 - 3\xi - 3\eta + 2\xi^2 + 2\eta^2 + 4\xi\eta$ $N_4(\xi, \eta) = 4\xi\eta$
$N_2(\xi, \eta) = 4\xi - 4\xi^2 - 4\xi\eta$ $N_5(\xi, \eta) = -\eta + 2\eta^2$
$N_3(\xi, \eta) = -\xi + 2\xi^2$ $N_6(\xi, \eta) = 4\eta - 4\xi - 4\eta^2$

Boom Elements

Two Nodes

Displacement functions
$u = a_1 + b_1\xi$
$v = a_2 + b_2\xi$
Degrees of freedom = 4

Three Nodes

Displacement functions
$u = a_1 + b_1\xi + c_1\xi^2$
$v = a_2 + b_2\xi + c_2\xi^2$
Degrees of freedom = 6

Two Nodes

Displacement functions
$u = u_1 L_1(\xi) + u_2 L_2(\xi)$
$v = v_1 L_1(\xi) + v_2 L_2(\xi)$
Degrees of freedom = 4

Beam Elements

Two Nodes

Displacement functions
$u = a_1 + b_1\xi$
$v = a_2 + b_2\xi + c_2\xi^2 + d_2\xi^3$
Degrees of freedom = 6

Three Nodes

Displacement functions
$u = a_1 + b_1\xi + c_1\xi^2$
$v = a_2 + b_2\xi + c_2\xi^2 + d_2\xi^3$
$\quad + e_2\xi^4 + f_2\xi^5$
Degrees of freedom = 9

Two Nodes

Displacement functions
$u = a_1 + b_1\xi + c_1\xi^2 + d_1\xi^3$
$v = a_2 + b_2\xi + c_2\xi^2 + d_2\xi^3$
$\quad + e_2\xi^4 + f_2\xi^5$
Degrees of freedom = 10

Linear and Quadratic Two-Dimensional Isoparametric Elements

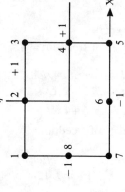

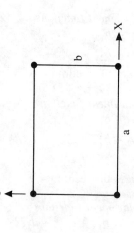

Linear shape functions:

$N_1 = \frac{1}{4}(1-\xi)(1+\eta)$
$N_2 = \frac{1}{4}(1+\xi)(1+\eta)$
$N_3 = \frac{1}{4}(1+\xi)(1-\eta)$
$N_4 = \frac{1}{4}(1-\xi)(1-\eta)$

mid-side nodes 2, 6 $(\xi_r = 0)$
$N_r = \frac{1}{2}(1-\xi^2)(1+\eta_0)$ $\Bigg\}$ Quadratic
mid-side nodes 4, 8 $(\eta_r = 0)$

Quadratic general shape function for the corner nodes:

$N_r = \frac{1}{4}(1+\xi_0)(1+\eta_0)(\xi_0+\eta_0+1)$ $(\xi_0 = \xi \xi_r,\ \eta_0 = \eta \eta_r)$

$$[B] = \sum_{r=1}^{4}[B]_r = \sum_{r=1}^{4} \begin{bmatrix} \frac{\partial N_r}{\partial y} & 0 \\ 0 & \frac{\partial N_r}{\partial y} \\ \frac{\partial N_r}{\partial y} & \frac{\partial N_r}{\partial x} \end{bmatrix} \qquad [J] = \begin{bmatrix} \frac{\partial X}{\partial \xi} & \frac{\partial Y}{\partial \xi} \\ \frac{\partial X}{\partial \eta} & \frac{\partial Y}{\partial \eta} \end{bmatrix} = \begin{bmatrix} \frac{a}{2} & 0 \\ 0 & \frac{b}{2} \end{bmatrix}$$

$$[B] = \begin{bmatrix} \dfrac{-(1+\eta)}{2a} & 0 & \dfrac{1-\xi}{2b} & 0 & \dfrac{(1+\eta)}{2a} & 0 & \dfrac{(1-\eta)}{2a} & 0 & \dfrac{-(1-\eta)}{2a} & 0 \\ 0 & \dfrac{1-\xi}{2b} & 0 & \dfrac{1+\xi}{2b} & 0 & \dfrac{(1+\xi)}{2b} & 0 & \dfrac{-(1+\xi)}{2b} & 0 & \dfrac{-(1-\xi)}{2b} \\ \dfrac{1-\xi}{2b} & \dfrac{-(1+\eta)}{2a} & \dfrac{1+\xi}{2b} & \dfrac{1+\eta}{2a} & \dfrac{1+\eta}{2a} & \dfrac{-(1+\xi)}{2b} & \dfrac{-(1-\xi)}{2b} & \dfrac{(1+\eta)}{2a} & \dfrac{-(1-\xi)}{2b} & \dfrac{-(1-\eta)}{2a} \end{bmatrix}$$

$$[K] = \int B^{T''} D B \, d\text{vol} = t \iint [B]^{T''}[D][B]|J| \, d\xi \, d\eta$$

Linear Strain Triangular Element

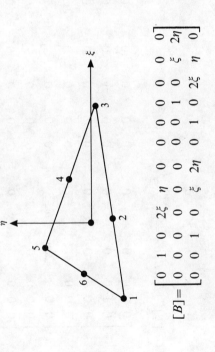

$$[B] = \begin{bmatrix} 0 & 1 & 0 & 2\xi & \eta & 0 & 0 & 0 & 0 & 0 & 0 & 0 \\ 0 & 0 & 0 & 0 & 0 & 0 & 0 & 0 & 1 & 0 & 2\eta & 0 \\ 0 & 0 & 1 & 0 & \xi & 2\eta & 0 & 1 & 0 & 2\xi & \eta & 0 \end{bmatrix}$$

$$[K] = [B]^{T''}[D][B]\,dA = \frac{\text{Area} \times E}{1-\nu^2}$$

$$\begin{bmatrix}
0 & 0 & 0 & 0 & 0 & 0 & 0 & 0 & 0 & 0 \\
0 & 1 & 0 & 0 & 0 & 0 & \nu & 0 & 0 & 0 \\
0 & 0 & d' & 0 & 0 & d' & 0 & 0 & 0 & 0 \\
0 & 0 & 0 & 4\overline{\xi\xi} & 2\overline{\xi\eta} & 0 & 0 & 0 & 2\nu\overline{\xi\xi} & 4\nu\overline{\xi\eta} \\
0 & 0 & 0 & 2\overline{\xi\eta} & d'\overline{\xi\xi}+\overline{\eta\eta} & 0 & 0 & G_1' & (\nu+d')\overline{\xi\eta} & 2\nu\eta\eta \\
0 & 0 & 0 & 0 & 2d'\overline{\xi\eta} & G' & 0 & G_2'' & 2d'\overline{\eta\eta} & 0 \\
0 & 0 & 0 & 0 & 0 & G'' & 0 & 0 & 0 & 0 \\
0 & 0 & d' & 0 & 0 & 0 & 0 & 1 & 0 & 0 \\
0 & \nu & 0 & 0 & 0 & G_2'' & 0 & 0 & G_1' & 2d'\overline{\xi\eta} \\
0 & 0 & 0 & 2d'\overline{\xi\xi} & (\nu+d')\overline{\xi\eta} & G_1' & 0 & 0 & G' & \overline{\xi\xi}+d'\overline{\eta\eta} \\
0 & 0 & 0 & 4d'\overline{\xi\eta} & 2\nu\eta\eta & 0 & 0 & 0 & 0 & 2\overline{\xi\eta} \\
\end{bmatrix}$$

$d' = (1-\nu)/2 \quad \overline{\xi\xi} = \frac{1}{12}(\xi_1^2 + \xi_3^2 + \xi_5^2) \quad \overline{\xi\eta} = \frac{1}{12}(\xi_1\eta_1 + \xi_3\eta_3 + \xi_5\eta_5)$

$\overline{\eta\eta} = \frac{1}{12}(\eta_1^2 + \eta_3^2 + \eta_5^2) \quad G' = 2d'\overline{\xi\eta} \quad G'' = 4d'\eta\eta$

$G_1' = 2d'\overline{\xi\xi} \quad G_2' = 4d'\overline{\xi\xi} \quad G_1'' = 4d'\overline{\xi\xi} \quad G_2'' = 4d'\overline{\xi\eta}$

Constant Stress/Strain [K] matrix

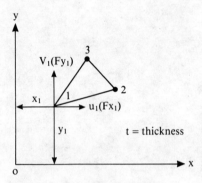

t = thickness

(A) Triangular

Plane stress:

$$\bar{d}_{11} = \bar{d}_{22} = E/(1 - v^2)$$
$$\bar{d}_{12} = \bar{d}_{21} = vE/(1 - v^2)$$
$$\bar{d}_{33} = E/2(1 + v)$$

Plane strain:

$$\bar{d}_{11} = \bar{d}_{22} = E(1 - v)/(1 + v)(1 - 2v)$$
$$\bar{d}_{12} = \bar{d}_{21} = vE/(1 + v)(1 - 2v)$$
$$\bar{d}_{33} = E/2(1 + v)$$

$$[K] = \frac{1}{4\Delta} \begin{bmatrix}
\bar{d}_{11}(y_2-y_3)^2 & \bar{d}_{12}(x_3-x_2)(y_2-y_3) & \bar{d}_{11}(y_2-y_3)(y_3-y_1) & \bar{d}_{12}(x_1-x_3)(y_2-y_3) & \bar{d}_{11}(y_2-y_3)(y_1-y_2) & \bar{d}_{12}(x_2-x_1)(y_2-y_3) \\
+\bar{d}_{33}(x_3-x_2)^2 & +\bar{d}_{33}(x_3-x_2)(y_2-y_3) & +\bar{d}_{33}(x_3-x_2)(x_1-x_3) & +\bar{d}_{33}(x_3-x_2)(y_3-y_1) & +\bar{d}_{33}(x_3-x_2)(x_2-x_1) & +\bar{d}_{33}(x_3-x_2)(y_1-y_2) \\
\bar{d}_{21}(x_3-x_2)(y_2-y_3) & \bar{d}_{22}(x_3-x_2)^2 & \bar{d}_{12}(y_3-y_1)(x_3-x_2) & \bar{d}_{22}(x_3-x_2)(x_1-x_3) & \bar{d}_{21}(x_3-x_2)(y_1-y_2) & \bar{d}_{22}(x_2-x_1)(x_3-x_2) \\
+\bar{d}_{33}(x_3-x_2)(y_2-y_3) & +\bar{d}_{33}(y_2-y_3)^2 & +\bar{d}_{33}(x_1-x_3)(y_2-y_3) & +\bar{d}_{33}(x_2-x_1)(y_2-y_3) & +\bar{d}_{33}(x_2-x_1)(y_2-y_3) & +\bar{d}_{33}(y_1-y_2)(y_2-y_3) \\
\bar{d}_{11}(y_2-y_3)(y_3-y_1) & \bar{d}_{12}(x_3-x_2)(y_3-y_1) & \bar{d}_{11}(y_3-y_1)^2 & \bar{d}_{12}(x_1-x_3)(y_3-y_1) & \bar{d}_{11}(y_3-y_1)(y_1-y_2) & \bar{d}_{12}(x_2-x_1)(y_3-y_1) \\
+\bar{d}_{33}(x_1-x_3)(x_3-x_2) & +\bar{d}_{33}(x_1-x_3)(y_2-y_3) & +\bar{d}_{33}(x_1-x_3)^2 & +\bar{d}_{33}(x_1-x_3)(y_3-y_1) & +\bar{d}_{33}(x_1-x_3)(x_2-x_1) & +\bar{d}_{33}(x_1-x_3)(y_1-y_2) \\
\bar{d}_{21}(x_1-x_3)(y_2-y_3) & \bar{d}_{22}(x_1-x_3)(x_3-x_2) & \bar{d}_{12}(x_1-x_3)(y_3-y_1) & \bar{d}_{22}(x_1-x_3)^2 & \bar{d}_{21}(x_1-x_3)(y_1-y_2) & \bar{d}_{22}(x_1-x_3)(x_2-x_1) \\
+\bar{d}_{33}(x_3-x_2)(y_3-y_1) & +\bar{d}_{33}(y_2-y_3)(y_3-y_1) & +\bar{d}_{33}(x_1-x_3)(y_3-y_1) & +\bar{d}_{33}(y_3-y_1)^2 & +\bar{d}_{33}(x_2-x_1)(y_3-y_1) & +\bar{d}_{33}(y_1-y_2)(y_3-y_1) \\
\bar{d}_{11}(y_1-y_2)(y_2-y_3) & \bar{d}_{12}(x_3-x_2)(y_1-y_2) & \bar{d}_{11}(y_1-y_2)(y_3-y_1) & \bar{d}_{12}(x_1-x_3)(y_1-y_2) & \bar{d}_{11}(y_1-y_2)^2 & \bar{d}_{12}(x_2-x_1)(y_1-y_2) \\
+\bar{d}_{33}(x_2-x_1)(x_3-x_2) & +\bar{d}_{33}(x_2-x_1)(y_2-y_3) & +\bar{d}_{33}(x_1-x_3)(x_2-x_1) & +\bar{d}_{33}(x_2-x_1)(y_3-y_1) & +\bar{d}_{33}(x_2-x_1)^2 & +\bar{d}_{33}(x_2-x_1)(y_1-y_2) \\
\bar{d}_{21}(x_2-x_1)(y_2-y_3) & \bar{d}_{22}(x_2-x_1)(x_3-x_2) & \bar{d}_{12}(x_2-x_1)(y_3-y_1) & \bar{d}_{22}(x_1-x_3)(x_2-x_1) & \bar{d}_{21}(x_2-x_1)(y_1-y_2) & \bar{d}_{22}(x_2-x_1)^2 \\
+\bar{d}_{33}(x_3-x_2)(y_1-y_2) & +\bar{d}_{33}(y_1-y_2)(y_2-y_3) & +\bar{d}_{33}(x_1-x_3)(y_1-y_2) & +\bar{d}_{33}(y_1-y_2)(y_3-y_1) & +\bar{d}_{33}(x_2-x_1)(y_1-y_2) & +\bar{d}_{33}(y_1-y_2)^2
\end{bmatrix}$$

(B) Rectangular

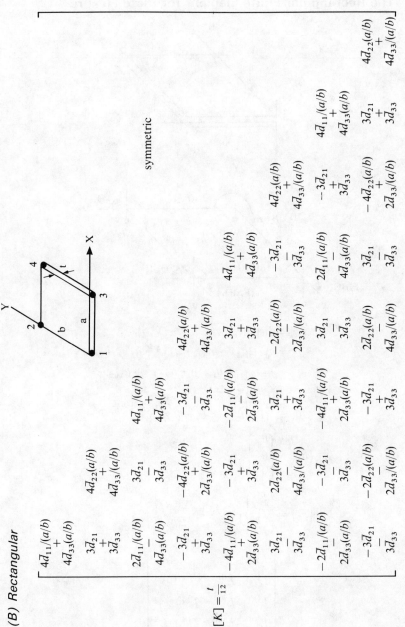

$$[K] = \frac{t}{12} \begin{bmatrix}
4\bar{d}_{11}/(a/b) \\
+ \\
4\bar{d}_{33}(a/b) \\[4pt]
\begin{array}{c}3\bar{d}_{21}\\+\\3\bar{d}_{33}\end{array} & 4\bar{d}_{22}(a/b)+4\bar{d}_{33}/(a/b) \\[4pt]
\begin{array}{c}2\bar{d}_{11}/(a/b)\\-\\4\bar{d}_{33}(a/b)\end{array} & \begin{array}{c}3\bar{d}_{21}\\-\\3\bar{d}_{33}\end{array} & \begin{array}{c}4\bar{d}_{11}/(a/b)\\+\\4\bar{d}_{33}(a/b)\end{array} \\[4pt]
\begin{array}{c}-3\bar{d}_{21}\\+\\3\bar{d}_{33}\end{array} & \begin{array}{c}-4\bar{d}_{22}(a/b)\\+\\2\bar{d}_{33}/(a/b)\end{array} & \begin{array}{c}-3\bar{d}_{21}\\-\\3\bar{d}_{33}\end{array} & 4\bar{d}_{22}(a/b)+4\bar{d}_{33}/(a/b) \\[4pt]
\begin{array}{c}-4\bar{d}_{11}/(a/b)\\+\\2\bar{d}_{33}(a/b)\end{array} & \begin{array}{c}-3\bar{d}_{21}\\+\\3\bar{d}_{33}\end{array} & \begin{array}{c}-2\bar{d}_{11}/(a/b)\\-\\2\bar{d}_{33}(a/b)\end{array} & \begin{array}{c}3\bar{d}_{21}\\+\\3\bar{d}_{33}\end{array} & \begin{array}{c}4\bar{d}_{11}/(a/b)\\+\\4\bar{d}_{33}(a/b)\end{array} \\[4pt]
\begin{array}{c}3\bar{d}_{21}\\-\\3\bar{d}_{33}\end{array} & \begin{array}{c}2\bar{d}_{22}(a/b)\\-\\4\bar{d}_{33}/(a/b)\end{array} & \begin{array}{c}3\bar{d}_{21}\\+\\3\bar{d}_{33}\end{array} & \begin{array}{c}-2\bar{d}_{22}(a/b)\\-\\2\bar{d}_{33}/(a/b)\end{array} & \begin{array}{c}-3\bar{d}_{21}\\-\\3\bar{d}_{33}\end{array} & 4\bar{d}_{22}(a/b)+4\bar{d}_{33}/(a/b) \\[4pt]
\begin{array}{c}-2\bar{d}_{11}/(a/b)\\-\\2\bar{d}_{33}(a/b)\end{array} & \begin{array}{c}-3\bar{d}_{21}\\-\\3\bar{d}_{33}\end{array} & \begin{array}{c}-4\bar{d}_{11}/(a/b)\\+\\2\bar{d}_{33}(a/b)\end{array} & \begin{array}{c}3\bar{d}_{21}\\-\\3\bar{d}_{33}\end{array} & \begin{array}{c}2\bar{d}_{11}/(a/b)\\-\\4\bar{d}_{33}(a/b)\end{array} & \begin{array}{c}-3\bar{d}_{21}\\+\\3\bar{d}_{33}\end{array} & \begin{array}{c}4\bar{d}_{11}/(a/b)\\+\\4\bar{d}_{33}(a/b)\end{array} \\[4pt]
\begin{array}{c}-3\bar{d}_{21}\\-\\3\bar{d}_{33}\end{array} & \begin{array}{c}-2\bar{d}_{22}(a/b)\\-\\2\bar{d}_{33}/(a/b)\end{array} & \begin{array}{c}-3\bar{d}_{21}\\+\\3\bar{d}_{33}\end{array} & \begin{array}{c}2\bar{d}_{22}(a/b)\\-\\4\bar{d}_{33}/(a/b)\end{array} & \begin{array}{c}3\bar{d}_{21}\\-\\3\bar{d}_{33}\end{array} & \begin{array}{c}-4\bar{d}_{22}(a/b)\\+\\2\bar{d}_{33}/(a/b)\end{array} & \begin{array}{c}3\bar{d}_{21}\\+\\3\bar{d}_{33}\end{array} & 4\bar{d}_{22}(a/b)+4\bar{d}_{33}/(a/b)
\end{bmatrix}$$

symmetric

The Rectangular Finite Element for Plate Flexure

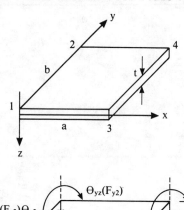

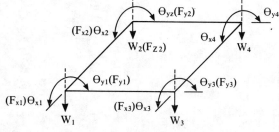

w = displacement
$$= a_1 + a_2 x + a_3 y + a_4 x^2 + a_5 xy + a_6 y^2$$
$$+ a_7 x^3 + a_8 x^2 y + a_9 xy^2 + a_{10} y^3$$
$$+ a_{11} x^3 y + a_{12} xy^3$$

$$\theta_x = -\partial w/\partial y \qquad \theta_y = \partial w/\partial x$$

$$\{\delta_1\} = \begin{Bmatrix} \theta_{x_1} \\ \theta_{y_1} \\ w_1 \end{Bmatrix} \text{ etc.}$$

$$\{F_1\} = \begin{Bmatrix} F_{x_1} \\ F_{y_1} \\ F_{z_1} \end{Bmatrix} \text{ etc.}$$

$$[K] = \begin{bmatrix} L^{\text{I}} & & & \\ L^{\text{II}} & L^{\text{III}} & & \\ L^{\text{IV}} & L^{\text{V}} & L^{\text{VI}} & \\ L^{\text{VII}} & L^{\text{VIII}} & L^{\text{IX}} & L^{\text{X}} \end{bmatrix}$$

$$[L^{\mathrm{I}}] = \begin{bmatrix} A & & \\ -B & C & \cdot \\ -D & E & F \end{bmatrix} \qquad [L^{\mathrm{II}}] = \begin{bmatrix} G & H & I \\ J & K & L \\ M & N & O \end{bmatrix}$$

$$[L^{\mathrm{III}}] = \begin{bmatrix} A & & \\ B & C & \\ D & E & F \end{bmatrix} \qquad [L^{\mathrm{IV}}] = \begin{bmatrix} P & Q & R \\ S & T & S' \\ X & Y & Z \end{bmatrix}$$

$$[L^{\mathrm{V}}] = \begin{bmatrix} G' & H' & I' \\ J' & K' & L' \\ M' & N' & O' \end{bmatrix} \qquad [L^{\mathrm{VI}}] = \begin{bmatrix} A & & \\ B & C & \\ -D & -E & F \end{bmatrix}$$

$$[L^{\mathrm{VII}}] = \begin{bmatrix} G' & H' & -I' \\ J' & K' & L' \\ M' & -N' & O' \end{bmatrix} \qquad [L^{\mathrm{VIII}}] = \begin{bmatrix} P & Q & -R \\ S & T & S' \\ -X & Y & Z \end{bmatrix}$$

$$[L^{\mathrm{IX}}] = \begin{bmatrix} G & H & I \\ J & K & -L \\ -M & -N & O \end{bmatrix} \qquad [L^{\mathrm{X}}] = \begin{bmatrix} A & & \\ -B & C & \\ D & -E & F \end{bmatrix}$$

$$H = 0 = J \qquad Q = O = S \qquad H' = 0 = J'$$

$A = 20a^2 D_y + 8b^2 D_{xy}$

$B = 15ab D_1$

$C = 20b^2 D_x + 8a^2 D_{xy}$

$D = 30\dfrac{a^2}{b} D_y + 15b D_1 + 6b D_{xy}$

$E = 30\dfrac{b^2}{a} D_x + 15a D_1 + 6a D_{xy}$

$F = 60\dfrac{b^2}{a^2} D_x + 60\dfrac{b^2}{a^2} D_y + 30 D_1 + 84 D_{xy}$

$G = 10a^2 D_y - 2b^2 D_{xy}$

$I = -30\dfrac{a^2}{b} D_y - 6b D_{xy}$

$K = 10b^2 D_x - 8a^2 D_{xy}$

$L = 15\dfrac{b^2}{a} D_x - 15a D_1 - 6a D_{xy}$

$O = 30\dfrac{b^2}{a} - 60\dfrac{a^2}{b^2} D_y - 30 D_1 - 84 D_{xy}$

$P = 10a^2 D_y - 8b^2 D_{xy}$

$R = -15\dfrac{a^2}{b} D_y + 15b D_1 + 6b D_{xy}$

$G' = 5a^2 D_y + 2b^2 D_{xy}$

$M' = 15\dfrac{a^2}{b} D_y - 6b D_{xy}$

$T = 10b^2 D_x - 2a^2 D_{xy}$

$Y = 30\dfrac{b^2}{a} D_x + 6a D_{xy}$

$K' = 5b^2 D_x + 2a^2 D_{xy}$

$N' = 15\dfrac{b^2}{a} D_x - 6a D_{xy}$

$Z = 60\dfrac{b^2}{a^2} D_x + 30\dfrac{a^2}{b^2} D_y - 30 D_1 - 84 D_{xy}$

$O' = -30\dfrac{b^2}{a^2} D_x - 30\dfrac{a^2}{b^2} D_y + 30 D_1 + 84 D_{xy}$

The Semiloof Shell Element

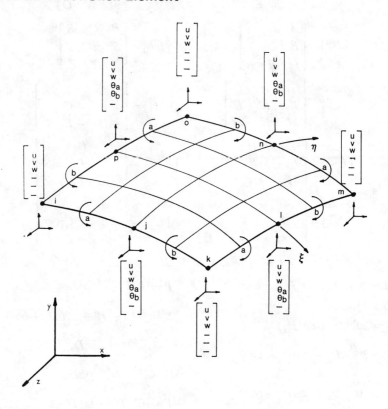

Loof Nodal Degrees of Freedom

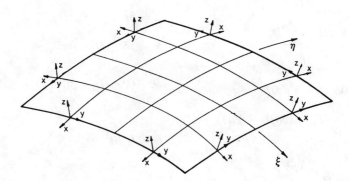

Bending Strain Nomenclature

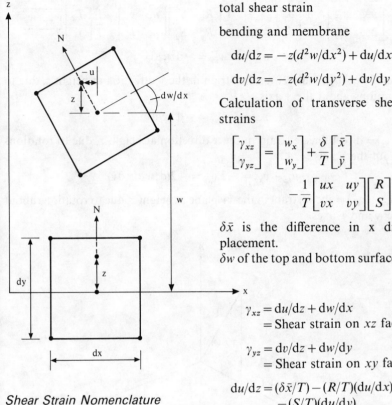

total shear strain

bending and membrane

$$du/dz = -z(d^2w/dx^2) + du/dx$$
$$dv/dz = -z(d^2w/dy^2) + dv/dy$$

Calculation of transverse shear strains

$$\begin{bmatrix} \gamma_{xz} \\ \gamma_{yz} \end{bmatrix} = \begin{bmatrix} w_x \\ w_y \end{bmatrix} + \frac{\delta}{T}\begin{bmatrix} \bar{x} \\ \bar{y} \end{bmatrix}$$

$$-\frac{1}{T}\begin{bmatrix} ux & uy \\ vx & vy \end{bmatrix}\begin{bmatrix} R \\ S \end{bmatrix}$$

$\delta\bar{x}$ is the difference in x displacement.
δw of the top and bottom surfaces.

$\gamma_{xz} = du/dz + dw/dx$
 = Shear strain on xz face

$\gamma_{yz} = dv/dz + dw/dy$
 = Shear strain on xy face

$du/dz = (\delta\bar{x}/T) - (R/T)(du/dx)$
$\qquad - (S/T)(du/dy)$

Shear Strain Nomenclature

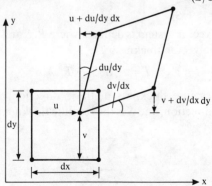

Note: x, y axes are the same as X, Y in the text.

Calculation of Equivalent Rotational Second Derivatives

$$\begin{bmatrix} u_{xz} & v_{xz} \\ u_{yz} & v_{yz} \end{bmatrix} = \frac{1}{T}\begin{bmatrix} \delta_x \\ \delta_y \end{bmatrix}\begin{bmatrix} \bar{x} & \bar{y} \end{bmatrix} - \frac{1}{T}\begin{bmatrix} R_x & S_x \\ R_y & S_y \end{bmatrix}\begin{bmatrix} u_x & v_x \\ u_y & v_y \end{bmatrix} + \frac{1}{T}\begin{bmatrix} T_x \\ T_y \end{bmatrix}\begin{bmatrix} w_x & w_y \end{bmatrix}$$

$$u_{xz} = -w_{xx} = -d^2w/dx^2$$

therefore $z(u_{xz})$ = direct tensile strain in the y direction at height z, due to rotations about the x axis.

$$v_{yz} = -w_{yy} = -d^2w/dy^2$$

$z(v_{yz})$ = direct tensile strain in the x direction at height z, due to rotations about the y axis.

$$u_{yz} + v_{xz} = -2w_{xy} = -2(d^2w/dx\,dy)$$

$z(u_{yz} + v_{xz})$ = shear strain in the xy plane at height z, due to rotations about the x and y axes.

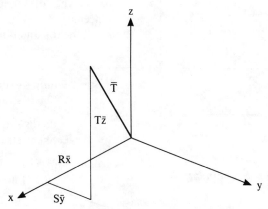

An orthogonal vector system is defined where R, S and T represent slope components of the vector thickness $\bar{T}$

$$\bar{T} = R_{\bar{x}} + S_{\bar{y}} + T_{\bar{z}}$$

where T is the resolute thickness in the z direction, R is the resolute thickness in the x direction, and S is the resolute thickness in the y direction

$$\bar{T} = \sum_{j=1}^{9} \bar{T}jL_j$$

where L_j is a Loof mapping function.

Resolution of a Thickness Vector

$$[D] = \frac{Et}{(1-v^2)}\begin{bmatrix} 1 & v & 0 & 0 & 0 & 0 \\ v & 1 & 0 & 0 & 0 & 0 \\ 0 & 0 & 1/2(1-v) & 0 & 0 & 0 \\ 0 & 0 & 0 & t^2/12 & vt^2/12 & 0 \\ 0 & 0 & 0 & vt^2/12 & t^2/12 & 0 \\ 0 & 0 & 0 & 0 & 0 & t^2(1-v)/24 \end{bmatrix}$$

Gap Element

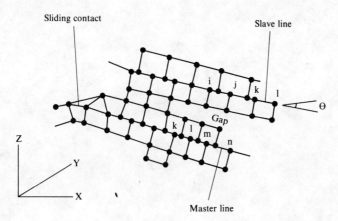

$$\{\vec{\bar{F}}_N\} = [K_{\gamma'\gamma'}]_i \{U_{\gamma'}\}_i = [\Sigma k]\{\Sigma \Delta_i, \Delta_j \ldots\} = \{\vec{\bar{F}}_{i,j}, \ldots\} + \{\pm \mu \vec{\bar{F}}_n \ldots \pm k\Delta_i \ldots\}$$

Δ_{sl} = distance of sliding

$$= (\Delta_j - \Delta_i) - \frac{\mu|\vec{\bar{F}}_n|}{[K_{\gamma'\gamma'}]}$$

μ = friction

$\{\vec{\bar{F}}_{SN}\} \leq \mu\{\vec{\bar{F}}_N\}$ no sliding

$\phantom{\{\vec{\bar{F}}_{SN}\}} \geq \mu\{\vec{\bar{F}}_N\}$ sliding

$\phantom{\{\vec{\bar{F}}_{SN}\}} = 0$ contact broken

$$\theta = \cos^{-1}\frac{X}{\gamma} \quad \text{or} \quad \sin^{-1}\frac{Y}{\gamma}$$

APPENDIX II

TRANSFORMATION MATRICES

STRESS AND STRAIN TRANSFORMATION MATRICES

$$\{T''_\sigma\}\{T''_\varepsilon\}$$

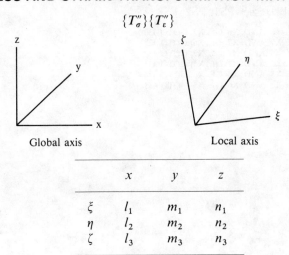

Global axis   Local axis

	x	y	z
ξ	l_1	m_1	n_1
η	l_2	m_2	n_2
ζ	l_3	m_3	n_3

Direction cosines of the two axes are given by:

$$l_1 = \cos(\xi, x) \quad m_1 = \cos(\xi, y) \quad n_1 = \cos(\xi, z)$$
$$l_2 = \cos(\eta, x) \quad m_2 = \cos(\eta, y) \quad n_2 = \cos(\eta, z)$$
$$l_3 = \cos(\zeta, x) \quad m_3 = \cos(\zeta, y) \quad n_3 = \cos(\zeta, z)$$

The following relationships can be written for local and global strain and stress vectors:

$$\{\varepsilon'_x\} = [T_\varepsilon]\{\varepsilon_x\} \qquad \{\sigma_x\} = [T_\varepsilon]^T\{\sigma'_x\}$$

and also

$$\{\sigma'_x\} = [T_\sigma]\{\sigma_x\} \qquad \{\varepsilon_x\} = [T_\sigma]^T\{\varepsilon'_x\}$$

$$[T''_\varepsilon] = \begin{bmatrix} l_1^2 & m_1^2 & n_1^2 & l_1 m_1 & m_1 n_1 & l_1 n_1 \\ l_2^2 & m_2^2 & n_2^2 & l_2 m_2 & m_2 n_2 & l_2 n_2 \\ l_3^2 & m_3^2 & n_3^2 & l_3 m_3 & m_3 n_3 & l_3 n_3 \\ 2l_1 l_2 & 2m_1 m_2 & 2n_1 n_2 & l_1 m_2 + l_2 m_1 & m_1 n_2 + m_2 n_1 & l_1 n_2 + l_2 n_1 \\ 2l_2 l_3 & 2m_2 m_3 & 2n_2 n_3 & l_2 m_3 + l_3 m_2 & m_2 n_3 + n_2 m_3 & l_2 n_3 + l_3 n_2 \\ 2l_1 l_3 & 2m_1 m_3 & 2n_1 n_3 & l_1 m_3 + m_1 l_3 & m_1 n_3 + m_3 n_1 & l_1 n_3 + n_1 l_3 \end{bmatrix}$$

$$[T''_\sigma] = \begin{bmatrix} l_1^2 & m_1^2 & n_1^2 & 2l_1 m_1 & 2m_1 n_1 & 2l_1 n_1 \\ l_2^2 & m_2^2 & n_2^2 & 2l_2 m_2 & 2m_2 n_2 & 2l_2 n_2 \\ l_3^2 & m_3^2 & n_3^2 & 2l_3 m_3 & 2m_3 n_3 & 2l_3 n_3 \\ l_1 l_2 & m_1 m_2 & n_1 n_2 & l_1 m_2 + l_2 m_1 & m_1 n_2 + n_1 m_2 & l_1 n_2 + l_2 n_1 \\ l_2 l_3 & m_2 m_3 & n_2 n_3 & l_2 m_3 + l_3 m_2 & m_2 n_3 + n_2 m_3 & l_2 n_3 + l_3 n_2 \\ l_1 l_3 & m_1 m_3 & n_1 n_3 & l_1 m_3 + l_3 m_1 & m_1 n_3 + m_3 n_1 & l_1 n_3 + n_1 l_3 \end{bmatrix}$$

Principal stresses and direction cosines D1, D2, D3 are the direction cosines of principal stresses PS1, PS2, PS3

```
      IF (X5 .GE. X6 .AND. X6 .GE. X7) GOTO 430
      IF (X5 .GE. X7 .AND. X7 .GE. X6) GOTO 431
      IF (X6 .GE. X5 .AND. X5 .GE. X7) GOTO 432
      IF (X6 .GE. X7 .AND. X7 .GE. X5) GOTO 433
      IF (X7 .GE. X5 .AND. X5 .GE. X6) GOTO 434
      IF (X7 .GE. X5 .AND. X6 .GE. X5) GOTO 435
  430 X1 = X5
      X2 = X6
      X3 = X7
      GOTO 438
  431 X1 = X5
      X2 = X7
      X3 = X6
      GOTO 438
  432 X1 = X6
      X2 = X5
      X3 = X7
      GOTO 438
  433 X1 = X6
      X2 = X7
      X3 = X5
      GOTO 438
  434 X1 = X7
      X2 = X5
      X3 = X6
      GOTO 438
  435 X1 = X7
      X2 = X6
      X3 = X5
  438 CONTINUE
```

PRINCIPAL STRESSES

```
      PS1(IPT) = X1
      PS2(IPT) = X2
      PS3(IPT) = X3
      DO 440 IS = 1,3
      GOTO (443,445,447),IS
  443 AS1 = G1 – X1
      AS2 = G2 – X1
      AS3 = G3 – X1
      GOTO 444
  445 AS1 = G1 – X2
      AS2 = G2 – X2
      AS3 = G3 – X2
      GOTO 444
  447 AS1 = G1 – X3
      AS2 = G2 – X3
      AS3 = G3 – X3
  444 CONTINUE
      AK = G4
      BK = G5
      CK = G6
      YAP1 = AS2*CK – BK*AK
      YAP2 = AK*AK – AS1*AS2
      IF (YAP1 .EQ. 0.0) YAP1 = 1.0
      IF (YAP2 .EQ. 0.0) YAP2 = 1.0
      BJM1 = (BK*BK – AS2*AS3)/YAP1
      BJM2 = (AS1*BK – AK*CK)/YAP2
      BJ1 = BJM1*BJM1
      BJ2 = BJM2*BJM2
      ZIP = DSQRT(BJ1 + BJ2 + 1.0)
```

APPENDIX III

MATERIAL AND CRACKING MATRICES

MATERIAL MATRICES [D] & [D*]
CRACKING MATRICES

Uncracked Material Matrix [D]

$$[D] = \begin{bmatrix} D_{11} & D_{12} & D_{13} & 0 & 0 & 0 \\ D_{21} & D_{22} & D_{23} & 0 & 0 & 0 \\ D_{31} & D_{32} & D_{33} & 0 & 0 & 0 \\ 0 & 0 & 0 & D_{44} & 0 & 0 \\ 0 & 0 & 0 & 0 & D_{55} & 0 \\ 0 & 0 & 0 & 0 & 0 & D_{66} \end{bmatrix}$$

Cracked Material Matrices [D*]

Cracks in Principal Direction 'One'

$D_{11}^* = D_{12}^* = D_{13}^* = D_{21}^* = D_{31}^* = 0$
$D_{22}^* = D_{22} - D_{12}D_{12}/D_{11}$
$D_{33}^* = D_{33} - D_{13}D_{13}/D_{11}$
$D_{23}^* = D_{23} - D_{12}D_{13}/D_{11}$
$D_{44}^* = \beta' D_{44}$
$D_{55}^* = D_{55}$
$D_{66}^* = \beta' D_{66}$

Cracks in Principal Direction 'Two'

$D_{22}^* = D_{12}^* = D_{21}^* = D_{23}^* = D_{32}^* = 0$
$D_{11}^* = D_{11} - D_{21}D_{21}/D_{22}$
$D_{33}^* = D_{33} - D_{23}D_{23}/D_{22}$
$D_{13}^* = D_{13} - D_{12}D_{23}/D_{22}$
$D_{31}^* = D_{13}^*$
$D_{44}^* = \beta' D_{44}$
$D_{55}^* = \beta' D_{55}$
$D_{66}^* = D_{66}$

Cracks in Principal Direction 'Three'

$$D_{33}^* = D_{13}^* = D_{31}^* = D_{23}^* = D_{31}^* = 0$$
$$D_{11}^* = D_{11} - D_{13}D_{13}/D_{33}$$
$$D_{22}^* = D_{22} - D_{23}D_{23}/D_{33}$$
$$D_{12}^* = D_{12} - D_{13}D_{23}/D_{33}$$
$$D_{21}^* = D_{12}^*$$
$$D_{44}^* = D_{44}$$
$$D_{55}^* = \beta' D_{55}$$
$$D_{66}^* = \beta' D_{66}$$

Cracks in Principal Directions 'One' and 'Two'

$$D_{11}^* = D_{22}^* = D_{12}^* = D_{21}^* = 0$$
$$D_{13}^* = D_{31}^* = D_{23}^* = D_{32}^* = 0$$
$$D_{33}^* = D_{33} - D_{13}D_{13}/D_{11} - D_{23}D_{23}/D_{22}$$
$$D_{44}^* = \beta' D_{44}$$
$$D_{55}^* = \beta' D_{55}$$
$$D_{66}^* = \beta' D_{66}$$

Cracks in Principal Directions 'Two' and 'Three'

$$D_{22}^* = D_{33}^* = D_{23}^* = D_{32}^* = 0$$
$$D_{13}^* = D_{31}^* = D_{12}^* = D_{21}^* = 0$$
$$D_{11}^* = D_{11} - D_{21}D_{21}/D_{22} - D_{13}D_{13}/D_{33}$$
$$D_{44}^* = \beta' D_{44}$$
$$D_{55}^* = \beta' D_{55}$$
$$D_{66}^* = \beta' D_{66}$$

Cracks in Principal Directions 'Three' and 'One'

$$D_{11}^* = D_{33}^* = D_{12}^* = D_{21}^* = 0$$
$$D_{13}^* = D_{31}^* = D_{23}^* = D_{31}^* = 0$$
$$D_{22}^* = D_{22} - D_{12}D_{12}/D_{11} - D_{23}D_{23}/D_{33}$$
$$D_{44}^* = \beta' D_{44}$$
$$D_{55}^* = \beta' D_{55}$$
$$D_{66}^* = \beta' D_{66}$$

Cracks in All Three Principal Directions

$$[D^*] = [0]$$

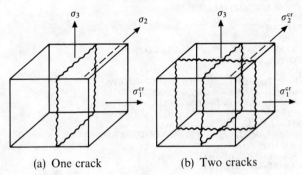

(a) One crack   (b) Two cracks

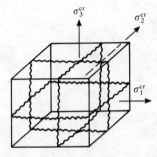

(c) Three cracks

Material Matrix for Reinforcement

```
      IMPLICIT REAL*8(A-H,O-Z)
      COMMON /MTMD3D/ D(6,6),STRESS(6),STRAIN(6),IPT,NEL
      DIMENSION PROP(1),DS(6,6),SIG(1),EPS(1),NCK(1),PS1(1),PS2(1),
     1           PS3(1),DC1(1),DC2(1),DC3(1)
      DO 111 II = 1,6
      DO 111 JJ = 1,6
  111 DS(II,JJ) = 0.0
      DS(1,1) = PROP(9)/PROP(6)*PROP(2)
      DS(2,2) = PROP(10)/PROP(7)*PROP(2)
      DS(3,3) = PROP(11)/PROP(8)*PROP(2)
      CALL TESTCK (PROP,SIG,EPS,NCK,PS1,PS2,PS3,DC1,DC2,DC3)
      IF (NCK(1) .EQ. 1 .OR. NCK(2) .EQ. 1 .OR. NCK(3) .EQ. 1)
     @   GOTO 220
      CALL DMAT(PROP,NCK)
  220 DO 222 III = 1,6
      DO 222 JJJ = 1,6
  222 D(III,JJJ) = D(III,JJJ) + DS(III,JJJ)
      RETURN
      END
```

Orthotropic Variable-Modulus Model for Concrete

```
      IMPLICIT REAL*8(A - H,O - Z)
      DIMENSION E(3),G(3,3),D(6,6),PROP(1)
      DO 222 II = 1,6
      DO 222 JJ = 1,6
  222 D(II,JJ) = 0.0
      AA = (1.0 - PROP(5))/(1.0 + PROP(5))*(1.0 - 2.0*PROP(5))
      BB = PROP(5)/(1.0 - PROP(5))

      E(1) = PROP(12)*PROP(1)*PROP(6) + PROP(2)*PROP(9)
      E(2) = PROP(12)*PROP(1)*PROP(7) + PROP(2)*PROP(10)
      E(3) = PROP(12)*PROP(1)*PROP(8) + PROP(2)*PROP(11)
      DO 7100 J = 1,3
      DO 7100 K = 1,3
 7100 G(J,K) = 0.25*(AA*(E(J) + E(K))) - 2.0*AA*BB*DSQRT(E(J)*E(K))
      D(1,1) = AA*E(1)
      D(1,2) = AA*BB*DSQRT(E(1)*E(2))
      D(1,3) = AA*BB*DSQRT(E(1)*E(3))
      D(2,1) = D(1,2)
      D(2,2) = AA*E(2)
      D(2,3) = BB*DSQRT(E(2)*E(3))
      D(3,1) = D(1,3)
      D(3,2) = D(2,3)
      D(3,3) = AA*E(3)
      D(4,4) = G(1,2)
      D(5,5) = G(1,3)
      D(6,6) = G(2,3)
      RETURN
      END
```

Ottoson Model

```
      IMPLICIT REAL*8(A - H,O - Z)
      COMMON /MTMD3D/ DEP(6,6),STRESS(6),STRAIN(6),IPT,NEL
      DIMENSION PAR(3,5),FS(6,6),FSTPOS(6,6),PROP(1),SIG(1),
     @          DVI1DS(6),DVJ2DS(6),DVJ3DS(6),DVTHDS(6)
      OPEN (UNIT = 5,FILE = 'PARAMETERS',STATUS = 'OLD')
      READ (5,*,END = 3700)((PAR(IF,JF),JF = 1,5),IF = 1,3)
 3700 CLOSE (5)
      PK = PROP(3)/PROP(4)
      IP = 0
      JP = 0
      IF (PK .LE. 0.08) IP = 1
      IF (PK .EQ. 0.10) IP = 2
      IF (PK .GE. 0.12) IP = 3
      IF (PK .LT. 0.10) JP = 1
      IF (PK .GT. 0.10) JP = 2
      IF (IP .EQ. 0) GOTO 3800
      A = PAR(IP,2)
      B = PAR(IP,3)
      PK1 = PAR(IP,4)
```

```
      PK2 = PAR(IP,5)
      GOTO 3900
3800  SUB1 = PK - PAR(JP,1)
      SUB2 = PAR(JP+1,1) - PAR(JP,1)
      A = SUB1*(PAR(JP+1,2) - PAR(JP,2))/SUB2 + PAR(JP,2)
      B = SUB1*(PAR(JP+1,3) - PAR(JP,3))/SUB2 + PAR(JP,3)
      PK1 = SUB1*(PAR(JP+1,4) - PAR(JP,4))/SUB2 + PAR(JP,4)
      PK2 = SUB1*(PAR(JP+1,5) - PAR(JP,5))/SUB2 + PAR(JP,5)
3900  VARI1 = SIG(1) + SIG(2) + SIG(3)
      VARJ2 = 1.0/6.0*((SIG(1) - SIG(2))**2 + (SIG(2) - SIG(3))**2 +
     @       (SIG(3) - SIG(1))**2) + SIG(4)**2 + SIG(5)**2 + SIG(6)**2
      VARI13 = VARI1/3.0
      VI131 = SIG(1) - VARI13
      VI132 = SIG(2) - VARI13
      VI133 = SIG(3) - VARI13
      VARJ3 = VI131*(VI132*VI133 - SIG(5)**2) - SIG(4)*(SIG(4)*VI133
     @       - SIG(5)*SIG(5)) + SIG(6)*(SIG(4)*SIG(5) - SIG(6)*VI132)
      VAR3TH = 1.5*3.0**(0.5)*VARJ3/VARJ2**1.5
      IF (VAR3TH .GE. 0.0) GOTO 4000
      ALAM = 22.0/21.0 - 1.0/3.0*ACOS( - PK2*VAR3TH)
      TOTLAM = PK1*COS(ALAM)
      DFD3TH = PK1*PK2*VARJ2**0.5*SIN(ALAM)/(3.0*PROP(4)*
     @       SIN(ACOS( - PK2*VAR3TH)))
      GOTO 4100
4000  ALAM = 1.0/3.0*ACOS(PK2*VAR3TH)
      TOTLAM = PK1*COS(ALAM)
      DFD3TH = PK1*PK2*VARJ2**0.5*SIN(ALAM)/(3.0*PROP(4)*
     @       SIN(ACOS(PK2*VAR3TH)))
4100  DFDI1 = B/PROP(4)
      DFDJ2 = A/PROP(4)**2 + TOTLAM/(PROP(4)*VARJ2**0.5)
      DVI1DS(1) = 1.0
      DVI1DS(2) = 1.0
      DVI1DS(3) = 1.0
      DVI1DS(4) = 0.0
      DVI1DS(5) = 0.0
      DVI1DS(6) = 0.0
      DVJ2DS(1) = 1.0/3.0*(2.0*SIG(1) - SIG(2) - SIG(3))
      DVJ2DS(2) = 1.0/3.0*(2.0*SIG(2) - SIG(1) - SIG(3))
      DVJ2DS(3) = 1.0/3.0*(2.0*SIG(3) - SIG(1) - SIG(2))
      DVJ2DS(4) = 2.0*SIG(4)
      DVJ2DS(5) = 2.0*SIG(5)
      DVJ2DS(6) = 2.0*SIG(6)
      DVJ3DS(1) = 1.0/3.0*(VI131*( - VI132 - VI133)) + 2.0*VI132*VI131 -
     @       2.0*SIG(5)**2 + SIG(4)**2 + SIG(6)**2
      DVJ3DS(2) = 1.0/3.0*(VI132*( - VI131 - VI133)) + 2.0*VI131*VI133 -
     @       2.0*SIG(6)**2 + SIG(4)**2 + SIG(5)**2
      DVJ3DS(3) = 1.0/3.0*(VI133*( - VI131 - VI132)) + 2.0*VI131*VI132 -
     @       2.0*SIG(4)**2 + SIG(5)**2 + SIG(6)**2
      DVJ3DS(4) = - 2.0*VI133*SIG(4) + 2.0*SIG(5)*SIG(6)
      DVJ3DS(5) = - 2.0*VI131*SIG(5) + 2.0*SIG(4)*SIG(6)
      DVJ3DS(6) = - - 2.0*VI132*SIG(6) + 2.0*SIG(4)*SIG(5)
      CONVJ2 = 3.0*3.0**0.5/(2.0*VARJ2*1.2)
      VJ3J2 = VARJ3/VARJ2**0.5
      DVTHDS(1) = CONVJ2*( - 0.5*VJ3J2*(2.0*SIG(1) - SIG(2) - SIG(3)) +
     @       DVJ3DS(1))
      DVTHDS(2) = CONVJ2*( - 0.5*VJ3J2*(2.0*SIG(2) - SIG(1) - SIG(3)) +
     @       DVJ3DS(2))
      DVTHDS(3) = CONVJ2*( - 0.5*VJ3J2*(2.0*SIG(3) - SIG(1) - SIG(2)) +
     @       DVJ3DS(3))
      DVTHDS(4) = CONVJ2*( - 3.0*VJ3J2*SIG(4) + DVJ3DS(4))
```

```
      DVTHDS(5) = CONVJ2*(-3.0*VJ3J2*SIG(5) + DVJ3DS(5))
      DVTHDS(6) = CONVJ2*(-3.0*VJ3J2*SIG(6) + DVJ3DS(6))
      DO 4200 IS = 1,6
      FS(IS,1) = DFDI1*DVI1DS(IS) + DFDJ2*DVJ2DS(IS) +
     1           DFD3TH*DVTHDS(IS)
 4200 FSTPDS(1,IS) = FS(IS,1)
      RETURN
      END
```

APPENDIX IV

GAUSSIAN QUADRATURE

THE GAUSS–LEGENDRE QUADRATURE

Integration Points, Co-ordinates and Weighting Factors

Typical Integral

$$\int_a^b F(x)\,dx = \int_{-1}^{+1} F(x)\,dx = \sum_{i=1}^n A_i f(x_i)$$

n	Point ξ_i	Weights w_i
1-point formula	0·000 000 000 0	2 × 10¹⁰
2-point formula	±0·577 350 269 2	1 × 10¹⁰
3-point formula	0·000 000 000 0	0·888 888 888 9
	±0·774 596 669 2	0·555 555 555 5
4-point formula	±0·339 981 043 5	0·652 145 154 8
	±0·861 136 311 6	0·347 854 845 1
5-point formula	0·000 000 000 0	0·568 888 888 9
	±0·538 469 310 1	0·478 628 670 5
	±0·906 179 845 9	0·236 926 885 0
6-point formula	±0·238 619 186 1	0·467 913 934 6
	±0·661 209 386 5	0·360 761 573 0
	±0·932 469 514 2	0·171 324 492 4

Typical Case: 20-Noded Isoparametric Element

$$\int_{-1}^{+1}\int_{-1}^{+1}\int_{-1}^{+1} F(\xi,\eta,\zeta)\,d\xi\,d\eta\,d\zeta$$

$$= B_6\{F(-b,0,0) + F(b,0,0) + F(0,-b,0) + F(0,b,0)$$
$$+ F(0,0,b) + F(0,0,b)\}$$
$$+ C_8\{F(-c,-c,-c) + F(c,-c,-c) + F(c,c,-c)$$
$$+ F(c,c,c) + F(-c,c,-c) + F(-c,-c,c)$$
$$+ F(-c,c,c) + F(c,-c,c)\}$$

$B_6 = 0·886\,426\,593 \quad C_8 = 0·335\,180\,055$
$b = 0·795\,822\,426$
$c = 0·758\,786\,911$

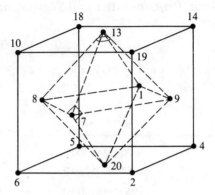

Typical locations of integration points (example: twenty-noded element).

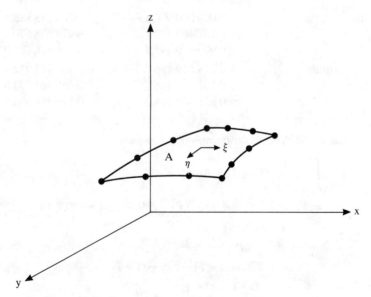

Surface area in curvilinear co-ordinates.

APPENDIX V

ACCELERATION AND CONVERGENCE

CRITERIA FOR CONVERGENCE AND ACCELERATION

Convergence Criteria

To ensure convergence to the correct solution by finer sub-division of the mesh, the assumed displacement function must satisfy the convergence criteria given below:

(a) Displacements must be continuous over element boundaries.
(b) Rigid body movements should be possible without straining.
(c) A state of constant strain should be reproducible.

Euclidean norm $\psi_i/R_i \leq C$. The term ψ_i represents the unbalanced forces and the norm of the residuals. With the aid of the iterative scheme described above, the unbalanced forces due to the initial stresses $\{\sigma_0\}$ become negligibly small. As a measure of their magnitude, the norm of the vector $\|\psi_i\|$ is used. The Euclidean norm and the absolute value of the largest component of the vector are written as

$$\|\psi_i\| = (|\psi_1|^2 + \cdots + |\psi_n|^2)^{1/2} \qquad \text{(AV.1)}$$
$$\|R_i\| = (|\{R_i\}^T\{R_i\}|)^{1/2}$$

the convergence criterion adopted is

$$\|\psi\| = \max_i |\psi_i| < C = 0 \cdot 001 \qquad \text{(AV.2)}$$

Uniform Acceleration

Various procedures are available for accelerating the convergence of the modified Newton–Raphson iterations. Figure AV.1 shows the technique of computing individual acceleration factors, δ_1 and δ_2 are known. Then, assuming a constant slope of the response curve, and from similar triangles, the value of δ_3 is computed.

$$\frac{\delta_1}{\delta_2} = \frac{\delta_2}{\delta_3} \qquad \delta_3 = \delta_2 \frac{\delta_2}{\delta_1} \qquad \text{(AV.3)}$$

When δ_3 is added to δ_2, then the accelerated displacement δ'_2 is expressed as

$$\delta'_2 = \delta_2 + \delta_3 = \delta_2\left(1 + \frac{\delta_2}{\delta_1}\right) = \alpha\delta_2 \qquad \text{(AV.4)}$$

where the acceleration factor α is

$$\alpha = 1 + \frac{\delta_2}{\delta_1} \qquad \text{(AV.5)}$$

Generally the range of α is between 1 and 2. The value of $\alpha = 1$ for zero

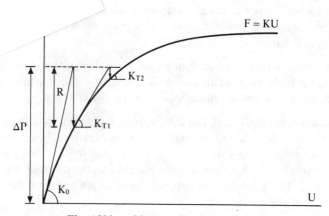

Fig. AV.1a. Newton–Raphson method.

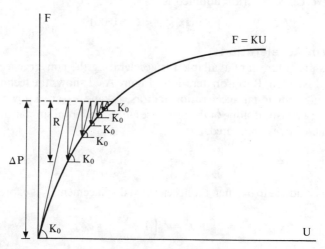

Fig. AV.1b. Initial stress method.

Note: ΔP is a specific value of F.

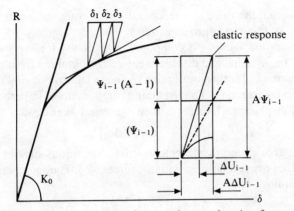

Fig. AV.1c. Technique of computing acceleration factors.

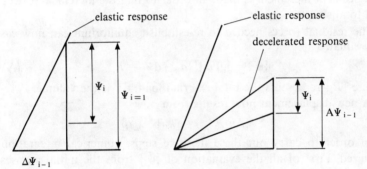

Fig. AV.1d. Grapical representation.

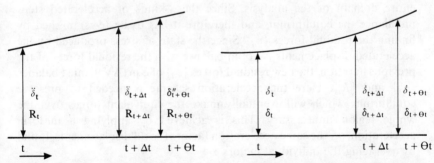

Fig. AV.1e. Linear acceleration and load assumptions of the Wilson θ method.

Fig. AV.1f. Quadratic and cubic variation of velocity and displacement assumptions of the Wilson θ method.

acceleration, and the value of α reaches the maximum value of 2 when the slope of the δ–R curve approaches zero.

The acceleration factor α is computed individually for every degree of freedom of the system. The displacement vector obtained from the linear stiffness matrix $[k_0]$ is then multiplied by the $[\alpha]$ matrix having the above constants on its diagonals. The remaining components of $[\alpha]$ are zero. The accelerated displacement vector is then expressed as follows:

$$\{\Delta u'_i\} = [a_{i-1}]\{\Delta u_i\} \tag{AV.6}$$

From these accelerated displacements $\{\Delta u'_i\}$, the initial stresses $\{\sigma_0\}$ are found and they are equilibrated with the forces $\{\psi_i\}$. They are then used for the next solution

$$\{\Delta \bar{u}_i\} = [k_0]^{-1}\{\psi_i\} \tag{AV.7}$$

which results in a new set of acceleration factors. Now an estimate for the displacement increment is made in order to find the incremental stresses and total stresses.

The residual forces needed to re-establish equilibrium can now easily be evaluated

$$\{\hat{\psi}_i\} = \int_v [B]^T \{\sigma_{0_T}\} \, dV - \{R_i\} \tag{AV.8}$$

where $\{R_i\}$ represents the total external load; dV is the volume.

A new displacement now results from

$$\{\Delta u_{i+1}\} = -[k_0]^{-1}\{\hat{\psi}_i\} \tag{AV.9}$$

In order to carry out these iterative steps, numerical integration is required. First of all the evaluation of $\{\hat{\psi}_i\}$ from the initial stresses is required, and this requires integration over the elastic–plastic region only. The value of $\{\hat{\psi}_i\}$ is computed by carrying out the integration over the entire domain of the analysis. Since these kinds of accelerated steps unbalance the equilibrium, and therefore it has to be re-established by finding the residual forces $\{\hat{\psi}_i\}$. Since the state of stress produced by the accelerated displacements is not in balance with the residual forces of the previous iteration, the new residual forces $\{\hat{\psi}_i\}$ of eqn (AV.9) must balance $\{\sigma_T\}$ and $\{R_i\}$. Here the acceleration scheme is needed to preserve equilibrium, which will eventually make the equivalent forces over the whole region unnecessary. This is achieved by applying a uniform acceleration, i.e. the same acceleration factor $\bar{A}$ to all displacements, found by averaging the individual factors α_i

$$\bar{A} = \frac{1}{n} \sum_{i=1}^{n} \alpha_i \tag{AV.10}$$

The force–displacement equation is then written by multiplying both sides with the scalar quantity $\bar{A}$ without disturbing the equilibrium.

$$\bar{A}\{\Delta u_i\} = [k_0]^{-1}\bar{A}\{\psi_i\} \quad (AV.11)$$

Now to evaluate $\{\psi_{i+1}\}$, the previous value of $\{\psi_i\}$ must be multiplied by $\bar{A}$, and the previously accelerated forces from the initial stresses $\{\sigma_0\}$ must be included such that

$$\{\psi_{i+1}\} = \int_V [B]^T \{\sigma_0\}\,dV - (A-1)\{\psi_{i-1}\} \quad (AV.12)$$

APPENDIX VI

*CONCRETE CUBES AND STEEL BARS—
EXPERIMENTAL AND FINITE ELEMENT RESULTS*

PREPARATION OF CONCRETE

Cement 43 lb
Sand 86 lb
Concentrated aggregate 172 lb
} 1:2:4 mix
Water 26 lb

The concrete is mixed in a laboratory batch mixer.

TESTING OF CONCRETE

Concrete is poured into test cubes and cylindrical moulds to test the strength.

Results

7 days: Cube 1
Cube 2

28 days: Cube 3 } Dry
Cube 2
Cube 3 } Wet
Cube 4
} ultimate loads and strength

Concrete cube strengths and the yield stress of steel reinforcement are as shown in Table AVI.1.

Two air-dried and two water-cured test cubes were tested for strength.

Table AVI.1

	Test cube			
	Air-dried		Water-cured	
	1	2	3	4
Concrete strength (kN)	1 197	1 150	1 291	1 286
Average strength (kN)	(1 090[a])		(1 270[a])	
	1 173·5		1 288·5	

[a] Results from finite element incremental analysis.

Size of test cube = $150 \times 150 \times 150$ mm
Area of test cube = $150 \times 150 = 2\,2500$ mm^2
Average concrete strength $f_{cu} = 54·71$ N/mm^2
Characteristic of reinforcement $f_y = 410$ N/mm^2

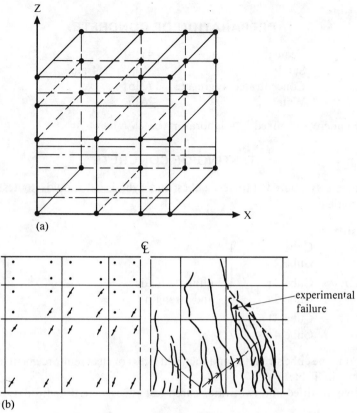

Fig. AVI. 1. (a) Finite element mesh; (b) computed crack patterns.

TENSION STRENGTH OF STEEL BARS

Area of 10 mm diameter bar

$$A = \pi(10^2/4)$$
$$= 78 \cdot 5 \, mm^2$$

Bar diameter (mm)	Area (mm)	Yield load (kN)	Failure load (kN)	Yield stress (N/mm²)
10	78·5	29 (31)[a]	39 (42)[a]	369·4 (410)

[a] Results from finite element incremental analysis.

Figure AVI.1 shows the finite element mesh and computed crack patterns. For cases for comparison refer to Figs I.5, I.10, I.16–I.19.

APPENDIX VII

STRESS–STRAIN CURVES FOR REINFORCEMENTS AND PRESTRESSING TENDONS (DATA FOR THE PROGRAM ISOPAR)

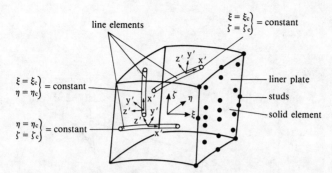

Finite element modelling of liner–stud interaction.

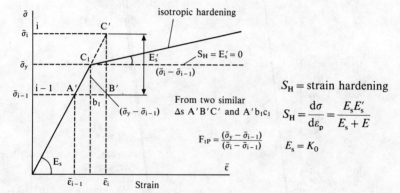

S_H = strain hardening

$$S_H = \frac{d\sigma}{d\varepsilon_p} = \frac{E_s E'_s}{E_s + E}$$

$$F_{tP} = \frac{(\bar{\sigma}_y - \bar{\sigma}_{i-1})}{(\bar{\sigma}_i - \bar{\sigma}_{i-1})} \qquad E_s = K_0$$

Equivalent stress–strain curve for steel. E_s = initial yield modulus; E'_s = post-yield modulus.

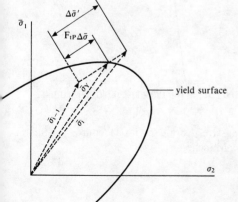

Transitional factor, elastic to plastic.
From the figure $\bar{\sigma}_{i-1} + F_{TR}\Delta\bar{\sigma}' = \bar{\sigma}_Y$

$$F_{tP} = \frac{\bar{\sigma}_Y - \bar{\sigma}_{i-1}}{\Delta\bar{\sigma}'} = \frac{\bar{\sigma}_Y - \bar{\sigma}_i}{\bar{\sigma}_i - \bar{\sigma}_Y}$$

$\bar{\sigma} = \sigma_{eq}$

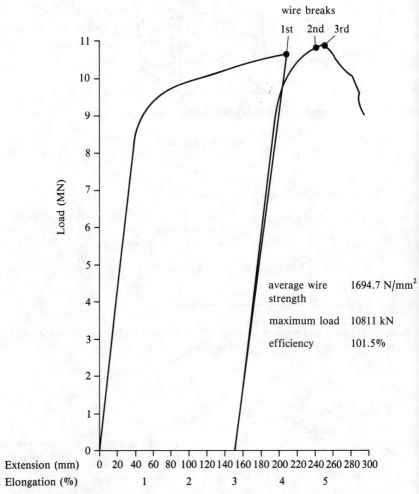

BBRV straight tendons 163/7 system—stress–strain test no. 11 (courtesy of Simon-Cave, Manchester).

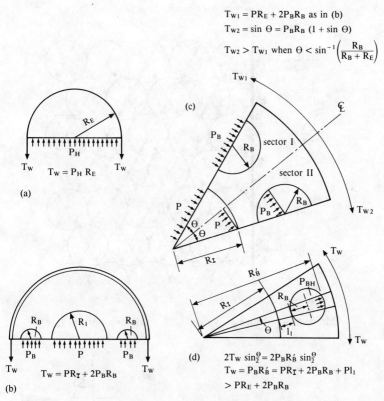

Wire/strand load versus uniform pressure. (a) Single cavity; (b) half main cavity with two boiler holes; (c) two sectors of multi-cavity; (d) a sector of main cavity with one boiler hole.

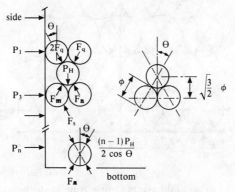

Contact forces from wires and strands.

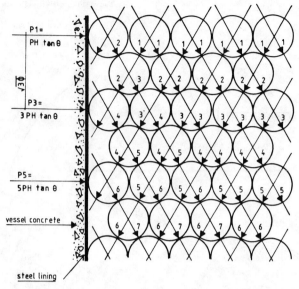

Odd number of layers—values $\times\ F_q$ (excluding bottom layer).

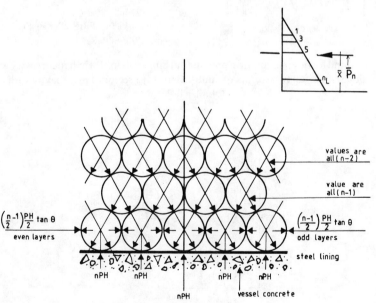

Even and odd layers (centre zone at the base).

Appendix VII

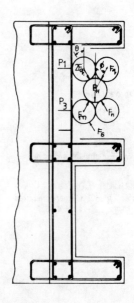

Wires

Even wires

No. of Layers $= \frac{1}{2}[n_L \cdot l/\phi + n_L(l/\phi - 1)]$

$P_H = \sum P_H$ Circumferential force

$ = \dfrac{n T_W}{l}\left[\dfrac{1}{R_1 + n\phi}\right]$

P_n = nTh layer pressure
$ = n P_H \tan\theta$

For example

$F_q = \frac{1}{2} P_H \sec\theta$

$P_3 = 2 F_q \sin\theta + 2 P_H \tan\theta$

$F_m = F_q + \frac{1}{2} P_H \sec\theta$

$F_s F_q + \frac{3}{2} P_H \sec\theta$

Strands

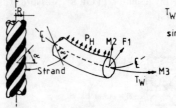

$P_H = -\left[\dfrac{\pi E \phi^4 \nu \sin\alpha \cos^4\alpha}{64 R^3 (1+\nu)} - \right.$

$\left. -\dfrac{\pi \phi^4}{4R}\sigma_t(\sin\alpha \cos\alpha + n_1 \sin\alpha \cos^4\alpha)\right]$

$T_W = \sigma_t \dfrac{\pi \phi^2}{4}(1 + n_1 \cos\alpha \times (1 + n_1 \cos^3\alpha))$

$\sin\phi_1 = \dfrac{1}{\sqrt{\left[1 + \dfrac{\sin^2\alpha}{\tan^2\left(\frac{\pi}{2} - \frac{\pi}{n}\right)}\right]}}$

Strength Reduction & Ultimate Tensile Stresses

$\sigma_{ut_1} = \bar{k}\{\sigma_1 - \nu(\sigma_2 - \sigma_3)\}$

$\sigma_{ut_1} = \bar{k}\{\sigma_3 - \nu(\sigma_1 - \sigma_3)\}$

$\sigma_{ut_1} = \bar{k}\{\sigma_3 - \nu(\sigma_1 - \sigma_3)\}$

$\bar{k}$ = Strength Reduction factor issued later on with equations for wires/strands

Load from wires/strands.

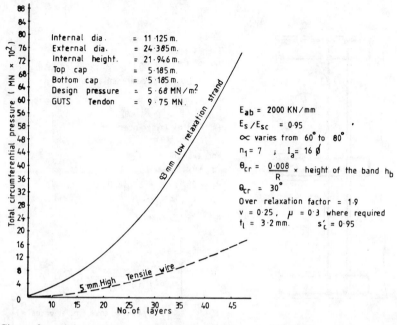

Circumferential pressure versus number of layers (HTGCR pressure vessel parameters).

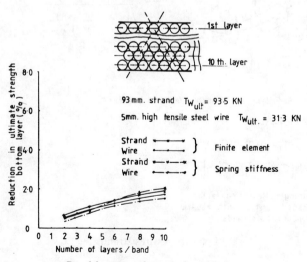

Band layers and strength reduction.

Summary of Circumferential and Vertical Tendon Losses Excluding Friction

Type of Loss	Walls Tons (MN)					Cap Tons MN	
	Circumferential				Vertical	Circumferential	
	2	3	4	7	1	5	6
Relaxation	22·0 (0·219)	30·8 (0·306)	28·6 (0·284)	25·0	29·4 (0·249)	19·1 (0·190 3)	22·8 (0·227)
Elastic	18·4 (0·183)	27·0 (0·269)	12·7 (0·126)	10·1 (0·10)	19·1 (0·19)	10·3 (0·102)	13·2 (0·131)
Creep	28·5 (0·283)	65·3 (0·65)	52·5 (0·523)	29·5 (0·293)	104·9 (1·045)	25·3 (0·252)	43·3 (0·431)
Pressure	−19·5 (−0·128)	−15·9 (−0·158)	−0·4 (−0·003 9)	−1·3 (−0·012 9)	−9·1 (−0·09)	−5·3 (−0·052)	−9·5 (−0·094)
Thermal	−12·9 (−0·128)	+14·6 (−0·145)	+9·1 (−0·09)	+2·2 (−0·021)	−28·9 −0·289	−10·6 (−0·105)	−4·0 (−0·039)
Total Tons	36·5 (0·363)	121·8 (1·213)	102·5 (1·021)	65·5 (0·652)	116·4 (1·159)	38·8 (0·386)	65·3 (0·650)
% Loss	5·0	16·5	14·0	8·9	15·9	5·3	9·0
Average % Loss/ Group of Tendons		9·0			15·9	7·8	

Geometry C₅—Type Tendons

Total tendon length = $2(l_1 + l'_2 + l'_3)$

$\cos\theta = \dfrac{R_2}{R_E}$ $\theta = 50°\,49'$

$\sin\gamma = \dfrac{l_2}{R_E} = 0\cdot5801$ $\gamma = 35°\,27'$

$\cos\theta_1 = \dfrac{R_2 - l'''}{R_E} = 0\cdot6004$ $\theta_1 = 53°\,6'$

$l_1 + l'_2 = R_E \sin\theta_1$ $36\cdot1864\,\text{ft}\ (11\cdot036\,\text{m})$

$l'_5 = \dfrac{2\pi \times R_2(90° + \theta_2)}{360°} = 46\cdot7611\,\text{ft}\ (14\cdot26\,\text{m})$

$\theta = 90° - (\theta + \gamma) = 90° - \left\{\cos^{-1}\dfrac{R_2}{R_E} + \sin^{-1}\dfrac{R_2}{R_E}\right\} = 3°\,44'$

Total Tendon length = $165\cdot8950\,\text{ft}\ (50\cdot59\,\text{m})$
 = $(2(l_1 + l'_3 + l'_5))$

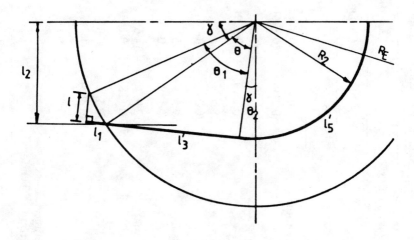

Appendix VII

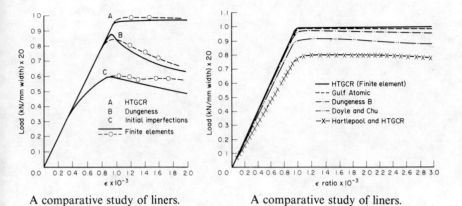

A comparative study of liners.

A comparative study of liners.

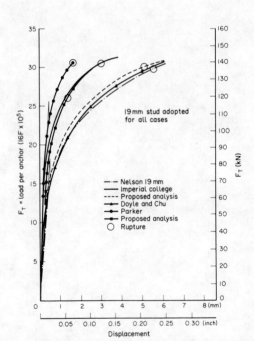

A comparative study of stud loads to failure.

APPENDIX VIII

CONCRETE CRACKING—STEP-BY-STEP FORMULATION AND COMPUTER SUB-ROUTINES

GENERAL STEPS OF FLOW AND CRACK CALCULATIONS

(1) The load increment $\{\Delta P_n\}$ is applied where n is the load increment.

(2) The total is accumulated as $\{P_n\} = \{P_{n-1}\} + \{\Delta P_n\}$, and $\{R\} = \{\Delta P_n\}$, where $\{R\}$ is the residual load vector.

(3) Incremental displacement is computed as $\{\Delta U_i\} = [K]_e^{-1}\{R\}$, where i is the iteration.

(4) Total displacements are now accumulated in the following form: $\{U_i\} = \{U_{i-1}\} + \{\Delta U_i\}$.

(5) Strain increments are calculated from step 4 as $\{\Delta\varepsilon_i\} = [B]\{\Delta U_i\}$. The accumulated strains at this stage would then be written as

$$\{\varepsilon_i\} = \{\varepsilon_{i-1}\} + \{\Delta\varepsilon_i\}$$

(6) The stress increments are calculated using the current nonlinear constitutive matrices of various models described earlier: $\{\Delta\sigma_i\} = \{f(\sigma)\}\{\Delta\varepsilon\}$. The accumulated stresses are computed as $\{\sigma_i\} = \{\sigma_{i-1}\} + \{\Delta\sigma_i\}$. In order to differentiate stresses at elastic and plastic conditions, a stress point indicator I_p is introduced.

$I_p = 0$ (elastic point)
$ = 1$ (plastic point)
$ = 2$ (unloading from plastic state)

(7) The stress increment is calculated using the elastic material matrix as $\{\Delta\sigma_i'\} = [D']\{\Delta\varepsilon\}$. Total stresses are given as $\{\sigma_i'\}_T = \{\sigma_{i-1}\} - \{\Delta\sigma_i'\}$.

(8) The stress $\{\sigma_i\}$ is now calculated using step 7: $\{\sigma_i\} = \{f(\sigma_i')\}$, $\{\sigma_{i-1}\} = \{f(\sigma_{i-1})\}$ – any yield criterion required.

(9) If a plastic point is obtained, step 11 should be considered.

(10) If $\sigma_i \geq \sigma_y$—plastic point ($I_P = 1$), transition from elastic to plastic, calculate factor F_{TP} using Figs VI.3 and VI.4:

$$F_{TP} = \left(\frac{\sigma_y - \bar{\sigma}_{i-1}}{\sigma_i - \sigma_{i-1}}\right)$$

The stress at the yield surface $\{\sigma_i\}_y^* = \{\sigma_{i-1}\} + F_{TP}^*\{\Delta\sigma_i\}$. Elasto-plastic stress increments are calculated as $\{\Delta\sigma_i\} = [D]_{ep}\{\sigma_i\}^*(1 - F_{TP})\{\Delta\varepsilon\}$. Total stress $\{\sigma_i\}_T = \{\sigma_i\} + \{\Delta\sigma_i\}$.

(11) Plastic point from steps 9 and 10, check for unloading, i.e. if $\bar{\sigma} \geq \sigma_y$ it is necessary to proceed to step 12. For the unloading case at this point, set $I_p = 2$, total stress $\{\sigma_i\}_T$ is then given by $\{\sigma_i\} = \{\sigma_{i-1}\} + \{\Delta\sigma_i\}$. Set $\{\sigma_y\} = \{\bar{\sigma}_{i-1}\}$, and the procedure is repeated for the additional increments.

(12) Loading at this point $\{\Delta\sigma_i\} = [D]_{ep}\{\bar{\sigma}_{i-1}\}\{\Delta\varepsilon\}$. Total stress $\{\sigma_i\}_T = \{\bar{\sigma}_{i-1}\} + \{\Delta\varepsilon\}$.

(13) Sometimes it is necessary to correct stresses from the equivalent stress–strain curve: $\sigma_{corr} = \sigma_{i-1} + S_H \Delta\varepsilon_p$, where $\Delta\varepsilon_p = \sqrt{\frac{2}{3}\Delta\varepsilon^p_{ij}\Delta\varepsilon^p_{ij}}$ = equivalent plastic strain increment. S_H is the strain hardening parameter, such that $\Delta\bar{\varepsilon}_p = \lambda$.

Equivalent stress, calculated from the current stress state, is given by $\{\sigma_i\} = F_{TP}\{(\sigma_i)\}$. The correct stress state, which is on the yield surface, will therefore be given as $\{\sigma_i\} = F_{TP}\{\sigma_i\}$. The total stresses are converted into equivalent nodal loads from $\int_v [B]^T \{\sigma_i\} \, d\,vol$, and the residual load vector is calculated from $\{R\} = \{F_n\} - \int_v [B]^T \{\sigma_i\} \, d\,vol$.

(14) Check for convergence. For detailed information see Appendix V.

$$(\|R\|/\|F\|) \leq \text{TOL} \qquad (\|\Delta U\|/\|U\|) \leq \text{TOL}$$

where TOL is chosen from 0·01 to 0·001; $\|R\| = \sqrt{R_i^{T''} R_i}$ is the Euclidean norm of the residuals; $\|F\| = \sqrt{P^T P}$ is the Euclidean norm of the externally applied load; and $\|\Delta U\| = \sqrt{\Delta U_i^{T''} U_i}$ is the Euclidean norm.

If convergence is not achieved, step 3 is invoked and all the steps repeated for the next iteration. If convergence is achieved, then proceed with the next load increment.

Appendix VIII

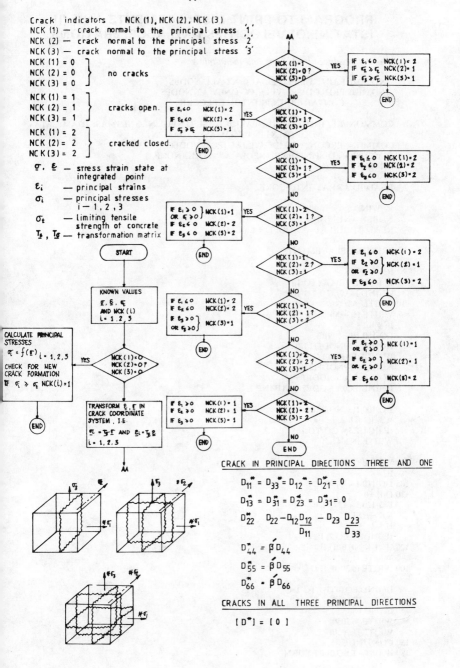

CRACK IN PRINCIPAL DIRECTIONS THREE AND ONE

$$D_{11}^{*} = D_{33}^{*} = D_{12}^{*} = D_{21}^{*} = 0$$

$$D_{13}^{*} = D_{31}^{*} = D_{23}^{*} = D_{31}^{*} = 0$$

$$D_{22}^{*} = D_{22} - D_{12}\frac{D_{12}}{D_{11}} - D_{23}\frac{D_{23}}{D_{33}}$$

$$D_{44}^{*} = \beta' D_{44}$$

$$D_{55}^{*} = \beta' D_{55}$$

$$D_{66}^{*} = \beta' D_{66}$$

CRACKS IN ALL THREE PRINCIPAL DIRECTIONS

$$[D^{*}] = [0]$$

PROGRAM TO PRINT DISPLACEMENTS AND (IF ISTAT.NE.0) VELOCITIES AND ACCELERATIONS

(Jointly developed by J. Tang and the author.)

```
      IND.EQ.0, PRINT DISPL/VEL/ACC AT ALL NODES
      IND.NE.0, PRINT DISPL/VEL/ACC ONLY AT NODES
               CONTAINED IN PRINTOUT BLOCKS

      COMMON /EL/ IND,ICOUNT,NPAR(20),NUMEG,NEGL,NEGNL,IMASS,IDAMP,ISTAT
     1            ,NDOFDM,KLIN,IEIG,IMASSN,IDAMPN
      COMMON /PRCON/ NPB,IDC,IVC,IAC,IPC,IPNODE(2,8)
      DIMENSION DISP(NEQ),VEL(NEQ),ACC(NEQ),ID(NDOF,1)
      DIMENSION D(6)

      READ ID ARRAY INTO CORE

      REWIND 8
      NDBLK = NUMNP
      READ (8) ((ID(I,J),I = 1,NDOF),J = 1,NUMNP)

      PRINT DISPLACEMENTS

      IC = 4
      IF (IND.EQ.0) GO TO 10
      IF (IND.EQ.0) GO TO 180
   10 WRITE (6,2000)
      WRITE (1,2000)
      IC = IC + 5
      DO 150 IB = 1,NPB
      NODE1 = IPNODE(1,IB)
      IF (NODE1.EQ.0) GO TO 150
      NODE2 = IPNODE(2,IB)
      IF (IND.EQ.0) NODE1 = 1
      IF (IND.EQ.0) NODE2 = NUMNP

      DO 100 II = NODE1,NODE2
      IC = IC + 1
      IF (IC.LT.56) GO TO 105
      WRITE (6,2045)

      WRITE (1,2045)
      IC = 4
  105 DO 110 I = 1,6
  110 D(I) = 0.
      DO 120 I = 1,NDOF
      KK = ID(I,II)
      IL = I
      IF (NDOF.EQ.2) IL = I + 1
  120 IF (KK.NE.0) D(IL) = DISP(KK)
      WRITE (1,2010) II,D
  100 WRITE (6,2010) II,D

      IF (IND.EQ.0) GO TO 180
      IF (IC.GE.55) GO TO 150
      IC = IC + 1
      WRITE (6,2050)
      WRITE (1,2050)
  150 CONTINUE
  180 IF (ISTAT.EQ.0) RETURN
```

PRINT VELOCITIES

```
    IF (IND.EQ.0) GO TO 201
    IF (IVC.EQ.0) GO TO 280
201 IC = IC + 5 + IDC
    IF (IDC.NE.0) WRITE (6,2050)
    IF (IDC.NE.0) WRITE (1,2050)
    IF (IC.GE.54) GO TO 205
    WRITE (6,2020)
    WRITE (1,2020)
    GO TO 206
205 WRITE (6,2022)
    WRITE (1,2022)
    IC = 4
206 DO 250 IB = 1,NPB
    NODE1 = IPNODE(1,IB)
    IF (NODE1.EQ.0) GO TO 250
    NODE2 = IPNODE(2,IB)
    IF (IND.EQ.0) NODE1 = 1
    IF (IND.EQ.0) NODE2 = NUMNP

    DO 200 II = NODE1,NODE2
    IC = IC + 1
    IF (IC.LT.56) GO TO 207
    WRITE (6,2022)
    WRITE (1,2022)
    IC = 4
207 DO 210 I = 1,6
210 D(I) = 0.
    DO 220 I = 1,NDOF
    KK = ID(I,II)
    IL = I
    IF (NDOF.EQ.2) IL = I + 1
220 IF (KK.NE.0) D(IL) = VEL(KK)
    WRITE (1,2010) II,D
200 WRITE (6,2010) II,D

    IF (IND.EQ.0) GO TO 280
    IF (IC.GE.55) GO TO 250
    IC = IC + 1
    WRITE (6,2050)
    WRITE (1,2050)

250 CONTINUE
```

PRINT ACCELERATIONS

```
280 IF (IND.EQ.0) GO TO 290
    IF (IAC.EQ.0) RETURN
    IF (IDC.EQ.0 .AND. IVC.EQ.0) GO TO 305
290 IC = IC + 6
    IF (IC.GE.54) GO TO 303
    WRITE (6,2050)
    WRITE (1,2050)
    WRITE (6,2030)
    WRITE (1,2030)
    GO TO 308
303 WRITE (6,2032)
    WRITE (1,2032)
    IC = 4
    GO TO 308
```

```
305 IC = IC + 5
    WRITE (6,2030)
    WRITE (1,2030)
308 DO 350 IB = 1,NPB
    NODE1 = IPNODE(1,IB)
    IF (NODE1.EQ.0) GO TO 350
    NODE2 = IPNODE(2,IB)
    IF (IND.EQ.0) NODE1 = 1
    IF (IND.EQ.0) NODE2 = NUMNP

    DO 300 II = NODE1,NODE2
    IC = IC + 1
    IF (IC.LT.56) GO TO 307
    WRITE (6,2032)
    WRITE (1,2032)
    IC = 4
307 DO 310 I = 1,6
310 D(I) = 0.
    DO 320 I = 1,NDOF
    KK = ID(I,II)
    IL = I
    IF (NDOF.EQ.2) IL = I + 1
320 IF (KK.NE.0) D(IL) = ACC(KK)
    WRITE (1,2010) II,D
300 WRITE (6,2010) II,D

    IF (IND.EQ.0) RETURN
    IF (IC.GE.55) GO TO 350
    IC = IC + 1
    WRITE (6,2050)
    WRITE (1,2050)
350 CONTINUE

    RETURN
```

SUBROUTINE CRACKD (PROP,NCK,SIG,EPS,PS1,PS2, PS3,DC1,DC2,DC3) IMPLICIT REAL*8(A−H,O−Z)

SET UP MATERIAL MATRICES FOR CRACKED CONCRETE

```
    COMMON /MTMD3D/ D(6,6),STRESS(6),STRAIN(6),IPT,NEL
    DIMENSION DD(6,6),PROP(1),NCK(1),SIG(1),EPS(1),
   @            PS1(1),PS2(1),PS3(1),DC1(1),DC2(1),DC3(1)
    CALL PRINCL (IPT,STRESS,PS1,PS2,PS3,DC1,DC2,DC3)
    CALL RCMOD (PROP,D)
    CALL DMAT (PROP,NCK)
    DO 222 I = 1,6

    DO 222 J = 1,6
222 DD(I,J) = 0.0
    JJJ = 1
    LL = 0
    IF (NCK(1).EQ.1) LL = 1
    IF (NCK(2).EQ.1) LL = 2
    IF (NCK(3).EQ.1) LL = 3
    IF (NCK(1).EQ.1.AND.NCK(2).EQ.1) LL = 4
    IF (NCK(2).EQ.1.AND.NCK(3).EQ.1) LL = 5
    IF (NCK(1).EQ.1.AND.NCK(3).EQ.1) LL = 6
```

IF (NCK(1).EQ.1.AND.NCK(2).EQ.1.AND.NCK(3).EQ1) LL = 7
IF (LL.EQ.7) GOTO 99
IF (JJJ.EQ.0) GOTO 200
IF (LL.EQ.0) GOTO 999
GOTO (113,114,115,116,117,118),LL

ONLY ONE DIRECTION CRACKED

113 CONTINUE

CRACK IN DIRECTION 1

DD(1,1) = 0.0
DD(1,2) = 0.0
DD(1,3) = 0.0
DD(2,1) = 0.0
DD(2,2) = D(2,2) − D(1,2)∗D(1,2)/D(1,1)
DD(2,3) = D(2,3) − D(1,3)∗D(1,2)/D(1,1)
DD(3,1) = 0.0
DD(3,2) = DD(2,3)
DD(3,3) = D(3,3) − D(1,3)∗D(1,3)/D(1,1)
DD(4,4) = PROP(12)∗D(4,4)
DD(5,5) = DD(5,5)
DD(6,6) = PROP(12)∗D(6,6)
GOTO 121
114 CONTINUE

CRACK IN DIRECTION 2

DD(1,1) = D(1,1) − D(2,1)∗D(2,1)/D(2,2)
DD(1,2) = 0.0
DD(1,3) = D(1,3) − D(1,2)∗D(2,3)/D(2,2)
DD(2,1) = 0.0
DD(2,2) = 0.0
DD(2,3) = 0.0
DD(3,1) = DD(1,3)
DD(3,2) = 0.0
DD(3,3) = D(3,3) − D(2,3)∗D(2,3)/D(2,2)
DD(4,4) = PROP(12)∗D(4,4)
DD(5,5) = PROP(12)∗D(5,5)
DD(6,6) = D(6,6)
GOTO 121
115 CONTINUE

CRACK IN DIRECTION 3

DD(1,1) = D(1,1) − D(1,3)∗D(1,3)/D(3,3)
DD(1,2) = D(1,2) − D(1,3)∗D(2,3)/D(3,3)
DD(1,3) = 0.0
DD(2,1) = DD(1,2)
DD(2,2) = D(2,2) − D(2,3)∗D(2,3)/D(3,3)
DD(2,3) = 0.0
DD(3,1) = 0.0
DD(3,2) = 0.0
DD(3,3) = 0.0
DD(4,4) = D(4,4)
DD(5,5) = D(5,5)∗PROP(12)
DD(6,6) = D(6,6)∗PROP(12)
GOTO 121
116 CONTINUE

CRACKS IN TWO DIRECTIONS
CRACKS IN DIRECTIONS 1 & 2

```
      DENOM = D(1,1)*D(2,2) - D(1,2)*D(2,1)
      DD(1,1) = 0.0
      DD(1,2) = 0.0
      DD(1,3) = 0.0
      DD(2,1) = 0.0
      DD(2,2) = 0.0
      DD(2,3) = 0.0
      DD(3,1) = 0.0
      DD(3,2) = 0.0
      DD(3,3) = D(3,3)
     1         - D(3,1)*(D(2,2)*D(1,3) - D(1,2)*D(2,3))/DENOM
     2         - D(3,2)*(D(1,1)*D(2,3) - D(2,1)*D(3,1))/DENOM
      DD(4,4) = PROP(12)*D(4,4)
      DD(5,5) = PROP(12)*D(5,5)
      DD(6,6) = PROP(12)*D(6,6)
      GOTO 121
  117 CONTINUE
```

CRACKS IN DIRECTIONS 3 & 2

```
      DENOM = D(2,2)*D(3,3) - D(2,3)*D(3,2)
      DD(1,1) = D(1,1)
     1         - D(1,2)*(D(3,3)*D(2,1) - D(3,1)*D(2,3))/DENOM
     2         - D(1,3)*(D(2,2)*D(3,1) - D(2,1)*D(3,2))/DENOM
      DD(1,2) = 0.0
      DD(1,3) = 0.0
      DD(2,1) = 0.0
      DD(2,2) = 0.0
      DD(2,3) = 0.0
      DD(3,1) = 0.0
      DD(3,2) = 0.0
      DD(3,3) = 0.0
      DD(4,4) = PROP(12)*D(4,4)
      DD(5,5) = PROP(12)*D(5,6)
      DD(6,6) = PROP(12)*D(6,6)
      GOTO 121
  118 CONTINUE
```

CRACKS IN DIRECTION 1 & 3

```
      DENOM = D(1,1)*D(3,3) - D(3,2)*D(1,3)
      DD(1,1) = 0.0
      DD(1,2) = 0.0
      DD(1,3) = 0.0
      DD(2,1) = 0.0
      DD(2,2) = D(2,2)
     1         - D(2,1)*(D(3,3)*D(1,2) - D(3,2)*D(1,3))/DENOM
     2         - D(2,3)*(D(1,1)*D(3,2) - D(3,1)*D(1,2))/DENOM
      DD(2,3) = 0.0
      DD(3,1) = 0.0
      DD(3,2) = 0.0
      DD(3,3) = 0.0
      DD(4,4) = PROP(12)*D(4,4)
      DD(5,5) = PROP(12)*D(5,5)
      DD(6,6) = PROP(12)*D(6,6)
  121 CONTINUE
      GOTO 99
```

```
200 CONTINUE
    IF (LL. .EQ. 0) GOTO 999
    GOTO (1,2,3,4,5,6),LL
  1 CONTINUE
    DD(2,2) = D(2,2)
    DD(2,3) = D(2,3)
    DD(3,2) = DD(2,3)
    DD(3,3) = D(3,3)
    DD(4,4) = PROP(12)*D(4,4)
    DD(5,5) = PROP(12)*D(5,5)
    DD(6,6) = PROP(12)*D(6,6)
    GOTO 99
  2 CONTINUE
    DD(1,1) = D(1,1)
    DD(1,3) = D(2,3)
    DD(3,1) = D(1,3)
    DD(3,3) = D(3,3)
    DD(4,4) = PROP(12)*D(4,4)
    DD(5,5) = PROP(12)*D(5,5)
    DD(6,6) = D(6,6)
    GOTO 99
  3 CONTINUE
    DD(1,1) = D(1,1)
    DD(2,2) = D(2,2)
    DD(1,2) = D(1,2)
    DD(3,3) = D(3,3)
    DD(2,1) = DD(1,2)
    DD(4,4) = D(4,4)
    DD(5,5) = PROP(12)*D(5,5)
    DD(6,6) = PROP(12)*D(6,6)
    GOTO 99
  4 CONTINUE
    DD(3,3) = D(3,3)
    DD(4,4) = PROP(12)*D(4,4)
    DD(5,5) = PROP(12)*D(5,5)
    DD(6,6) = PROP(12)*D(6,6)
    GOTO 99
  5 CONTINUE
    DD(1,1) = D(1,1)
    DD(4,4) = PROP(12)*D(4,4)
    DD(5,5) = PROP(12)*D(5,5)
    DD(6,6) = PROP(12)*D(6,6)
    GOTO 99
  6 CONTINUE
    DD(2,2) = D(2,2)
    DD(4,4) = D(4,4)
    DD(5,5) = D(5,5)
    DD(6,6) = D(6,6)
 99 CONTINUE

    CRACKS IN ALL THREE DIRECTIONS
    TRANSFER DD TO D

    DO 101 J = 1,6
    DO 101 K = 1,6
    D(J,K) = DD(J,K)
101 CONTINUE
999 CONTINUE
    RETURN
    END
```

APPENDIX IX

DATA AND RESULTS FOR OFFSHORE STRUCTURES

APPENDIX

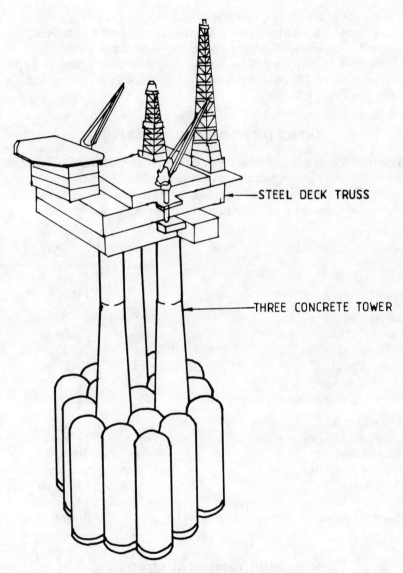

Fig. AIX.I. Brent C concrete gravity platform.

BRENT C PRODUCTION PLATFORM (Fig. AIX.1)

The caisson of the Brent C platform is made up of 36 square concrete cells. The external walls of the outer cells are semi-circular so as to resist any external overpressure, and the roofs of the cells are in the shape of truncated pyramids. Four of these cells rise to form four towers where the utilities are kept. The base slab is 4 m thick to give stability to the platform when it is being towed out.

BRENT D PRODUCTION PLATFORM

This platform consists of a base with three concrete towers. Some utilities are kept in these towers as well as in the depth of the deck truss, the rest are contained in the modules mounted on the steel deck truss assembly. No anchors or piles are needed when on location on the seabed.

	Brent D	Brent C
Environment		
Water depth	142 m	141 m
Substructure		
Caisson shape	hexagon	square
Caisson base area	6 360 m^2	10 340 m^2
Caisson height	58 m	57 m
Number of legs	3	4
Weight of sub-structure in air	175 000 tons	283 000 tons
Superstructure		
Height of deck above water	25·2 m	23·5 m
Area of deck	3 400 m^2	4 000 m^2
Type of deck construction	plate girder	lattice girder
Deck weight	3 200 tons	6 400 tons
Weight of equipment within deck	1 864 tons	3 420 tons
Weight of modules and equipment on deck	—	9 860 tons

ENVIRONMENTAL LOADS

Wave loads, current loads and wind loads are the main loads acting on gravity platforms. The other loads imposed are due to ice and tidal

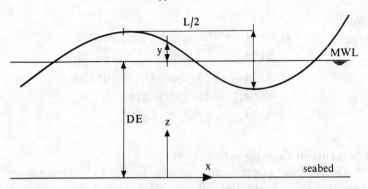

Fig. AIX.2. Wave description.

fluctuations. Stokes nonlinear fifth order wave theory is adopted for wave crest and incidental wave potential as shown in Fig. AIX.2.

The potential function assumed is

$$\frac{\beta_{w\phi}}{(CB)} = \frac{2(\pi)(\phi)}{L(CB)} = (\lambda A_{11} + \lambda^3 A_{13} + \lambda^5 A_{15})(\cosh(\beta_w(z)))\sin\theta$$
$$+ (\lambda^2 A_{22} + \lambda^4 A_{24})(\cosh(2\beta_w(z)))\sin(2\theta)$$
$$+ (\lambda^3 A_{33} + \lambda^5 A_{35})(\cosh(3\beta_w(z)))\sin(3\theta)$$
$$+ (\lambda^4 A_{44})(\cosh(4\beta_w(z)))\sin(4\theta)$$
$$+ (\lambda^5 A_{55})(\cosh(5\beta_w(z)))\sin(5\theta) \quad (IX.1)$$

where $\beta_w = 2\pi/\lambda$; λ is a parameter; and A_{ij} are constants depending on the ratio of water depth to wave length.

The final velocity and acceleration components adopted are given below.

Velocity Components
Horizontal

$$U_1 = (CB)((\lambda A_{11} + \lambda^3 A_{13} + \lambda^5 A_{15})\cosh(\beta_w P_N)\cos\theta$$
$$+ 2(\lambda^2 A_{22} + \lambda^4 A_{24})\cosh(2\beta_w P_N)\cos 2\theta$$
$$+ 3(\lambda^3 A_{33} + \lambda^5 A_{35})\cosh(3\beta_w P_N)\cos 3\theta$$
$$+ 4(\lambda^4 A_{44})\cosh(4\beta_w P_N)\cos 4\theta$$
$$+ 5(\lambda^5 A_{55})\cosh(5\beta_w P_N)\cos 5\theta) \quad (IX.2)$$

Vertical

$$\begin{aligned}W_1 = &(CB)(\lambda A_{11} + \lambda^5 A_{15})\sinh(\beta_w P_N)\sin\theta \\ &+ 2(\lambda^2 A_{22} + \lambda^4 A_{24})\sinh(2\beta_w P_N)\sin(2\theta) \\ &+ 3(\lambda^3 A_{33} + \lambda^5 A_{35})\sinh(3\beta_w P_N)\sin(3\theta) \\ &+ 4(\lambda^4 A_{44})\sinh(4\beta_w P_N)\sin(4\theta) \\ &+ 5(\lambda^5 A_{44})\sinh(5\beta_w P_N)\sin(5\theta)\end{aligned} \quad (IX.3)$$

Acceleration Components

The acceleration components can be found by differentiating the expressions for the horizontal and vertical velocities. These are given below.

Horizontal

$$\begin{aligned}U_2 = &\beta_w(CB)^2(\lambda A_{11} + \lambda^3 A_{13} + \lambda^5 A_{15})\cosh(\beta_w P_N)\sin\theta \\ &+ 4(\lambda^2 A_{22} + \lambda^4 A_{24})\cosh(2\beta_w P_N)\sin(2\theta) \\ &+ 9(\lambda^3 A_{33} + \lambda^5 A_{35})\cosh(3\beta_w P_N)\sin(3\theta) \\ &+ 16(\lambda^4 A_{44})\cosh(4\beta_w P_N)\sin(4\theta) \\ &+ 25(\lambda^5 A_{55})\cosh(5\beta_w P_N)\sin(5\theta)\end{aligned} \quad (IX.4)$$

Vertical

$$\begin{aligned}W_2 = &(-\beta_w(CB)^2(\lambda A_{11} + \lambda^3 A_{13} + \lambda^5 A_{15})\sinh(\beta_w P_N)\cos\theta \\ &+ 4(\lambda^2 A_{22} + \lambda^4 A_{24})\sinh(2\beta_w P_N)\cos(2\theta) \\ &+ 9(\lambda^3 A_{33} + \lambda^5 A_{55})\sinh(3\beta_w P_N)\cos(3\theta) \\ &+ 16(\lambda^4 A_{44}\sinh(4\beta_w P_N)\cos(4\theta) \\ &+ 25\lambda^5 A_{55}\sinh(5\beta_w P_N)\cos(5\theta)\end{aligned} \quad (IX.5)$$

The equations for the vertical and horizontal components of velocity and acceleration will be used together with current loads and Morrison's formula, taking into account diffraction of waves on a large base to calculate the wave forces.

Modified Morrison's Equation

The Morrison equation, expressing force as the sum of a velocity-dependent term (drag) and an acceleration-dependent term (inertia), is given by

$$F_M = \tfrac{1}{2}(P_w)(CD)(D_0)(U_1)|U_1| + (P_w)(CM)\pi(D_0)^2(U_2)/4 \quad (IX.6)$$

Appendix IX

The appropriate criterion for the neglect of the drag is $D_0/H > 0.2$. Diffraction is added if the ratio of the diameter of the main sections of the platform to the design wave length is > 0.2. For the Condeep platform having three small circular towers and a large hexagonal-shaped base formed from caissons, diffraction is assumed.

The diffraction coefficient CH is given by

$$CH = (F_u + F_d)/F_u \tag{IX.7}$$

where $F_u =$ contribution to the undisturbed incident wave, known as the Froude–Krylor component (Morrison's equation); and $F_d =$ contribution due to disturbance.

$$F = \text{inertia coefficient} = (CH)F_u \tag{IX.8}$$

The program ISOPAR includes two diffraction-coefficient formulae and they are given below.

Maccamy and Fuchs

$$CH = \frac{4(DL)^2}{\pi^3(DIA)^2[j_1'(x)^2 + Y_1'(x)^2]} \tag{IX.9}$$

when

$$X = (\pi)(DL)(DIA) \qquad j_1' = \tfrac{1}{2} \qquad Y_1' = \frac{2}{\pi x^2}$$

Hobgen and Standing

$$CH = 1 + (0.75)\left(\frac{h_c}{2R_e}\right)^{1/3}(1 - (0.3)\left(\frac{2\pi}{DL}\right)^2 R_e^2 \tag{IX.10}$$

where $h_c =$ caisson height; $R_e =$ equivalent radius of the caisson base; and $DL =$ design wave length.

Current Loads

Currents in conjunction with the wave loads normally result in stresses and overturning moments on the gravity platforms. The forces due to currents acting on the structure are obtained by the formula:

$$F_{DC} = (\tfrac{1}{2})(p_w)(CD)(CU)^2(D_0) \tag{IX.11}$$

where D_0 is a specific value of ϕ.

Due to lack of observational data on the existing offshore structure, a conservative approach is adopted by defining the current velocity (CU) at any elevation (RL) above the seabed as:

For RL less than 15 m

$$CU = 0.24 + (0.3)(RL)/15 \, (\text{m/s}) \qquad (\text{IX.12a})$$

For RL greater than or equal to 15 m

$$CU = 0.54 + (0.84)(RL - 15)/155 \, (\text{m/s}) \qquad (\text{IX.12b})$$

In addition velocity is 1·38 m/s at the water surface and 0·24 m/s at the bottom.

Wind Loads
The wind load adopted is based on Refs [1], [2] and [19].

Storm Wind
The maximum one-minute sustained wind velocity of 125 mph is used in conjunction with a storm wind.

Operating Winds and Gusts
A maximum instantaneous gust velocity of 160 mph is used on the deck.

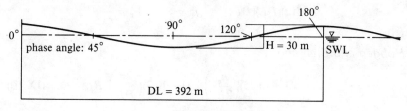

Wave profile

Appendix IX 623

AIRY WAVE THEORY

This is a programme to analyze the effects of surface wave on offshore structures by means of Airy wave theory, which is related to the surface data for the water velocity, acceleration, and pressure beneath the wave.

Notation

H	Wave height
D	Water depth
X	Horizontal direction
Y	Vertical direction
N	Wave profile
U	Horizontal velocity
V	Vertical velocity
W	Frequency
DL	Wave length
K	Wavenumber
T	Period
G	Gravitational acceleration
DEN	Density of water
AX	Horizontal acceleration
AY	Vertical Acceleration
P	Gauge pressure
C	Wave speed
FWAV	Wave force
FD	Frictional force from drag
FI	Inertia force due to water particle acceleration
CD	Drag coefficient
CM	Coefficient of virtual mass
A	Projected area of the structure
DIA	Diameter of the structure
MOM	Total moment
MD	Moment arising from drag force
MI	Moment arising from inertia force
CI	Inertia coefficient
CU	Current velocity
FCU	Current force
FENV	Environmental force

```
WRITE (6,'(A)') ' PLEASE INPUT THE WATER DEPTH (m):'
READ (5,*)D
WRITE (6,'(A)') ' PLEASE INPUT THE WAVE HEIGHT (m):'
READ (5,*)H
WRITE (6,'(A)') ' PLEASE INPUT THE PERIODIC TIME (sec):'
```

This is a sub-routine to determine the wave force
by Morrison's equation

```
SUBROUTINE WAVEFO (Y,DEN,DIA,U,PI,AX,FWAV)
IMPLICIT REAL (A-Z)
DIMENSION FD(10000),FI(10000),FWAV(10000),U(10000),AX(10000)
CD = 0.60
Cm = 2.0
FD(Y+1) = (0.5*CD*DEN*DIA*U(Y+1)*ABS(U(Y+1)))/1000.0
FI(Y+1) = (0.25*CM*DEN*PI*DIA**2.0*AX(Y+1))/1000.0
FWAV(Y+1) = FD(Y+1) + FI(Y+1)
RETURN
END
```

This is a sub-routine to determine the moments
arising from the drag and inertia forces

```
SUBROUTINE MOMENT (J,K,Y,D,DEN,DIA,W,H,THETA,MOM)
IMPLICIT REAL (A-Z)
DIMENSION MOM(10000)
CD = 0.60
CI = 2.0
IF (J.EQ.Y) THEN
   MOM(Y+1) = 0.0
   GOTO 1234
     ENDIF
     Q1A = 2.0*(K*(J-Y))*SINH(2.0*(K*(J-Y)))
     Q1B = COSH(2.0*K*(J-Y)) - 2.0*(K*(J-Y))**2.0 - 1.0
     Q1C = (SINH(K*D))**2.0
     Q1 = (Q1A - Q1B)/Q1C
------------------------------------------------
     Q2A = K*Y*SINH(K*(J-Y))
     Q2B = COSH(K*(J-Y)) - 1.0
     Q2C = SINH(K*D)
     Q2 = (Q2A - Q2B)/Q2C
------------------------------------------------
   MCOM = DEN*DIA*W**2.0*H/(8.0*K**2.0)
   MD = MCOM*CD*H*Q1*ABS(COS(THETA))*COS(THETA)/8.0
   MD = (DEN*CD*DIA)/(64.0*K**2.0)*(W*H)**2.0*Q1
   *    *ABS(COS(THETA))*COS(THETA)
   MI = (DEN*CI*PI*DIA**2.0)/(2.0*K*2.0*4.0)
   *    *W**2.0*H*Q2*SIN(THETA)
   MOM(Y+1) = (MD + MI)/1000.0
1234 RETURN
   END
```

This is a sub-routine to find the current velocity

```
SUBROUTINE CURVEL (Y,CU)
IMPLICIT REAL (A-Z)
DIMENSION CU(10000)
CU(Y+1) = 0.54 + (0.84)*(Y-15.0)/155.0
RETURN
END
```

This is a sub-routine to determine the current force.

```
SUBROUTINE CURRFO (Y,DEN,CU,DIA,FCU)
    IMPLICIT REAL (A−Z)
    DIMENSION FCU (10000), CU(10000)
    CD = 0.60
    FCU(Y + 1) = (0.5*DEN*CD*CU(Y + 1)**2*DIA)/1000
    RETURN
    END
```

This is a sub-routine to find the environmental force.

```
SUBROUTINE ENVOFO (Y,FWAV,FCU,FENV)
    IMPLICIT REAL (A−Z)
    DIMENSION FENV(10000),FWAV(10000),FCU(10000)
    FENV(Y + 1) = FWAV(Y + 1) + FCU(Y + 1)
    RETURN
    END
```

Sub-routine for drilling tower with caisson base.

```
SUBROUTINE DTOWER (DIA,Y)
    IMPLICIT REAL (A−Z)
    IF (Y.LT.65.0)                        DIA = 110.0
    IF ((Y.GE.65.0).AND.(Y.LT.72.0))      DIA = 2.0*11.758
    IF ((Y.GE.72.0).AND.(Y.LT.82.0))      DIA = 2.0*11.067
    IF ((Y.GE.82.0).AND.(Y.LT.92.0))      DIA = 2.0*10.108
    IF ((Y.GE.92.0).AND.(Y.LT.102.0))     DIA = 2.0* 9.150
    IF ((Y.GE.102.0).AND.(Y.LT.112.0))    DIA = 2.0* 8.192
    IF ((Y.GE.112.0).AND.(Y.LT.122.0))    DIA = 2.0* 7.233
    IF (Y.GE.122.0)                       DIA = 2.0* 6.275
    RETURN
    END
```

Sub-routine for utility tower with caisson base.

```
SUBROUTINE UTOWER (DIA,Y)
    IMPLICIT REAL (A−Z)
    IF (Y.LT.65.0)                        DIA = 110.0
    IF ((Y.GE.65.0).AND.(Y.LT.72.0))      DIA = 2.0*11.708
    IF ((Y.GE.72.0).AND.(Y.LT.82.0))      DIA = 2.0*10.967
    IF ((Y.GE.82.0).AND.(Y.LT.92.0))      DIA = 2.0* 9.908
    IF ((Y.GE.92.0).AND.(Y.LT.102.0))     DIA = 2.0* 8.850
    IF ((Y.GE.102.0).AND.(Y.LT.112.0))    DIA = 2.0* 7.792
    IF ((Y.GE.112.0).AND.(Y.LT.122.0))    DIA = 2.0* 6.735
    IF ((Y.GE.122.0)                      DIA = 2.0* 5.675
    RETURN
    END
```

A subroutine to find the diffraction coefficient by Maccamy and Fuchs

```
SUBROUTINE MACFUC (Y,DIA,PI,DL,CHF)
    IMPLICIT REAL (A−Z)
    DIMENSION CHF(10000)
    DIJ = 0.5
    DIY = 2.0/PI/(PI*DIA*DL)**2
    CHF(Y + 1) = 4.0*DL**2/(PI**3*DIA**2*(DIJ**2 + DIY**2)**0.5)
    RETURN
    END
```

A subroutine to find the diffraction coefficient by Hobgen and Standing

```
SUBROUTINE HOBSTA (Y,PI,DL,DIA,HC,CHH)
IMPLICIT REAL (A-Z)
DIMENSION CHH(10000)
HC=60.0
RC=DIA/2.0
PW=1.0/3.0
CHH(Y+1)=(1.0+0.75*(HC/2.0/RC)**PW*
*      (1.0-0.3*(2.0*PI/DL)**2*RC**2))
RETURN
END
```

A subroutine to find the total wave force by Morrison's equation and Maccamy and Fuchs coefficient

```
SUBROUTINE TOTFOM (Y,FWAV,CHF,FOM)
IMPLICIT REAL (A-Z)
DIMENSION FOM(10000),FWAV(10000),CHF(10000)
FOM(Y+1)=FWAV(Y+1)*CHF(Y+1)
RETURN
END
```

A subroutine to find the total wave force by Morrison's equation and Hobgen and Standing coefficient

```
SUBROUTINE TOTFOH (Y,FWAV,CHH,FOH)
IMPLICIT REAL (A-Z)
DIMENSION FOH(10000),FWAV(10000),CHH(10000)
FOH(Y+1)=FWAV(Y+1)*CHH(Y+1)
RETURN
END
```

Appendix IX

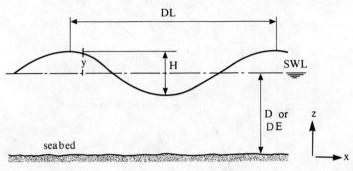

Wave definition sketch.

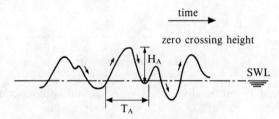

Typical sea surface record.

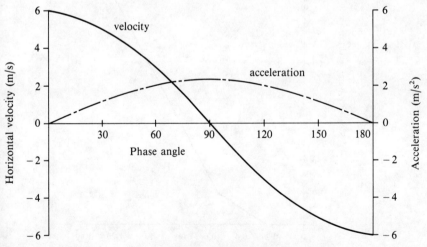

The variation of horizontal velocity and acceleration with respect to phase angle at a water depth of 145 m.

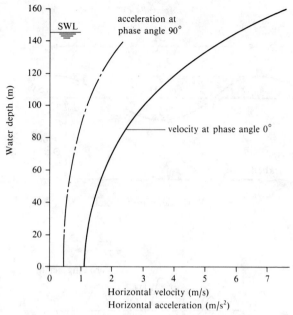

Variation of horizontal velocity and acceleration with water depth.

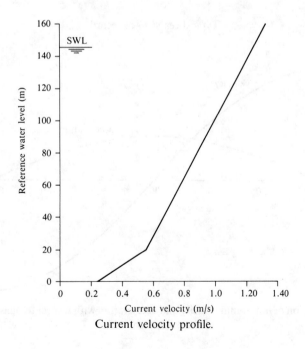

Current velocity profile.

Appendix IX 629

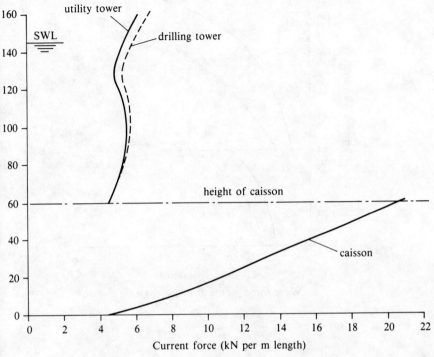

The distribution of current force/unit length with water level.

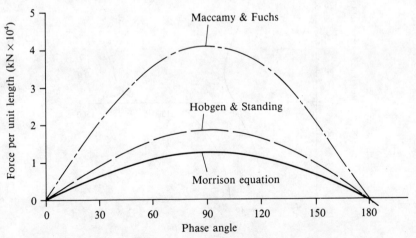

The variation of wave force with phase angle at a water depth of 55 m.

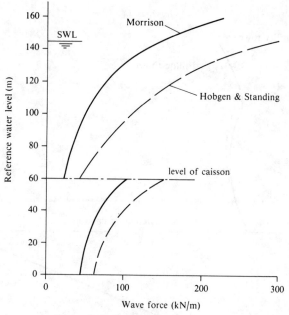

Wave force profile at phase angle 0°.

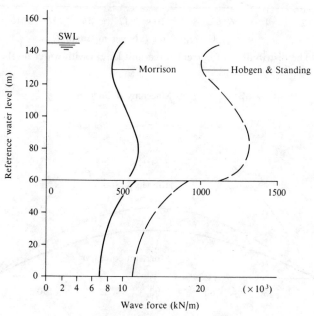

Wave force profile at phase angle 90°.

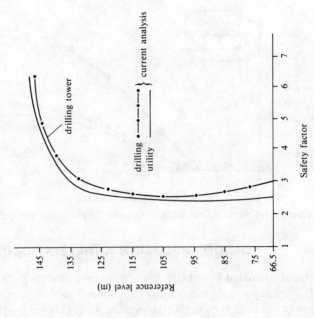
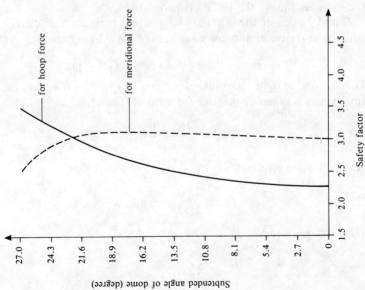

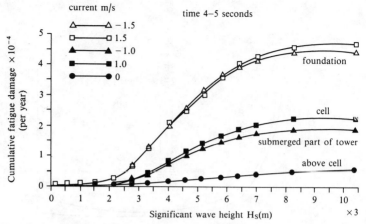

Cumulative damage with significant wave height in the presence of current.

OLSON IMPLOSION ANALYSIS [1029]

General equations have been reproduced for cylinders from Olson's work [1029]. They are then modelled in a special subprogram IMPLOSION given in this section by the Author and A. Aziz. This work is then compared in the main text using the ISOPAR program.

If P_i and P_0 denote the internal and external pressure respectively, the conditions at the outer and inner surfaces of the cylinder are (Fig. AIX.3)

$$(\sigma_r)_{r=R_0} = -P_0 \qquad (\sigma_r)_{r=R_0-t} = P_i = -P_0$$

The sign on the right-hand side of each equation is negative because the normal stress is taken as positive for tension. Therefore

$$P_{\text{imp}} = \frac{t(2R_0 - t)}{2R_0^2} \sigma_{\text{max}}$$

In terms of outer diameter D_0

$$P_{\text{imp}} = \frac{2t(D_0 - t)}{D_0^2} \sigma_{\text{max}}$$

In terms of mean radius R

$$P_{\text{imp}} = 4 \frac{t}{\left(2 + \dfrac{t}{R}\right)^2} \sigma_{\text{max}}$$

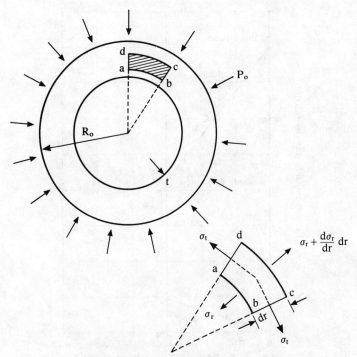

Fig. AIX. 3. Cylinder under loads.

where
$$R = \frac{D_0 - t}{2}$$

The separation point between thin-walled and thick-walled cylinders is selected at $t/D_0 = 0.063$, while the separation point for short, moderately long and long thin-walled cylinders is given in Fig. AIX. 4.

A   thick-walled cylinders of all lengths
B   thin-walled cylinders
B_1 short cylinder
B_2 moderately long cylinder
B_3 long cylinder

Thick-Walled Cylinders

The implosion pressure is developed by using an average wall hoop stress equation and a strength increase factor K to account for the difference in

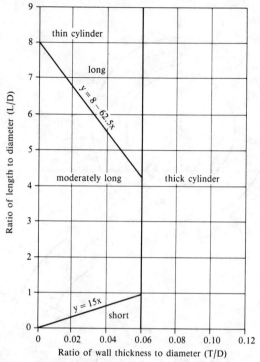

Fig. AIX. 4. Cylindrical structures of different geometries) [1029].

concrete length under multiaxial loading. For a cylindrical structure the ratio of hoop to axial stress is nearly 2:1. From studies of biaxial loading of concrete, for the loading case where principal stresses are in this ratio, the strength increase factor $K = 1.25$. The expression to predict implosion strength is developed as follows.

The average wall stress equation is

$$\sigma_{\text{imp}} = \frac{P_{\text{imp}} \times R_0}{t}$$

where

$$\sigma_{\text{imp}} = K f'_c \quad \text{and} \quad R_0 = \frac{D_0}{2}$$

therefore

$$P_{\text{imp}} = 2K \frac{t}{D_0} f'_c$$

Thin-Walled Cylinders
Short Cylinders
Short thin-walled cylinders are prevented from buckling by the influence of end closures. When the cylinder wall stress in the hoop direction equals the ultimate strength of the concrete, a material failure will cause implosion.

$$\sigma_{max} = f'_c \qquad P_{imp} = 4\left[\frac{\dfrac{t}{R}}{\left(2 + \dfrac{t}{R}\right)}\right] f'_c$$

Moderately Long Cylinders
For thin-walled cylinders of moderate length, end closures influence the behaviour of the cylindrical shell by restraining radial displacement. An expression to predict the implosion pressure of moderately long cylinders can be written

$$P_{imp} = 2330\eta \left[\frac{\left(\dfrac{t}{R}\right)^{5/2}\left(\dfrac{R}{L}\right)}{\left(2 + \dfrac{t}{R}\right)^2}\right] f'_c$$

A factor β is introduced in this equation depending on the type of end condition. In the computer program the value of β is taken as 1·2 assuming the cylinders are fixed at the end.

$$P_{imp} = 2330 \times \beta \times \eta \left[\frac{\left(\dfrac{t}{R}\right)^{5/2}\left(\dfrac{R}{L}\right)}{\left(2 + \dfrac{t}{R}\right)^2}\right] f'_c$$

η = plasticity reduction factor; E_i = initial tangential modulus of elasticity.

The plasticity reduction factor accounts for the inelastic material behaviour in the elastic buckling analysis. σ_{max}/f'_c is the ratio of wall stress at implosion (experimental) to wall stress at implosion (analytical).

Because the plasticity reduction factor is a function of modulus of elasticity, the η curve was selected at $E_i/f'_c = 667$ (Fig. AIX.5). This value was obtained from the relationship

$$E_i = \frac{2f'_c}{\varepsilon_{ult}} \qquad \varepsilon_{ult} = 0·003$$

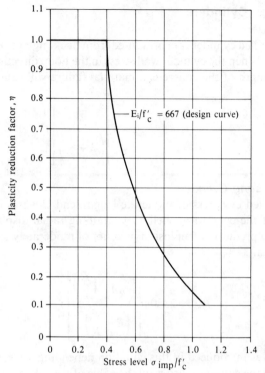

Fig. AIX.5. η curve [1029].

Long Cylinders

A long thin-walled cylinder has its mid-length portion unaffected by the end closure conditions. For this case the elliptical out-of-round shape, where the number of buckles is two, will be the geometry at implosion.

Using the same substitution $E_i/f_c' = 667$ and $v = 0.18$ the equation obtained is

$$\frac{\sigma_{max}}{f_c'} = 172\left(\frac{t}{R}\right)^2 \eta$$

The implosion pressure of a long cylinder can be written as

$$P_{imp} = 686 \times \eta \times \left[\frac{\left(\frac{t}{R}\right)^3}{\left(2 + \frac{t}{R}\right)^2}\right] \times f_c'$$

Appendix IX

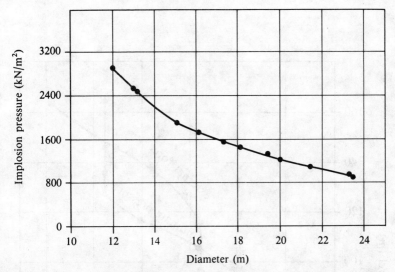

Implosion pressure versus diameter (utility and drilling tower). Results from ISOPAR (Bangash). (See Implosion program in this Section.)

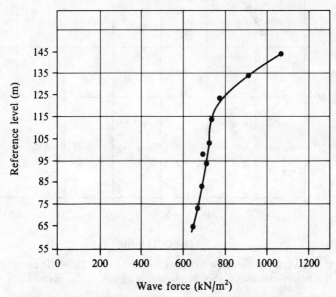

Reference level versus wave force (utility tower). Results from ISOPAR (Bangash). (See Implosion program in this Section.)

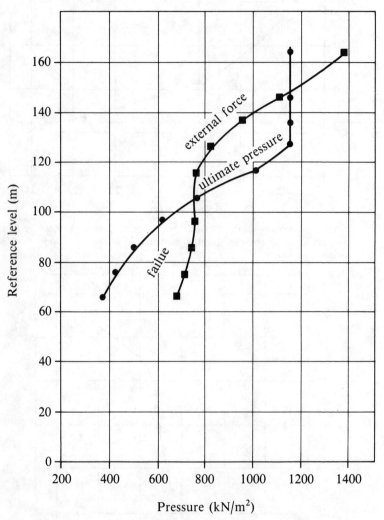

Reference level versus pressure (utility tower). Results from ISOPAR using Olson's equations (Bangash). (See Implosion program in this Section.)

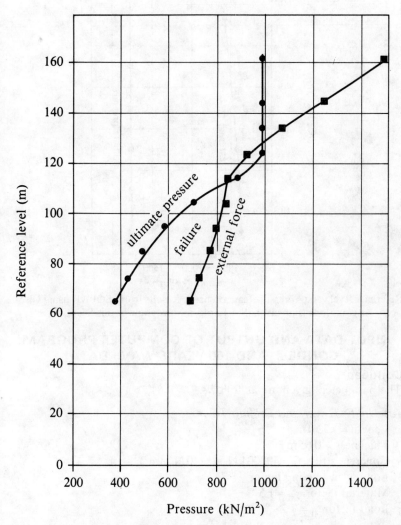

Reference level versus pressure (drilling tower). Results from ISOPAR using Olson's equations (Bangash). (See Implosion program in this Section.)

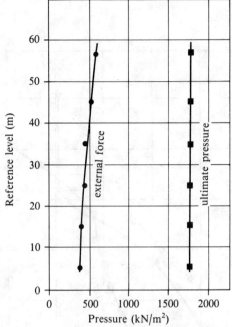

Reference level versus pressure (caisson base). Results from ISOPAR using Olson's equations (Bangash). (See Implosion program in this Section.)

INPUT DATA AND OUTPUT OF COMPUTER PROGRAM: CONDEEP AND 100 YEARS WAVE DATA

Condeep

The parameters given are as follows.

(a) Utility Tower and Drilling Tower
Length = 172·0 m
Thickness = 0·65 m
Concrete stress $f_c = 40000 \text{ kN/m}^2$ (40 N/mm²)
Long-term loading factor $\lambda = 0·7$
Material factor $\gamma_m = 1·5$
Load factor $\gamma_f = 1·2$

(b) Caisson Base
Length = 56·5 m
Thickness = 1·2 m

The other data are similar to those for the tower.

One Hundred Years Wave Data
Wave period = 16 s
Water depth (from mean water level) = 145 m
Wave height = 30 m
Density of seawater = 10·1 kN/m^3
C_D = 0·6
C_I = 2·0

(appendix overleaf)

```
C        *****************************************
*        *    PROGRAM TO PREDICT IMPLOSION       *
*        *         IN GRAVITY PLATFORM           *
*        *****************************************
*
* ==============================================================================
* I                                                                            I
* I   NOMENCLATURE :                                                           I
* I   ============                                                             I
* I                                                                            I
* I   UT   = UTILITY TOWER              LW    = WAVE LENGTH                    I
* I   DT   = DRILLING TOWER             K     = WAVE NUMBER                    I
* I   C    = CAISSON BASE               W     = WAVE FREQUENCY                 I
* I   D    = DIAMETER                   FD    = DRAG FORCE                     I
* I   R    = MEAN RADIUS                FI    = INERTIA FORCE                  I
* I   PIMP = IMPLOSION PRESSURE         FDI   = DRAG & INERTIA FORCE           I
* I   PUL  = ULT. LIMIT PRESSURE        FCU   = CURRENT FORCE                  I
* I   ITA  = PLASTICITY FACTOR          FDIFF = DIFFRACTION FORCE              I
* I   L    = LENGTH                     CI    = COEFFICIENT OF INERTIA         I
* I   RL   = REDUCED LEVEL              CD    = COEFFICIENT OF DRAG            I
* I   FM   = MAX FORCE                  P     = DENSITY                        I
* I                                                                            I
* ==============================================================================

         PARAMETER ( N=9,M=6 )
         REAL DUT(N),DDT(N),DC(N),ITA(N),PIMP(N),PIMP1(N),PIMP2(N),
        *PIMP3(N),PIMP4(N),PIMP5(N),PIMP6(N),PIMP7(N),PIMP8(N),PIMP9(N),
        *PIMP10(N),PIMP11(N),PIMP12(N),RLUT(N),RLDT(N),RLC(N),YUT(N),
        *YDT(N),YC(N),XUT(N),XDT(N),XC(N),DUT1(N),DDT1(N),DC1(M),RUT1(N),
        *RUT2(N),RUT3(N),RDT1(N),RDT2(N),RDT3(N),RC1(N),RC2(N),RC3(N),
        *RLUT1(N),RLUT2(N),RLUT3(N),RLUT4(N),RLDT1(N),RLDT2(N),RLDT4(N),
        *RLC1(N),RLC2(N),RLC3(N),RLC4(N),RL1(N),RL2(N),RL3(N),RL4(N),
        *A(N),B(N),D(N),D11(N),RC(N),RDT(N),RUT(N),YUT1(N),YUT2(N),
        *YUT3(N),YDT1(N),YDT2(N),YDT3(N),LUT,LDT,LC,YC1(N),YC2(N),YC3(N),
        *X2(N),Y(N),RL(N),RLDT3(N),ITA1(N),ITA2(N),FA1(9),FA2(9),FA3(9),
        *FA4(9),FA5(9),FA6(9),Z(9),GA1(9),GA2(9),GA3(9),GA4(9),GA5(9),
        *GA6(9),GA7(9),GA8(9),GA9(9),FI(N),FD(N),Y4(N),CIV(N),C(N),E(N),
        *E1(N),DIFFRC(N),DIFFRU(N),DIFFRD(N),F1(N),RATIO(N),FCU(N),CU(N),
        *ANGLEU(N),ANGLED(N),ANGLEC(N),T1(N),F14(N),F23(N),
        *V1(N),V2(N),V3(N),V4(N),V5(N),U1(N),U2(N),U3(N),U4(N),U5(N),R1(N),
        *THITA(180),FD1(180),FI1(180),ANGLE(180),F32(N),F41(N),
        *R2(N),R3(N),R4(N),R5(N),K,K1,K0,LW,NITA,NITAC,NITAT,Y5(N),Y6(N),
        *FDIFF(N),FDIFFU(N),FDIFFD(N),FDIFFC(N),AUT(N),BUT(N),CUT(N),
        *D11UT(N),ADT(N),BDT(N),CDT(N),D11DT(N),AC(N),BC(N),CC(N),
        *D11C(N),ITA3(N),ANGLE1(N),YL(N),X(N),FD2(180),FD3(180),FD4(180)
         REAL FD5(180),F(180),FTOTAL(N),FT(N),PULA(N),PULB(N),PULC(N),
        *FIUT(N),FDDT(N),FIC(N),FIDT(N),FDUT(N),FDC(N),FCUDT(N),FCUDT(N),
        *FCUC(N),PULD(N),FM1(180),FM2(180),F1A(180),F2A(180),F3A(180),
        *F4A(180),F1B(180),F1C(180),PUL1(N),PUL2(N),PUL3(N),PUL4(N),
        *PUL5(N),PUL6(N),PUL7(N),PUL8(N),PUL9(N),PUL10(N),PUL11(N),
        *FDI(N),FDIUT(N),FDIDT(N),FDIC(N),FDI1(180),D1(N),PUL12(N)
         DATA TDT,TUT,TC,LDT,LUT,LC/0.65,0.65,1.2,172.0,172.0,56.5/
         DATA FC,GAMA,GAMAM,GAMAF,LU,LD,LC1/40000.0,0.7,1.5,1.2,1,2,3/
         DATA DIUT,DIDT,DIC/12.0,13.0,0.0/
         OPEN(20,FILE='DIAMRL',STATUS='OLD')
         READ(20,*)(RLUT(I),DUT(I),RLDT(I),DDT(I),I=1,N)
         OPEN(20,FILE='DIARLC')
         READ(20,*)(RLC(I),DC(I),I=1,M)
         OPEN(20,FILE='SLUT')
```

```
      READ(20,*)(Y4(I),Y5(I),I=1,N)
      OPEN(20,FILE='SLC')
      READ(20,*)(Y6(I),I=1,M)
      CALL CALC (DUT,LUT,TUT,RUT,YUT1,YUT2,YUT3,XUT)
      CALL CALC (DDT,LDT,TDT,RDT,YDT1,YDT2,YDT3,XDT)
      CALL CALC (DC,LC,TC,RC,YC1,YC2,YC3,XC)
      CALL CALC1(LU,LUT,YUT1,YUT2,YUT3,XUT,DUT,RUT,RLUT,NORLU1,
     *    NORLU2,NORLU3,NORLU4,RUT1,RUT2,RUT3,DUT1,A1UT,B1UT,
     *         C1UT,D12UT,AUT,BUT,CUT,D11UT,RLUT1,RLUT2,RLUT3,RLUT4)
      CALL CALC1(LD,LDT,YDT1,YDT2,YDT3,XDT,DDT,RDT,RLDT,NORLD1,NORLD2,
     *         NORLD3,NORLD4,RDT1,RDT2,RDT3,DDT1,A1DT,B1DT,C1DT,D12DT,
     *         ADT,BDT,CDT,D11DT,RLDT1,RLDT2,RLDT3,RLDT4)
      CALL CALC1(LC1,LC,YC1,YC2,YC3,XC,DC,RC,RLC,NORLC1,NORLC2,NORLC3,
     *    NORLC4,RC1,RC2,RC3,DC1,A1C,B1C,C1C,D12C,AC,BC,CC,D11C,
     *             RLC1,RLC2,RLC3,RLC4)
      PRINT1
      PRINT2
      PRINT5
      PRINT4, (RLUT(I),DUT(I),RLDT(I),DDT(I),I=1,N)
      PRINT7
      PRINT6,(RLC(I),DC(I),I=1,M)
      PRINT50
50    FORMAT(/9X,'IMPLOSION PRESSURE AND ULTIMATE PRESSURE :',/9X,42
     *         ('='))
      PRINT8
      CALL THICK(DUT1,NORLU1,LUT,TUT,FC,GAMA,GAMAM,GAMAF,PUL1,PIMP1)
      CALL THIN1(RUT1,NORLU2,LUT,TUT,FC,GAMA,GAMAM,GAMAF,PUL4,PIMP4)
      CALL THIN2(RUT2,NORLU3,LUT,TUT,FC,GAMA,GAMAM,GAMAF,ITA,
     *           PUL7,PIMP7)
      CALL THIN3(RUT3,NORLU4,LUT,TUT,FC,GAMA,GAMAM,GAMAF,ITA2,
     *           PUL10,PIMP10)
      IF(AUT(1).EQ.1)THEN
      PRINT23
      PRINT27
      PRINT11,(RLUT1(I),PIMP1(I),PUL1(I),I=1,NORLU1)
      ELSEIF(BUT(1).EQ.1)THEN
      PRINT24
      PRINT27
      PRINT14,(RLUT2(I),PIMP4(I),PUL4(I),I=1,NORLU2)
      ELSEIF(CUT(1).EQ.1)THEN
      PRINT25
      PRINT27
      PRINT17,(RLUT3(I),PIMP7(I),PUL7(I),ITA(I),I=1,NORLU3)
      ELSEIF(D11UT(1).EQ.1)THEN
      PRINT26
      PRINT27
      PRINT20,(RLUT4(I),PIMP10(I),PUL10(I),ITA2(I),I=1,NORLU4)
      END IF
      IF(A1UT.EQ.1.AND.AUT(1).NE.1)THEN
      PRINT23
      PRINT11,(RLUT1(I),PIMP1(I),PUL1(I),I=1,NORLU1)
      ELSEIF(B1UT.EQ.1.AND.BUT(1).NE.1)THEN
      PRINT24
      PRINT14,(RLUT2(I),PIMP4(I),PUL4(I),I=1,NORLU2)
      ELSEIF(C1UT.EQ.1.AND.CUT(1).NE.1)THEN
      PRINT25
      PRINT17,(RLUT3(I),PIMP7(I),PUL7(I),ITA(I),I=1,NORLU3)
      ELSEIF(D12UT.EQ.1.AND.D11UT(1).NE.1)THEN
      PRINT26
```

```
      PRINT20,(RLUT4(I),PIMP10(I),PUL10(I),ITA2(I),I=1,NORLU4)
      ENDIF
      PRINT10
      CALL THICK(DDT1,NORLD1,LDT,TDT,FC,GAMA,GAMAM,GAMAF,PUL2,PIMP2)
      CALL THIN1(RDT1,NORLD2,LDT,TDT,FC,GAMA,GAMAM,GAMAF,PUL5,PIMP5)
      CALL THIN2(RDT2,NORLD3,LDT,TDT,FC,GAMA,GAMAM,GAMAF,ITA,
     *           PUL8,PIMP8)
      CALL THIN3(RDT3,NORLD4,LDT,TDT,FC,GAMA,GAMAM,GAMAF,ITA,
     *           PUL11,PIMP11)
      IF(ADT(1).EQ.1)THEN
      PRINT23
      PRINT27
      PRINT12,(RLDT1(I),PIMP2(I),PUL2(I),I=1,NORLD1)
      ELSEIF(BDT(1).EQ.1)THEN
      PRINT24
      PRINT27
      PRINT15,(RLDT2(I),PIMP5(I),PUL5(I),I=1,NORLD2)
      ELSEIF(CDT(1).EQ.1)THEN
      PRINT25
      PRINT27
      PRINT18,(RLDT3(I),PIMP8(I),PUL8(I),ITA(I),I=1,NORLD3)
      ELSEIF(D11DT(1).EQ.1)THEN
      PRINT26
      PRINT27
      PRINT21,(RLDT4(I),PIMP11(I),PUL11(I),ITA(I),I=1,NORLD4)
      END IF
      IF(A1DT.EQ.1.AND.ADT(1).NE.1)THEN
      PRINT23
      PRINT12,(RLDT1(I),PIMP2(I),PUL2(I),I=1,NORLD1)
      ELSEIF(B1DT.EQ.1.AND.BDT(1).NE.1)THEN
      PRINT24
      PRINT15,(RLDT2(I),PIMP5(I),PUL5(I),I=1,NORLD2)
      ELSEIF(C1DT.EQ.1.AND.CDT(1).NE.1)THEN
      PRINT25
      PRINT18,(RLDT3(I),PIMP8(I),PUL8(I),ITA(I),I=1,NORLD3)
      ELSEIF(D12DT.EQ.1.AND.D11DT(1).NE.1)THEN
      PRINT26
      PRINT21,(RLDT4(I),PIMP11(I),PUL11(I),ITA(I),I=1,NORLD4)
      END IF
      PRINT9
      CALL THICK(DC1,NORLC1,LC,TC,FC,GAMA,GAMAM,GAMAF,PUL3,PIMP3)
      CALL THIN1(RC1,NORLC2,LC,TC,FC,GAMA,GAMAM,GAMAF,PUL6,PIMP6)
      CALL THIN2(RC2,NORLC3,LC,TC,FC,GAMA,GAMAM,GAMAF,ITA1,
     *           PUL9,PIMP9)
      CALL THIN3(RC3,NORLC4,LC,TC,FC,GAMA,GAMAM,GAMAF,ITA3,
     *           PUL12,PIMP12)
      IF(AC(1).EQ.1)THEN
      PRINT23
      PRINT27
      PRINT13,(RLC1(I),PIMP3(I),PUL3(I),I=1,NORLC1)
      ELSEIF(BC(1).EQ.1)THEN
      PRINT24
      PRINT27
      PRINT16,(RLC2(I),PIMP6(I),PUL6(I),I=1,NORLC2)
      ELSEIF(CC(1).EQ.1)THEN
      PRINT25
      PRINT27
```

```
      PRINT19,(RLC3(I),PIMP9(I),PUL9(I),ITA1(I),I=1,NORLC3)
      ELSEIF(D11C(1).EQ.1)THEN
      PRINT26
      PRINT27
      PRINT22,(RLC4(I),PIMP12(I),PUL12(I),ITA3(I),I=1,NORLC4)
      END IF
      IF(A1C.EQ.1.AND.AC(1).NE.1)THEN
      PRINT23
      PRINT13,(RLC1(I),PIMP3(I),PUL3(I),I=1,NORLC1)
      ELSEIF(B1C.EQ.1.AND.BC(1).NE.1)THEN
      PRINT24
      PRINT16,(RLC2(I),PIMP6(I),PUL6(I),I=1,NORLC2)
      ELSEIF(C1C.EQ.1.AND.CC(1).NE.1)THEN
      PRINT25
      PRINT19,(RLC3(I),PIMP9(I),PUL9(I),ITA1(I),I=1,NORL3)
      ELSEIF(D12C.EQ.1.AND.D11C(1).NE.1)THEN
      PRINT26
      PRINT22,(RLC4(I),PIMP12(I),PUL12(I),ITA3(I),I=1,NORLC4)
      END IF
  1   FORMAT(4(/),19X,'5.3  RESULT OF PROGRAM'/19X,22('='))
  2   FORMAT(/12X,'UTILITY TOWER',8X,'DRILLING TOWER'/12X,13('='),8X,
     *       14('='))
  5   FORMAT(11X,'RL',9X,'D',11X,'RL',9X,'D'/11X,2('='),9X,1('='),
     *       11X,2('='),9X,1('='))
  4   FORMAT(9X,F6.2,4X,F6.2,7X,F6.2,4X,F6.2)
  7   FORMAT(/12X,'CAISSON BASE'/12X,12('='),/11X,'RL',9X,'D',/11X,
     *       2('='),9X,1('='))
  6   FORMAT(9X,F6.2,4X,F6.2)
  8   FORMAT(/18X,'UTILITY TOWER'/18X,13('='))
 10   FORMAT(/18X,'DRILLING TOWER'/18X,14('='))
  9   FORMAT(///18X,'CAISSON BASE'/18X,12('='))
 11   FORMAT(9X,F6.2,4X,F7.2,4X,F7.2)
 12   FORMAT(9X,F6.2,4X,F7.2,4X,F7.2)
 13   FORMAT(9X,F6.2,4X,F7.2,4X,F7.2)
 14   FORMAT(9X,F6.2,4X,F7.2,4X,F7.2)
 15   FORMAT(9X,F6.2,4X,F7.2,4X,F7.2)
 16   FORMAT(9X,F6.2,4X,F7.2,4X,F7.2)
 17   FORMAT(9X,F6.2,6X,F7.2,10X,F7.2,7X,F5.2)
 18   FORMAT(9X,F6.2,6X,F7.2,10X,F7.2,7X,F5.2)
 19   FORMAT(9X,F6.2,6X,F7.2,10X,F7.2,7X,F5.2)
 20   FORMAT(9X,F6.2,6X,F7.2,10X,F7.2,7X,F5.2)
 21   FORMAT(9X,F6.2,6X,F7.2,10X,F7.2,7X,F5.2)
 22   FORMAT(9X,F6.2,6X,F7.2,10X,F7.2,7X,F5.2)
 23   FORMAT(9X,'THICK-WALLED CYLINDER'/9X,21('='))
 24   FORMAT(9X,'THIN-WALLED SHORT CYLINDER'/9X,26('='))
 25   FORMAT(9X,'THIN-WALLED MODERATELY LONG'/9X,27('='))
 26   FORMAT(9X,'THIN-WALLED LONG CYLINDER'/9X,25('='))
 27   FORMAT(10X,'RL(m)',4X,'PIMP(kn/m~2)',4X,'ULT.PRES(kn/m~2)',3X,
     *       'ITA',/10X,5('='),4X,12('='),4X,16('='),3X,3('='))
      CALL WAVE(DUT,DIUT,LUT,Y4,FDIUT,FIUT,ANGLEU,RATIOU,DIFFRU,
     *          FMU,FDIFFU,NODIFU,NODFFU,NODU,NOIU,NOINTU,NODRAU,YUT7)
      CALL WAVE(DDT,DIDT,LDT,Y5,FDIDT,FIDT,ANGLED,RATIOD,DIFFRD,FMD,
     *          FDIFFD,NODIFD,NODFFD,NODD,NOID,NOINTD,NODRAD,YDT7)
      CALL WAVE(DC,DIC,LC,Y6,FDIC,FIC,ANGLEC,RATIOC,DIFFRC,FMC,
     *          FDIFFC,NODIFC,NODFFC,NODC,NOIC,NOINTC,NODRAC,YC7)
      PRINT30
 30   FORMAT(/9X,'PRESSURE LOADING DUE TO WAVE :',/9X,30('='))
```

```
      PRINT31
31    FORMAT(/18X,'UTILITY TOWER',/18X,13('='))
      CALL PRINTF(RLUT,NODFFU,FDIFFU,NODIFU,NOINTU,NODU,NOIU,FIUT,FDUT
     *            ,ANGLEU)
      PRINT100,FMU,YUT7
100   FORMAT(/7X,'MAX FORCE =',F7.2,' (KN/M~2) OCCUR @ R. LEVEL=',F6.2,
     *       /7X,51('='))
      PRINT32
32    FORMAT(/18X,'DRILLING TOWER',/18X,14('='))
      CALL PRINTF(RLDT,NODFFD,FDIFFD,NODIFD,NOINTD,NODD,NOID,FIDT,FDDT
     *            ,ANGLED)
      PRINT102,FMD,YDT7
102   FORMAT(/7X,'MAX FORCE =',F7.2,' (KN/M~2) OCCUR @ R. LEVEL=',F6.2,
     *       /7X,51('='))
      PRINT33
33    FORMAT(14(/),18X,'CAISSON BASE',/18X,12('='))
      CALL PRINTF(RLC,NODFFC,FDIFFC,NODIFC,NOINTC,NODC,NOIC,FIC,FDC,
     *            ANGLEC)
      PRINT34
      CALL FCUR(DUT,LUT,Y4,FCUUT)
      CALL FCUR(DDT,LDT,Y5,FCUDT)
      CALL FCUR(DC,LC,Y6,FCUC)
34    FORMAT(/9X,'CURRENT FORCE:',/9X,14('='))
      PRINT35
35    FORMAT(/18X,'UTILITY TOWER',/18X,13('='))
      CALL PRINTC(RLUT,FCUUT,LUT)
      PRINT36
36    FORMAT(/18X,'DRILLING TOWER',/18X,14('='))
      CALL PRINTC(RLDT,FCUDT,LDT)
      PRINT37
37    FORMAT(/18X,'CAISSON BASE',/18X,12('='))
      CALL PRINTC(RLC,FCUC,LC)
      PRINT38
38    FORMAT(5(/),9X,'TO CHECK WHETHER THE STRUCTURE IS SAFE',
     *       /9X,38('='))
      PRINT39
39    FORMAT(/18X,'UTILITY TOWER',/18X,13('='))
      DO 70 I=1,N
      IF(NODFFU.NE.1)FDIFFU(I)=0.0
      IF(NOINTU.NE.1)FIUT(I)=0.0
      IF(NODRAU.NE.1)FDIUT(I)=0.0
      FTOTAL(I)=FDIUT(I)+FIUT(I)+FDIFFU(I)+FCUUT(I)
70    CONTINUE
      CALL FTOTL(LUT,RLUT,FTOTAL,PUL1,PUL4,PUL7,PUL10)
      PRINT40
40    FORMAT(/18X,'DRILLING TOWER',/18X,14('='))
      DO 80 I=1,N
      IF(NODFFD.NE.1)FDIFFD(I)=0.0
      IF(NOINTD.NE.1)FIDT(I)=0.0
      IF(NODRAD.NE.1)FDIDT(I)=0.0
      FTOTAL(I)=FDIDT(I)+FIDT(I)+FDIFFD(I)+FCUDT(I)
80    CONTINUE
      CALL FTOTL(LDT,RLDT,FTOTAL,PUL2,PUL5,PUL8,PUL11)
      PRINT41
41    FORMAT(/18X,'CAISSON BASE',/18X,12('='))
      DO 60 I=1,M
      IF(NODFFC.NE.1)FDIFFC(I)=0.0
```

```
      END IF
      END IF
      RETURN
      END
C ========================================================================
*          ***************************                                   *
*          * 1. THICK-WALLED CYLINDER *                                  *
*          * ------------------------ *                                  *
*          ***************************                                   *
* ========================================================================
      SUBROUTINE THICK(D1,NORL1,L,T,FC,GAMA,GAMAM,GAMAF,PUL,PIMP)
      PARAMETER ( N=9 )
      REAL D1(N),PIMP(N),PUL(N),K,L
      K=1.25
      DO 3 I=1,NORL1
      PIMP(I)=2*K*(T/D1(I))*FC
      PUL(I)=GAMA/(GAMAM*GAMAF)*PIMP(I)
    3 CONTINUE
      RETURN
      END

C ========================================================================
*                    2. THIN-WALLED CYLINDER.                            *
* ========================================================================
*                                                                        *
*              ***************                                           *
*              * 2.1 SHORT   *                                           *
*              * ----------  *                                           *
*              ***************                                           *
* ========================================================================
      SUBROUTINE THIN1(R1,NORL2,L,T,FC,GAMA,GAMAM,GAMAF,PUL,PIMP)
      PARAMETER ( N=9 )
      REAL R1(N),PIMP(N),PUL(N),L
      DO 1 I=1,NORL2
      PIMP(I)=4*FC*(T/R1(I))/(2+T/R1(I))**2
      PUL(I)=GAMA/(GAMAM*GAMAF)*PIMP(I)
    1 CONTINUE
      RETURN
      END

C ========================================================================
*          **************************                                    *
*          * 2.2 MODERATELY LONG    *                                    *
*          * --------------------   *                                    *
*          **************************                                    *
* ========================================================================
      SUBROUTINE THIN2(R,NORL,L,T,FC,GAMA,GAMAM,GAMAF,ITA,PUL,PIMP)
      PARAMETER ( N=9 )
      REAL R(N),PIMP(N),ITA(N),Y(N),PUL(N),T,L
      CALL CALCI(NORL ,R ,T,L,ITA,Y)
      BETA=1.2
      DO 1 I=1,NORL
      PIMP(I)=2330*ITA (I)*FC*BETA*((T/R(I))**2.5*(R(I)/L)/
```

```
      *((2+T/R(I))**2))
      PUL(I)=GAMA/(GAMAM*GAMAF)*PIMP(I)
    1 CONTINUE
      RETURN
      END
```

```
C ======================================================================
*                    **************                                    *
*                    *  2.3 LONG  *                                    *
*                    *  --------  *                                    *
*                    **************                                    *
* ======================================================================
```

```
      SUBROUTINE THIN3(R,NORL,L,T,FC,GAMA,GAMAM,GAMAF,ITA,PUL,PIMP)
      PARAMETER ( N=9 )
      REAL R(N),PIMP(N),ITA(N),Y(N),PUL(N),T,L
      CALL CALCI1(NORL,R,T,L,ITA)
      DO 1 I=1,NORL
      PIMP(I)=686*ITA(I)*FC*(T/R(I))**3/(2+T/R(I))**2
      PUL(I)=GAMA/(GAMAM*GAMAF)*PIMP(I)
    1 CONTINUE
      RETURN
      END
```

```
C ======================================================================
*        ********************************************************      *
*        *   SUBROUTINE TO CALCULATE ITA [ PLASTICITY FACTOR ]  *      *
*        *          FOR MODERATELY LONG CYLINDER.               *      *
*        *   -------------------------------------------------- *      *
*        ********************************************************      *
* ======================================================================
```

```
      SUBROUTINE CALCI(NORL,R,T,L,ITA,Y)
      PARAMETER ( N=9 )
      REAL R(N),ITA(N),X2(N),Y(N),T,L
      DO 1 I=1,NORL
      DO 2 X1=0.4,1.10,1D-5
      ITA(I)=1.0-0.35*(10*X1-4)+0.15/2*(10*X1-4)*(10*X1-5)
     *       -0.05/6*(10*X1-4)*(10*X1-5)*(10*X1-6)
     *       -0.03/24*(10*X1-4)*(10*X1-5)*(10*X1-6)*(10*X1-7)
     *       +0.1/120*(10*X1-4)*(10*X1-5)*(10*X1-6)*(10*X1-7)*(10*X1-8)
     *       -0.15/720*(10*X1-4)*(10*X1-5)*(10*X1-6)*(10*X1-7)*(10*X1-8)
     *       *(10*X1-9)+0.15/5040*(10*X1-4)*(10*X1-5)*(10*X1-6)*(10*X1-7)
     *       *(10*X1-8)*(10*X1-9)*(10*X1-10)
      X2(I)=583*(T/R(I))**1.5*(R(I)/L)*ITA(I)
      Y(I)=INT(X2(I))-INT(X0)
      IF(ABS(Y(I)).EQ.0.0)GOTO 1
    2 CONTINUE
    1 CONTINUE
      RETURN
      END
```

APPENDIX X

GEOMETRY AND FORCES FOR DOMES FOR OFFSHORE CELLS AND LNG TANKS

GEOMETRY AND FORCES OF ELLIPTICAL DOMES

Meridional thrust due to wave/current loads

$$MTL = \frac{wla^2}{2b} Q$$

Hoop thrust due to wave/current loads

$$HTL = \frac{wla^2}{2b} \frac{\{(2(y/b)^2 - 1)\}}{Q}$$

where

W_v = wave/current load on dome surface per unit area
$Q = \sqrt{(1 - (a^2 - b^2)^2/a^2)(1 - (y/b)^2)}$

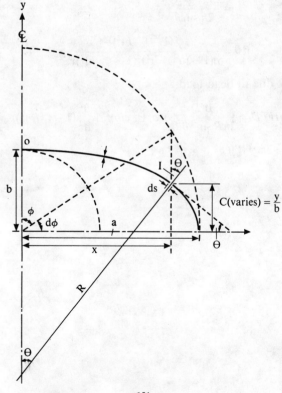

External Applied Forces

$$H = N_\phi = \frac{W_v r_0}{2 \sin \phi} = \frac{W_v r_0}{2}$$

$$T = N_\theta = W_v \left(r_2 - \frac{r_2^2}{2r_1} \right)$$

where

$$r_1 = \frac{a^2}{(a^2 \sin^2 \phi + b^2 \cos^2 \phi)^{1/2}}$$

Resisting Forces
Meridional and hoop forces

$$M_{TD} = \text{meridional thrust due to deep load}$$

$$= \frac{2\pi a^2 W_s C}{2\pi \sin \theta}$$

$$\sin \theta = \frac{b(1 - (y/b)^2)^{1/2}}{a(1 - (a^2 - b^2)^2(1 - (y/b)^2))^{1/2}}$$

Hoop thrust due to dead load

$$HTD = \frac{-WD}{2\pi R \sin^2 \theta} + wa^2 y((y/b)^2 + \frac{b^2}{a^2}(1 - (y/b)^2))$$

where $WD = 2\pi a^2 W_s C$.

Index

Aas–Jacobson model (for fatigue), 136
ACE formulae (for penetration/perforation), 87–8
ACI–NBC allowable load formulae, 271
Advanced gas-cooled reactors, 294
Ahmed's constitutive model (for bond and bond–slip), 123, 128–34
Ahmlink elements, 123, 204–5
  high-temperature gas-cooled reactor, 341, 344
  prestressed concrete slab, 260
Airburst (bomb)
  blast wave from, 400–7
  meaning of term, 400
Aircraft impact
  containment vessel affected by, 368, 373–5
  structural response to, 180
Airy's earth pressure formulae, 419, 420
Airy's wave theory, 623–6
Akosombo (Ghana), hydro-electric tunnels, 282, 285
American Association of State Highway and Transportation Officials (AASHTO), fatigue model, 139
American Concrete Institute (ACI)
  creep model, 97–8
  shrinkage model, 107–8
  silo calculations, 422, 430

American Concrete Institute (ACI)
  —contd.
  Young's modulus, 38
American Defense Agency, nuclear shelter study, 413
Anchor studs, analytical formulation of, 171
Argyris three-parameter criterion, 29, 45
Atom bomb
  characteristic blast wave, 400–7, 408
  see also Blast wave
  crater size caused by, 406
  energy released from, 400
  shelter design, 413–18

Baker's failure model, 26–8
Bandama (Ivory Coast), hydro-electric tunnels, 282, 289
Bangash incremental creep formulation, 102–3
Bangash yield criterion, 166
Barr–Carter–Howe–Neilson formulae (for perforation), 89
BBRV prestressing system, 119, 120, 596
Beam elements
  2-noded, 549
  3-noded, 549
  shape functions for, 549
Bechtel formula (for scabbing), 86–7

Bellefonte reactor, containment vessel
  finite element mesh scheme for, 370
  plan and section of dome, 371, 372
  safety margin for, 372
  size of, 362
Bersimis (Canada), hydro-electric
  tunnels, 283
Biaxial state curves, simulation of, 185
Biaxial stress behaviour, 11–19
Biaxial stress envelope, isoparametric
  elements for, 186–91
Biaxial stress state
  octahedral stress approach, 16–19
  ultimate uniaxial cylinder crushing
    strength for, 12–13
Birkinhead (South Australia), silo, 430
Blast wave (from atom bomb)
  arrival times of, 403, 408
  building types affected by, 412–13
  characteristics in air, 400–7, 408
  diffraction loading due to, 409,
    411–12
  drag loading due to, 412
  duration of, 403–4, 408
  dynamic pressure of, 402–3
  ground shock due to, 406
  loadings due to, 409–12
  overpressure of, 400–1
  pressure variation with time, 401–2
  reflection at surface, 404–5
  suction phase of, 401
  transient wind caused by, 402
  underpressure of, 401
Blunt crack band propagation, 80–4
  formulation of cracking directions
    for, 82–4
Body forces, 153
Boeing 707–320 aircraft, impact by,
  374, 380
Boeing 747 aircraft, impact by, 373,
  374, 380
Boiling-water reactors, containment
  vessels, 358
Bombs
  blast wave from, 400–7, 408
  see also Blast wave
  structural response to, 180

Bond
  Ahmed's constitutive model for,
    123, 128–34
  Brice theory for, 115–16
  BS8110 model for, 113
  Marshall–Krishnamurthy model
    for, 118
  models for, 118–34
  Saliger theory for, 114–16
  simple methods for, 113–16
  Somayaji–Shah model for, 116–17
  variables affecting, 112
  Wastlund theory for, 114–16
Bond linkage element, nonlinear,
  204–5
Bond stress–slip
  constitutive models for, 127–34
  curves
    prestressed concrete slab, 264
    prestressing wire (5 mm diameter),
      125, 248
    prestressing wire (7 mm diameter),
      126, 249
Bond–slip
  Ahmed's constitutive model for,
    123, 128–34
  curves
    deformed bar, 121
    plain wire, 121
    strands of different diameters, 122
  linear interpretation of nonlinear
    curve, 205
  models for, 118–34
  Ngo–Scordelis model for, 123,
    127–34
  test specimen, 244–50
    experimental details, 244–6
    experimental results, 247–9
    finite element analysis of, 246–7
Bond-linkage element, 123, 130
Bond-linkage stiffness matrix, 132
  load vector for, 133
Bonded prestressed concrete slab
  experimental details of, 250–6
  loading system for, 253–5
  reinforcement of, 250, 251
  strain gauge positions for, 252, 253

Bonded prestressed concrete
  specimens, 244–50
Boom elements
  2-noded, 549
  3-noded, 549
  shape functions for, 549
Brent C production platform, 617, 618
Brent D production platform, 618
Brice theory, 115–16
British Standards
  bond model, 113
  silo calculations, 421
  stress–strain curves for uniaxial compression, 6, 8
  Young's modulus calculations, 38
Buckling analysis, 181
Bugey reactor
  containment vessel, size of, 362
  pressure vessel, 30, 32
Bulk moduli models, 66–9
  finite element analysis used with, 206
Bunkers, 418–30
  formulae for, 420–1
  see also Silos
Buoyancy effects, 365
Burger (creep) model, 105
Burrendong (Australia), hydro-electric tunnels, 283
Buyukozturk strain-hardening model, 217–18

C_5 tendons, 600
CCL prestressing system, 119, 120, 250
CEA–EDF formula, 89
CEB–FIP methods/models
  creep, 98–9
  fatigue, 138
  shrinkage, 108–9
Cement powder, material properties of, 421
Chain rule of partial differentiation, 196, 201, 537
Chen–Chen strain-hardening model, 216–17

Chinon reactor, pressure vessel, 30
Circular tubes, empirical formulae for, 272
CKW–BRL formula (for penetration and perforation), 88
Coal, material properties of, 421
Common points, meaning of term, 7, 10
Concentrated loads, 153
Concrete
  sample preparation for, 589
  strength of, 589
  testing of, 589
Condeep platform
  currents data for, 438
  data for, 433, 436–8
  design of, 431–6
  implosion analysis for, 447, 449, 637–40
  implosion program, 642–8
  layout of, 432
  ship impact analysis for, 448–50
  soil data for, 437
  tide data for, 438
  wave data for, 437
  wind data for, 437
Constant stress–strain matrix
  rectangular, 554–5
  triangular, 553
Constitutive laws, multiaxial stress state, 19–36
Containment vessels, 356–99
  aircraft impact on, 368, 373–5
  analysis of, 368–99
  available data for, 362
  loads on, 364–8
  model of, 387, 388
    finite element mesh used, 395
    reinforcement in, 389–90
  overpressurisation analysis of, 388–99
    model used, 387, 388
    comparison of studies, 391–4
    crack zones for, 397
    final modes for, 396
    finite element mesh scheme for, 395

Containment vessels—*contd.*
 overpressurisation analysis of—*contd.*
  model used—*contd.*
   prestressing tendons in, 359–61
   purpose of, 356
   seismic analysis of, 375, 377, 381–6
   types of, 356, 358
Convection, virtual work associated with, 209
Convergence criteria, 581
Coulomb's criterion, 24
Crack
 calculations
  algorithm for, 607
  general steps of, 605–6
 closure, 77
  criterion, 79
 control theory, 453, 455
 displacement, tangent shear modulus affected by, 83
 opening, 79
 prevention concept, 455
 tip solid element, shape functions for, 545
Cracking
 matrices for, 569–71
 models for, 74–5
 step-by-step formulation for, 77–80
Creep, 91–106
 literature survey on, 91–6
 loads, 152
 models, 96–106
  ACI209 method, 97–8
  Bangash incremental formulation for, 102–3
  CEB–FIB method, 98–9
  Dibsi strain-hardening, 100
  finite element analysis used with, 202–3
  Hansen's formula, 99–100
  Hilsdorf–Muller method, 97
  Indian Standard Code IS 1343, 96–7
  three-dimensional thermal creep model, 103–6
  Zienkiewicz–Watson formulation, 100–1

Creep—*contd.*
 Poisson ratio, 202
 temperature effects on, 91–2, 94
 visco-elastic models for, 104–5
Creys–Malville reactor, containment vessel
 safety margin for, 372
 size of, 362
Cryogenic temperatures, material properties at, 38, 457, 462–3
Cubic integration, 175, 177
Currents loading, offshore structures, 438, 621–2
CVTR reactor containment vessel
 safety margin for, 372
 size of, 362
Cyclic loading stress–strain curves, 7, 9, 10

Damage classification, atom bomb blast, 413
Darwin–Pecknold constitutive model, 6, 9, 13, 214–15
Decomposition parameters, 187
Derivatives
 line isoparametric element, 162–3
 membrane isoparametric element, 159–61, 541, 542, 543
 solid isoparametric element, 148–50, 538, 540
Dibsi strain-hardening model, 100
Differentiation, partial, chain rule for, 196, 201, 537
Diffraction loading
 bomb blast response, 409, 411–12
 offshore structures, 621
Dilatancy, meaning of term, 14
Direction cosines, 58, 59, 158–9, 163
Discretisation, time domain, 173–4
Distributed loads, 153
Diversion tunnels (hydro-electric scheme), data for, 285–91
Doel (Belgium) reactor, containment vessel
 safety margin for, 372
 size of, 362

Index

Drag loading
  bomb blast response, 412
  offshore structures, 620, 621
Drucker–Prager criterion, 44, 45, 48, 53
Dungeness B Power Station
  liner study, 601
  pressure vessel, 30, 32, 296
    crack patterns, 319–22
    design analysis for, 309–23
    finite element mesh scheme used, 306
    flexural failure of caps, 316
    prestressing layout for, 301
    prestressing systems used, 302
    principal stresses in, 313, 317
    temperature distribution in, 353
Dykman LNG tank, 454
Dynamic analysis, 173–80
  nonlinear transient, 173–4
  reduced linear transient, 174–5
Dynamic finite element analysis
  nuclear shelters, 416
  offshore structures, 442, 444–51
Dynamic pressure (of atom bomb blast), 402–3
  relation to overpressure, 403, 408

Earthquakes
  loads due to, 365, 366
  nuclear reactor containment vessels affected by, 383–5
EDF-3 reactor, pressure vessel, 32, 294
EDF-4 reactor, pressure vessel, 32, 295
EGCR reactor, containment vessel
  safety margin for, 372
  size of, 362
Elastic strain softening material, concrete as, 74, 75
Elasticity, modulus of, 38–9
Elastoplastic straining, stresses during, 166–71
Element
  displacement vectors, 150
  stiffness matrix, 151, 152

Elements
  beam, 549
  boom, 549
  crack tip solid, 545
  gap, 563
  line isoparametric, 161–6
  membrane isoparametric, 157–61, 541–3
  prism, 546
  semiloof shell, 558–561
  shell, 550
  solid isoparametric, 145–57, 538–40
  tetrahedral, 547
  three-dimensional reinforced concrete solid, 544
  triangular, 548
  two-dimensional isoparametric, 552
  see also Finite element analysis; Isoparametric elements
Elliptical domes
  forces on, 652
  geometry of, 651
Elwi–Murray constitutive model, 215–16
Elwi–Murray decomposition parameters, 188
Empirical design methods, LNG tanks, 453
Endochronic model, 69–74
  coefficients for, 73
  constitutive equations of, 70
  containment vessel, 377, 388
  finite element analysis used with, 206–7
  hydro-electric tunnels, 291
  prestressed concrete beam, 229–30
  reinforced concrete slabs, 234, 235
Enel/Is pressure vessel, 32
Enricu Fermi reactor, containment vessel, size of, 362
Envelope curve, meaning of term, 7, 10
Environmental loads
  nuclear reactor containment vessels, 364–8
  offshore structures, 431, 433, 618–22
Equivalent uniaxial strain, 5–6

European Concrete Committee (CEB)
  stress–strain curves, 5
Young's modulus, 38
Explosion, structural response to, 180
Extension test, 34, 35
Extreme loads, examples of, 365

F-16 aircraft, impact by, 373, 375, 376
Failure criteria, 20–36
Failure models, 41–74
  1-parameter model, 41–2
  2-parameter model, 42–4
  3-parameter model, 45–7
  4-parameter model, 47–51
  5-parameter model, 51–4
Fatigue
  Aas–Jacobson model for, 136
  AASHTO model for, 139
  CEB–FIP model for, 138
  durability factor, 141
  JSCE model for, 137–8, 140
  Kakuta model for, 136, 140
  loading, types of, 135
  Miner model for, 136–7
  models for, 136–41
  NCHR model for, 139
  numerical modelling for, 135–41
  Waagaard model for, 137
FB-111 aircraft, impact by, 373, 380
Fessenheim reactor, containment vessel
  safety margin for, 372
  size of, 362
Finite element analysis
  bunkers, 430
  containment vessels, 368–99
  hydro-electric tunnels, 291
  LNG tanks, 463–9
  material modelling applied to, 182–218
  nuclear shelters, 413–18
  offshore structures, 442, 444–51
  prestressed concrete beams, 229
  prestressed concrete nuclear reactor vessels, 306–8

Finite element analysis—contd.
  prestressed concrete slabs, 260, 262, 263, 264, 265, 267
  reinforced concrete beams, 226
  reinforced concrete slabs, 236
  silos, 430
  thermal stress–strain model, 204
  tubular columns, 276, 279
Finite element modelling, 145–66
Finite elements
  8-noded, 147
  20-noded, 147
  32-noded, 147
  see also Elements; Isoparametric elements
Five-parameter model, 51–4
  nuclear shelter, 416
  prestressed concrete beam, 228, 229–30
  reinforced concrete beam, 226
  reinforced concrete slab, 234, 235
  simulation procedure for, 201–2
Fixed crack approach, 75
Flow calculations, 605–6
Fort Peck (USA), hydro-electric tunnels, 282
Fort St Vrain reactor, pressure vessel, 30, 295
Four-parameter model, 47–51
  containment vessel, 377
  prestressed concrete beam, 229–30
  simulation procedure for, 195–200
French code, silo calculations, 422
Froude–Krylor component (Morrison's equation), 621

Gap element, 561
Garrison (USA), hydro-electric tunnels, 282
Gauss point locations, high-temperature gas-cooled reactor, 340
Gauss–Legrendre quadrature, 577–8
Gaussian integration rule, 150
Gaussian quadrature, 575–8

Gentilly reactor, containment vessel
  safety margin for, 372
  size of, 362
German Specifications
  silo calculations, 422
  Young's modulus calculations, 38
Ginna reactor, containment vessel
  safety margin for, 372
  size of, 362
Gravitational forces, 153

Haig–Westergaard co-ordinate system, 46
Hannant–Frederick failure criterion, 26
Hansen's formula (for creep), 99–100
Hartlepool Power Station
  liner study, 601
  pressure vessel, 30, 296
  crack patterns in, 345, 349
Heat-conduction models
  finite element analysis used with, 207–12
  steady-state, 208
  transient, 208–12
Heysham Power Station, 296
High-temperature gas-cooled reactor
  liner study, 601
  pressure vessel
    circumferential pressure data, 598
    crack patterns in, 345–51
    design analysis of, 334–51, 355
    finite element mesh schemes for, 308, 334
    layout of, 304–5
    principal stresses in, 336–7
    requirements for, 293
    temperature distribution in, 354, 355
    ultimate load tests on model, 352
    Young's moduli for, 355
Hilsdorf model (for creep), 107
Hilsdorf–Muller method (for creep analysis), 97
Hinkley B Power Station, pressure vessel, 30, 32, 296

Hobbs triaxial test, 33–4, 35
Hobgen–Standing diffraction coefficient formula, 621
Hoyer's method (for transmission length), 117
Hunterston B Power Station, pressure vessel, 32, 296
Hurricanes
  loads due to, 367
  meaning of term, 367
Hydro-electric tunnels, 280–93
  classification of, 280
  components of, 280
  diversion tunnels, 285–91
  finite element analysis of, 291
  finite element mesh used, 290
  hydraulic criteria for, 281, 284
  loading of, 284–5
  parameters used in recent designs, 282–3
  reinforcement layouts for, 287, 288, 292
  shape of lining in, 280–1
  structural criteria for, 284
  thickness of lining in, 281
Hydrostatic pressure loads, 365
Hydrostatic stress axis, meaning of term, 42
Hypoelastic concept, 33
Hypoelastic model, 55–66
  reinforced concrete slabs, 234, 235
Hypoelasticity
  isotropic case, 63–5
  one-dimensional tension/compression cases, 65–6
  orthotropic case, 60–3
  stress–strain criteria using, 55–60

Impact, structural response to, 180
Impact loads
  load–time function for, 240
  missiles used, 240
  reinforced concrete slabs under, 238–44

Implosion analysis, 632–40
  see also Olson, implosion analysis
Indian Standards
  creep model used, 96–7
  Young's modulus calculation, 38
Inelastic dilatancy variable, 72
Integration
  cubic, 175, 177
  quadratic, 175, 177
Interpolation function, 148
IRS formulae (for penetration), 87
ISOPAR program
  diffraction-coefficient formulae in, 621
  flexibility of, 231
  impact loading of reinforced concrete slab, 243
  limitations of, 247
  offshore structures, 431, 444
  overpressurisation analysis, 388
  seismic analysis, 377
Isoparametric elements, 145–66
  line, 161–6
  membrane, 157–61
  solid (three-dimensional), 145–57
  see also Line isoparametric elements; Membrane isoparametric elements; Solid isoparametric elements; Two-dimensional isoparametric elements
Isotropicity of concrete, 59–60, 63–5
Iteration methods
  acceleration of convergence of, 581–5
  convergence criteria for, 581

Jacobian matrix, 146, 148–50
Janssen's formulae, 417, 420
Japan Society of Civil Engineers (JSCE), fatigue model, 137–8, 140

Kakuta model (for fatigue), 136, 140
Kar formulae (for penetration), 89–90

Kelvin (elastic) model, 105
Kloppel–Godr–Furlong formulae, 272
Knowles–Park formulae, 271
Kronecker delta, 57, 67, 172, 202
Kupfer model, 49, 50, 190

Launay–Gachon failure criterion, 29, 31
Line isoparametric elements, 161–6
  derivatives of, 162–3
  four-noded, 147, 161, 163
  shape functions of, 162–3
  solid element containment of, 164–5
  three-noded, 147, 161, 163
  two-noded, 147, 161, 162
Liners
  comparative study of, 601
  data on existing designs, 362
  finite element modelling of interaction with studs, 593
Liquefied natural gas (LNG) tanks, 451–69
  concrete material characteristics for, 457, 462–3
  corrosion of, 455
  crack patterns in, 466–7
  data for, 456–63
  design considerations for, 455–6
  design methods used, 453, 455
  elliptical domes
    forces on, 652
    geometry of, 651
  finite element analysis of, 463–9
  finite element mesh scheme for, 464
  first developed, 452
  leakage of, 455
  life span of, 455
  positioning of prestressing tendons, 460
  prestressing steel used
    design parameters for, 459, 461
    material properties of, 457
    positioning of, 460
    stresses in, 461–2
  principal stresses in, 466
  seismic analysis of, 468, 469

Liu–Nilson–Slate constitutive model, 212–14
Load vectors, 154–5
Locus of common points, 7, 9, 10
Loof mapping function, 560
Losenhausen compressive machine, 273

Maccamy–Fuchs diffraction coefficient formula, 621
Mach region, 404
Mangla (Pakistan), hydro-electric tunnels, 283
Marcoule (France), reactor pressure vessel, 294
Marshall–Krishnamurthy model (for bond), 118
Marshall–Marshall bond stress formula, 118
Material matrices
  constant Young's modulus, constant Poisson's ratio, 533
  cracked material, 569–71
  orthotropic variable-modulus model, 572
  Ottoson model, 573
  reinforcement, 572
  shear and torsion included, 536
  uncracked material, 569
  variable Young's modulus
    constant Poisson's ratio, 531
    variable Poisson's ratio, 532
Material modelling simulation, 182–218
Material properties, 3–40
Maxwell (flow) model, 104, 105
Membrane isoparametric elements, 157–61
  4-noded
    derivatives of, 541
    shape functions of, 541
  8-noded
    derivatives of, 542
    shape functions of, 542
  8-noded solid element used, 147

Membrane isoparametric elements
  —contd.
  12-noded
    derivatives of, 543
    shape functions of, 543
  20-noded solid element used, 147
  32-noded solid element used, 147
    derivatives of, 159–61, 543, 544, 545
    shape functions of, 541, 542, 543
Micro-concrete slabs, patch-loaded, 237–8
Mig-23 aircraft, impact by, 373, 380
Miner's Rule, 136–7
Mirage 2000 aircraft, impact by, 373, 376, 380
Missile impact, nuclear plant affected by, 367–8
Mode frequency analysis, 175–6
Mohr–Coulomb criterion, 20, 24, 43, 193
Mohr criterion, 24
Mol (Belgium) reactor, containment vessel
  safety margin of, 372
  size of, 362
Morrison's equation, 437, 620–1
MRCA aircraft, impact by, 373, 374, 375, 376, 378, 379, 380
Multi-cavity (nuclear reactor) vessels, 294, 296
  components of, 297
  design analysis of, 299–356
  design objective for, 298–9
  finite element mesh schemes used, 308
  liners of, 298
  prestressing loads in, 297
  problems associated with, 296–9
  reinforcements in, 298
Multiaxial stress state, constitutive laws for, 19–36
Multiple cracks, step-by-step analysis of, 78–80

National Cooperative Highway Research (NCHR), fatigue model, 139

National Defense Research
 Committee (NDRC) formula,
 85–6
Newton–Raphson iteration method,
 145, 236
 acceleration of convergence of,
 581–5
 prestressed concrete slab, 265
Ngo–Scorcelis model (for
 bond/bond–slip), 123, 127–8
Nodal displacement vector, 149, 150
Nodal forces, 152
Nonlinear incremental elastic models,
 66
Nonlinear transient dynamic analysis,
 173–4
N-SARVE program, 246
Nuclear reactor
 containment vessels, 356–99
 loss-of-coolant accident design
  criteria, 383
 pressure vessels, 293–356
 see also Containment vessels;
  Prestressed concrete nuclear
  reactor vessels
Nuclear shelters, 399–418
 damage *post mortem* for, 418
 dynamic finite element analysis of,
  416
 finite element analysis of, 413–18
 finite element mesh scheme for, 416
 general arrangement of, 414
 principal stresses for, 417
 reinforcement in, 415
Numerical modelling
 creep, 91–106
 failure, 41–90
 fatigue, 135–41
 temperature effects, 109–11

Oahe (USA), hydro-electric tunnels,
 282
Octagonal prestressed concrete slab,
 250–70
Octahedral shear stress concept, 23,
 42

Octahedral stress approach
 biaxial compression stress state,
  17–18
 biaxial stress state, 16–19
 biaxial tension and combined
  compression–tension stress
  states, 18–19
Offshore structures, 430–51
 concrete gravity platforms
  advantages of, 430–1
  currents data for, 438
  data on typical design, 436–8
  design of, 431–6
  dynamic finite element analysis
   of, 442, 444–51
  examples of, 431, 617–18
  finite element mesh scheme for,
   446
  implosion analysis for, 436, 444,
   447, 449, 632–40
  implosion program for, 642–8
  ship impact on, 438–42
  soil data for, 437
  tide data for, 438
  wave data for, 437
  wind data for, 437
  cracking of concrete in, 435
  data for, 617–41
  design requirements for, 433–6
  dynamic finite element analysis of,
   442, 444–52
  effects of surface waves on, 623–7
  elliptical domes
   forces on, 652
   geometry of, 651
  materials used, 434
  ship impact on, 438–42
  structural requirements for, 434–5
  typical design, 431–6
   data on, 436–8
Oldbury Power Station pressure
 vessel, 30, 32, 295
 circumferential stresses in, 329
 crack patterns in, 332–4
 design analysis of, 323–34
 layout of prestressing cables, 303
 principal stresses in, 330–1

Oldbury Power Station pressure
   vessel—*contd.*
   temperature distribution in, 354
   top-slab deflections in, 324, 328
   wall deflections in, 324, 327
One-parameter model, 41–2
   simulation procedure for, 191–2
Olson inplosion analysis, 623–40
   Condeep platform, 637–40
   offshore structures, 436, 444, 447,
      449, 632–40
   program, 642–8
   thick-walled cylinders, 633–4
   thin-walled cylinders
      long cylinders, 633, 636
      moderately long cylinders, 633, 635
      short cylinders, 633, 635
Operating basis earthquake loads, 365
Operating pressure loads, 365
Operating temperature loads, 365
Orthotropic constitutive models,
      212–18
   Darwin–Pecknold model, 6, 9, 13,
      214–15
   Elwi–Murray model, 215–16
   Liu–Nilson–Slate model, 212–14
Orthotropic variable-modulus model,
   material matrices for, 572
Orthotropicity of concrete, 59, 60–3
Ottoson model, 47, 48, 53, 195
   material matrices for, 573
Overpressurisation analysis
   containment vessels, 388–99
   model used, 387, 388

Palisades reactor, containment vessel
   safety margin for, 372
   size of, 362
Peak overpressure
   meaning of term, 401
   time variation, 404, 408
   wind velocity relation, 403, 408
Peekel strain indicator, 245
Penetration
   ACE formulae for, 87–8
   CKW–BRL formula for, 88

Penetration—*contd.*
   design formulae for, 85–6, 87–90
   IRS formulae for, 87
   Kar formulae for, 89
   meaning of term, 84
   modified NDRC formula for, 85–6
   US Army Corps of Engineers
      formula for, 85–6
Perforation
   ACE formulae for, 87–8
   Barr–Carter–Howe–Neilson
      formulae for, 89
   CEA–EDF formula for, 89
   CKW–BRL formula for, 88
   design formulae for, 85, 87–8
   meaning of term, 84
   modified NDRC formula for, 85
Phantom aircraft, impact by, 373, 374,
      376, 378, 379, 380
Piping equipment reaction loads, 365
Plastic flow rule, 166–71, 218
   examples of, 172
Plate flexure, rectangular finite
      element for, 556–7
Poisson's ratio
   cryogenic temperatures, 38
   data on, 36–8
Pressurised-water reactors,
   containment vessels, 356, 358
Prestressed concrete beams, 227–31
   collapse mode of, 228
   experimental details, 227
   experimental results, 228–30
   finite element analysis of, 229
   load–displacement relation for, 229
   strain distribution along depth of,
      228
Prestressed concrete nuclear reactor
   vessels, 293–356
   advantages of, 293
   design analysis of, 299–356
   finite element mesh schemes used,
      306–8
   historical development of, 294–6
   methods of analysis used, 300
   model testing for, 300, 305, 309
   service conditions for, 299–300

Prestressed concrete nuclear reactor vessels—*contd.*
  shape of, 294
  thermal analysis of, 355–6
  ultimate conditions for, 300
  vessel layouts for, 301–5
Prestressed concrete pull-out specimen, 244–50
Prestressed concrete slabs
  bonded slab, finite element analysis of, 260, 262
  crack pattern for, 258–9
  crack patterns in, 268–70
  finite element analysis of, 259–70
  load–displacement relations for, 256, 265
  loading system for, 253–5
  perfectly bonded slab, finite element analysis of, 262, 263
  pull-out specimens for, 124
  reinforcement of, 250, 251
  strain gauge positions for, 252, 253
  unbonded slab, finite element analysis of, 262, 264, 265, 267
  variation of steel stress with load, 267
Prestressed concrete tanks, 452
Prestressing tendons, stress–strain curves for, 594
Prism element, shape functions for, 546
Programs
  accelerations, printing of, 608–10
  cracked concrete sub-routine, 610–13
  displacements, printing of, 608–10
  velocities, printing of, 608–10
  see also **ISOPAR** program; **N-SARVE** program
Pseudo-time, meaning of term, 104
Pull-out test, 112
  bond–slip studied by, 118–19, 122
  prestressed concrete slabs, 124
  prestressed concrete specimen, 244–50

Quadratic integration, 175, 177

Rama-ganga (India), hydro-electric tunnels, 282
Rancho Seco reactor, containment vessel
  safety margin for, 372
  size of, 362
Rectangular finite element (plate flexure), 556–7
Reduced linear transient dynamic analysis, 174–5
Reimann criterion, 20, 29, 45
Reinforced concrete beams, 221–7
  comparison of crack patterns, 226–7
  cracking of, 225
  experimental details, 222
  experimental results, 222–7
  finite element analysis of, 226
  load–displacement relation for, 224, 226
  strain distributions along depth of, 222–4
Reinforced concrete slabs, 231–44
  concentrated loading of, 231–7
    crack pattern for, 233, 236
    finite element mesh used, 236
    load–displacement relation for, 235
  impact loading of, 238–44
  patch-loading of, 237–8
Ringhals reactor containment vessel
  safety margin for, 372
  size of, 362
Roselend (France), hydro-electric tunnels, 283
Rotating crack model, 75

Safe shut-down earthquake, 366
St Laurent reactor, pressure vessel, 30
Saliger theory, 114, 116
San Onofre reactor, containment vessel, size of, 362
Scabbing
  Bechtel formula for, 86–7
  design formulae for, 86–7, 88
  meaning of term, 84

Scabbing—*contd.*
  Stone–Webster formula for, 88
Seawaves. *See* Wave
Seebrook reactor, containment vessel
  safety margin for, 372
  size of, 362
Seismic analysis
  containment vessels, 375, 377, 381–6
  LNG tanks, 468, 469
Semiloof shell element, 558–61
  bending strain nomenclature for, 559
  equivalent rotational second derivatives calculated for, 560
  Loof nodal degrees of freedom, 558
  resolution of thickness vector for, 561
  shear strain nomenclature for, 559
Shape functions
  crack tip solid element, 545
  line elements, 162–3
  membrane isoparametric elements, 541, 542, 543
  solid isoparametric elements, 146–8, 538, 539, 540
  tetrahedral element, 547
  three-dimensional reinforced concrete solid element, 544
  two-dimensional isoparametric elements, 550
Shear moduli models, 66–9
  finite element analysis used with, 206
Shell element, shape functions for, 548
Ship impact
  data used, 439–40, 441, 442, 443
  dynamic finite element analysis for, 449–51
  impact mechanics for, 440, 442
  offshore concrete gravity platforms, 438–42
Shrinkage, 106–9
  ACI209 analysis method for, 107–8
  CEB–FIP 70 analysis method for, 108–9
  Hilsdorf model for, 107
Silos, 418–30
  Airy's formulae for, 419, 420–1

Silos—*contd.*
  bin dimensions of, 423
  British code formulae for, 421
  buttress construction in, 425
  duct layout of, 425
  finite element analysis of, 430
  finite element mesh scheme for, 427
  flow patterns in, 423
  Janssen's formulae for, 419, 420
  layout of, 424
  loads on, 419
  material properties for, 419
  pressures calculated by various codes, 422
  reinforcement in, 425–6
  wall construction in, 426
Simulation procedures
  1-parameter model, 191–2
  2-parameter model, 192–4
  3-parameter model, 194–5
  4-parameter model, 195–200
  5-parameter model, 200–2
Single cracks, step-by-step analysis of, 78–80
Sizewell B reactor
  containment vessel
    finite element mesh scheme for, 369
    overpressurisation analysis of, 398–9
    plan of, 359
    prestressing tendons in, 359–61
    safety margin for, 372
    seismic analysis of, 384, 386
    size of, 360
  reactor building, 357
Smeared crack concept, 80–1
Snow loads, 364–5
Softening behaviour, 3, 4
Soil data, offshore structures, 437
Soil pressure, 365
Solid isoparametric elements, 145–57
  8-noded, 260, 291, 463–4
    derivatives of, 538
    shape functions of, 538
  20-noded, 239–40, 274, 430
    derivatives of, 543

Solid isoparametric elements—*contd.*
  20-noded—*contd.*
    Gauss–Legrendre quadrature for, 577–8
    shape functions of, 539
  32-noded
    derivatives of, 540
    shape functions of, 540
    derivatives of, 148–50, 538, 539, 540
    line elements in, 164–5
    shape functions of, 538, 539, 540
Somayaji–Shah model (for bond), 116–17
Soviet code, silo calculations, 422
Spalling, meaning of term, 84
Specific creep, 93–4, 95
Specific heat, virtual work associated with, 210
Specific thermal creep, 93–4
Spectrum analysis, 178–80
Square tubes, empirical formulae for, 272
Steady-state heat-conduction models, 208
Steel
  stress–strain curves for, 593
  tensile strength of, 590
Step-by-step integration method, 176–7
Stiffness matrix, 151–2
Stokes nonlinear fifth-order wave theory, 619
Stone–Webster formula (for scabbing), 88
Stone–Webster reactor, containment vessel
  safety margin for, 372
  size of, 362
Storm wind, meaning of term, 437 622
Strain–displacement relation, line elements, 163–4
Strain-hardening models, 216–18
  Buyukozturk model, 217–18
  Chen–Chen model, 216–17
Strain-softening rule, 75–7
Strains, determination of, 150–1

Stress–strain curves
  equivalent uniaxial, 6, 7
  solution technique for, 183, 184–6
  steel, 593
  uniaxial, 182–6
    compression, 3–11
    tension, 4, 11
Stresses, determination of, 151
Substructuring, 155–7
  thermal model, 208
Superelement, 155
Surface burst (of bomb)
  blast from, 405
  crater size caused by, 405, 406
  meaning of term, 400

Tangent shear modulus, 82
  crack displacement effects on, 83–4
Temperature
  creep affected by, 91–2, 94
  numerical modelling for effects of, 109–11
  Young's modulus affected by, 39–40
Tetrahedral elements
  4-noded, 547
  10-noded, 547
  shape functions for, 547
Tetrahedral model (of failure), 26–8
Thermal analysis, prestressed concrete nuclear reactor vessels, 355–6
Thermal damping
  matrix, 211–12
  virtual work associated with, 210
Thermal expansion coefficients, LNG tank materials, 457, 463
Thermal loads, 154
Thermal stress–strain model, finite element analysis used with, 204
Three Mile Island reactor, 362
Three-dimensional reinforced concrete solid element, shape functions for, 544
Three-dimensional thermal creep model, 103–6
Three-parameter model, 45–7
  LNG tank, 469
  simulation procedure for, 194–5

Time-dependent strain, 93
Tornado aircraft, impact by, 373, 376, 378, 379
Tornado generated missiles, 368
Tornado loads, 366–7
Transformation matrices, stress and strain, 565–6
Transient, heat-conduction model, 208–12
Tresca criterion, 20, 21–2, 24, 41
Triangular elements
　3-noded, 548
　6-noded, 548
　linear strain, 551
　shape functions for, 548
Triaxial compression, stress–strain curves for, 33–4
Trojan reactor
　containment vessel
　　safety margin for, 372
　　size of, 362
Tsunami (ocean waves), 367
　factors affecting, 367
Tubular columns
　concrete-filled, 270–9
　empirical formulae used, 271–2
　experimental testing of, 272–5
　failure mode of, 276
　finite element analysis for, 276, 279
　finite element mesh scheme for, 275
　load–displacement curves for, 278–9
　load–strain curves for, 275–7
Tumut, hydro-electric tunnels, 283
Tunnels, 280–93
　see also Hydro-electric tunnels
Turning points, meaning of term, 10
Two-dimensional isoparametric elements, shape functions for, 550
Two-parameter model, 42–4
　simulation procedure for, 192–4

Underpressure (of bomb blast), 401
Uniaxial stress–strain curves, 182–6
Uniaxial tension, 11

US Army Corps of Engineers formula, 85–6
US Bureau of Reclamation, Young's modulus, 38

Von Mises criterion, 20, 23, 41, 45, 48, 53, 275
Von Mises theory, 22–3
Von Mises yield surface, 21, 42, 172
VSL prestressing system, 119, 120, 430

Waagaard model (for fatigue), 137
Warsak (Pakistan), hydro-electric tunnels, 283
Wastlund Theory, 114, 116
Wave
　data
　　offshore structures, 437
　　100-year wave data, 641
　　definitions for, 627
　loads
　　acceleration components of, 620, 627, 628
　　offshore structures, 619
　　velocity components of, 619–20
Westinghouse reactor, containment vessel
　safety margin for, 372
　size of, 362
Wheat, material properties of, 421
Willam three-parameter model, 45, 51, 53
Willam–Wanke five-parameter criterion, 29
Willam–Wanke three-parameter model, 45–7, 194
Wilson θ method, 583
Wind data, offshore structures, 437
Wind loads
　nuclear reactor containment vessels, 364
　offshore structures, 622
Wylfa Power Station, pressure vessel, 30, 295

Yamuna (India), hydro-electric
    tunnels, 283
Young's modulus
    data on, 38–9
    high-temperature gas-cooled reactor
        pressure vessel, 355

Young's modulus—*contd.*
    temperature variation of, 39–40

Zienkiewicz–Watson creep
    formulation, 100–1